Basic College Mathematics:
An Applied Approach

Seventh Edition

Basic College Mathematics: An Applied Approach

Richard N. Aufmann
Palomar College, California

Vernon C. Barker
Palomar College, California

Joanne S. Lockwood
Plymouth State College, New Hampshire

HOUGHTON MIFFLIN COMPANY
Boston   New York

Editor-in-Chief: *Jack Shira*
Sponsoring Editor: *Lynn Cox*
Senior Development Editor: *Dawn Nuttall*
Editorial Assistant: *Melissa Parkin*
Senior Project Editor: *Tamela Ambush*
Senior Production/Design Coordinator: *Carol Merrigan*
Manufacturing Manager: *Florence Cadran*
Marketing Manager: *Ben Rivera*

Cover Photographer: Harold Burch, Harold Burch Design/NYC.

Printed in the U.S.A.

Library of Congress Control Number: 2001099313

ISBN
Student Text: 0-618-202854
Instructor's Annotated Edition: 0-618-202862

3456789-VH-06 05 04 03 02

Contents

4 Ratio and Proportion   167

5 Percents   195

6 Applications for Business and Consumers   227

7 Statistics and Probability   287

8 U.S. Customary Units of Measurement   331

9 The Metric System of Measurement   361

12 Geometry 499

Preface

The seventh edition of *Basic College Mathematics: An Applied Approach* provides mathematically sound and comprehensive coverage of the topics considered essential in a basic college mathematics course. The text has been designed not only to meet the needs of the traditional college student, but also to serve the needs of returning students whose mathematical proficiency may have declined during years away from formal education.

In this new edition of *Basic College Mathematics: An Applied Approach*, we have continued to integrate approaches suggested by AMATYC. Each chapter opens by illustrating and referencing a mathematical application within the chapter. At the end of each section there are "Applying the Concepts" exercises that include writing, synthesis, critical thinking, and challenge problems. At the end of each chapter there is a "Focus on Problem Solving" that introduces students to various problem-solving strategies. This is followed by "Projects and Group Activities" that can be used for cooperative learning activities.

One of the main challenges for students at this level is the ability to translate verbal phrases into mathematical expressions. One reason for this difficulty is that students are not exposed to verbal phrases until later in most texts. In *Basic College Mathematics: An Applied Approach*, we introduce verbal phrases for operations as we introduce the operation. For instance, after addition concepts have been presented, we provide exercises which say "Find the sum of...." or "What is 6 more than 7?" In this way, students are constantly confronted with verbal phrases and must make a mathematical connection between the phrase and a mathematical operation.

NEW! Changes to this Edition

In response to user requests, we have integrated topics from geometry through the early chapters of the text. In Chapters 1, 2, and 3, exercise sets include problems on finding the perimeter of a triangle and of a square, as well as finding the area of a square and of a rectangle. This is done very informally, without formulas—for example, students are presented with a verbal explanation, such as "The perimeter of a triangle is the sum of the lengths of the three sides of the triangle." Formal presentation of these topics remains in Chapter 12, *Geometry*.

In Chapter 6, *Applications for Business and Consumers*, the coverage of simple interest has been expanded and now includes determining the maturity value for simple interest loans and calculating the monthly payment on a simple interest loan. Also, a new objective on calculating finance charges on a credit card bill has been added. The objective on compound interest has been expanded; this presentation of the material should provide students with a better understanding of the topic.

Because of the growing popularity of 15-year mortgages, factors necessary to compute monthly payments on 15-year mortgages were added to the Monthly Payment Table, and exercises in Chapter 6 now include calculating monthly payments on 15-year mortgages.

New to this edition is Section 7.5, *Introduction to Probability*, in which students are provided instruction on calculating the probability of simple events.

Also in Chapter 7 is a presentation of box-and-whiskers plots. This material has been expanded in order to give students a better understanding of the use of this statistical tool to represent data.

Throughout the text, developmental exercises have been added to the exercise sets. The intent of these exercises is to reinforce the concepts underlying the skills presented in the lesson. For example, in the objective on multiplying whole numbers, students are asked to write $2 + 2 + 2 + 2 + 2 + 2$ as a product.

Throughout the text, data problems have been updated to reflect current data and trends. These application problems will demonstrate to students the variety of problems in real life that require mathematical analysis. Instructors will find that many of these problems may lead to interesting class discussions.

Many of the activities in the *Projects and Group Activities* feature are new to this edition. Some of those from the previous edition were revised to provide more structure. These projects include applications of mathematics to a variety of fields, for example, music, the consumer price index, buying a car, and symmetry.

Another new feature of this edition is *AIM for Success,* which explains what is required of a student to be successful and how this text has been designed to foster student success. *AIM for Success* can be used as a lesson on the first day of class or as a project for students to complete to strengthen their study skills. There are suggestions for teaching this lesson in the *Instructor's Resource Manual.*

Related to *AIM for Success* are *Prep Tests* which occur at the beginning of each chapter. These tests focus on the particular prerequisite skills that will be used in the upcoming chapter. The answers to these questions can be found in the Answer Appendix along with a reference (except for Chapter 1) to the objective from which the question was taken. Students who miss a question are encouraged to review the objective from which the question was taken.

The *Go Figure* problem that follows the *Prep Test* is a puzzle problem designed to engage students in problem solving.

Chapter Opening Features

NEW! Chapter Opener

New, motivating chapter opener photos and captions have been added, illustrating and referencing a specific application from the chapter.

The at the bottom of the page lets students know of additional on-line resources at math.college.hmco.com/students.

Objective-Specific Approach

Each chapter begins with a list of learning objectives which form the framework for a complete learning system. The objectives are woven throughout the text (i.e. Exercises, Prep Tests, Chapter Reviews, Chapter Tests, Cumulative Reviews) as well as through the print and multimedia ancillaries. This results in a seamless learning system delivered in one consistent voice.

Page 288

Chapter **5** **Percents**

Objectives

Section 5.1
A To write a percent as a fraction or a decimal
B To write a fraction or a decimal as a percent

Section 5.2
A To find the amount when the percent and the base are given
B To solve application problems

Section 5.3
A To find the percent when the base and amount are given
B To solve application problems

Section 5.4
A To find the base when the percent and amount are given
B To solve application problems

Section 5.5
A To solve percent problems using proportions
B To solve application problems

The number of children in this kindergarten class represents a percent of the population of a school district. Local populations and estimates of growth or decline are used to predict future job markets in that area. For example, population growth may indicate a need for more teachers in the local schools. The demography **project on page 219** describes different segments of the U.S. population as percents of the total population.

Need help? For on-line student resources, such as section quizzes, visit this textbook's web site at **math.college.hmco.com/students.**

Page 195

Prep Test

1. Bill-related mail accounted for 49 billion of the 102 billion pieces of first-class mail handled by the U.S. Postal Service during a recent year. (Source: USPS) What percent of the pieces of first-class handled by the U.S. Postal Service was bill-related mail? Round to the nearest tenth of a percent.

2. The table at the right shows the estimated costs of funding an education at a public college. Between which two enrollment years is the increase in cost greatest? What is the increase between these two years?

Enrollment Year	Cost of Public College
2005	$70,206
2006	$74,418
2007	$78,883
2008	$83,616
2009	$88,633
2010	$93,951

Source: The College Board's Annual Survey of Colleges

3. During the 1924 Summer Olympics in Paris, France, the United States won 45 gold medals, 27 silver medals, and 27 bronze medals. (Source: The Ultimate Book of Sports Lists)
   a. Find the ratio of gold medals won by the United States to silver medals won by the United States during the 1924 Summer Olympics. Write the ratio as a fraction in simplest form.
   b. Find the ratio of silver medals won by the United States to bronze medals won by the United States during the 1924 Summer Olympics. Write the ratio using a colon.

4. The table below shows the number of television viewers, in millions, who watch pay-cable channels, such as HBO and Showtime, each night of the week. (Source: Neilsen Media Research analyzed by Initiative Media North America)

Mon	Tue	Wed	Thu	Fri	Sat	Sun
3.9	4.5	4.2	3.9	5.2	7.1	5.5

   a. Arrange the numbers in the table from smallest to largest.
   b. Find the average number of viewers per night.

5. Approximately 90,000 women serve in the U.S. military. Five percent of these women serve in the Marine Corps. (Source: U.S. Department of Defense)
   a. Approximately how many women are in the Marine Corps?
   b. What fractional amount of women in the military are in the Marine Corps?

Go Figure

I have 2 brothers and 1 sister. My father's parents have 10 grandchildren. My mother's parents have 11 grandchildren. If no divorces or remarriages occurred, how many first cousins do I have?

NEW! Prep Test and Go Figure

Prep Tests occur at the beginning of each chapter and test students on previously covered concepts that are required in the coming chapter. Answers are provided in the Answer Appendix. Objective references are also provided if a student needs to review specific concepts.

The **Go Figure** problem that follows the *Prep Test* is a playful puzzle problem designed to engage students in problem solving.

Aufmann Interactive Method (AIM)

Page 253

Page S16

An Interactive Approach

Basic College Mathematics: An Applied Approach uses an interactive style that provides a student with an opportunity to try a skill as it is presented. Each section is divided into objectives, and every objective contains one or more sets of matched-pair examples. The first example in each set is worked out; the second example, called "You Try It," is for the student to work. By solving this problem, the student actively practices concepts as they are presented in the text.

There are complete worked-out solutions to these examples in an appendix. By comparing their solution to the solution in the appendix, students obtain immediate feedback on, and reinforcement of, the concept.

NEW! AIM for Success Student Preface

This new student 'how to use this book' preface explains what is required of a student to be successful and how this text has been designed to foster student success, including the Aufmann Interactive Method (AIM). *AIM for Success* can be used as a lesson on the first day of class or as a project for students to complete to strengthen their study skills. There are suggestions for teaching this lesson in the *Instructor's Resource Manual*.

Page xxvii

Problem Solving

Focus on Problem Solving

At the end of each chapter is a Focus on Problem Solving feature which introduces the student to various successful problem-solving strategies. Strategies such as drawing a diagram, applying solutions to other problems, working backwards, inductive reasoning, and trial and error are some of the techniques that are demonstrated.

Focus on Problem Solving

Introduction to Problem Solving

You encounter problem-solving situations every day. Some problems are easy to solve, and you may mentally solve these problems without considering the steps you are taking in order to draw a conclusion. Others may be more challenging and may require more thought and consideration.

Suppose a friend suggests that you both take a trip over spring break. You'd like to go. What questions go through your mind? You might ask yourself some of the following questions:

How much will the trip cost? What will be the cost for travel, hotel rooms, meals, and so on?

Are some costs going to be shared by both me and my friend?

Can I afford it?

How much money do I have in the bank?

How much more money than I have now do I need?

How much time is there to earn that much money?

How much can I earn in that amount of time?

How much money must I keep in the bank in order to pay the next tuition bill (or some other expense)?

These questions require different mathematical skills. Determining the cost of the trip requires **estimation;** for example, you must use your knowledge of air fares or the cost of gasoline to arrive at an estimate of these costs. If some of the costs are going to be shared, you need to **divide** those costs by 2 in order to determine your share of the expense. The question regarding how much more money you need requires **subtraction:** the amount needed minus the amount currently in the bank. To determine how much money you can earn in the given amount of time requires **multiplication**—for example, the amount you earn per week times the number of weeks to be worked. To determine if the amount you can earn in the given amount of time is sufficient, you need to use your knowledge of **order relations** to compare the amount you can earn with the amount needed.

Facing the problem-solving situation described above may not seem difficult to you. The reason may be that you have faced similar situations before and, therefore, know how to work through this one. You may feel better prepared to deal with a circumstance such as this one because you know what questions to ask. An important aspect of learning to solve problems is learning what questions to ask. As you work through application problems in this text, try to become more conscious of the mental process you are going through. You might begin the process by asking yourself the following questions whenever you are solving an application problem.

1. Have I read the problem enough times to be able to understand the situation being described?

2. Will restating the problem in different words help me to understand the problem situation better?

3. What facts are given? (You might make a list of the information contained in the problem.)

4. What information is being asked for?

Page 53

Page 86

Problem-Solving Strategies

The text features a carefully developed approach to problem-solving that emphasizes the importance of *strategy* when solving problems. Students are encouraged to develop their own strategies—to draw diagrams, to write out the solution steps in words—as part of their solution to a problem. In each case, model strategies are presented as guides for students to follow as they attempt the "You Try It" problem. Having students provide strategies is a natural way to incorporate writing into the math curriculum.

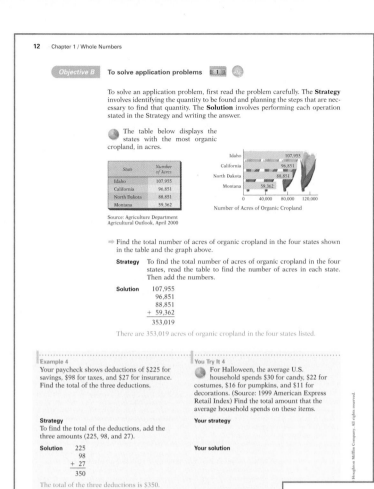

Real Data and Applications

Applications

One way to motivate an interest in mathematics is through applications. Wherever appropriate, the last objective of a section presents applications that require the student to use problem-solving strategies, along with the skills covered in that section, to solve practical problems. This carefully integrated applied approach generates student awareness of the value of arithmetic as a real-life tool.

Applications are taken from many disciplines including agriculture, business, carpentry, chemistry, construction, Earth science, education, manufacturing, nutrition, real estate, and sociology.

Page 24

Page 12 content:

Objective B   To solve application problems

To solve an application problem, first read the problem carefully. The **Strategy** involves identifying the quantity to be found and planning the steps that are necessary to find that quantity. The **Solution** involves performing each operation stated in the Strategy and writing the answer.

The table below displays the states with the most organic cropland, in acres.

State	Number of Acres
Idaho	107,955
California	96,851
North Dakota	88,851
Montana	59,362

Source: Agriculture Department Agricultural Outlook, April 2000

Idaho 107,955
California 96,851
North Dakota 88,851
Montana 59,362

0   40,000   80,000   120,000
Number of Acres of Organic Cropland

→ Find the total number of acres of organic cropland in the four states shown in the table and the graph above.

Strategy   To find the total number of acres of organic cropland in the four states, read the table to find the number of acres in each state. Then add the numbers.

Solution
```
   107,955
    96,851
    88,851
 +  59,362
   353,019
```

There are 353,019 acres of organic cropland in the four states listed.

Example 4

Your paycheck shows deductions of $225 for savings, $98 for taxes, and $27 for insurance. Find the total of the three deductions.

Strategy

To find the total of the deductions, add the three amounts (225, 98, and 27).

Solution
```
  225
   98
 + 27
  350
```

The total of the three deductions is $350.

You Try It 4

For Halloween, the average U.S. household spends $30 for candy, $22 for costumes, $16 for pumpkins, and $11 for decorations. (Source: 1999 American Express Retail Index) Find the total amount that the average household spends on these items.

Your strategy

Your solution

Page 12

Real Data

Real data examples and exercises, identified by , ask students to analyze and solve problems taken from actual situations. Students are often required to work with tables, graphs, and charts drawn from a variety of disciplines.

Page 24 content:

112. The leader in pass receiving for the 1990s in the National Football League was Jerry Rice with 12,078 yards. His closest competitor was Michael Irvin with 10,872 yards. (Source: NFL) How many more yards did Jerry Rice gain than Michael Irvin?

113. Use the graph at the right to find the difference between the maximum height to which Great Fountain geyser erupts and the maximum height to which Valentine erupts.

114. According to the graph at the right, how much higher is the eruption of the Giant than that of Old Faithful?

The Maximum Heights of the Eruptions of Six Geysers at Yellowstone National Park

115. The top six live acts of the 1990s are shown in the table at the right.
a. How much more did the Rolling Stones gross from their concerts in the 1990s than the Grateful Dead?
b. How much more did Neil Diamond gross from his concerts in the 1990s than Jimmy Buffett?

Live Acts	Gross
Rolling Stones	$751,000,000
Grateful Dead	$285,000,000
U2	$283,000,000
Eagles	$197,000,000
Neil Diamond	$183,000,000
Jimmy Buffett	$167,000,000

Source: Amusement Business BOXSCORE

The graph at the right shows the expected U.S. population aged 100 and over, every two years from 2010 to 2020. Use this information for Exercises 116 and 117.

116. What is the expected growth in the population aged 100 and over during the 10-year period?

117. a. Which 2-year period has the smallest increase in the number of people aged 100 and over?
b. Which 2-year period has the greatest increase?

Expected U.S. Population Aged 100 and Over
Source: Census Bureau

APPLYING THE CONCEPTS

118. Answer true or false.
a. The phrases "the difference between 9 and 5" and "5 less than 9" mean the same thing.
b. $9 - (5 - 3) = (9 - 5) - 3$.
c. Subtraction is an associative operation. *Hint:* See part b of this exercise.

119. Make up a word problem for which the difference between 15 and 8 is the answer.

Student Pedagogy

Icons

The at each objective head remind students that both a video and a tutorial lesson are available for that objective.

Key Terms and Concepts

Key terms, in bold, emphasize important terms. The key terms are also provided in a **Glossary** at the back of the text.

Key concepts are presented in purple boxes in order to highlight these important concepts and to provide for easy reference.

Take Note

These margin notes alert students to a point requiring special attention or are used to amplify the concept under discussion.

6.3 Interest

Objective A   To calculate simple interest

When you deposit money in a bank—for example, in a savings account—you are permitting the bank to use your money. The bank may use the deposited money to lend customers the money to buy cars or make renovations on their homes. The bank pays you for the privilege of using your money. The amount paid to you is called **interest.** If you are the one borrowing money from the bank, the amount you pay for the privilege of using that money is also called interest.

The original amount deposited or borrowed is called the **principal.** The amount of interest paid is usually given as a percent of the principal. The percent used to determine the amount of interest is the **interest rate.**

TAKE NOTE
If you deposit $1000 in a savings account paying 5% interest, the $1000 is the principal and 5% is the interest rate.

Interest paid on the original principal is called **simple interest.** To calculate simple interest, multiply the principal by the interest rate per period by the number of time periods. In this objective, we are working with annual interest rates, so the time periods are years. The simple interest formula for an annual interest rate is given below.

> **Simple Interest Formula for Annual Interest Rates**
> Principal $\times$ annual interest rate $\times$ time (in years) $=$ interest

Page 243

Page 10

CALCULATOR NOTE

A scientific calculator is a useful tool in mathematical computation. To add $24 + 71$ with your calculator, enter the following:

$24 \boxed{+} 71 \boxed{=}$

➡ What is the sum of 24 and 71?

$$\begin{array}{r} 24 \\ + 71 \\ \hline 95 \end{array}$$

• The phrase *the sum of* means to add.

The sum of 24 and 71 is 95.

When the sum of the digits in a column exceeds 9, the addition will involve "carrying."

➡ Add: $487 + 369$

$$\begin{array}{r} \overset{1}{4}\ 8\ 7 \\ + 3\ 6\ 9 \\ \hline 6 \end{array}$$

• Add the ones column.
$7 + 9 = 16$ (1 ten + 6 ones).
Write the 6 in the ones column and carry the 1 ten to the tens column.

$$\begin{array}{r} \overset{1}{4}\ \overset{1}{8}\ 7 \\ + 3\ 6\ 9 \\ \hline 5\ 6 \end{array}$$

• Add the tens column.
$1 + 8 + 6 = 15$ (1 hundred + 5 tens).
Write the 5 in the tens column and carry the 1 hundred to the hundreds column.

$$\begin{array}{r} \overset{1}{4}\ \overset{1}{8}\ 7 \\ + 3\ 6\ 9 \\ \hline 8\ 5\ 6 \end{array}$$

• Add the hundreds column.
$1 + 4 + 3 = 8$ (8 hundreds).
Write the 8 in the hundreds column.

Calculator Note

These margin notes provide suggestions for using a calculator in certain situations.

Annotated Examples

Examples indicated by ➡ use annotations in blue to explain what is happening in key steps of the complete, worked-out solutions.

Page 50

Point of Interest

Prime numbers are an important part of cryptology, the study of secret codes. To make it less likely that codes can be broken, cryptologists use prime numbers that have hundreds of digits.

A number is a **prime number** if its only whole-number factors are 1 and itself. 7 is prime because its only factors are 1 and 7. If a number is not prime, it is called a **composite number.** Because 6 has factors of 2 and 3, 6 is a composite number. The number 1 is not considered a prime number; therefore it is not included in the following list of prime numbers less than 50.

2, 3, 5, 7, 11, 13, 17, 19, 23, 29, 31, 37, 41, 43, 47

Point of Interest

These margin notes contain interesting sidelights about mathematics, its history, or its application.

Exercises and Projects

82. 206)3097   **83.** 504)6504   **84.** 654)1217   **85.** 546)2344

86. Find the quotient of 5432 and 21.   **87.** Find the quotient of 8507 and 53.

88. What is 37,294 divided by 72?   **89.** What is 76,788 divided by 46?

90. Find 23,457 divided by 43. Round to the nearest hundred.   **91.** Find 341,781 divided by 43. Round to the nearest ten.

 Estimate and then use your calculator to divide.

92. 76)389,804   **93.** 53)117,925   **94.** 29)637,072   **95.** 67)738,072

96. 38)934,648   **97.** 34)906,304   **98.** 309)876,324   **99.** 642)323,568

100. 209)632,016   **101.** 614)332,174   **102.** 179)5,734,444   **103.** 374)7,712,254

Objective D *Application Problems*

The graph at the right shows the annual expenditures, in a recent year, of the average household in the United States. Use this information for Exercises 104 to 106. Round answers to the nearest whole number.

104. What is the total amount spent annually by the average household in the United States?

105. What is the average monthly expense for housing?

106. What is the difference between the average monthly expense for food and the average monthly expense for health care?

Other $5366
Entertainment $1746
Housing $11,713
Health Care $1903
Insurance $3381
Food $4810
Transportation $6616

Average Annual Household Expenses
Source: Bureau of Labor Statistics Consumer Expenditure Survey

Page 43

Exercises

The exercise sets of *Basic College Mathematics: An Applied Approach* emphasize skill building, skill maintenance, and applications. Concept-based writing or developmental exercises have been integrated with the exercise sets. Icons identify appropriate writing , data analysis , and calculator exercises.

Included in each exercise set are **Applying the Concepts** that present extensions of topics, require analysis, or offer challenge problems. The writing exercises ask students to explain answers, write about a topic in the section, or research and report on a related topic.

APPLYING THE CONCEPTS

The table at the right shows how each dollar of projected spending by the federal government for a recent year was distributed. The costs for Social Security, health, income security, net interest, and the category education, veterans, and transportation are fixed expenditures. Defense and "Other" represent the only discretionary spending by the government. The projected budget for the year was $1977.6 billion.

72. a. Is at least one-half of federal spending discretionary spending?
b. Using a colon, write the ratio of the expenditures for "health" to the discretionary spending.
c. Find the amount of the budget to be spent on fixed expenditures.
d. Find the amount of the budget to be spent on Social Security.

73. A survey of voters in a city claimed that 2 people of every 5 who voted cast a ballot in favor of city amendment A and that 3 people of every 4 who voted cast a ballot against amendment A. Is this possible? Explain your answer.

How Your Federal Tax Dollar Is Spent	
Social Security	23 cents
Health	21 cents
Income security	14 cents
Net interest	11 cents
Edu., Vets, Transp.	10 cents
Defense	16 cents
Other	5 cents

Source: Tax Foundation, Office of Management and Budget

Page 184

Projects and Group Activities

The Projects and Group Activities feature at the end of each chapter can be used as extra credit or for cooperative learning activities. The projects cover various aspects of mathematics, including the use of calculators, collecting data from the Internet, data analysis, and extended applications.

Page 279

Credit Card Charges All credit card companies assess a finance charge when a credit card balance is not paid within a certain number of days of the *billing date* (the date the credit card bill is sent). There may also be an annual fee.

The table below shows the charges for five banks in September of 2000. (If you have Internet access, you can get current rates by visiting **http://www.bankrate.com**.)

Bank	Annual Interest Rate	Annual Fee
Pulaski Bank & Trust Co.	8.75%	$35
Wachovia Bank Card Services	9.50%	$88
USAA Savings Bank	10.50%	$45
Metropolitan National Bank	10.98%	$25
First Internet Bank of Indiana	12.00%	$0

The amount of the monthly finance charge is usually based on the average daily balance. For instance, say your credit card was issued by the USAA Savings Bank. If your average daily balance for one month was $375.89, then the monthly finance charge is calculated, using the simple interest formula, as follows:

TAKE NOTE
The monthly interest rate is the annual interest rate divided by 12.

$$\boxed{\text{Interest}} = \boxed{\text{principal}} \times \boxed{\text{monthly interest rate}} \times \boxed{\text{time (in months)}}$$

$$= \quad 375.89 \quad \times \quad \frac{0.1050}{12} \quad \times \quad 1$$

$$\approx 3.29$$

The finance charge is $3.29.

1. Suppose you have an average daily balance of $421.65 for 12 months. With which bank will your *annual* charges (the finance charges and the annual fee) be least?

2. Suppose you have an average daily balance of $1175.91 for 12 months. With which bank will your *annual* charges (the finance charges and the annual fee) be least?

End of Chapter

Chapter Summary

At the end of each chapter there is a Chapter Summary that includes Key Words and Essential Rules that were covered in the chapter. These chapter summaries provide a single point of reference as the student prepares for a test. Each concept references the page number from the lesson where the concept is introduced.

Chapter Summary

Key Words
Quantities such as 8 feet and 60 miles are number quantities written with *units*. [p. 169]

A *ratio* is a comparison of two quantities that have the same units. [p. 169]

A ratio is in *simplest form* when the two numbers that make up the ratio have no common factors. [p. 169]

A *rate* is a comparison of two quantities that have different units. [p. 173]

A rate is in *simplest form* when the numbers that make up the rate have no common factors. [p. 173]

A *unit rate* is a rate in which the number in the denominator is 1. [p. 173]

A *proportion* is an expression of the equality of two ratios or rates. [p. 177]

A proportion is true if the fractions are equal when written in lowest terms. In any true proportion, the "cross products" are equal. [p. 177]

Essential Rules
Ways to Express a Ratio
A ratio can be written in three different ways:

1. As a fraction
2. As two numbers separated by a colon (:)
3. As two numbers separated by the word *to* [p. 169]

To Find Unit Rates
To find unit rates, divide the number in the numerator of the rate by the number in the denominator of the rate. [p. 173]

To Solve a Proportion
One of the numbers in a proportion may be unknown. To solve a proportion, find a number to replace the unknown so that the proportion is true. [p. 178]

Page 188

Page 189

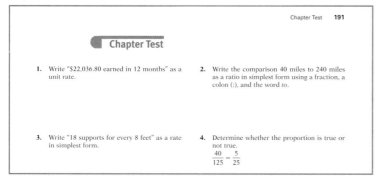

Chapter Review   Chapter Review   **189**

1. Determine whether the proportion is true or not true.
$$\frac{2}{9} = \frac{10}{45}$$

2. Write the comparison 32 dollars to 80 dollars as a ratio in simplest form using a fraction, a colon (:), and the word *to*.

3. Write "250 miles in 4 hours" as a unit rate.

4. Determine whether the proportion is true or not true.
$$\frac{8}{15} = \frac{32}{60}$$

Chapter Review

Review exercises are found at the end of each chapter. These exercises are selected to help the student integrate all of the topics presented in the chapter.

Page 191

Chapter Test   Chapter Test   **191**

1. Write "$22,036.80 earned in 12 months" as a unit rate.

2. Write the comparison 40 miles to 240 miles as a ratio in simplest form using a fraction, a colon (:), and the word *to*.

3. Write "18 supports for every 8 feet" as a rate in simplest form.

4. Determine whether the proportion is true or not true.
$$\frac{40}{125} = \frac{5}{25}$$

Chapter Test

The Chapter Test exercises are designed to simulate a possible test of the material in the chapter.

Cumulative Review

Cumulative Review exercises, which appear at the end of each chapter (beginning with Chapter 2), help students maintain skills learned in previous chapters.

The answers to all Chapter Review exercises, all Chapter Test exercises, and all Cumulative Review exercises are given in the Answer Section. Along with the answer, there is a reference to the objective that pertains to each exercise.

Page 193

Cumulative Review   **193**

Cumulative Review

1. Subtract: 20,095
           − 10,937

2. Write $2 \cdot 2 \cdot 2 \cdot 2 \cdot 3 \cdot 3 \cdot 3$ in exponential notation.

3. Simplify: $4 - (5 - 2)^2 \div 3 + 2$

4. Find the prime factorization of 160.

Supplements for the Instructor

Basic College Mathematics: An Applied Approach has a complete set of teaching aids for the instructor.

Instructor's Annotated Edition   This edition contains a replica of the student text and additional items just for the instructor. These include: *Instructor Notes, Transparency Master icons, In-Class Examples, Concept Checks, Discuss the Concepts, New Vocabulary/Symbols, etc.,Vocabulary/Symbols, etc. to Review, Optional Student Activities, Quick Quizzes, Answers to Writing Exercises,* and *Suggested Assignments.* Answers to all exercises are also provided.

Instructor's Solutions Manual   The *Instructor's Solutions Manual* contains worked-out solutions for all exercises in the text.

Instructor's Answer Booklet   The *Instructor's Answer Booklet* contains final answers for all exercises in the text.

Instructor's Resource Manual with Chapter Tests   The *Instructor's Resource Manual* contains ready-to-use printed Chapter Tests, which is the first of three sources of testing material. Eight printed tests (in two formats—free response and multiple choice) are provided for each chapter. Cumulative tests and final exams are also provided. These tests are available on the *Class Prep* CD or can be downloaded from our web site at math.college.hmco.com/instructors. The tests are in Microsoft Word® format and can be edited to suit the needs of the instructor. The *Instructor's Resource Manual* also includes transparency masters.

Printed Test Bank   The *Printed Test Bank,* the second component of the testing material, offers a printout of the items in the *HM Testing* computerized testing software, including an example of each algorithmic item. All multiple-choice items are carefully worded to make them appropriate to use as free-response items also.

NEW! ***HM Testing***   *HM Testing,* our computerized test generator, is our third source of testing material. The database contains test items and algorithms and is designed to produce an unlimited number of tests for each chapter of the text, including cumulative tests and final exams. These questions are unique to *HM Testing* and do not repeat items provided in the Chapter Tests of the *Instructor's Resource Manual.* It is available for Microsoft Windows® and the Macintosh. Both versions provide **on-line testing** and **gradebook** functions.

HM³ Tutorial (Instructor version)   This state-of-the-art tutorial CD software package for Microsoft Windows® was written and developed by the authors specifically for use with this text. Every objective is supported by the HM³ Tutor; exercises and quizzes are algorithmically generated; solution steps are animated; lessons and exercises are presented in a colorful, lively manner; and an integrated classroom management system tracks student performance.

NEW! ***WebCT ePacks***   *WebCT ePacks* provide instructors with a flexible, Internet-based education platform providing multiple ways to present learning materials. The *WebCT ePacks* come with a full array of features to enrich the on-line learning experience.

NEW! ***Blackboard Cartridges***   The *Houghton Mifflin Blackboard cartridge* allows flexible, efficient, and creative ways to present learning materials and opportunities. In addition to course management benefits, instructors may make use of an electronic grade book, receive papers from students enrolled in the course via the Internet, and track student use of the communication and collaboration functions.

NEW! *HM ClassPrep* These CD-ROMs contain a multitude of text-specific resources for instructors to use to enhance the classroom experience. The resources, or assets, are available as pdf and/or customizable Microsoft Word® files and include transparency masters and Chapter Tests from the IRM, to name only a few. These resources can be easily accessed from the CD-ROM by chapter or resource type. The CD can also link you to the text's web site.

NEW! *Instructor Text-Specific Web Site* The resources available on the *Class Prep* CD are also available on the instructor web site at math.college.hmco.com/ instructors. Appropriate items will be password-protected. Instructors also have access to the student part of the text's web site.

Supplements for the Student

Student Solutions Manual The *Student Solutions Manual* contains complete solutions to all odd-numbered exercises in the text.

Math Study Skills Workbook *by Paul D. Nolting* This workbook is designed to reinforce skills and minimize frustration for students in any math class, lab, or study skills course. It offers a wealth of study tips and sound advice on note taking, time management, and reducing math anxiety. In addition, numerous opportunities for self assessment enable students to track their own progress.

NEW! *HM eduSpace* eduSpace is a new content delivery system, combining an algorithmic tutorial program, on-line delivery of course materials, and classroom management functions. The interactive on-line content correlates directly to this text and can be accessed 24 hours a day.

HM³ Tutorial (Student version) This state-of-the-art tutorial CD software package for Microsoft Windows® was written and developed by the authors specifically for use with this text. *HM³* is an interactive tutorial containing lessons, exercises, and quizzes for every section of the text—lessons are presented in a colorful, lively manner and solution steps are animated. Exercises and quizzes are algorithmically generated. The lessons provide additional instruction and practice and can be used in several ways: (1) to cover material that was missed because of absence from class; (2) to reinforce instruction on a concept that has not yet been mastered; and (3) to review material in preparation for an examination. Following each lesson are practice exercises for the student to try. These exercises are created by carefully constructed algorithms that enable a student to practice a variety of exercise types. Because the exercises are created algorithmically, each time the student uses the tutorial, the student is presented with different problems. Following the practice exercises is an algorithmically-generated quiz. Next to every objective head in the text, the serves as a reminder that there is an *HM³* tutorial lesson corresponding to that objective.

NEW! *SMARTHINKING™ Live, On-Line Tutoring* Houghton Mifflin has partnered with *SMARTHINKING* to provide an easy-to-use, effective, on-line tutorial service. **Whiteboard Simulations** and **Practice Area** promote real-time visual interaction.

Three levels of service are offered.

- **Text-Specific Tutoring** provides real-time, one-on-one instruction with a specially qualified 'e-structor.'

- **Questions Any Time** allows students to submit questions to the tutor outside the scheduled hours and receive a reply within 24 hours.
- **Independent Study Resources** connect students with around-the-clock access to additional educational services, including interactive web sites, diagnostic tests, and Frequently Asked Questions posed to *SMARTHINKING* e-structors.

NEW! *Videos* This edition offers brand new text-specific videos, hosted by Dana Mosely, covering all sections of the text. These videos, professionally produced specifically for this text, offer a valuable resource for further instruction and review. Next to every objective head, the serves as a reminder that the objective is covered in a video lesson.

NEW! *Student Text-Specific Web Site* On-line student resources, such as section quizzes, can be found at this text's web site at math.college.hmco.com/students.

Acknowledgments

The authors would like to thank the people who have reviewed this manuscript and provided many valuable suggestions.

Barry Baron, *Baker College of Cadillac, MI*
Allan Berele, *De Paul University, IL*
Walter E. Burlage, *Lake City Community College, FL*
John Cabell Church, *Longview Community College, MO*
Dan Cortney, *Glendale Community College, AZ*
Sharon Dunn, *Southside Virginia Community College*
Anne Haney
Heidi A. Howard, *Florida Community College at Jacksonville, South Campus*
Domingo Javier Litong, *South Texas Community College*
John Searcy
Jackie Wing, *Angelica College, TX*

AIM for Success

Welcome to *Basic College Mathematics: An Applied Approach*. As you begin this course we know two important facts: (1) We want you to succeed. (2) You want to succeed. To do that requires an effort from each of us. For the next few pages, we are going to show you what is required of you to achieve that success and how you can use the features of this text to be successful.

Motivation

One of the most important keys to success is motivation. We can try to motivate you by offering interesting or important ways mathematics can benefit you. But, in the end, motivation must come from you. On the first day of class it is easy to be motivated. Eight weeks into the term, it is harder to keep that motivation.

To stay motivated, there must be outcomes from this course that are worth your time, money, and energy.

List some reasons you are taking this course.

_____

_____

_____

TAKE NOTE
Motivation alone will not lead to success. For instance, suppose a person who cannot swim is placed in a boat, taken out to the middle of a lake, and then thrown overboard. That person has a lot of motivation but there is a high likelihood the person will drown without some help. Motivation gives us the desire to learn but is not the same as learning.

Although we hope that one of the reasons you listed was an interest in mathematics, we know that many of you are taking this course because it is required to graduate, it is a prerequisite for a course you must take, or because it is required for your major. Although you may not agree that this course should be necessary, it is! If you are motivated to graduate or to complete the requirements for your major, then use that motivation to succeed in this course. Do not become distracted from your goal to complete your education!

Commitment

To be successful, you must make a commitment to succeed. This means devoting time to math so that you achieve a better understanding of the subject.

List some activities (sports, hobbies, talents such as dance, art or music) that you enjoy and at which you would like to become better.

ACTIVITY	TIME SPENT	TIME WISHED SPENT
_____	_____	_____
_____	_____	_____
_____	_____	_____

Thinking about these activities, put the number of hours that you spend each week practicing these activities next to the activity. Next to that number, indicate the number of hours a week you would like to spend on these activities.

Whether you listed surfing or sailing, aerobics or restoring cars, or any other activity you enjoy, note how many hours a week you spend doing it. To succeed in math, you must be willing to commit the same amount of time. Success requires some sacrifice.

The "I Can't Do Math" Syndrome

There may be things you cannot do, for instance, lift a two-ton boulder. You can, however, do math. It is much easier than lifting the two-ton boulder. When you first learned the activities you listed above, you probably could not do them well. With practice, you got better. With practice, you will be better at math. Stay focused, motivated, and committed to success.

It is difficult for us to emphasize how important it is to overcome the "I Can't Do Math Syndrome." If you listen to interviews of very successful athletes after a particularly bad performance, you will note that they focus on the positive aspect of what they did—not the negative. Sports psychologists encourage athletes to always be positive—to have a "Can Do" attitude. Develop this attitude toward math.

Strategies for Success

Textbook Reconnaissance   Right now, do a 15-minute "textbook reconnaissance" of this book. Here's how:

First, read the table of contents. Do it in three minutes or less. Next, look through the entire book, page by page. Move quickly. Scan titles, look at pictures, notice diagrams.

A textbook reconnaissance shows you where a course is going. It gives you the big picture. That's useful because brains work best when going from the general to the specific. Getting the big picture before you start makes details easier to recall and understand later on.

Your textbook reconnaissance will work even better if, as you scan, you look for ideas or topics that are interesting to you. List three facts, topics, or problems that you found interesting during your textbook reconnaissance.

_____

_____

_____

The idea behind this technique is simple: It's easier to work at learning material if you know it's going to be useful to you.

Not all the topics in this book will be "interesting" to you. But that is true of any subject. Surfers find that on some days the waves are better than others, musicians find some music more appealing than other music, computer gamers find some computer games more interesting than others, car enthusiasts find some cars more exciting than others. Some car enthusiasts would prefer to have a completely restored 1957 Chevrolet than a new Ferrari.

Know the Course Requirements   To do your best in this course, you must know exactly what your instructor requires. Course requirements may be stated in a *syllabus*, which is a printed outline of the main topics of the course, or they may be presented orally. When they are listed in a syllabus or on other printed pages, keep them in a safe place. When they are presented orally, make sure to take complete notes. In either case, it is important that you understand them completely and follow them exactly. Be sure you know the answer to each of the following questions.

1. What is your instructor's name?
2. Where is your instructor's office?
3. At what times does your instructor hold office hours?
4. Besides the textbook, what other materials does your instructor require?
5. What is your instructor's attendance policy?
6. If you must be absent from a class meeting, what should you do before returning to class? What should you do when you return to class?

7. What is the instructor's policy regarding collection or grading of homework assignments?

8. What options are available if you are having difficulty with an assignment? Is there a math tutoring center?

9. If there is a math lab at your school, where is it located? What hours is it open?

10. What is the instructor's policy if you miss a quiz?

11. What is the instructor's policy if you miss an exam?

12. Where can you get help when studying for an exam?

Remember: Your instructor wants to see you succeed. If you need help, ask! Do not fall behind. If you were running a race and fell behind by 100 yards, you might be able to catch up, but it would require more effort than had you not fallen behind.

Time Management   We know that there are demands on your time. Family, work, friends, and entertainment all compete for your time. We do not want to see you receive poor job evaluations because you are studying math. However, it is also true that we do not want to see you receive poor math test scores because you devoted too much time to work. When several competing and important tasks require your time and energy, the only way to manage the stress of being successful at both is to manage your time efficiently.

Instructors often advise students to spend twice the amount of time outside of class studying as they spend in the classroom. Time management is important to accomplishing this goal and being successful in school. The following activity is intended to help you structure your time more efficiently.

List the name of each course you are taking this term, the number of class hours each course meets, and the number of hours you should spend studying each subject outside of class. Then fill in a weekly schedule like the one printed below. Begin by writing in the hours spent in your classes, the hours spent at work (if you have a job), and any other commitments that are not flexible with respect to the time that you do them. Then begin to write down the commitments which are more flexible, including hours spent studying. Remember to reserve time for activities such as meals and exercise. You should also schedule free time.

TAKE NOTE

Besides time management, there must be realistic ideas of how much time is available. There are very few people who can *successfully* work full-time and go to school full-time. If you work 40 hours a week, take 15 units, spend the recommended study time given at the right, and sleep 8 hours a day, you will use over 80% of the available hours in a week. That leaves less than 20% of the hours in a week for family, friends, eating, recreation, and other activities.

	Monday	Tuesday	Wednesday	Thursday	Friday	Saturday	Sunday
7–8 a.m.							
8–9 a.m.							
9–10 a.m.							
10–11 a.m.							
11–12 p.m.							
12–1 p.m.							
1–2 p.m.							
2–3 p.m.							
3–4 p.m.							
4–5 p.m.							
5–6 p.m.							
6–7 p.m.							
7–8 p.m.							
8–9 p.m.							
9–10 p.m.							
10–11 p.m.							
11–12 a.m.							

We know that many of you must work. If that is the case, realize that a part-time job working 10 hours a week is equivalent to taking a three unit class. If you must work, consider letting your education progress at a slower rate to allow you to be successful at both work and school. There is no rule that says you must finish school in a certain time frame.

Schedule Study Time   As we encouraged you to do by filling out the time management form above, schedule a certain time to study. You should think of this time like being at work or in class. Reasons for missing "study time" should be as compelling as reasons for missing work or class. "I just didn't feel like it." is not a good reason to miss your scheduled study time.

Although this may seem like an obvious exercise, list a few reasons you might study.

_____

_____

_____

Of course we have no way of knowing the reasons you listed but from our experience one reason given quite frequently is, "To pass the course." There is nothing wrong with that reason. If that is the most important reason for you to study, then use it to stay focused.

One method of keeping to a study schedule is to form a ***study group.*** Look for people who are committed to learning, who pay attention in class, and who are punctual to class. Ask them to join your group. Choose people with similar educational goals but different methods of learning. You can gain insight from seeing the material from a new perspective. Limit groups to four or five people; larger groups are unwieldy.

There are many ways to conduct a study group. Begin with the following suggestions and see what works best for your group.

1. Test each other by asking questions. Each group member might bring two or three sample test questions to each meeting.
2. Practice teaching each other. Many of us that are teachers learned a lot about our subject when we had to explain it to someone else.
3. Compare class notes. You might ask other students about material in your notes that is difficult for you to understand.
4. Brainstorm test questions.
5. Set an agenda for each meeting. Set approximate time limits for each agenda item and determine a quitting time.

And now, probably the most important aspect of studying is that it should be done in relatively small chunks. If you can only study three hours a week for this course (probably not enough for most people), do it in blocks of one hour on three separate days, preferably after class. Three hours of studying on a Sunday is not as productive as three hours of paced study.

Text Features That Promote Success

There are 12 chapters in this text. Each chapter is divided into sections, and each section is subdivided into learning objectives. Each learning objective is labeled with a letter from A to D.

Preparing for a Chapter   Before you begin a new chapter, you should take some time to review previously learned skills. There are two ways to do this. The first is to complete the *Cumulative Review* which occurs after every chapter (except Chapter 1). For instance, turn to page 225. The questions in this review are taken from the previous chapters. The answers for all these exercises can be found on page A11. Turn to that page now and locate the answers for the Chapter 5 Cumulative Review. After the answer to the first exercise, which is 4, you will see the objective reference [1.6B]. This means that this question was taken from Chapter 1, Section 6, Objective B. If you missed this question, you should return to that objective and restudy the material.

A second way of preparing for a new chapter is to complete the ***Prep Test.*** This test focuses on the particular skills that will be required for the new chapter. Turn to page 196 to see a Prep Test. The answers for the Prep Test are the first set of answers in the answer section for a chapter. Turn to page A10 to see the answers for the Prep Test for Chapter 5. Note that an objective reference is given for each question. If you answer a question incorrectly, restudy the objective from which the question was taken.

Before the class meeting in which your professor begins a new section, you should read each objective statement. Next browse through the objective material, being sure to note each word in bold type. These words indicate important concepts that you must know to learn the material. Do not worry about trying to understand all the material. Your professor is there to assist you with that endeavor. The purpose of browsing through the material is that when it is presented to you, your brain will be prepared to accept and organize the new information.

Turn to page 3. Write down the title of the first objective in Section 1.1. Under the title of the objective, write down the words on the page that are in bold print. It is not necessary that you understand the meaning of these words. You are in this class to learn their meaning.

_____   _____

_____   _____

_____   _____

Math is Not a Spectator Sport   To learn mathematics you must be an active participant. Listening and watching your professor do mathematics is not enough. Mathematics requires that you interact with the lesson you are studying. If you filled in the blanks above, you were being interactive. There are other ways this textbook has been designed so that you can be an active learner.

Annotated Examples   An orange arrow indicates an example with explanatory remarks to the right of the work. Using paper and pencil, you should work along as you go through the example.

When you complete the example, get a clean sheet of paper. Write down the problem and then try to complete the solution without referring to your notes or the book. When you can do that, move on to the next part of the objective.

Leaf through the book now and write down the page numbers of two other occurrences of an arrowed example.

$$\frac{5}{6} = \frac{10}{12}$$
$$-\frac{1}{4} = \frac{3}{12}$$
$$\frac{7}{12}$$

Objective B   **To subtract fractions with unlike denominators**

To subtract fractions with unlike denominators, first rewrite the fractions as equivalent fractions with a common denominator. As with adding fractions, the common denominator is the LCM of the denominators of the fractions.

⇒ Subtract: $\frac{5}{6} - \frac{1}{4}$

The common denominator is the LCM of 6 and 4. The LCM = 12.	Write equivalent fractions using the LCM.	Subtract the fractions.
	$\frac{5}{6} = \frac{10}{12}$ $-\frac{1}{4} = \frac{3}{12}$	$\frac{5}{6} = \frac{10}{12}$ $-\frac{1}{4} = \frac{3}{12}$ $\frac{7}{12}$

You Try Its   One of the key instructional features of this text is the paired examples. Notice that in each example box, the example on the left is completely worked out and the "You Try It" example on the right is not. Study the worked-out example carefully by working through each step. Then work the You Try It. If you get stuck, refer to the page number at the end of the example which directs you to the page on which the You Try It is solved—a complete worked-out solution is provided. Try to use the given solution to get a hint for the step you are stuck on. Then try to complete your solution.

Example 2

Subtract: $\frac{11}{16} - \frac{5}{12}$

Solution

$\frac{11}{16} = \frac{33}{48}$ • LCM = 48

$-\frac{5}{12} = \frac{20}{48}$

$\frac{13}{48}$

You Try It 2

Subtract: $\frac{13}{18} - \frac{7}{24}$

Your solution

$\frac{13}{18} = \frac{52}{72}$

$-\frac{7}{24} = \frac{21}{72}$

$\frac{31}{72}$

Solution on p. S5

When you have completed your solution, check your work against the solution we provided. (Turn to page S5 to see the solution of You Try It 2.) Be aware that frequently there is more than one way to solve a problem. Your answer, however, should be the same as the given answer. If you have any question as to whether your method will "always work," check with your instructor or with someone in the math center.

Browse through the text book and write down the page numbers of two other occurrences of the paired example feature.

Remember: Be an active participant in your learning process. When you are sitting in class watching and listening to an explanation, you may think that you understand. However, until you actually try to do it, you will have no confirmation of the new knowledge or skill. Most of us have had the experience of sitting in class thinking we knew how to do something only to get home and realize we didn't.

Word Problems   Word problems are difficult because we must read the problem, determine the quantity we must find, think of a method to do that, and then actually solve the problem. A short summary of this process is to formulate a *strategy* to solve the problem and then devise a *solution*.

Note in the paired example below that part of every word problem is a strategy and a solution. The strategy is a written description of how we will solve the problem. In the corresponding You Try It, you are asked to formulate a strategy. Do not skip this step and be sure to write it out.

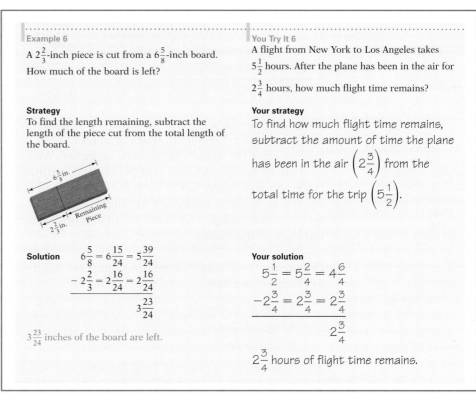

Example 6

A $2\frac{2}{3}$-inch piece is cut from a $6\frac{5}{8}$-inch board. How much of the board is left?

Strategy

To find the length remaining, subtract the length of the piece cut from the total length of the board.

Solution

$$6\frac{5}{8} = 6\frac{15}{24} = 5\frac{39}{24}$$
$$-2\frac{2}{3} = 2\frac{16}{24} = 2\frac{16}{24}$$
$$\overline{\hspace{2cm}3\frac{23}{24}}$$

$3\frac{23}{24}$ inches of the board are left.

You Try It 6

A flight from New York to Los Angeles takes $5\frac{1}{2}$ hours. After the plane has been in the air for $2\frac{3}{4}$ hours, how much flight time remains?

Your strategy

To find how much flight time remains, subtract the amount of time the plane has been in the air $\left(2\frac{3}{4}\right)$ from the total time for the trip $\left(5\frac{1}{2}\right)$.

Your solution

$$5\frac{1}{2} = 5\frac{2}{4} = 4\frac{6}{4}$$
$$-2\frac{3}{4} = 2\frac{3}{4} = 2\frac{3}{4}$$
$$\overline{\hspace{2cm}2\frac{3}{4}}$$

$2\frac{3}{4}$ hours of flight time remains.

Page 86

Rule Boxes   Pay special attention to rules placed in boxes. These rules give you the reasons certain types of problems are solved the way they are. When you see a rule, try to rewrite the rule in your own words.

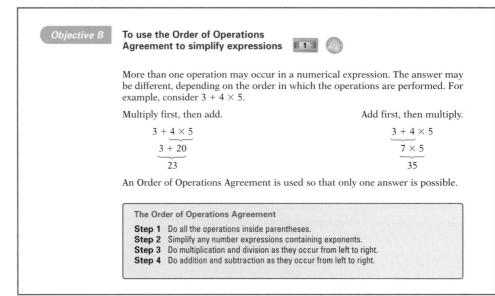

Parentheses
Exponents
Multiplication and division
Addition and subtraction

Page 46

TAKE NOTE

If you are working at home and need assistance, there is on-line help available at math.college.hmco.com/ students, at this text's web site.

Chapter Exercises   When you have completed studying an objective, do the exercises in the exercise set that correspond with that objective. The exercises are labeled with the same letter as the objective. Math is a subject that needs to be learned in small sections and practiced continually in order to be mastered. Doing all of the exercises in each exercise set will help you to master the problem-solving techniques necessary for success. As you work through the exercises for an objective, check your answers to the odd-numbered exercises with those in the back of the book.

Preparing for a Test   There are important features of this text that can be used to prepare for a test.

- Chapter Summary
- Chapter Review
- Chapter Test

After completing a chapter, read the Chapter Summary. (See pages 113–114 for the Chapter 2 Summary.) This summary highlights the important topics covered in the chapter. The page number following each topic refers you to the page in the text on which you can find more information about the concept.

Following the Chapter Summary are a Chapter Review (see page 115) and a Chapter Test (see page 117). Doing the review exercises is an important way of testing your understanding of the chapter. The answer to each review exercise is given at the back of the book along with its objective reference. After checking your answers, restudy any objective from which a question you missed was taken. It may be helpful to retry some of the exercises for that objective to reinforce your problem-solving techniques.

The Chapter Test should be used to prepare for an exam. We suggest that you try the Chapter Test a few days before your actual exam. Take the test in a quiet place and try to complete the test in the same amount of time you will be allowed for your exam. When taking the Chapter Test, practice the strategies of successful test takers: 1) scan the entire test to get a feel for the questions; 2) read the directions carefully; 3) work the problems that are easiest for you first; and perhaps most importantly, 4) try to stay calm.

When you have completed the Chapter Test, check your answers. If you missed a question, review the material in that objective and rework some of the exercises from that objective. This will strengthen your ability to perform the skills in that objective.

Your major goes here.⟶

Is it difficult to be successful? YES! Members of any successful music group, any successful artist, any professional athlete, any successful chef, and successful _____ have worked very hard to achieve their goals. They focused on their goals and ignored distractions. The things we ask you to do to achieve success take time and a commitment by you. We are confident that if you follow our suggestions, you will succeed.

Index of Applications

Basic College Mathematics:
An Applied Approach

1

Whole Numbers

When a contractor prepares an estimate for a job, many factors are considered: the number of workers needed, their hourly wages, how many hours they will work, the cost of the materials. **Exercises 103-105 on page 32** illustrate how a contractor would use operations on whole numbers to determine costs associated with a construction job.

Need help? For on-line student resources, such as section quizzes, visit this textbook's web site at **math.college.hmco.com/students**.

1. Name the number of ♦s shown below.

   ♦ ♦ ♦ ♦ ♦ ♦ ♦ ♦

2. Write the numbers from 1 to 10.

   1 ___ ___ ___ ___ ___ ___ ___ ___ 10

3. Match the number with its word form.
a. 4		**A.** five
b. 2		**B.** one
c. 5		**C.** zero
d. 1		**D.** four
e. 3		**E.** two
f. 0		**F.** three

Go Figure

Five adults and two children want to cross a river in a rowboat. The boat can hold one adult or two children or one child. Everyone is able to row the boat. What is the minimum number of trips that will be necessary for everyone to get to the other side?

1.1 Introduction to Whole Numbers

Objective A   **To identify the order relation between two numbers**

The **whole numbers** are 0, 1, 2, 3, 4, 5, 6, 7, 8, 9, 10, 11, 12, 13, 14, ...

The three dots mean that the list continues on and on and that there is no largest whole number.

Just as distances are associated with the markings on the edge of a ruler, the whole numbers can be associated with points on a line. This line is called the **number line.** The arrow on the number line below indicates that there is no largest whole number.

The **graph of a whole number** is shown by placing a heavy dot on the number line directly above the number. Here is the graph of 7 on the number line:

The number line can be used to show the order of whole numbers. A number that appears to the left of a given number is **less than** the given number. The symbol for "is less than" is <. A number that appears to the right of a given number is **greater than** the given number. The symbol for "is greater than" is >.

Four is less than seven.
4 < 7

Twelve is greater than seven.
12 > 7

Example 1   Graph 11 on the number line.

Solution

You Try It 1   Graph 6 on the number line.

Your solution

Example 2   Place the correct symbol, < or >, between the two numbers.
a. 39   24
b. 0   51

Solution   **a.** 39 > 24
b. 0 < 51

You Try It 2   Place the correct symbol, < or >, between the two numbers.
a. 45   29
b. 27   0

Your solution   **a.**
b.

Solutions on p. S1

Objective B To write whole numbers in words and in standard form

Point of Interest

The Babylonians had a place-value system based on 60. Its influence is still with us in angle measurement and time: 60 seconds in 1 minute, 60 minutes in 1 hour. It appears that the earliest record of a base-10 place-value system for natural numbers dates from the 8th century.

When a whole number is written using the digits 0, 1, 2, 3, 4, 5, 6, 7, 8, and 9, it is said to be in **standard form.** The position of each digit in the number determines the digit's **place value.** The diagram below shows a **place-value chart** naming the first twelve place values. The number 37,462 is in standard form and has been entered in the chart.

In the number 37,462, the position of the digit 3 determines that its place value is ten-thousands.

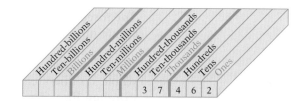

When a number is written in standard form, each group of digits separated by a comma is called a **period.** The number 3,786,451,294 has four periods. The period names are shown in red in the place-value chart above.

To write a number in words, start from the left. Name the number in each period. Then write the period name in place of the comma.

3,786,451,294 is read "three billion seven hundred eighty-six million four hundred fifty-one thousand two hundred ninety-four."

To write a whole number in standard form, write the number named in each period, and replace each period name with a comma.

Four million sixty-two thousand five hundred eighty-four is written 4,062,584. The zero is used as a place holder for the hundred-thousands place.

Example 3 Write 25,478,083 in words.

Solution Twenty-five million four hundred seventy-eight thousand eighty-three

You Try It 3 Write 36,462,075 in words.

Your solution

Example 4 Write three hundred three thousand three in standard form.

Solution 303,003

You Try It 4 Write four hundred fifty-two thousand seven in standard form.

Your solution

Solutions on p. S1

Objective C To write whole numbers in expanded form

The whole number 26,429 can be written in **expanded form** as

20,000 + 6000 + 400 + 20 + 9

The place-value chart can be used to find the expanded form of a number.

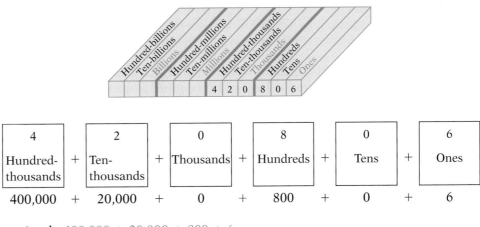

2		6		4		2		9
Ten-thousands	+	Thousands	+	Hundreds	+	Tens	+	Ones
20,000	+	6000	+	400	+	20	+	9

The number 420,806 is written in expanded form below. Note the effect of having zeros in the number.

4		2		0		8		0		6
Hundred-thousands	+	Ten-thousands	+	Thousands	+	Hundreds	+	Tens	+	Ones
400,000	+	20,000	+	0	+	800	+	0	+	6

or simply 400,000 + 20,000 + 800 + 6.

Example 5 Write 23,859 in expanded form.

Solution 20,000 + 3000 + 800 + 50 + 9

You Try It 5 Write 68,281 in expanded form.

Your solution

Example 6 Write 709,542 in expanded form.

Solution 700,000 + 9000 + 500 + 40 + 2

You Try It 6 Write 109,207 in expanded form.

Your solution

Solutions on p. S1

Objective D   **To round a whole number to a given place value**

When the distance to the moon is given as 240,000 miles, the number represents an approximation to the true distance. Giving an approximate value for an exact number is called **rounding.** A rounded number is always rounded to a given place value.

37 is closer to 40 than it is to 30. 37 rounded to the nearest ten is 40.

673 rounded to the nearest ten is 670. 673 rounded to the nearest hundred is 700.

A whole number is rounded to a given place value without using the number line by looking at the first digit to the right of the given place value.

If the digit to the right of the given place value is less than 5, that digit and all digits to the right are replaced by zeros.

⇒ Round 13,834 to the nearest hundred.

Given place value

13,834

3 < 5

13,834 rounded to the nearest hundred is 13,800.

If the digit to the right of the given place value is greater than or equal to 5, increase the digit in the given place value by 1, and replace all other digits to the right by zeros.

⇒ Round 386,217 to the nearest ten-thousand.

Given place value

386,217

6 > 5

386,217 rounded to the nearest ten-thousand is 390,000.

Example 7   Round 525,453 to the nearest ten-thousand.

Solution

Given place value

525,453

5 = 5

525,453 rounded to the nearest ten-thousand is 530,000.

You Try It 7   Round 368,492 to the nearest ten-thousand.

Your solution

Example 8   Round 1972 to the nearest hundred.

Solution

Given place value

1972

7 > 5

1972 rounded to the nearest hundred is 2000.

You Try It 8   Round 3962 to the nearest hundred.

Your solution

Solutions on p. S1

1.1 Exercises

Objective A

Graph the number on the number line.

1. 3

0 1 2 3 4 5 6 7 8 9 10 11 12

2. 5

0 1 2 3 4 5 6 7 8 9 10 11 12

3. 9

0 1 2 3 4 5 6 7 8 9 10 11 12

4. 0

0 1 2 3 4 5 6 7 8 9 10 11 12

Place the correct symbol, < or >, between the two numbers.

5. 37   49

6. 58   21

7. 101   87

8. 16   5

9. 245   158

10. 2701   2071

11. 0   45

12. 107   0

13. 815   928

Objective B

Name the place value of the digit 3.

14. 83,479

15. 3,491,507

16. 2,634,958

17. 76,319,204

Write the number in words.

18. 2675

19. 3790

20. 42,928

21. 58,473

22. 356,943

23. 498,512

24. 3,697,483

25. 6,842,715

Write the number in standard form.

26. Eighty-five

27. Three hundred fifty-seven

28. Three thousand four hundred fifty-six

29. Sixty-three thousand seven hundred eighty

30. Six hundred nine thousand nine hundred forty-eight

31. Seven million twenty-four thousand seven hundred nine

Objective C

Write the number in expanded form.

32. 5287   **33.** 6295   **34.** 58,943   **35.** 453,921

36. 200,583   **37.** 301,809   **38.** 403,705   **39.** 3,000,642

Objective D

Round the number to the given place value.

40. 926   Tens   **41.** 845   Tens

42. 1439   Hundreds   **43.** 3973   Hundreds

44. 43,607   Thousands   **45.** 52,715   Thousands

46. 647,989   Ten-thousands   **47.** 253,678   Ten-thousands

APPLYING THE CONCEPTS

Answer true or false for Exercise 48a and 48b. If the answer is false, give an example to show that it is false.

48.   **a.** If you are given two distinct whole numbers, then one of the numbers is always greater than the other number.
      b. A rounded-off number is always less than its exact value.

49. What is the largest three-digit whole number? What is the smallest five-digit whole number?

50. If 3846 is rounded to the nearest ten and then that number is rounded to the nearest hundred, is the result the same as what you get when you round 3846 to the nearest hundred? If not, which of the two methods is correct for rounding to the nearest hundred?

1.2 Addition of Whole Numbers

Objective A   **To add whole numbers**

Addition is the process of finding the total of two or more numbers.

Point of Interest

The first use of the plus sign appeared in 1489 in *Mercantile Arithmetic*. It was used to indicate a surplus and not as the symbol for addition. That use did not appear until around 1515.

By counting, we see that the total of $3 and $4 is $7.

$3 + $4 = $7

Addend   Addend   Sum

Addition can be illustrated on the number line by using arrows to represent the addends. The size, or magnitude, of a number can be represented on the number line by an arrow.

The number 3 can be represented anywhere on the number line by an arrow that is 3 units in length.

To add on the number line, place the arrows representing the addends head to tail, with the first arrow starting at zero. The sum is represented by an arrow starting at zero and stopping at the tip of the last arrow.

$$3 + 4 = 7$$

More than two numbers can be added on the number line.

$$3 + 2 + 4 = 9$$

Some special properties of addition that are used frequently are given below.

Addition Property of Zero

Zero added to a number does not change the number.

$$4 + 0 = 4$$
$$0 + 7 = 7$$

Commutative Property of Addition

Two numbers can be added in either order; the sum will be the same.

$$4 + 8 = 8 + 4$$
$$12 = 12$$

Associative Property of Addition

TAKE NOTE
This is the same addition problem shown on the number line above.

Grouping the addition in any order gives the same result. The parentheses are grouping symbols and have the meaning "Do the operations inside the parentheses first."

$$(3 + 2) + 4 = 3 + (2 + 4)$$
$$5 + 4 = 3 + 6$$
$$9 = 9$$

The number line is not useful for adding large numbers. The basic addition facts for adding one digit to one digit should be memorized. Addition of larger numbers requires the repeated use of the basic addition facts.

To add large numbers, begin by arranging the numbers vertically, keeping the digits of the same place value in the same column.

➡ Add: 321 + 6472

```
        THOUSANDS
          HUNDREDS
            TENS
              ONES

          3 : 2 : 1
      + 6 : 4 : 7 : 2
        6 : 7 : 9 : 3
```

• Add the digits in each column.

There are several words or phrases in English that indicate the operation of addition. Here are some examples:

added to	3 added to 5	5 + 3
more than	7 more than 5	5 + 7
the sum of	the sum of 3 and 9	3 + 9
increased by	4 increased by 6	4 + 6
the total of	the total of 8 and 3	8 + 3
plus	5 plus 10	5 + 10

CALCULATOR NOTE

A scientific calculator is a useful tool in mathematical computation. To add 24 + 71 with your calculator, enter the following:

24 [+] 71 [=]

➡ What is the sum of 24 and 71?

```
    24
  + 71
    95
```

• The phrase *the sum of* means to add.

The sum of 24 and 71 is 95.

When the sum of the digits in a column exceeds 9, the addition will involve "carrying."

➡ Add: 487 + 369

```
      HUNDREDS
        TENS
          ONES

          1
      4 : 8 : 7
    + 3 : 6 : 9
          : 6
```

• Add the ones column.
  7 + 9 = 16 (1 ten + 6 ones).
  Write the 6 in the ones column and carry the 1 ten to the tens column.

```
    1   1
    4   8   7
  + 3   6   9
        5   6
```

• Add the tens column.
  1 + 8 + 6 = 15 (1 hundred + 5 tens).
  Write the 5 in the tens column and carry the 1 hundred to the hundreds column.

```
    1   1
    4   8   7
  + 3   6   9
    8   5   6
```

• Add the hundreds column.
  1 + 4 + 3 = 8 (8 hundreds).
  Write the 8 in the hundreds column.

Example 1 Find the total of 17, 103, and 8.

Solution

$$
\begin{array}{r}
{\scriptstyle 1} \\
17 \\
103 \\
+\ \ \ 8 \\
\hline
128
\end{array}
$$

You Try It 1 What is 347 increased by 12,453?

Your solution

Example 2 Add: 89 + 36 + 98

Solution

$$
\begin{array}{r}
{\scriptstyle 2} \\
89 \\
36 \\
+\ 98 \\
\hline
223
\end{array}
$$

You Try It 2 Add: 95 + 88 + 67

Your solution

Example 3 Add:

$$
\begin{array}{r}
41{,}395 \\
4{,}327 \\
497{,}625 \\
+\ 32{,}991
\end{array}
$$

Solution

$$
\begin{array}{r}
{\scriptstyle 1\ 1\ 2\ \ 2\ 1} \\
41{,}395 \\
4{,}327 \\
497{,}625 \\
+\ 32{,}991 \\
\hline
576{,}338
\end{array}
$$

You Try It 3 Add:

$$
\begin{array}{r}
392 \\
4{,}079 \\
89{,}035 \\
+\ 4{,}992
\end{array}
$$

Your solution

Solutions on p. S1

E S T I M A T I O N

Estimation and Calculators

At some places in the text, you will be asked to use your calculator. Effective use of a calculator requires that you estimate the answer to the problem. This helps ensure that you have entered the numbers correctly and pressed the correct keys.

For example, if you use your calculator to find 22,347 + 5896 and the answer in the calculator's display is 131,757,912, you should realize that you have entered some part of the calculation incorrectly. In this case, you pressed ☒ instead of ⊞. By estimating the answer to a problem, you can help ensure the accuracy of your calculations. The symbol $\approx$ is used to denote **is approximately equal to.**

For example, to estimate the answer to 22,347 + 5896, round each number to the same place value. In this case, we will round to the nearest thousand. Then add.

$$
\begin{array}{r}
22{,}347 \approx\ \ \ \ 22{,}000 \\
+\ \ 5{,}896 \approx\ +\ \ 6{,}000 \\
\hline
28{,}000
\end{array}
$$

The sum 22,347 + 5896 is approximately 28,000. Knowing this, you would know that 131,757,912 is much too large and is therefore incorrect.

To estimate the sum of two numbers, first round each whole number to the same place value and then add. Compare this answer with the calculator's answer.

Objective B   **To solve application problems**

To solve an application problem, first read the problem carefully. The **Strategy** involves identifying the quantity to be found and planning the steps that are necessary to find that quantity. The **Solution** involves performing each operation stated in the Strategy and writing the answer.

The table below displays the states with the most organic cropland, in acres.

State	Number of Acres
Idaho	107,955
California	96,851
North Dakota	88,851
Montana	59,362

Source: Agriculture Department
Agricultural Outlook, April 2000

Number of Acres of Organic Cropland

⇒ Find the total number of acres of organic cropland in the four states shown in the table and the graph above.

Strategy   To find the total number of acres of organic cropland in the four states, read the table to find the number of acres in each state. Then add the numbers.

Solution   107,955
     96,851
     88,851
+ 59,362
   353,019

There are 353,019 acres of organic cropland in the four states listed.

Example 4

Your paycheck shows deductions of $225 for savings, $98 for taxes, and $27 for insurance. Find the total of the three deductions.

Strategy

To find the total of the deductions, add the three amounts (225, 98, and 27).

Solution   225
     98
+ 27
  350

The total of the three deductions is $350.

You Try It 4

For Halloween, the average U.S. household spends $30 for candy, $22 for costumes, $16 for pumpkins, and $11 for decorations. (Source: 1999 American Express Retail Index) Find the total amount that the average household spends on these items.

Your strategy

Your solution

Solution on p. S1

1.2 Exercises

Objective A

Add.

1.   17
    + 11

2.   25
    + 63

3.   83
    + 42

4.   63
    + 94

5.   77
    + 25

6.   63
    + 49

7.   56
    + 98

8.   86
    + 68

9.   658
    + 831

10.   842
    + 936

11.   735
    + 93

12.   189
    + 50

13.   859
    + 725

14.   637
    + 829

15.   470
    + 749

16.   427
    + 690

17.   36,925
    + 65,392

18.   56,772
    + 51,239

19.   50,873
    + 28,453

20.   34,872
    + 46,079

21.   878
    737
    + 189

22.   768
    461
    + 669

23.   319
    348
    + 912

24.   292
    579
    + 315

25.   9409
    3253
    + 7078

26.   8188
    8020
    + 7104

27.   2038
    2243
    + 3139

28.   4252
    6882
    + 5235

29.   67,428
    32,171
    + 20,971

30.   52,801
    11,664
    + 89,638

31.   76,290
    43,761
    + 87,402

32.   43,901
    98,301
    + 67,943

Add.

33. 20,958 + 3218 + 42

34. 80,973 + 5168 + 29

35. 392 + 37 + 10,924 + 621

36. 694 + 62 + 70,129 + 217

37. 294 + 1029 + 7935 + 65

38. 692 + 2107 + 3196 + 92

39. 97 + 7234 + 69,532 + 276

40. 87 + 1698 + 27,317 + 727

41. What is 9874 plus 4509?

42. What is 7988 plus 5678?

43. What is 3487 increased by 5986?

44. What is 99,567 added to 126,863?

45. What is 23,569 more than 9678?

46. What is 7894 more than 45,872?

47. What is 479 added to 4579?

48. What is 23,902 added to 23,885?

49. Find the total of 659, 55, and 1278.

50. Find the total of 4561, 56, and 2309.

51. Find the sum of 34, 329, 8, and 67,892.

52. Find the sum of 45, 1289, 7, and 32,876.

 Estimate by rounding to the nearest hundred. Then use your calculator to add.

53. 1234 + 9780 + 6740

54. 919 + 3642 + 8796

55. 241 + 569 + 390 + 1672

56. 107 + 984 + 1035 + 2904

 Estimate by rounding to the nearest thousand. Then use your calculator to add.

57.	**58.**	**59.**	**60.**
32,461	29,036	25,432	66,541
9,844	22,904	62,941	29,365
+ 59,407	+ 7,903	+ 70,390	+ 98,742

Estimate by rounding to the nearest ten-thousand. Then use your calculator to add.

61.   67,421
      82,984
      66,361
      10,792
   + 34,037

62.   21,896
       4,235
      62,544
      21,892
    + 1,334

63.   281,421
        9,874
      34,394
    526,398
  + 94,631

64.   542,698
      97,327
       7,235
      73,667
  + 173,201

 Estimate by rounding to the nearest million. Then use your calculator to add.

65.   28,627,052
        983,073
  + 3,081,496

66.   1,792,085
   29,919,301
  + 3,406,882

67.   12,377,491
    3,409,723
    7,928,026
 + 10,705,682

68.   46,751,070
   6,095,832
     280,011
 + 1,563,897

Objective B   *Application Problems*

69. In a recent year, according to the U.S. Department of Health and Human Services, there were 110,670 twin births in this country, 6919 triplet births, 627 quadruplet deliveries, and 79 quintuplet and other higher-order multiple births. Find the total number of multiple births during the year.

70. The Census Bureau estimates that the U.S. population will grow by 296 million people from 2000 to 2100. Given that the U.S. population in 2000 was 281 million, find the Census Bureau's estimate of the U.S. population in 2100.

The graph at the right shows the domestic box-office income from the *Star Wars* movies. Use this information for Exercises 71 to 73.

71. Estimate the total income from the four *Star Wars* movies.

72. Find the total income from the four *Star Wars* movies.

73. a. Find the total income from the two movies with the lowest box-office incomes.
   b. Does the total income from the two movies with the lowest box-office incomes exceed the income from the 1977 *Star Wars* production?

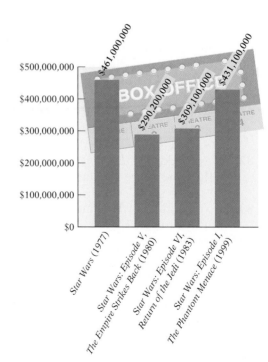

Source: www.worldwideboxoffice.com

74. The perimeter of a triangle is the sum of the lengths of the three sides of the triangle. Find the perimeter of a triangle that has sides that measure 12 inches, 14 inches, and 17 inches.

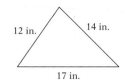

75. The odometer on a moving van reads 68,692. The driver plans to drive 515 miles the first day, 492 miles the second day, and 278 miles the third day.
    a. How many miles will be driven during the three days?
    b. What will the odometer reading be at the end of the trip?

The accompanying table shows the average amounts of money that all Americans have invested in selected assets and the average amounts invested for Americans between the ages of 16 and 34.

76. What is the total average amount for all Americans in checking accounts, savings accounts, and U.S. Savings Bonds?

77. What is the total average amount for Americans aged 16 to 34 in checking accounts, savings accounts, and U.S. Savings Bonds?

78. What is the total average amount for Americans aged 16 to 34 in all categories except home equity and retirement?

	All Americans	Ages 16 to 34
Checking accounts	$487	$375
Savings accounts	$3,494	$1,155
U.S. Savings Bonds	$546	$266
Money market	$10,911	$4,427
Stock/mutual funds	$4,510	$1,615
Home equity	$43,070	$17,184
Retirement	$9,016	$4,298

79. Is the sum of the average amounts invested in home equity and retirement for all Americans greater than or less than that same sum for Americans between the ages of 16 and 34?

APPLYING THE CONCEPTS

80. How many two-digit numbers are there? How many three-digit numbers are there?

81. If you roll two ordinary six-sided dice and add the two numbers that appear on top, how many different sums are possible?

82. If you add two *different* whole numbers, is the sum always greater than either one of the numbers? If not, give an example.

83. If you add two whole numbers, is the sum always greater than either one of the numbers? If not, give an example. (Compare this with the previous exercise.)

84. Make up a word problem for which the answer is the sum of 34 and 28.

85. Call a number "lucky" if it ends in a 7. How many lucky numbers are less than 100?

1.3 Subtraction of Whole Numbers

Objective A **To subtract whole numbers without borrowing**

Subtraction is the process of finding the difference between two numbers.

Point of Interest

The use of the minus sign dates from the same period as the plus sign, around 1515.

By counting, we see that the difference between $8 and $5 is $3.

$8 – $5 = $3

Minuend   Subtrahend   Difference

The difference 8 − 5 can be shown on the number line.

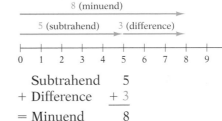

Note from the number line that addition and subtraction are related.

Subtrahend   5
+ Difference   + 3
= Minuend   8

The fact that the sum of the subtrahend and the difference equals the minuend can be used to check subtraction.

To subtract large numbers, begin by arranging the numbers vertically, keeping the digits that have the same place value in the same column. Then subtract the digits in each column.

➡ Subtract 8955 − 2432 and check.

```
THOUSANDS
  HUNDREDS
    TENS
      ONES
  8 | 9 | 5 | 5          Check:   Subtrahend      2432
− 2 | 4 | 3 | 2                 + Difference   + 6523
  6 | 5 | 2 | 3                 = Minuend        8955
```

Example 1 Subtract 6594 − 3271 and check.

Solution
```
  6594    Check:    3271
− 3271            + 3323
  3323              6594
```

You Try It 1 Subtract 8925 − 6413 and check.

Your solution

Example 2 Subtract 15,762 − 7541 and check.

Solution
```
  15,762    Check:    7,541
−  7,541            + 8,221
   8,221             15,762
```

You Try It 2 Subtract 17,504 − 9302 and check.

Your solution

Solutions on p. S1

Objective B | **To subtract whole numbers with borrowing**

In all the subtraction problems in the previous objective, for each place value the lower digit was not larger than the upper digit. When the lower digit *is* larger than the upper digit, subtraction will involve "borrowing."

➡ Subtract: 692 − 378

$$
\begin{array}{r}
6\ \overset{8+1}{9}\ 2 \\
-\ 3\ 7\ 8 \\
\hline
\end{array}
\qquad
\begin{array}{r}
6\ \overset{8+1}{9}\ \overset{10}{2} \\
-\ 3\ 7\ 8 \\
\hline
\end{array}
\qquad
\begin{array}{r}
6\ \overset{8}{9}\ \overset{12}{2} \\
-\ 3\ 7\ 8 \\
\hline
\end{array}
\qquad
\begin{array}{r}
6\ \overset{8}{9}\ \overset{12}{2} \\
-\ 3\ 7\ 8 \\
\hline
3\ \ 1\ \ 4
\end{array}
$$

Because 8 > 2, borrowing is necessary. 9 tens = 8 tens + 1 ten.

Borrow 1 ten from the tens column and write 10 in the ones column.

Add the borrowed 10 to 2.

Subtract the digits in each column.

The phrases below are used to indicate the operation of subtraction. An example is shown at the right of each phrase.

minus	8 minus 5	8 − 5
less	9 less 3	9 − 3
less than	2 less than 7	7 − 2
the difference between	the difference between 8 and 2	8 − 2
decreased by	5 decreased by 1	5 − 1

➡ Find the difference between 1234 and 485 and check.

"The difference between 1234 and 485" means 1234 − 485.

$$
\begin{array}{r}
1\ 2\ \overset{2}{3}\ \overset{14}{4} \\
-\ \ \ 4\ 8\ 5 \\
\hline
9
\end{array}
\quad
\begin{array}{r}
1\ \overset{1}{2}\ \overset{12}{3}\ \overset{14}{4} \\
-\ \ \ 4\ 8\ 5 \\
\hline
4\ 9
\end{array}
\quad
\begin{array}{r}
\overset{0}{1}\ \overset{11}{2}\ \overset{12}{3}\ \overset{14}{4} \\
-\ \ \ 4\ 8\ 5 \\
\hline
7\ 4\ 9
\end{array}
\quad
Check:\
\begin{array}{r}
^{1\,1} \\
485 \\
+\ 749 \\
\hline
1234
\end{array}
$$

Subtraction with a zero in the minuend involves repeated borrowing.

➡ Subtract: 3904 − 1775

$$
\begin{array}{r}
3\ \overset{8}{9}\ \overset{10}{0}\ 4 \\
-\ 1\ 7\ 7\ 5 \\
\hline
\end{array}
\qquad
\begin{array}{r}
\overset{\ \ \ 9}{}\ \\
3\ \overset{8}{9}\ \overset{10}{0}\ \overset{14}{4} \\
-\ 1\ 7\ 7\ 5 \\
\hline
\end{array}
\qquad
\begin{array}{r}
\overset{\ \ \ 9}{}\ \\
3\ \overset{8}{9}\ \overset{10}{0}\ \overset{14}{4} \\
-\ 1\ 7\ 7\ 5 \\
\hline
2\ 1\ 2\ 9
\end{array}
$$

5 > 4
There is a 0 in the tens column. Borrow 1 hundred (= 10 tens) from the hundreds column and write 10 in the tens column.

Borrow 1 ten from the tens column and add 10 to the 4 in the ones column.

Subtract the digits in each column.

Example 3 Subtract 4392 − 678 and check.

Solution

$$
\begin{array}{c}
\overset{3}{\cancel{4}}\ \overset{13}{\cancel{3}}\ \overset{8}{\cancel{9}}\ \overset{12}{\cancel{2}} \\
-\ \ \ 6\ \ 7\ \ 8 \\
\hline
3\ \ 7\ \ 1\ \ 4
\end{array}
\qquad
\begin{array}{r}
Check: \quad 678 \\
+\ 3714 \\
\hline
4392
\end{array}
$$

You Try It 3 Subtract 3481 − 865 and check.

Your solution

Example 4 Find 23,954 less than 63,221 and check.

Solution

$$
\begin{array}{c}
\overset{5}{\cancel{6}}\ \overset{12}{\cancel{3}},\ \overset{11}{\cancel{2}}\ \overset{11}{\cancel{2}}\ \overset{11}{\cancel{1}} \\
-\ 2\ 3,9\ 5\ 4 \\
\hline
3\ 9,2\ 6\ 7
\end{array}
\qquad
\begin{array}{r}
Check: \quad 23,954 \\
+39,267 \\
\hline
63,221
\end{array}
$$

You Try It 4 Find 54,562 decreased by 14,485 and check.

Your solution

Example 5 Subtract 46,005 − 32,167 and check.

You Try It 5 Subtract 64,003 − 54,936 and check.

Solution

$$
\begin{array}{c}
\overset{5\ \ 10}{46,\cancel{0}05} \\
-32,167
\end{array}
$$

• There are two zeros in the minuend. Borrow 1 thousand from the thousands column and write 10 in the hundreds column.

$$
\begin{array}{c}
\overset{9}{} \\
\overset{5\ \ \cancel{10}10}{46,\cancel{0}\cancel{0}5} \\
-32,167
\end{array}
$$

• Borrow 1 hundred from the hundreds column and write 10 in the tens column.

$$
\begin{array}{c}
\overset{9\ \ 9}{} \\
\overset{5\ \ \cancel{10}\cancel{10}15}{46,\cancel{0}\cancel{0}\cancel{5}} \\
-32,167 \\
\hline
13,838
\end{array}
$$

• Borrow 1 ten from the tens column and add 10 to the 5 in the ones column.

$$
\begin{array}{r}
Check: \quad 32,167 \\
+\ 13,838 \\
\hline
46,005
\end{array}
$$

Your solution

Solutions on p. S1

ESTIMATION

Estimating the Difference Between Two Whole Numbers

Estimate and then use your calculator to find 323,502 − 28,912.

To estimate the difference between two numbers, round each number to the same place value. In this case we will round to the nearest ten-thousand. Then subtract. The estimated answer is 290,000.

$$
\begin{array}{r}
323,502 \approx \ \ \ 320,000 \\
-\ \ 28,912 \approx -\ \ 30,000 \\
\hline
290,000
\end{array}
$$

Now use your calculator to find the exact difference. The exact answer is 294,590.

323502 ⊟ 28912 ⊜ 294590

Objective C   **To solve application problems**

The table at the right shows the number of personnel on active duty in the U.S. branches of the military in 1940 and 1945. Use this table for Example 6 and You Try It 6.

Branch	1940	1945
U.S. Army	267,767	8,266,373
U.S. Navy	160,997	3,380,817
U.S. Air Force	51,165	2,282,259
U.S. Marine Corps	28,345	474,680

Source: Dept. of the Army, Dept. of the Navy, Air Force Dept., Dept. of the Marines, U.S. Dept. of Defense

Example 6

Find the difference between the number of U.S. Army personnel on active duty in 1940 and in 1945.

Strategy

To find the difference, subtract the number of U.S. Army personnel on active duty in 1940 (267,767) from the number on active duty in 1945 (8,266,373).

Solution
$$\begin{array}{r} 8,266,373 \\ -\ \ \ 267,767 \\ \hline 7,998,606 \end{array}$$

There were 7,998,606 more personnel on active duty in the U.S. Army in 1945 than in 1940.

You Try It 6

Find the difference between the number of personnel on active duty in the Navy and in the Air Force in 1945.

Your strategy

Your solution

Example 7

You had a balance of $415 on your student debit card. You then used the card, deducting $197 for books, $48 for art supplies, and $24 for theater tickets. What is your new student debit card balance?

Strategy

To find your new debit card balance:

• Add to find the total of the three deductions (197 + 48 + 24).
• Subtract the total of the three deductions from the old balance (415).

Solution

$$\begin{array}{r} 197 \\ 48 \\ +\ 24 \\ \hline 269 \end{array} \text{ total deductions} \qquad \begin{array}{r} 415 \\ -269 \\ \hline 146 \end{array}$$

Your new debit card balance is $146.

You Try It 7

Your total weekly salary is $638. Deductions of $127 for taxes, $18 for insurance, and $35 for savings are taken from your pay. Find your weekly take-home pay.

Your strategy

Your solution

Solutions on pp. S1–S2

1.3 Exercises

Objective A

Subtract.

1. $\begin{array}{r} 9 \\ -5 \\ \hline \end{array}$ **2.** $\begin{array}{r} 8 \\ -7 \\ \hline \end{array}$ **3.** $\begin{array}{r} 8 \\ -4 \\ \hline \end{array}$ **4.** $\begin{array}{r} 7 \\ -3 \\ \hline \end{array}$ **5.** $\begin{array}{r} 10 \\ -0 \\ \hline \end{array}$

6. $\begin{array}{r} 11 \\ -4 \\ \hline \end{array}$ **7.** $\begin{array}{r} 12 \\ -8 \\ \hline \end{array}$ **8.** $\begin{array}{r} 19 \\ -8 \\ \hline \end{array}$ **9.** $\begin{array}{r} 15 \\ -6 \\ \hline \end{array}$ **10.** $\begin{array}{r} 16 \\ -7 \\ \hline \end{array}$

11. $\begin{array}{r} 25 \\ -3 \\ \hline \end{array}$ **12.** $\begin{array}{r} 55 \\ -4 \\ \hline \end{array}$ **13.** $\begin{array}{r} 68 \\ -8 \\ \hline \end{array}$ **14.** $\begin{array}{r} 77 \\ -3 \\ \hline \end{array}$ **15.** $\begin{array}{r} 89 \\ -23 \\ \hline \end{array}$

16. $\begin{array}{r} 54 \\ -21 \\ \hline \end{array}$ **17.** $\begin{array}{r} 88 \\ -57 \\ \hline \end{array}$ **18.** $\begin{array}{r} 1202 \\ -701 \\ \hline \end{array}$ **19.** $\begin{array}{r} 1305 \\ -404 \\ \hline \end{array}$ **20.** $\begin{array}{r} 1763 \\ -801 \\ \hline \end{array}$

21. $\begin{array}{r} 1497 \\ -706 \\ \hline \end{array}$ **22.** $\begin{array}{r} 8974 \\ -3972 \\ \hline \end{array}$ **23.** $\begin{array}{r} 2836 \\ -1711 \\ \hline \end{array}$ **24.** $\begin{array}{r} 8976 \\ -7463 \\ \hline \end{array}$ **25.** $\begin{array}{r} 9273 \\ -6142 \\ \hline \end{array}$

26. $77 - 36$ **27.** $129 - 82$ **28.** $132 - 61$ **29.** $969 - 44$ **30.** $1347 - 103$

31. $4865 - 304$ **32.** $1525 - 702$ **33.** $9999 - 6794$ **34.** $7806 - 3405$ **35.** $8843 - 7621$

36. What is 3795 minus 1092?   **37.** What is 9071 minus 6050?

38. Find the difference between 9763 and 541.   **39.** Find the difference between 6094 and 3072.

40. What is 3701 less than 6932?   **41.** What is 2031 less than 5071?

42. Find 6509 decreased by 3102.   **43.** Find 7994 decreased by 7782.

44. Find 23,907 less 12,705.   **45.** Find 65,986 less 5741.

Objective B

Subtract.

46. $\begin{array}{r} 71 \\ -18 \\ \hline \end{array}$

47. $\begin{array}{r} 93 \\ -28 \\ \hline \end{array}$

48. $\begin{array}{r} 47 \\ -18 \\ \hline \end{array}$

49. $\begin{array}{r} 44 \\ -27 \\ \hline \end{array}$

50. $\begin{array}{r} 37 \\ -29 \\ \hline \end{array}$

51. $\begin{array}{r} 50 \\ -27 \\ \hline \end{array}$

52. $\begin{array}{r} 70 \\ -33 \\ \hline \end{array}$

53. $\begin{array}{r} 993 \\ -537 \\ \hline \end{array}$

54. $\begin{array}{r} 250 \\ -192 \\ \hline \end{array}$

55. $\begin{array}{r} 840 \\ -783 \\ \hline \end{array}$

56. $\begin{array}{r} 768 \\ -194 \\ \hline \end{array}$

57. $\begin{array}{r} 770 \\ -395 \\ \hline \end{array}$

58. 674 − 337

59. 3526 − 387

60. 1712 − 289

61. 4350 − 729

62. 1702 − 948

63. 1607 − 869

64. 5933 − 3754

65. 7293 − 3748

66. 9407 − 2918

67. 3706 − 2957

68. 8605 − 7716

69. 8052 − 2709

70. 80,305 − 9176

71. 70,702 − 4239

72. 10,004 − 9306

73. 80,009 − 63,419

74. 70,618 − 41,213

75. 80,053 − 27,649

76. 70,700 − 21,076

77. 80,800 − 42,023

78. $\begin{array}{r} 2600 \\ -1972 \\ \hline \end{array}$

79. $\begin{array}{r} 8400 \\ -3762 \\ \hline \end{array}$

80. $\begin{array}{r} 9003 \\ -2471 \\ \hline \end{array}$

81. $\begin{array}{r} 6004 \\ -2392 \\ \hline \end{array}$

82. $\begin{array}{r} 8202 \\ -3916 \\ \hline \end{array}$

83. $\begin{array}{r} 7050 \\ -4137 \\ \hline \end{array}$

84. $\begin{array}{r} 7015 \\ -2973 \\ \hline \end{array}$

85. $\begin{array}{r} 4207 \\ -1624 \\ \hline \end{array}$

86. $\begin{array}{r} 7005 \\ -1796 \\ \hline \end{array}$

87. $\begin{array}{r} 8003 \\ -2735 \\ \hline \end{array}$

88. $\begin{array}{r} 20,005 \\ -9,627 \\ \hline \end{array}$

89. $\begin{array}{r} 80,004 \\ -8,237 \\ \hline \end{array}$

90. Find 10,051 less 9027.

91. Find 17,031 less 5792.

92. Find the difference between 1003 and 447.

93. What is 29,874 minus 21,392?

94. What is 29,797 less than 68,005?

95. What is 69,379 less than 70,004?

96. What is 25,432 decreased by 7994?

97. What is 86,701 decreased by 9976?

Use the relationship between addition and subtraction to complete the statement.

98. ___ + 39 = 104   **99.** 67 + ___ = 90   **100.** ___ + 497 = 862   **101.** 253 + ___ = 4901

 Estimate by rounding to the nearest ten-thousand. Then use your calculator to subtract.

102. 80,032
       − 19,605

103. 90,765
       − 60,928

104. 32,574
       − 10,961

105. 96,430
       − 59,762

106. 567,423
       − 208,444

107. 300,712
       − 198,714

Objective C   *Application Problems*

108. You have $304 in your checking account. If you write a check for $139, how much is left in your checking account?

109. In a recent year, the U.S. Patent and Trademark Office granted 151,024 patents. IBM received 2682 of those patents, more than any other company. Canon, with 1934 patents, received the second highest number. (Source: IFI/Plenum Data) How many more patents was IBM granted than Canon?

110. The tennis coach at a high school purchased a video camera that costs $1079 and made a down payment of $180. Find the amount that remains to be paid.

111. Rod Guerra, an engineer, purchased a used car that cost $11,225 and made a down payment of $950. Find the amount that remains to be paid.

112. The leader in pass receiving for the 1990s in the National Football League was Jerry Rice with 12,078 yards. His closest competitor was Michael Irvin with 10,872 yards. (Source: NFL) How many more yards did Jerry Rice gain than Michael Irvin?

113. Use the graph at the right to find the difference between the maximum height to which Great Fountain geyser erupts and the maximum height to which Valentine erupts.

114. According to the graph at the right, how much higher is the eruption of the Giant than that of Old Faithful?

The Maximum Heights of the Eruptions of Six Geysers at Yellowstone National Park

115. The top six live acts of the 1990s are shown in the table at the right.

a. How much more did the Rolling Stones gross from their concerts in the 1990s than the Grateful Dead?

b. How much more did Neil Diamond gross from his concerts in the 1990s than Jimmy Buffett?

Live Acts	Gross
Rolling Stones	$751,000,000
Grateful Dead	$285,000,000
U2	$283,000,000
Eagles	$197,000,000
Neil Diamond	$183,000,000
Jimmy Buffett	$167,000,000

Source: Amusement Business
BOXSCORE

 The graph at the right shows the expected U.S. population aged 100 and over, every two years from 2010 to 2020. Use this information for Exercises 116 and 117.

116. What is the expected growth in the population aged 100 and over during the 10-year period?

117. a. Which 2-year period has the smallest increase in the number of people aged 100 and over?
b. Which 2-year period has the greatest increase?

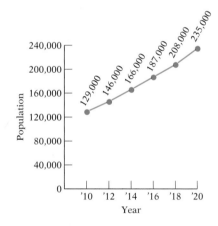

Expected U.S. Population Aged 100 and Over

Source: Census Bureau

APPLYING THE CONCEPTS

118. Answer true or false.
a. The phrases "the difference between 9 and 5" and "5 less than 9" mean the same thing.
b. $9 - (5 - 3) = (9 - 5) - 3$.
c. Subtraction is an associative operation. *Hint:* See part b of this exercise.

119. Make up a word problem for which the difference between 15 and 8 is the answer.

1.4 Multiplication of Whole Numbers

Objective A   **To multiply a number by a single digit**

Six boxes of toasters are ordered. Each box contains eight toasters. How many toasters are ordered?

This problem can be worked by adding 6 eights.

$8 + 8 + 8 + 8 + 8 + 8 = 48$

This problem involves repeated addition of the same number and can be worked by a shorter process called **multiplication.** Multiplication is the repeated addition of the same number.

$8 + 8 + 8 + 8 + 8 + 8 = 48$

The numbers that are multiplied are called **factors.** The answer is called the **product.**

or

$6 \quad \times \quad 8 \qquad = 48$

Factor   Factor   Product

The product of 6×8 can be represented on the number line. The arrow representing the whole number 8 is repeated 6 times. The result is the arrow representing 48.

The times sign "×" is one symbol that is used to mean multiplication. Another common symbol used for this purpose is a dot placed between the numbers.

$$7 \times 8 = 56 \qquad 7 \cdot 8 = 56$$

As with addition, there are some useful properties of multiplication.

Multiplication Property of Zero

The product of a number and zero is zero.

$0 \times 4 = 0$
$7 \times 0 = 0$

Multiplication Property of One

The product of a number and one is the number.

$1 \times 6 = 6$
$8 \times 1 = 8$

Commutative Property of Multiplication

Two numbers can be multiplied in either order. The product will be the same.

$4 \times 3 = 3 \times 4$
$12 = 12$

Associative Property of Multiplication

Grouping the numbers to be multiplied in any order gives the same result. Do the multiplication inside the parentheses first.

$(4 \times 2) \times 3 = 4 \times (2 \times 3)$
$8 \quad \times 3 = 4 \times \quad 6$
$24 = 24$

The basic facts for multiplying one-digit numbers should be memorized. Multiplication of larger numbers requires the repeated use of the basic multiplication facts.

➡ Multiply: 37 × 4

$$
\begin{array}{r}
{\scriptstyle 2} \\
3\ 7 \\
\times\ \ \ \ 4 \\
\hline
8
\end{array}
$$

- $4 \times 7 = 28$ (2 tens + 8 ones).
  Write the 8 in the ones column and
  carry the 2 to the tens column.

$$
\begin{array}{r}
{\scriptstyle 2} \\
3\ 7 \\
\times\ \ \ \ 4 \\
\hline
14\ 8
\end{array}
$$

- The 3 in 37 is 3 tens.

$$
\begin{array}{r}
4 \times 3 \text{ tens} =\ \ 12 \text{ tens} \\
\text{Add the carry digit.}\ \ +\ 2 \text{ tens} \\
\hline
14 \text{ tens}
\end{array}
$$

- Write the 14.

The phrases below are used to indicate the operation of multiplication. An example is shown at the right of each phrase.

times	7 times 3	7 · 3
the product of	the product of 6 and 9	6 · 9
multiplied by	8 multiplied by 2	2 · 8

Example 1   Multiply: 735 × 9

Solution
$$
\begin{array}{r}
{\scriptstyle 3\ 4} \\
735 \\
\times\ \ \ \ 9 \\
\hline
6615
\end{array}
$$

You Try It 1   Multiply: 648 × 7

Your solution

Solution on p. S2

Objective B   **To multiply larger whole numbers**

Note the pattern when the following numbers are multiplied.

Multiply the nonzero part of the factors.

Now attach the same number of zeros to the product as the total number of zeros in the factors.

$$
\begin{array}{l}
4 \times 2 \\
4 \times 20 = 80 \quad \text{1 zero}
\end{array}
$$

$$
\begin{array}{l}
4 \times 2 \\
4 \times 200 = 800 \quad \text{2 zeros}
\end{array}
$$

$$
\begin{array}{l}
4 \times 2 \\
40 \times 200 = 8000 \quad \text{3 zeros}
\end{array}
$$

$$
\begin{array}{l}
12 \times 5 \\
12 \times 5000 = 60{,}000 \quad \text{3 zeros}
\end{array}
$$

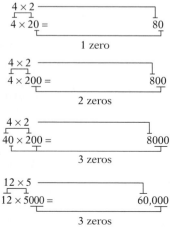

➡ Find the product of 47 and 23.

Multiply by the ones digit.	Multiply by the tens digit.	Add.

$$\begin{array}{r} 47 \\ \times\ 23 \\ \hline 141 \end{array} \ (=47\times 3)$$

$$\begin{array}{r} 47 \\ \times\ 23 \\ \hline 141 \\ 940 \end{array} \ (=47\times 20)$$

$$\begin{array}{r} 47 \\ \times\ 23 \\ \hline 141 \\ 940 \\ \hline 1081 \end{array}$$

Writing the 0 is optional.

The place-value chart illustrates the placement of the products.

Note the placement of the products when we are multiplying by a factor that contains a zero.

➡ Multiply: 439×206

$$\begin{array}{r} 439 \\ \times\ 206 \\ \hline 2634 \\ 000 \quad 0\times 439 \\ 878 \\ \hline 90{,}434 \end{array}$$

When working the problem, we usually write only one zero. Writing this zero ensures the proper placement of the products.

$$\begin{array}{r} 439 \\ \times\ 206 \\ \hline 2634 \\ 8780 \\ \hline 90{,}434 \end{array}$$

Example 2   Find 829 multiplied by 603.

Solution
$$\begin{array}{r} 829 \\ \times\ 603 \\ \hline 2487 \\ 49740 \\ \hline 499{,}887 \end{array}$$

You Try It 2   Multiply: 756×305

Your solution

Solution on p. S2

E S T I M A T I O N

Estimating the Product of Two Whole Numbers

Estimate and then use your calculator to find 3267×389.

To estimate the product, round each number so that all the digits are zero except the first digit. Then multiply. The estimated answer is 1,200,000.

$$\begin{array}{r} 3267 \approx\quad 3000 \\ \times\ 389 \approx\ \times\ 400 \\ \hline 1{,}200{,}000 \end{array}$$

Now use your calculator to find the exact answer. The exact answer is 1,270,863.

3267 ☒ 389 ⬜ 1270863

Objective C **To solve application problems**

Example 3

An auto mechanic receives a salary of $1050 each week. How much does the auto mechanic earn in 4 weeks?

Strategy

To find the mechanic's earnings for 4 weeks, multiply the weekly salary (1050) by the number of weeks (4).

Solution

```
  1050
×    4
  4200
```

The mechanic earns $4200 in 4 weeks.

You Try It 3

A new-car dealer receives a shipment of 37 cars each month. Find the number of cars the dealer will receive in 12 months.

Your strategy

Your solution

Example 4

A press operator earns $640 for working a 40-hour week. This week the press operator also worked 7 hours of overtime at $26 an hour. Find the press operator's total pay for the week.

Strategy

To find the press operator's total pay for the week:

• Find the overtime pay by multiplying the hours of overtime (7) by the overtime rate of pay (26).
• Add the weekly salary (640) to the overtime pay.

Solution

```
  26              640
×  7            +182
 182 overtime pay  822
```

The press operator earned $822 this week.

You Try It 4

The buyer for Ross Department Store can buy 80 men's suits for $4800. Each sports jacket will cost the store $23. The manager orders 80 men's suits and 25 sports jackets. What is the total cost of the order?

Your strategy

Your solution

Solutions on p. S2

1.4 Exercises

Objective A

Write each of the following as a product.

1. $2 + 2 + 2 + 2 + 2 + 2$ **2.** $4 + 4 + 4 + 4 + 4$ **3.** $7 + 7 + 7 + 7$ **4.** $18 + 18 + 18$

Multiply.

5. $\begin{array}{r} 3 \\ \times\, 4 \\ \hline \end{array}$

6. $\begin{array}{r} 2 \\ \times\, 8 \\ \hline \end{array}$

7. $\begin{array}{r} 5 \\ \times\, 7 \\ \hline \end{array}$

8. $\begin{array}{r} 6 \\ \times\, 4 \\ \hline \end{array}$

9. $\begin{array}{r} 5 \\ \times\, 5 \\ \hline \end{array}$

10. $\begin{array}{r} 7 \\ \times\, 7 \\ \hline \end{array}$

11. $\begin{array}{r} 0 \\ \times\, 7 \\ \hline \end{array}$

12. $\begin{array}{r} 8 \\ \times\, 0 \\ \hline \end{array}$

13. $\begin{array}{r} 8 \\ \times\, 9 \\ \hline \end{array}$

14. $\begin{array}{r} 7 \\ \times\, 6 \\ \hline \end{array}$

15. $\begin{array}{r} 66 \\ \times\, 3 \\ \hline \end{array}$

16. $\begin{array}{r} 70 \\ \times\, 4 \\ \hline \end{array}$

17. $\begin{array}{r} 67 \\ \times\, 5 \\ \hline \end{array}$

18. $\begin{array}{r} 127 \\ \times\, 9 \\ \hline \end{array}$

19. $\begin{array}{r} 623 \\ \times\, 4 \\ \hline \end{array}$

20. $\begin{array}{r} 802 \\ \times\, 5 \\ \hline \end{array}$

21. $\begin{array}{r} 607 \\ \times\, 9 \\ \hline \end{array}$

22. $\begin{array}{r} 300 \\ \times\, 5 \\ \hline \end{array}$

23. $\begin{array}{r} 600 \\ \times\, 7 \\ \hline \end{array}$

24. $\begin{array}{r} 906 \\ \times\, 8 \\ \hline \end{array}$

25. $\begin{array}{r} 703 \\ \times\, 9 \\ \hline \end{array}$

26. $\begin{array}{r} 127 \\ \times\, 5 \\ \hline \end{array}$

27. $\begin{array}{r} 632 \\ \times\, 3 \\ \hline \end{array}$

28. $\begin{array}{r} 559 \\ \times\, 4 \\ \hline \end{array}$

29. $\begin{array}{r} 632 \\ \times\, 8 \\ \hline \end{array}$

30. $\begin{array}{r} 524 \\ \times\, 4 \\ \hline \end{array}$

31. $\begin{array}{r} 337 \\ \times\, 5 \\ \hline \end{array}$

32. $\begin{array}{r} 841 \\ \times\, 6 \\ \hline \end{array}$

33. $\begin{array}{r} 6709 \\ \times\, 7 \\ \hline \end{array}$

34. $\begin{array}{r} 3608 \\ \times\, 5 \\ \hline \end{array}$

35. $\begin{array}{r} 8568 \\ \times\, 7 \\ \hline \end{array}$

36. $\begin{array}{r} 5495 \\ \times\, 4 \\ \hline \end{array}$

37. $\begin{array}{r} 4780 \\ \times\, 4 \\ \hline \end{array}$

38. $\begin{array}{r} 3690 \\ \times\, 5 \\ \hline \end{array}$

39. $\begin{array}{r} 9895 \\ \times\, 2 \\ \hline \end{array}$

40. Find the product of 5, 7, and 4.

41. Find the product of 6, 2, and 9.

42. Find the product of 5304 and 9.

43. Find the product of 458 and 8.

44. What is 3208 multiplied by 7?

45. What is 5009 multiplied by 4?

46. What is 3105 times 6?

47. What is 8957 times 8?

Objective B

Multiply.

48. 16
× 21

49. 18
× 24

50. 35
× 26

51. 27
× 72

52. 693
× 91

53. 581
× 72

54. 419
× 80

55. 727
× 60

56. 8279
× 46

57. 9577
× 35

58. 6938
× 78

59. 8875
× 67

60. 7035
× 57

61. 6702
× 48

62. 3009
× 35

63. 6003
× 57

64. 809
× 530

65. 607
× 460

66. 800
× 325

67. 700
× 274

68. 987
× 349

69. 688
× 674

70. 312
× 134

71. 423
× 427

72. 379
× 500

73. 684
× 700

74. 985
× 408

75. 758
× 209

76. 3407
× 309

77. 5207
× 902

78. 4258
× 986

79. 6327
× 876

80. What is 5763 times 45?

81. What is 7349 times 27?

82. Find the product of 2, 19, and 34.

83. Find the product of 6, 73, and 43.

84. What is 376 multiplied by 402?

85. What is 842 multiplied by 309?

86. Find the product of 3005 and 233,489.

87. Find the product of 9007 and 34,985.

 Estimate and then use your calculator to multiply.

88.   8745
        ×   63

89.   4732
        ×   93

90.   2937
        ×   206

91.   8941
        ×   726

92.   3097
        × 1025

93.   6379
        × 2936

94.   32,508
        ×   591

95.   62,504
        ×   923

Objective C   *Application Problems*

96. Rob Hill owns a compact car that averages 43 miles on 1 gallon of gas. How many miles could the car travel on 12 gallons of gas?

97. A plane flying from Los Angeles to Boston uses 865 gallons of jet fuel each hour. How many gallons of jet fuel were used on a 6-hour flight?

98. The perimeter of a square is equal to four times the length of a side of the square. Find the perimeter of a square whose side measures 16 miles.

16 mi

99. To find the area of a square, multiply the length of one side of the square times itself. What is the area of a square whose side measures 16 miles? The area of the square will be in square miles.

100. The area of a rectangle is equal to the product of the length of the rectangle times its width. Find the area of a rectangle that has a length of 24 meters and a width of 15 meters. The area will be in square meters.

15 m

24 m

A lighting consultant to a bank suggests that the bank lobby contain 43 can lights, 15 high-intensity lights, 20 fire safety lights, and one chandelier. The table at the right gives the costs for each type of light from two companies. Use this table for Exercises 101 and 102.

	Company A	Company B
Can lights	$2 each	$3 each
High-intensity	$6 each	$4 each
Fire safety	$12 each	$11 each
Chandelier	$998 each	$1089 each

101. Which company offers the lights for the lower total price?

102. How much can the lighting consultant save by purchasing the lights from the company that offers the lower total price?

The table at the right shows the hourly wages of four different job classifications at a small construction company. Use this table for Exercises 103 to 105.

Type of Work	Wage per Hour
Electrician	$34
Plumber	$30
Clerk	$16
Bookkeeper	$20

103. The owner of this company wants to provide the electrical installation for a new house. On the basis of the architectural plans for the house, it is estimated that it will require 3 electricians each working 50 hours to complete the job. What is the estimated cost for the electricians' labor?

104. Carlos Vasquez, a plumbing contractor, hires 4 plumbers from this company at the hourly wage given in the table. If each plumber works 23 hours, what are the total wages paid by Carlos?

105. The owner of this company estimates that remodeling a kitchen will require 1 electrician working 30 hours and 1 plumber working 33 hours. This project also requires 3 hours of clerical work and 4 hours of bookkeeping. What is the total cost for these four components of this remodeling?

APPLYING THE CONCEPTS

106. Determine whether each of the following statements is always true, sometimes true, or never true.
a. A whole number times zero is zero.
b. A whole number times one is the whole number.
c. The product of two whole numbers is greater than either one of the whole numbers.

107. According to the National Safety Council, in a recent year a death resulting from an accident occurred at the rate of 1 every 5 minutes. At this rate, how many accidental deaths occurred each hour? Each day? Throughout the year? Explain how you arrived at your answers.

108. According to the Population Reference Bureau, in the world today, 261 people are born every minute and 101 people die every minute. Using this statistic, what is the increase in the world's population every hour? Every day? Every week? Every year? Use a 365-day year. Explain how you arrived at your answers.

109. Pick your favorite number between 1 and 9. Multiply the number by 3. Multiply that product by 37,037. How is the product related to your favorite number? Explain why this works. (*Suggestion*: Multiply 3 and 37,037 first.)

110. There are quite a few tricks based on whole numbers. Here's one about birthdays. Write down the month in which you were born. Multiply by 5. Add 7. Multiply by 20. Subtract 100. Add the day of the month on which you were born. Multiply by 4. Subtract 100. Multiply by 25. Add the year you were born. Subtract 3400. The answer is the month/day/year of your birthday.

1.5 Division of Whole Numbers

Objective A   **To divide by a single digit with no remainder in the quotient**

Division is used to separate objects into equal groups.

A store manager wants to display 24 new objects equally on 4 shelves. From the diagram, we see that the manager would place 6 objects on each shelf.

The manager's division problem can be written as follows:

Number of shelves — Divisor
Number on each shelf — **Quotient**
Number of objects — **Dividend**

$$4\overline{)24}\ \ \ \ \ \begin{array}{c}6\end{array}$$

Note that the quotient multiplied by the divisor equals the dividend.

$$\overset{6}{4\overline{)24}} \quad \text{because} \quad \boxed{\underset{\text{Quotient}}{6}} \times \boxed{\underset{\text{Divisor}}{4}} = \boxed{\underset{\text{Dividend}}{24}}$$

$$\overset{6}{9\overline{)54}} \quad \text{because} \quad 6 \quad \times \quad 9 \quad = \quad 54$$

$$\overset{5}{8\overline{)40}} \quad \text{because} \quad 5 \quad \times \quad 8 \quad = \quad 40$$

Here are some important quotients and the properties of zero in division:

Important Quotients

Any whole number, except zero, divided by itself is 1.

$$\overset{1}{8\overline{)8}} \qquad \overset{1}{14\overline{)14}} \qquad \overset{1}{10\overline{)10}}$$

Any whole number divided by 1 is the whole number.

$$\overset{9}{1\overline{)9}} \qquad \overset{27}{1\overline{)27}} \qquad \overset{10}{1\overline{)10}}$$

Properties of Zero in Division

Zero divided by any other whole number is zero.

$$\overset{0}{7\overline{)0}} \qquad \overset{0}{13\overline{)0}} \qquad \overset{0}{10\overline{)0}}$$

Division by zero is not allowed.

$$\overset{?}{0\overline{)8}}$$

There is no number whose product with 0 is 8.

When the dividend is a larger whole number, the digits in the quotient are found in steps.

➡ Divide $4\overline{)3192}$ and check.

$$
\begin{array}{r}
7 \\
4\overline{)\;3192} \\
-28 \\
\hline
39
\end{array}
$$

- Think $4\overline{)31}$.
- Subtract 7×4.
- Bring down the 9.

$$
\begin{array}{r}
79 \\
4\overline{)\;3192} \\
-28 \\
\hline
39 \\
-36 \\
\hline
32
\end{array}
$$

- Think $4\overline{)39}$.
- Subtract 9×4.
- Bring down the 2.

$$
\begin{array}{r}
798 \\
4\overline{)\;3192} \\
-28 \\
\hline
39 \\
-36 \\
\hline
32 \\
-32 \\
\hline
0
\end{array}
$$

- Think $4\overline{)32}$.
- Subtract 8×4.

Check:
$$
\begin{array}{r}
798 \\
\times\quad 4 \\
\hline
3192
\end{array}
$$

The place-value chart can be used to show why this method works.

$$
\begin{array}{r}
\text{HUNDREDS} \quad \text{TENS} \quad \text{ONES} \\
7\;\;9\;\;8 \\
4\overline{)\;3\;\;1\;\;9\;\;2} \\
-2\;8\;0\;0 \quad \text{7 hundreds} \times 4 \\
\hline
3\;9\;2 \\
-3\;6\;0 \quad \text{9 tens} \times 4 \\
\hline
3\;2 \\
-3\;2 \quad \text{8 ones} \times 4 \\
\hline
0
\end{array}
$$

There are other ways of expressing division.

54 divided by 9 equals 6.

54   ÷   9 equals 6.

$\dfrac{54}{9}$ equals 6.

Example 1   Divide 7)56 and check.

Solution   8
                7)56

Check: 8 × 7 = 56

You Try It 1   Divide 9)63 and check.

Your solution

Example 2   Divide 2808 ÷ 8 and check.

Solution   351
                8) 2808
                   −24
                    40
                   −40
                    08
                   − 8
                     0

Check: 351 × 8 = 2808

You Try It 2   Divide 4077 ÷ 9 and check.

Your solution

Example 3   Divide 7)2856 and check.

Solution   408
                7) 2856
                   −28
                    05   • **Think 7)5. Place 0 in quotient.**
                   −0   **Subtract 0 × 7.**
                    56   • **Bring down the 6.**
                   −56
                     0

Check: 408 × 7 = 2856

You Try It 3   Divide 9)6345 and check.

Your solution

Solutions on p. S2

Objective B   **To divide by a single digit with a remainder in the quotient**

Sometimes it is not possible to separate objects into a whole number of equal groups.

A baker has 14 muffins to pack into 3 boxes. Each box holds 4 muffins. From the diagram, we see that after the baker places 4 muffins in each box, there are 2 left over. The 2 is called the **remainder.**

The clerk's division problem could be written

$$
\begin{array}{r}
\textbf{Quotient} \\
\text{(Number in each box)} \\
4 \\
\textbf{Divisor} \longrightarrow 3\overline{)\,14} \longleftarrow \textbf{Dividend} \\
\text{(Number of boxes)} \quad -12 \quad \text{(Total number of objects)} \\
\hline
2 \longleftarrow \textbf{Remainder} \\
\text{(Number left over)}
\end{array}
$$

The answer to a division problem with a remainder is frequently written

$$3\overline{)14}^{\,4\ \text{r2}}$$

Note that $\boxed{\underset{\text{Quotient}}{4} \times \underset{\text{Divisor}}{3}} + \boxed{\underset{\text{Remainder}}{2}} = \boxed{\underset{\text{Dividend}}{14}}$

Example 4 Divide $4\overline{)2522}$ and check.

Solution

$$
\begin{array}{r}
630 \text{ r2} \\
4\overline{)\,2522} \\
-24 \\
\hline
12 \\
-12 \\
\hline
02 \\
-\ 0 \\
\hline
2
\end{array}
$$

- Think $4\overline{)2}$. Place 0 in quotient.
- Subtract 0×4.

Check: $(630 \times 4) + 2 =$
$2520\ + 2 = 2522$

You Try It 4 Divide $6\overline{)5225}$ and check.

Your solution

Example 5 Divide $9\overline{)27,438}$ and check.

Solution

$$
\begin{array}{r}
3,048 \text{ r6} \\
9\overline{)\,27,438} \\
-27 \\
\hline
04 \\
-\ 0 \\
\hline
43 \\
-36 \\
\hline
78 \\
-72 \\
\hline
6
\end{array}
$$

- Think $9\overline{)4}$.
- Subtract 0×9.

Check: $(3048 \times 9) + 6 =$
$27,432\ + 6 = 27,438$

You Try It 5 Divide $7\overline{)21,409}$ and check.

Your solution

Solutions on p. S2

Objective C   **To divide by larger whole numbers**

When the divisor has more than one digit, estimate at each step by using the first digit of the divisor. If that product is too large, lower the guess by 1 and try again.

⇒ Divide $34\overline{)1598}$ and check.

$$\begin{array}{r} 5 \\ 34\overline{)\ 1598} \\ -170 \end{array}$$

• Think $3\overline{)15}$.
• Subtract 5×34.

$$\begin{array}{r} 4 \\ 34\overline{)\ 1598} \\ -136 \\ \hline 238 \end{array}$$

• Subtract 4×34.

170 is too large. Lower the guess by 1 and try again.

$$\begin{array}{r} 47 \\ 34\overline{)\ 1598} \\ -136 \\ \hline 238 \\ -238 \\ \hline 0 \end{array}$$

• Think $3\overline{)23}$.
• Subtract 7×34.

Check:
$$\begin{array}{r} 47 \\ \times 34 \\ \hline 188 \\ 141\ \ \\ \hline 1598 \end{array}$$

The phrases below are used to indicate the operation of division. An example is shown at the right of each phrase.

| the quotient of | the quotient of 9 and 3 | $9 \div 3$ |
| divided by | 6 divided by 2 | $6 \div 2$ |

Example 6   Find 7077 divided by 34 and check.

Solution

$$\begin{array}{r} 208\ \text{r}5 \\ 34\overline{)\ 7077} \\ -68\ \ \ \\ \hline 27\ \ \\ -\ 0\ \ \\ \hline 277 \\ -272 \\ \hline 5 \end{array}$$

• Think $34\overline{)27}$.
• Place 0 in the quotient.
• Subtract 0×34.

Check: $(208 \times 34) + 5 =$
$\qquad\quad 7072\ \ \ + 5 = 7077$

You Try It 6   Divide $4578 \div 42$ and check.

Your solution

Solution on p. S3

Example 7 Find the quotient of 21,312 and 56 and check.

Solution

$$
\begin{array}{r}
380 \text{ r32} \\
56\overline{)21{,}312} \\
-16\ 8 \\
\hline
4\ 51 \\
-4\ 48 \\
\hline
32 \\
-\ 0 \\
\hline
32
\end{array}
$$

Think $5\overline{)21}$.
4×56 is too large. Try 3.

Check: $(380 \times 56) + 32 =$
$21{,}280 \quad + 32 = 21{,}312$

You Try It 7 Divide $18{,}359 \div 39$ and check.

Your solution

Example 8 Divide $427\overline{)24{,}782}$ and check.

Solution

$$
\begin{array}{r}
58 \text{ r16} \\
427\overline{)24{,}782} \\
-21\ 35 \\
\hline
3\ 432 \\
-3\ 416 \\
\hline
16
\end{array}
$$

Check: $(58 \times 427) + 16 =$
$24{,}766 \quad + 16 = 24{,}782$

You Try It 8 Divide $534\overline{)33{,}219}$ and check.

Your solution

Example 9 Divide $386\overline{)206{,}149}$ and check.

Solution

$$
\begin{array}{r}
534 \text{ r25} \\
386\overline{)206{,}149} \\
-193\ 0 \\
\hline
13\ 14 \\
-11\ 58 \\
\hline
1\ 569 \\
-1\ 544 \\
\hline
25
\end{array}
$$

Check: $(534 \times 386) + 25 =$
$206{,}124 \quad + 25 = 206{,}149$

You Try It 9 Divide $515\overline{)216{,}848}$ and check.

Your solution

Solutions on p. S3

ESTIMATION

Estimating the Quotient of Two Whole Numbers

Estimate and then use your calculator to find 36,936 ÷ 54.

To estimate a quotient, round each number so that all the digits are zero except the first digit. Then divide.

$$36,936 ÷ 54 ≈$$
$$40,000 ÷ 50 = 800$$

The estimated answer is 800.

Now use your calculator to find the exact quotient. The exact quotient is 684.

36936 ÷ 54 = 684

Objective D   **To solve application problems**

The *average* of several numbers is the sum of all the numbers divided by the number of numbers.

$$\text{Average test score} = \frac{81 + 87 + 80 + 85 + 79 + 86}{6} = \frac{498}{6} = 83$$

The table at the right shows the seating capacity of the 8 stadiums used for major-league soccer games. Find the average seating capacity of these stadiums. Round to the nearest whole number.

Stadium	Seating Capacity
Columbus Crew Stadium	30,000
Giants Stadium	79,649
Mile High Stadium	76,098
Raymond James Stadium	66,321
Ericsson Stadium	73,250
Legion Field	83,000
Peoria Stadium	10,123
The Rose Bowl	92,542

Source: www.stadianet.com

Strategy

To find the average seating capacity:

- Add to find the sum of the seating capacities of the 8 stadiums.
- Divide the sum by 8.

Solution

```
  30,000        63,872
  79,649     8) 510,983
  76,098       −48
  66,321       ────
  73,250        30
  83,000       −24
  10,123       ───
+ 92,542        6 9
  ──────       −6 4
  510,983      ────
                 58
                −56
                ───
                 23
                −16
                ───
                  7
```

Sum of all the seating capacities

- When rounding to the nearest whole number, compare twice the remainder to the divisor. If twice the remainder is less than the divisor, drop the remainder. If twice the remainder is greater than or equal to the divisor, add 1 to the units digit of the quotient.

- Twice the remainder is 2 × 7 = 14. Because 14 > 8, add 1 to the units digit of the quotient.

The average seating capacity is 63,873 seats.

Example 10

Ngan Hui, a freight supervisor, shipped 192,600 bushels of wheat in 9 railroad cars. Find the amount of wheat shipped in each car.

Strategy

To find the amount of wheat shipped in each car, divide the number of bushels (192,600) by the number of cars (9).

Solution

$$
\begin{array}{r}
21{,}400 \\
9\overline{)192{,}600} \\
-18 \\
\hline
12 \\
-9 \\
\hline
36 \\
-36 \\
\hline
0
\end{array}
$$

Each car carried 21,400 bushels of wheat.

You Try It 10

Suppose a Michelin retail outlet can store 270 tires on 15 shelves. How many tires can be stored on each shelf?

Your strategy

Your solution

Example 11

The used car you are buying costs $11,216. A down payment of $2000 is required. The remaining balance is paid in 48 equal monthly payments. What is the monthly payment?

Strategy

To find the monthly payment:

• Find the remaining balance by subtracting the down payment (2000) from the total cost of the car (11,216).
• Divide the remaining balance by the number of equal monthly payments (48).

Solution

$$
\begin{array}{r}
11{,}216 \\
-\ 2{,}000 \\
\hline
9{,}216
\end{array}
$$
Remaining balance

$$
\begin{array}{r}
192 \\
48\overline{)9216} \\
-48 \\
\hline
441 \\
-432 \\
\hline
96 \\
-96 \\
\hline
0
\end{array}
$$

The monthly payment is $192.

You Try It 11

A soft-drink manufacturer produces 12,600 cans of soft drink each hour. Cans are packed 24 to a case. How many cases of soft drink are produced in 8 hours?

Your strategy

Your solution

Solutions on p. S3

1.5 Exercises

Objective A

Divide.

1. $4\overline{)8}$ **2.** $3\overline{)9}$ **3.** $6\overline{)36}$ **4.** $9\overline{)81}$

5. $7\overline{)49}$ **6.** $5\overline{)80}$ **7.** $6\overline{)96}$ **8.** $6\overline{)480}$

9. $4\overline{)840}$ **10.** $3\overline{)690}$ **11.** $7\overline{)308}$ **12.** $7\overline{)203}$

13. $9\overline{)6327}$ **14.** $4\overline{)2120}$ **15.** $8\overline{)7280}$ **16.** $9\overline{)8118}$

17. $3\overline{)64,680}$ **18.** $4\overline{)50,760}$ **19.** $6\overline{)21,480}$ **20.** $5\overline{)18,050}$

21. Find the quotient of 1446 and 3.   **22.** Find the quotient of 4123 and 7.

23. What is 7525 divided by 7?   **24.** What is 32,364 divided by 4?

Use the relationship between multiplication and division to complete the multiplication problem.

25. ___ × 7 = 364   **26.** 8 × ___ = 376   **27.** 34 × ___ = 170   **28.** ___ × 14 = 322

Objective B

Divide.

29. $4\overline{)9}$ **30.** $2\overline{)7}$ **31.** $5\overline{)27}$ **32.** $9\overline{)88}$ **33.** $3\overline{)40}$

34. $6\overline{)97}$ **35.** $8\overline{)83}$ **36.** $5\overline{)54}$ **37.** $7\overline{)632}$ **38.** $4\overline{)363}$

39. $4\overline{)921}$ **40.** $7\overline{)845}$ **41.** $8\overline{)1635}$ **42.** $5\overline{)1548}$ **43.** $7\overline{)9432}$

44. $7\overline{)8124}$ **45.** $3\overline{)5162}$ **46.** $5\overline{)3542}$ **47.** $8\overline{)3274}$

48. $4\overline{)15{,}301}$ **49.** $7\overline{)43{,}500}$ **50.** $8\overline{)72{,}354}$ **51.** $5\overline{)43{,}542}$

52. Find the quotient of 3107 and 8.

53. Find the quotient of 8642 and 8.

54. What is 45,738 divided by 4? Round to the nearest ten.

55. What is 37,896 divided by 9? Round to the nearest hundred.

56. What is 3572 divided by 7? Round to the nearest ten.

57. What is 78,345 divided by 4? Round to the nearest hundred.

Objective C

Divide.

58. $27\overline{)96}$ **59.** $44\overline{)82}$ **60.** $42\overline{)87}$ **61.** $67\overline{)93}$

62. $41\overline{)897}$ **63.** $32\overline{)693}$ **64.** $23\overline{)784}$ **65.** $25\overline{)772}$

66. $74\overline{)600}$ **67.** $92\overline{)500}$ **68.** $70\overline{)329}$ **69.** $50\overline{)467}$

70. $36\overline{)7225}$ **71.** $44\overline{)8821}$ **72.** $19\overline{)3859}$ **73.** $32\overline{)9697}$

74. $88\overline{)3127}$ **75.** $92\overline{)6177}$ **76.** $33\overline{)8943}$ **77.** $27\overline{)4765}$

78. $22\overline{)98{,}654}$ **79.** $77\overline{)83{,}629}$ **80.** $64\overline{)38{,}912}$ **81.** $78\overline{)31{,}434}$

82. $206\overline{)3097}$ **83.** $504\overline{)6504}$ **84.** $654\overline{)1217}$ **85.** $546\overline{)2344}$

86. Find the quotient of 5432 and 21.   **87.** Find the quotient of 8507 and 53.

88. What is 37,294 divided by 72?   **89.** What is 76,788 divided by 46?

90. Find 23,457 divided by 43. Round to the nearest hundred.   **91.** Find 341,781 divided by 43. Round to the nearest ten.

 Estimate and then use your calculator to divide.

92. $76\overline{)389,804}$ **93.** $53\overline{)117,925}$ **94.** $29\overline{)637,072}$ **95.** $67\overline{)738,072}$

96. $38\overline{)934,648}$ **97.** $34\overline{)906,304}$ **98.** $309\overline{)876,324}$ **99.** $642\overline{)323,568}$

100. $209\overline{)632,016}$ **101.** $614\overline{)332,174}$ **102.** $179\overline{)5,734,444}$ **103.** $374\overline{)7,712,254}$

Objective D   *Application Problems*

The graph at the right shows the annual expenditures, in a recent year, of the average household in the United States. Use this information for Exercises 104 to 106. Round answers to the nearest whole number.

104. What is the total amount spent annually by the average household in the United States?

105. What is the average monthly expense for housing?

106. What is the difference between the average monthly expense for food and the average monthly expense for health care?

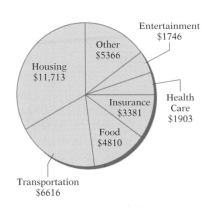

Average Annual Household Expenses

Source: Bureau of Labor Statistics Consumer Expenditure Survey

The table at the right shows the sources of laptop insurance claims in a recent year. Claims have been rounded to the nearest ten thousand dollars. Use this information for Exercises 107 and 108.

107. What was the average monthly claim for theft?

108. For all sources combined, find the average claims per month.

Source	Claims
Accidents	$560,000
Theft	$300,000
Power surge	$80,000
Lightning	$50,000
Transit	$20,000
Water/flood	$20,000
Other	$110,000

Source: Safeware, The Insurance Company

The table at the right shows, for different countries, the average number of hours per year that employees work. Use this information for Exercises 109 and 110. Use a 50-week year. Round answers to the nearest whole number.

109. What is the average number of hours worked per week by employees in Britain?

110. On average, how many more hours per week do employees in the United States work than employees in France?

Country	Annual Number of Hours Worked
Britain	1731
France	1656
Japan	1889
Norway	1399
United States	1966

Source: International Labor Organization

111. The U.S. Mint estimates that about 114,000,000,000 of the 312,000,000,000 pennies it has minted over the last 30 years are in active circulation. That works out to how many pennies in circulation for each of the 280,000,000 people living in the United States?

APPLYING THE CONCEPTS

112. A palindromic number is a whole number that remains unchanged when its digits are written in reverse order. For instance, 292 is a palindromic number. Find the smallest three-digit palindromic number that is divisible by 4.

113. Find the smallest four-digit palindromic number (see Exercise 112) that is divisible by 8. (*Helpful fact:* A whole number whose last three digits form a number that is divisible by 8 is itself divisible by 8. For example, because 408 is divisible by 8, the number 3408 is divisible by 8.)

114. Determine whether each of the following statements is true or false.
   a. Any whole number divided by zero is zero.
   b. $\dfrac{0}{0} = 1$
   c. Zero divided by any whole number, except zero, is zero.

Exponential Notation and the Order of Operations Agreement

Objective A   **To simplify expressions that contain exponents**

Repeated multiplication of the same factor can be written in two ways:

$$3 \cdot 3 \cdot 3 \cdot 3 \cdot 3 \quad \text{or} \quad 3^5 \leftarrow \textbf{Exponent}$$

The exponent indicates how many times the factor occurs in the multiplication. The expression 3^5 is in **exponential notation.**

It is important to be able to read numbers written in exponential notation.

$6 = 6^1$ is read "six to the first power" or just "six." Usually the exponent 1 is not written.

$6 \cdot 6 = 6^2$ is read "six squared" or "six to the second power."

$6 \cdot 6 \cdot 6 = 6^3$ is read "six cubed" or "six to the third power."

$6 \cdot 6 \cdot 6 \cdot 6 = 6^4$ is read "six to the fourth power."

$6 \cdot 6 \cdot 6 \cdot 6 \cdot 6 = 6^5$ is read "six to the fifth power."

Each place value in the place-value chart can be expressed as a power of 10.

$$
\begin{array}{rcccl}
\text{Ten} = & 10 & = & 10 & = 10^1 \\
\text{Hundred} = & 100 & = & 10 \cdot 10 & = 10^2 \\
\text{Thousand} = & 1000 & = & 10 \cdot 10 \cdot 10 & = 10^3 \\
\text{Ten-Thousand} = & 10,000 & = & 10 \cdot 10 \cdot 10 \cdot 10 & = 10^4 \\
\text{Hundred-Thousand} = & 100,000 & = & 10 \cdot 10 \cdot 10 \cdot 10 \cdot 10 & = 10^5 \\
\text{Million} = & 1,000,000 & = & 10 \cdot 10 \cdot 10 \cdot 10 \cdot 10 \cdot 10 & = 10^6 \\
\end{array}
$$

To simplify a numerical expression containing exponents, write each factor as many times as indicated by the exponent and carry out the indicated multiplication.

$$4^3 = 4 \cdot 4 \cdot 4 = 64$$
$$2^2 \cdot 3^4 = (2 \cdot 2) \cdot (3 \cdot 3 \cdot 3 \cdot 3) = 4 \cdot 81 = 324$$

Example 1   Write $3 \cdot 3 \cdot 3 \cdot 5 \cdot 5$ in exponential notation.

Solution $3 \cdot 3 \cdot 3 \cdot 5 \cdot 5 = 3^3 \cdot 5^2$

You Try It 1   Write $2 \cdot 2 \cdot 2 \cdot 2 \cdot 3 \cdot 3 \cdot 3$ in exponential notation.

Your solution

Example 2   Write as a power of 10:
$10 \cdot 10 \cdot 10 \cdot 10$

Solution $10 \cdot 10 \cdot 10 \cdot 10 = 10^4$

You Try It 2   Write as a power of 10:
$10 \cdot 10 \cdot 10 \cdot 10 \cdot 10 \cdot 10 \cdot 10$

Your solution

Example 3   Simplify $3^2 \cdot 5^3$.

Solution $3^2 \cdot 5^3 = (3 \cdot 3) \cdot (5 \cdot 5 \cdot 5)$
    $= 9 \cdot 125 = 1125$

You Try It 3   Simplify $2^3 \cdot 5^2$.

Your solution

Solutions on p. S3

Objective B | To use the Order of Operations Agreement to simplify expressions

More than one operation may occur in a numerical expression. The answer may be different, depending on the order in which the operations are performed. For example, consider $3 + 4 \times 5$.

Multiply first, then add.

$$3 + \underbrace{4 \times 5}$$
$$\underbrace{3 + 20}$$
$$23$$

Add first, then multiply.

$$\underbrace{3 + 4} \times 5$$
$$\underbrace{7 \times 5}$$
$$35$$

An Order of Operations Agreement is used so that only one answer is possible.

The Order of Operations Agreement

Step 1 Do all the operations inside parentheses.
Step 2 Simplify any number expressions containing exponents.
Step 3 Do multiplication and division as they occur from left to right.
Step 4 Do addition and subtraction as they occur from left to right.

⇒ Simplify $3 \times (2 + 1) - 2^2 + 4 \div 2$ by using the Order of Operations Agreement.

$$3 \times \underbrace{(2 + 1)} - 2^2 + 4 \div 2$$
$$\underbrace{3 \times 3} - \underbrace{2^2} + 4 \div 2$$
$$9 - 4 + \underbrace{4 \div 2}$$
$$\underbrace{9 - 4} + 2$$
$$\underbrace{5 + 2}$$
$$7$$

1. Perform operations in parentheses.

2. Simplify expressions with exponents.

3. Do multiplications and divisions as they occur from left to right.

4. Do additions and subtractions as they occur from left to right.

One or more of the above steps may not be needed to simplify an expression. In that case, proceed to the next step in the Order of Operations Agreement.

⇒ Simplify $5 + 8 \div 2$.

$$5 + \underbrace{8 \div 2}$$
$$\underbrace{5 + 4}$$
$$9$$

There are no parentheses or exponents. Proceed to Step 3 of the Agreement.

3. Do multiplication or division.

4. Do addition or subtraction.

Example 4 Simplify: $64 \div (8 - 4)^2 \cdot 9 - 5^2$

Solution
$$64 \div (8 - 4)^2 \cdot 9 - 5^2$$
$$= 64 \div 4^2 \cdot 9 - 5^2$$
$$= 64 \div 16 \cdot 9 - 25$$
$$= 4 \cdot 9 - 25$$
$$= 36 - 25$$
$$= 11$$

You Try It 4 Simplify: $5 \cdot (8 - 4)^2 \div 4 - 2$

Your solution

Solution on p. S3

1.6 Exercises

Objective A

Write the number in exponential notation.

1. $2 \cdot 2 \cdot 2$

2. $7 \cdot 7 \cdot 7 \cdot 7 \cdot 7$

3. $6 \cdot 6 \cdot 6 \cdot 7 \cdot 7 \cdot 7 \cdot 7$

4. $6 \cdot 6 \cdot 9 \cdot 9 \cdot 9 \cdot 9$

5. $2 \cdot 2 \cdot 2 \cdot 3 \cdot 3 \cdot 3$

6. $3 \cdot 3 \cdot 10 \cdot 10$

7. $5 \cdot 7 \cdot 7 \cdot 7 \cdot 7 \cdot 7$

8. $4 \cdot 4 \cdot 4 \cdot 5 \cdot 5 \cdot 5$

9. $3 \cdot 3 \cdot 3 \cdot 6 \cdot 6 \cdot 6 \cdot 6$

10. $2 \cdot 2 \cdot 5 \cdot 5 \cdot 5 \cdot 8$

11. $3 \cdot 3 \cdot 3 \cdot 5 \cdot 9 \cdot 9 \cdot 9$

12. $2 \cdot 2 \cdot 2 \cdot 4 \cdot 7 \cdot 7 \cdot 7$

Simplify.

13. 2^3

14. 2^6

15. $2^4 \cdot 5^2$

16. $2^6 \cdot 3^2$

17. $3^2 \cdot 10^2$

18. $2^3 \cdot 10^4$

19. $6^2 \cdot 3^3$

20. $4^3 \cdot 5^2$

21. $5 \cdot 2^3 \cdot 3$

22. $6 \cdot 3^2 \cdot 4$

23. $2^2 \cdot 3^2 \cdot 10$

24. $3^2 \cdot 5^2 \cdot 10$

25. $0^2 \cdot 4^3$

26. $6^2 \cdot 0^3$

27. $3^2 \cdot 10^4$

28. $5^3 \cdot 10^3$

29. $2^2 \cdot 3^3 \cdot 5$

30. $5^2 \cdot 7^3 \cdot 2$

31. $2 \cdot 3^4 \cdot 5^2$

32. $6 \cdot 2^6 \cdot 7^2$

33. $5^2 \cdot 3^2 \cdot 7^2$

34. $4^2 \cdot 9^2 \cdot 6^2$

35. $3^4 \cdot 2^6 \cdot 5$

36. $4^3 \cdot 6^3 \cdot 7$

37. $4^2 \cdot 3^3 \cdot 10^4$

Objective B

Simplify by using the Order of Operations Agreement.

38. $4 - 2 + 3$

39. $6 - 3 + 2$

40. $6 \div 3 + 2$

41. $8 \div 4 + 8$

42. $6 \cdot 3 + 5$

43. $5 \cdot 9 + 2$

44. $3^2 - 4$

45. $5^2 - 17$

46. $4 \cdot (5 - 3) + 2$ **47.** $3 + (4 + 2) \div 3$ **48.** $5 + (8 + 4) \div 6$ **49.** $8 - 2^2 + 4$

50. $16 \cdot (3 + 2) \div 10$ **51.** $12 \cdot (1 + 5) \div 12$ **52.** $10 - 2^3 + 4$ **53.** $5 \cdot 3^2 + 8$

54. $16 + 4 \cdot 3^2$ **55.** $12 + 4 \cdot 2^3$ **56.** $16 + (8 - 3) \cdot 2$ **57.** $7 + (9 - 5) \cdot 3$

58. $2^2 + 3 \cdot (6 - 2)^2$ **59.** $3^3 + 5 \cdot (8 - 6)^3$ **60.** $2^2 \cdot 3^2 + 2 \cdot 3$ **61.** $4 \cdot 6 + 3^2 \cdot 4^2$

62. $16 - 2 \cdot 4$ **63.** $12 + 3 \cdot 5$ **64.** $3 \cdot (6 - 2) + 4$

65. $5 \cdot (8 - 4) - 6$ **66.** $8 - (8 - 2) \div 3$ **67.** $12 - (12 - 4) \div 4$

68. $8 + 2 - 3 \cdot 2 \div 3$ **69.** $10 + 1 - 5 \cdot 2 \div 5$ **70.** $3 \cdot (4 + 2) \div 6$

71. $(7 - 3)^2 \div 2 - 4 + 8$ **72.** $20 - 4 \div 2 \cdot (3 - 1)^3$ **73.** $12 \div 3 \cdot 2^2 + (7 - 3)^2$

74. $(4 - 2) \cdot 6 \div 3 + (5 - 2)^2$ **75.** $18 - 2 \cdot 3 + (4 - 1)^3$ **76.** $100 \div (2 + 3)^2 - 8 \div 2$

APPLYING THE CONCEPTS

77. Memory in computers is measured in bytes. One kilobyte (1 K) is 2^{10} bytes. Write this number in standard form.

78. Explain the difference in the order of operations for **a.** $\dfrac{14 - 2}{2} \div 2 \cdot 3$ and **b.** $\dfrac{14 - 2}{2} \div (2 \cdot 3)$. Work the two problems. What is the difference between the larger and the smaller answer?

79. If a number is divisible by 6 and 10, is the number divisible by 30? If so, explain why. If not, give an example.

1.7 Prime Numbers and Factoring

Objective A   **To factor numbers**

Whole-number **factors of a number** divide that number evenly (there is no remainder).

1, 2, 3, and 6 are whole-number factors of 6 because they divide 6 evenly.

$$\dfrac{6}{1\overline{)6}} \quad \dfrac{3}{2\overline{)6}} \quad \dfrac{2}{3\overline{)6}} \quad \dfrac{1}{6\overline{)6}}$$

Note that both the divisor and the quotient are factors of the dividend.

To find the factors of a number, try dividing the number by 1, 2, 3, 4, 5, ... Those numbers that divide the number evenly are its factors. Continue this process until the factors start to repeat.

→ Find all the factors of 42.

$42 \div 1 = 42$	1 and 42 are factors.
$42 \div 2 = 21$	2 and 21 are factors.
$42 \div 3 = 14$	3 and 14 are factors.
$42 \div 4$	Will not divide evenly
$42 \div 5$	Will not divide evenly
$42 \div 6 = 7$	6 and 7 are factors. ⎫ Factors are repeating; all the
$42 \div 7 = 6$	7 and 6 are factors. ⎭ factors of 42 have been found.

1, 2, 3, 6, 7, 14, 21, and 42 are factors of 42.

The following rules are helpful in finding the factors of a number.

2 is a factor of a number if the last digit of the number is 0, 2, 4, 6, or 8.

436 ends in 6; therefore, 2 is a factor of 436. ($436 \div 2 = 218$)

3 is a factor of a number if the sum of the digits of the number is divisible by 3.

The sum of the digits of 489 is $4 + 8 + 9 = 21$. 21 is divisible by 3. Therefore, 3 is a factor of 489. ($489 \div 3 = 163$)

5 is a factor of a number if the last digit of the number is 0 or 5.

520 ends in 0; therefore, 5 is a factor of 520. ($520 \div 5 = 104$)

Example 1   Find all the factors of 30.

Solution
$30 \div 1 = 30$
$30 \div 2 = 15$
$30 \div 3 = 10$
$30 \div 4$ Will not divide evenly
$30 \div 5 = 6$
$30 \div 6 = 5$

1 2, 3, 5, 6, 10, 15, and 30 are factors of 30.

You Try It 1   Find all the factors of 40.

Your solution

Solution on p. S3

Objective B · **To find the prime factorization of a number**

Point of Interest

Prime numbers are an important part of cryptology, the study of secret codes. To make it less likely that codes can be broken, cryptologists use prime numbers that have hundreds of digits.

A number is a **prime number** if its only whole-number factors are 1 and itself. 7 is prime because its only factors are 1 and 7. If a number is not prime, it is called a **composite number.** Because 6 has factors of 2 and 3, 6 is a composite number. The number 1 is not considered a prime number; therefore it is not included in the following list of prime numbers less than 50.

$$2, 3, 5, 7, 11, 13, 17, 19, 23, 29, 31, 37, 41, 43, 47$$

The **prime factorization** of a number is the expression of the number as a product of its prime factors. We use a "T-diagram" to find the prime factors of 60. Begin with the smallest prime number as a trial divisor, and continue with prime numbers as trial divisors until the final quotient is 1.

$$
\begin{array}{c|c}
\multicolumn{2}{c}{60} \\
\hline
2 & 30 \\
2 & 15 \\
3 & 5 \\
5 & 1 \\
\end{array}
\qquad
\begin{array}{l}
60 \div 2 = 30 \\
30 \div 2 = 15 \\
15 \div 3 = 5 \\
5 \div 5 = 1 \\
\end{array}
$$

The prime factorization of 60 is $2 \cdot 2 \cdot 3 \cdot 5$.

Finding the prime factorization of larger numbers can be more difficult. Try each prime number as a trial divisor. Stop when the square of the trial divisor is greater than the number being factored.

➡ Find the prime factorization of 106.

$$
\begin{array}{c|c}
\multicolumn{2}{c}{106} \\
\hline
2 & 53 \\
53 & 1 \\
\end{array}
$$

• 53 cannot be divided evenly by 2, 3, 5, 7, or 11. Prime numbers greater than 11 need not be tested because 11^2 is greater than 53.

The prime factorization of 106 is $2 \cdot 53$.

Example 2   Find the prime factorization of 315.

Solution

$$
\begin{array}{c|c}
\multicolumn{2}{c}{315} \\
\hline
3 & 105 \\
3 & 35 \\
5 & 7 \\
7 & 1 \\
\end{array}
$$

$315 = 3 \cdot 3 \cdot 5 \cdot 7$

You Try It 2   Find the prime factorization of 44.

Your solution

Example 3   Find the prime factorization of 201.

Solution

$$
\begin{array}{c|c}
\multicolumn{2}{c}{201} \\
\hline
3 & 67 \\
67 & 1 \\
\end{array}
$$

• Try only 2, 3, 5, 7, and 11 because $11^2 > 67$.

$201 = 3 \cdot 67$

You Try It 3   Find the prime factorization of 177.

Your solution

Solutions on p. S4

1.7 Exercises

Objective A

Find all the factors of the number.

1. 4	**2.** 6	**3.** 10	**4.** 20
5. 7	**6.** 12	**7.** 9	**8.** 8
9. 13	**10.** 17	**11.** 18	**12.** 24
13. 56	**14.** 36	**15.** 45	**16.** 28
17. 29	**18.** 33	**19.** 22	**20.** 26
21. 52	**22.** 49	**23.** 82	**24.** 37
25. 57	**26.** 69	**27.** 48	**28.** 64
29. 95	**30.** 46	**31.** 54	**32.** 50
33. 66	**34.** 77	**35.** 80	**36.** 100
37. 96	**38.** 85	**39.** 90	**40.** 101

Objective B

Find the prime factorization.

41. 6	**42.** 14	**43.** 17	**44.** 83
45. 24	**46.** 12	**47.** 27	**48.** 9

49. 36   **50.** 40   **51.** 19   **52.** 37

53. 90   **54.** 65   **55.** 115   **56.** 80

57. 18   **58.** 26   **59.** 28   **60.** 49

61. 31   **62.** 42   **63.** 62   **64.** 81

65. 22   **66.** 39   **67.** 101   **68.** 89

69. 66   **70.** 86   **71.** 74   **72.** 95

73. 67   **74.** 78   **75.** 55   **76.** 46

77. 120   **78.** 144   **79.** 160   **80.** 175

81. 216   **82.** 400   **83.** 625   **84.** 225

APPLYING THE CONCEPTS

85. Twin primes are two prime numbers that differ by 2. For instance, 17 and 19 are twin primes. Find three sets of twin primes, not including 17 and 19.

86. In 1742, Christian Goldbach conjectured that every even number greater than 2 could be expressed as the sum of two prime numbers. Show that this conjecture is true for 8, 24, and 72. (*Note*: Mathematicians have not yet been able to determine whether Goldbach's conjecture is true or false.)

87. Explain why 2 is the only even prime number.

88. Explain the method of finding prime numbers using the Sieve of Erathosthenes.

Focus on Problem Solving

Introduction to Problem Solving

You encounter problem-solving situations every day. Some problems are easy to solve, and you may mentally solve these problems without considering the steps you are taking in order to draw a conclusion. Others may be more challenging and may require more thought and consideration.

Suppose a friend suggests that you both take a trip over spring break. You'd like to go. What questions go through your mind? You might ask yourself some of the following questions:

How much will the trip cost? What will be the cost for travel, hotel rooms, meals, and so on?

Are some costs going to be shared by both me and my friend?

Can I afford it?

How much money do I have in the bank?

How much more money than I have now do I need?

How much time is there to earn that much money?

How much can I earn in that amount of time?

How much money must I keep in the bank in order to pay the next tuition bill (or some other expense)?

These questions require different mathematical skills. Determining the cost of the trip requires **estimation;** for example, you must use your knowledge of air fares or the cost of gasoline to arrive at an estimate of these costs. If some of the costs are going to be shared, you need to **divide** those costs by 2 in order to determine your share of the expense. The question regarding how much more money you need requires **subtraction:** the amount needed minus the amount currently in the bank. To determine how much money you can earn in the given amount of time requires **multiplication**—for example, the amount you earn per week times the number of weeks to be worked. To determine if the amount you can earn in the given amount of time is sufficient, you need to use your knowledge of **order relations** to compare the amount you can earn with the amount needed.

Facing the problem-solving situation described above may not seem difficult to you. The reason may be that you have faced similar situations before and, therefore, know how to work through this one. You may feel better prepared to deal with a circumstance such as this one because you know what questions to ask. An important aspect of learning to solve problems is learning what questions to ask. As you work through application problems in this text, try to become more conscious of the mental process you are going through. You might begin the process by asking yourself the following questions whenever you are solving an application problem.

1. Have I read the problem enough times to be able to understand the situation being described?

2. Will restating the problem in different words help me to understand the problem situation better?

3. What facts are given? (You might make a list of the information contained in the problem.)

4. What information is being asked for?

5. What relationship exists among the given facts? What relationship exists between the given facts and the solution?

6. What mathematical operations are needed in order to solve the problem?

Try to focus on the problem-solving situation, not on the computation or on getting the answer quickly. And remember, the more problems you solve, the better able you will be to solve other problems in the future, partly because you are learning what questions to ask.

Projects and Group Activities

Order of Operations

Does your calculator use the Order of Operations Agreement? To find out, try this problem:

$$2 + 4 \cdot 7$$

If your answer is 30, then the calculator uses the Order of Operations Agreement. If your answer is 42, it does not use that agreement.

Even if your calculator does not use the Order of Operations Agreement, you can still correctly evaluate numerical expressions. The parentheses keys, $\boxed{(}$ and $\boxed{)}$, are used for this purpose.

Remember that $2 + 4 \cdot 7$ means $2 + (4 \cdot 7)$ because the multiplication must be completed before the addition. To evaluate this expression, enter the following:

Enter:	2	$\boxed{+}$	$\boxed{(}$	4	$\boxed{\times}$	7	$\boxed{)}$	$\boxed{=}$
Display:	2	2	$\boxed{(}$	4	4	7	28	30

When using your calculator to evaluate numerical expressions, insert parentheses around multiplications and around divisions. This has the effect of forcing the calculator to do the operations in the order you want.

Evaluate.

1. $3 \cdot 8 - 5$

2. $6 + 8 \div 2$

3. $3 \cdot (8 - 2)^2$

4. $24 - (4 - 2)^2 \div 4$

5. $3 + (6 \div 2 + 4)^2 - 2$

6. $16 \div 2 + 4 \cdot (8 - 12 \div 4)^2 - 50$

7. $3 \cdot (15 - 2 \cdot 3) - 36 \div 3$

8. $4 \cdot 2^2 - (12 + 24 \div 6) + 5$

9. $16 \div 4 \cdot 3 + (3 \cdot 4 - 5) + 2$

10. $15 \cdot 3 \div 9 + (2 \cdot 6 - 3) + 4$

Patterns in Mathematics

For the circle at the left, use a straight line to connect each dot on the circle with every other dot on the circle. How many different straight lines are there?

Follow the same procedure for each of the circles shown below. How many different straight lines are there in each?

Find a pattern to describe the number of dots on a circle and the corresponding number of different lines drawn. Use the pattern to determine the number of different lines that would be drawn in a circle with 7 dots and in a circle with 8 dots.

Now use the pattern to answer the following question. You are arranging a tennis tournament with 9 players. How many singles matches will be played among the 9 players if each player plays each of the other players only once?

Search the World Wide Web

On the Internet, go to www.census.gov. Find the web page on the U.S. population. Use the data to estimate the following to the nearest thousand. Assume that the same rate of increase in population will continue into the future.

1. What is the population of the United States today?
2. Find the increase in population for a 24-hour period.
3. Using the answer for Exercise 2, find the increase in population for a 30-day period. Find the increase in population for 1 year (365 days).
4. Using the information for Exercise 3, estimate the population of the United States 50 years from now.

Now find the web page on world population. Use the data to estimate the following to the nearest million.

5. How much did the population increase in the last month? In the last year?
6. Using the answer for Exercise 5, estimate what the population of the world will be 10 years from now and 50 years from now.

Chapter Summary

Key Words

The *whole numbers* are 0, 1, 2, 3, 4, 5, 6, 7, 8, 9, 10, ... [p. 3]

The symbol for *is less than* is $<$. The symbol for *is greater than* is $>$. [p. 3]

When a whole number is written using the digits 0, 1, 2, 3, 4, 5, 6, 7, 8, and 9, it is said to be in *standard form*. The number 58,429 is written in standard form. The position of each digit in the number determines the digit's *place value*. The number 58,429 is written in *expanded form* as 50,000 + 8000 + 400 + 20 + 9. [p. 4]

Addition is the process of finding the total of two or more numbers. The answer to an addition problem is called the *sum*. [p. 9]

Subtraction is the process of finding the *difference* between two numbers. [p. 17]

Multiplication is the repeated addition of the same number. The numbers that are multiplied are called *factors*. The answer is the *product*. [p. 25]

Division is used to separate objects into equal groups. The answer to a division problem is called the *quotient*. [p. 33]

The expression 4^3 is in *exponential notation*. The *exponent*, 3, indicates how many times 4 occurs as a factor in the multiplication. [p. 45]

Factors of a number divide that number evenly (there is no remainder). [p. 49]

A number greater than 1 is *prime* if its only whole-number factors are 1 and itself. [p. 50]

The *prime factorization* of a number is the expression of the number as a product of its prime factors. [p. 50]

Essential Rules

The Addition Property of Zero
Zero added to a number does not change the number.
5 + 0 = 0 + 5 = 5 [p. 9]

The Commutative Property of Addition
Two numbers can be added in either order.
6 + 5 = 5 + 6 = 11 [p. 9]

The Associative Property of Addition
Grouping numbers to be added in any order gives the same result.
(2 + 3) + 5 = 2 + (3 + 5) = 10 [p. 9]

The Multiplication Property of Zero
The product of a number and zero is zero.
0 × 3 = 3 × 0 = 0 [p. 25]

The Multiplication Property of One
The product of a number and one is the number.
1 × 8 = 8 × 1 = 8 [p. 25]

The Commutative Property of Multiplication
Two numbers can be multiplied in any order.
5 × 2 = 2 × 5 = 10 [p. 25]

The Associative Property of Multiplication
Grouping numbers to be multiplied in any order gives the same result.
(3 × 2) × 5 = 3 × (2 × 5) = 30 [p. 25]

The Properties of Zero in Division
Zero divided by any other number is zero. Division by zero is not allowed. [p. 33]

Rounding a Number to a Given Place Value
If the digit to the right of the given place value is less than 5, replace that digit and all digits to the right by zeros. If the digit to the right of the given place value is greater than or equal to 5, increase the digit in the given place value by 1, and replace all other digits to the right by zeros. [p. 6]

Order of Operations Agreement [p. 46]
Step 1 Perform operations inside grouping symbols.
Step 2 Simplify expressions with exponents.
Step 3 Do multiplications and divisions as they occur from left to right.
Step 4 Do additions and subtractions as they occur from left to right.

Chapter Review

1. Simplify: $3 \cdot 2^3 \cdot 5^2$

2. Write 10,327 in expanded form.

3. Find all the factors of 18.

4. Find the sum of 5894, 6301, and 298.

5. Subtract: $\begin{array}{r} 4926 \\ -\ 3177 \\ \hline \end{array}$

6. Divide: $7\overline{)14{,}945}$

7. Place the correct symbol, $<$ or $>$, between the two numbers: 101   87

8. Write $5 \cdot 5 \cdot 7 \cdot 7 \cdot 7 \cdot 7 \cdot 7$ in exponential notation.

9. What is 2019 multiplied by 307?

10. What is 10,134 decreased by 4725?

11. Add: $\begin{array}{r} 298 \\ 461 \\ +\ 322 \\ \hline \end{array}$

12. Simplify: $2^3 - 3 \cdot 2$

13. Round 45,672 to the nearest hundred.

14. Write 276,057 in words.

15. Find the quotient of 109,763 and 84.

16. Write two million eleven thousand forty-four in standard form.

17. What is 3906 divided by 8?

18. Simplify: $3^2 + 2^2 \cdot (5 - 3)$

19. Simplify: $8 \cdot (6 - 2)^2 \div 4$

20. Find the prime factorization of 72.

21. What is 3895 minus 1762?

22. Multiply: 843
$$\times\ 27$$

23. Vincent Meyers, a sales assistant, earns $480 for working a 40-hour week. Last week Vincent worked an additional 12 hours at $24 an hour. Find Vincent's total pay for last week's work.

24. Louis Reyes, a sales executive, drove a car 351 miles on 13 gallons of gas. Find the number of miles driven per gallon of gasoline.

25. A car is purchased for $17,880, with a down payment of $3000. The balance is paid in 48 equal monthly payments. Find the monthly car payment.

26. An insurance account executive received commissions of $723, $544, $812, and $488 during a 4-week period. Find the total income from commissions for the 4 weeks.

27. You had a balance of $516 in your checking account before making deposits of $88 and $213. Find the total amount deposited, and determine your new account balance.

28. You have a car payment of $246 per month. What is the total of the car payments over a 12-month period?

The table at the right shows the average earnings, by level of education, for people 18 years of age and older. Use this information for Exercises 29 to 31.

29. What is the difference between the annual salary of a woman with an advanced degree and that of a woman with an associate's degree?

Level of Education	Men	Women
Some college or an associates degree	$33,830	$21,057
Bachelor's degree	$50,056	$30,119
Advanced degree	$86,617	$50,546

Source: Census Bureau, *Current Population Reports*, March 1998

30. Find the difference between the annual salary of a man and that of a woman if both have a bachelor's degree.

31. How much more does a man with an associate's degree earn in 5 years than a woman with the same degree?

Chapter Test

1. Simplify: $3^3 \cdot 4^2$

2. Write 207,068 in words.

3. Subtract: $\begin{array}{r} 17,495 \\ -\ 8,162 \\ \hline \end{array}$

4. Find all the factors of 20.

5. Multiply: $\begin{array}{r} 9736 \\ \times\ 704 \\ \hline \end{array}$

6. Simplify: $4^2 \cdot (4-2) \div 8 + 5$

7. Write 906,378 in expanded form.

8. Round 74,965 to the nearest hundred.

9. Divide: $97\overline{)108,764}$

10. Write $3 \cdot 3 \cdot 3 \cdot 7 \cdot 7$ in exponential form.

11. Find the sum of 8756, 9094, and 37,065.

12. Find the prime factorization of 84.

13. Simplify: $16 \div 4 \cdot 2 - (7-5)^2$

14. Find the product of 8 and 90,763.

15. Write one million two hundred four thousand six in standard form.

16. Divide: $7\overline{)60,972}$

17. Place the correct symbol, < or >, between the two numbers: 21   19

18. Find the quotient of 5624 and 8.

19. Add: 25,492
            +71,306

20. Find the difference between 29,736 and 9814.

The table at the right shows the projected enrollment in public and private elementary and secondary schools in the fall of 2000 and the fall of 2005. Use this information for Exercises 21 and 22.

Year	Kindergarten Through Grade 8	Grades 9 Through 12
2000	39,152,000	15,250,000
2005	39,437,000	16,434,000

Source: The National Center for Education Statistics

21. Find the difference between the total enrollment in 2000 and that projected for 2005.

22. Find the average projected enrollment in each of grades 9 through 12 in 2005.

23. A farmer harvested 48,290 pounds of lemons from one grove and 23,710 pounds of lemons from another grove. The lemons were packed in boxes with 24 pounds of lemons in each box. How many boxes were needed to pack the lemons?

24. An investor receives $237 each month from a corporate bond fund. How much will the investor receive over a 12-month period?

25. A family drives 425 miles the first day, 187 miles the second day, and 243 miles the third day of their vacation. The odometer read 47,626 miles at the start of the vacation.
   a. How many miles were driven during the 3 days?
   b. What is the odometer reading at the end of the 3 days?

Chapter 2 Fractions

Members of the Boston Pops follow conductor Keith Lockhart's movements to guarantee synchronized playing. By paying attention to both Lockhart and the time signature for each musical piece, the musicians play in time with each other. The time signature appears as a fraction at the beginning of a piece of music and tells them how many beats to play per measure. The **project on page 112** demonstrates how to interpret the time signature.

Need help? For on-line student resources, such as section quizzes, visit this textbook's web site at **math.college.hmco.com/students**.

For Exercises 1 to 6, add, subtract, multiply, or divide.

1. 4×5

2. $2 \cdot 2 \cdot 2 \cdot 3 \cdot 5$

3. 9×1

4. $6 + 4$

5. $10 - 3$

6. $63 \div 30$

7. Which of the following numbers divide evenly into 12?

   1 2 3 4 5 6 7 8 9 10 11 12

8. Simplify: $8 \times 7 + 3$

9. Complete: $8 = ? + 1$

10. Place the correct symbol, $<$ or $>$, between the two numbers.

    44   48

Go Figure

You and a friend are swimming laps in a pool. You swim one lap every 4 minutes. Your friend swims one lap every 5 minutes. If you start at the same time from the same end of the pool, in how many minutes will both of you be at the starting point again? How many times will you have passed each other in the pool prior to that time?

The Least Common Multiple and Greatest Common Factor

Objective A   **To find the least common multiple (LCM)**

The **multiples** of a number are the products of that number and the numbers 1, 2, 3, 4, 5, ...

$3 \times 1 = 3$
$3 \times 2 = 6$
$3 \times 3 = 9$
$3 \times 4 = 12$ The multiples of 3 are 3, 6, 9, 12, 15, ...
$3 \times 5 = 15$
.
.
.

A number that is a multiple of two or more numbers is a **common multiple** of those numbers.

The multiples of 4 are 4, 8, 12, 16, 20, 24, 28, 32, 36, ...
The multiples of 6 are 6, 12, 18, 24, 30, 36, 42, ...
Some common multiples of 4 and 6 are 12, 24, and 36.

The **least common multiple** (LCM) is the smallest common multiple of two or more numbers.

The least common multiple of 4 and 6 is 12.

Listing the multiples of each number is one way to find the LCM. Another way to find the LCM uses the prime factorization of each number.

To find the LCM of 450 and 600, find the prime factorization of each number and write the factorization of each number in a table. Circle the largest product in each column. The LCM is the product of the circled numbers.

	2	3	5
450 =	2	(3 · 3)	(5 · 5)
600 =	(2 · 2 · 2)	3	5 · 5

In the column headed by 5, the products are equal. Circle just one product.

The LCM is the product of the circled numbers.
The LCM = $2 \cdot 2 \cdot 2 \cdot 3 \cdot 3 \cdot 5 \cdot 5 = 1800$.

Example 1   Find the LCM of 24, 36, and 50.

Solution

	2	3	5
24 =	(2 · 2 · 2)	3	
36 =	2 · 2	(3 · 3)	
50 =	2		(5 · 5)

The LCM = $2 \cdot 2 \cdot 2 \cdot 3 \cdot 3 \cdot 5 \cdot 5$
        $= 1800$.

You Try It 1   Find the LCM of 50, 84, and 135.

Your solution

Solution on p. S4

Objective B   **To find the greatest common factor (GCF)**

Recall that a number that divides another number evenly is a factor of that number. The number 64 can be evenly divided by 1, 2, 4, 8, 16, 32, and 64, so the numbers 1, 2, 4, 8, 16, 32, and 64 are factors of 64.

A number that is a factor of two or more numbers is a **common factor** of those numbers.

The factors of 30 are 1, 2, 3, 5, 6, 10, 15, and 30.
The factors of 105 are 1, 3, 5, 7, 15, 21, 35, and 105.
The common factors of 30 and 105 are 1, 3, 5, and 15.

The **greatest common factor** (GCF) is the largest common factor of two or more numbers.

The greatest common factor of 30 and 105 is 15.

Listing the factors of each number is one way of finding the GCF. Another way to find the GCF uses the prime factorization of each number.

To find the GCF of 126 and 180, find the prime factorization of each number and write the factorization of each number in a table. Circle the smallest product in each column that does not have a blank. The GCF is the product of the circled numbers.

	2	3	5	7
126 =	②	(3 · 3)		7
180 =	2 · 2	3 · 3	5	

In the column headed by 3, the products are equal. Circle just one product.
Columns 5 and 7 have a blank, so 5 and 7 are not common factors of 126 and 180. Do not circle any number in these columns.

The GCF is the product of the circled numbers.
The GCF = 2 · 3 · 3 = 18.

Example 2   Find the GCF of 90, 168, and 420.

Solution

	2	3	5	7
90 =	②	3 · 3	5	
168 =	2 · 2 · 2	③		7
420 =	2 · 2	3	5	7

The GCF = 2 · 3 = 6.

You Try It 2   Find the GCF of 36, 60, and 72.

Your solution

Example 3   Find the GCF of 7, 12, and 20.

Solution

	2	3	5	7
7 =				7
12 =	2 · 2	3		
20 =	2 · 2		5	

Because no numbers are circled, the GCF = 1.

You Try It 3   Find the GCF of 11, 24, and 30.

Your solution

Solutions on p. S4

2.1 Exercises

Objective A

Find the LCM.

1. 5, 8   **2.** 3, 6   **3.** 3, 8   **4.** 2, 5   **5.** 5, 6

6. 5, 7   **7.** 4, 6   **8.** 6, 8   **9.** 8, 12   **10.** 12, 16

11. 5, 12   **12.** 3, 16   **13.** 8, 14   **14.** 6, 18   **15.** 3, 9

16. 4, 10   **17.** 8, 32   **18.** 7, 21   **19.** 9, 36   **20.** 14, 42

21. 44, 60   **22.** 120, 160   **23.** 102, 184   **24.** 123, 234   **25.** 4, 8, 12

26. 5, 10, 15   **27.** 3, 5, 10   **28.** 2, 5, 8   **29.** 3, 8, 12   **30.** 5, 12, 18

31. 9, 36, 64   **32.** 18, 54, 63   **33.** 16, 30, 84   **34.** 9, 12, 15

Objective B

Find the GCF.

35. 3, 5   **36.** 5, 7   **37.** 6, 9   **38.** 18, 24   **39.** 15, 25

40. 14, 49   **41.** 25, 100   **42.** 16, 80   **43.** 32, 51   **44.** 21, 44

45. 12, 80 **46.** 8, 36 **47.** 16, 140 **48.** 12, 76

49. 24, 30 **50.** 48, 144 **51.** 44, 96 **52.** 18, 32

53. 3, 5, 11 **54.** 6, 8, 10 **55.** 7, 14, 49 **56.** 6, 15, 36

57. 10, 15, 20 **58.** 12, 18, 20 **59.** 24, 40, 72 **60.** 3, 17, 51

61. 17, 31, 81 **62.** 14, 42, 84 **63.** 25, 125, 625 **64.** 12, 68, 92

65. 28, 35, 70 **66.** 1, 49, 153 **67.** 32, 56, 72 **68.** 24, 36, 48

APPLYING THE CONCEPTS

69. Define the phrase *relatively prime numbers*. List three pairs of relatively prime numbers.

70. Joe Salvo, a clerk, works three days and then has a day off. A friend works five days and then has a day off. How many days after Joe and his friend have a day off together will they have another day off together?

71. Find the LCM of each of the following pairs of numbers: 2 and 3, 5 and 7, and 11 and 19. Can you draw a conclusion about the LCM of two prime numbers? Suggest a way of finding the LCM of three prime numbers.

72. Find the GCF of each of the following pairs of numbers: 3 and 5, 7 and 11, and 29 and 43. Can you draw a conclusion about the GCF of two prime numbers? What is the GCF of three prime numbers?

73. Is the LCM of two numbers always divisible by the GCF of the two numbers? If so, explain why. If not, give an example.

74. Using the pattern for the first two triangles at the right, determine the center number of the last triangle.

2.2 Introduction to Fractions

Objective A   **To write a fraction that represents part of a whole**

A **fraction** can represent the number of equal parts of a whole. The shaded portion of the circle is represented by the fraction $\frac{4}{7}$. Four of the seven equal parts of the circle (that is, four-sevenths of it) are shaded.

Each part of a fraction has a name.

Point of Interest

The fraction bar was first used in 1050 by al-Hassar. It is also called a vinculum.

Fraction bar → $\dfrac{4}{7}$ ← **Numerator**
   ← **Denominator**

A **proper fraction** is a fraction less than 1. The numerator of a proper fraction is smaller than the denominator. The shaded portion of the circle can be represented by the proper fraction $\frac{3}{4}$.

A **mixed number** is a number greater than 1 with a whole-number part and a fractional part. The shaded portion of the circles can be represented by the mixed number $2\frac{1}{4}$.

An **improper fraction** is a fraction greater than or equal to 1. The numerator of an improper fraction is greater than or equal to the denominator. The shaded portion of the circles can be represented by the improper fraction $\frac{9}{4}$. The shaded portion of the square can be represented by $\frac{4}{4}$.

Example 1

Express the shaded portion of the circles as a mixed number.

Solution $3\frac{2}{5}$

You Try It 1

Express the shaded portion of the circles as a mixed number.

Your solution

Example 2

Express the shaded portion of the circles as an improper fraction.

Solution $\frac{17}{5}$

You Try It 2

Express the shaded portion of the circles as an improper fraction.

Your solution

Solutions on p. S4

Objective B   **To write an improper fraction as a mixed number or a whole number, and a mixed number as an improper fraction**

Note from the diagram that the mixed number $2\frac{3}{5}$ and the improper fraction $\frac{13}{5}$ both represent the shaded portion of the circles.

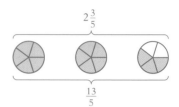

$$2\frac{3}{5} = \frac{13}{5}$$

An improper fraction can be written as a mixed number or a whole number.

➡ Write $\frac{13}{5}$ as a mixed number.

Point of Interest

Archimedes (c. 287–212 B.C.) is the person who calculated that $\pi \approx 3\frac{1}{7}$. He actually showed that $3\frac{10}{71} < \pi < 3\frac{1}{7}$. The approximation $3\frac{10}{71}$ is more accurate but more difficult to use.

Divide the numerator by the denominator.	To write the fractional part of the mixed number, write the remainder over the divisor.	Write the answer.
$\begin{array}{r} 2 \\ 5\overline{)\,13} \\ -10 \\ \hline 3 \end{array}$	$\begin{array}{r} 2\frac{3}{5} \\ 5\overline{)\,13} \\ -10 \\ \hline 3 \end{array}$	$\frac{13}{5} = 2\frac{3}{5}$

To write a mixed number as an improper fraction, multiply the denominator of the fractional part by the whole-number part. The sum of this product and the numerator of the fractional part is the numerator of the improper fraction. The denominator remains the same.

➡ Write $7\frac{3}{8}$ as an improper fraction.

$$7\frac{3}{8} = \frac{(8 \times 7) + 3}{8} = \frac{56 + 3}{8} = \frac{59}{8} \qquad 7\frac{3}{8} = \frac{59}{8}$$

Example 3

Write $\frac{21}{4}$ as a mixed number.

Solution

$\begin{array}{r} 5 \\ 4\overline{)\,21} \\ -20 \\ \hline 1 \end{array} \qquad \frac{21}{4} = 5\frac{1}{4}$

You Try It 3

Write $\frac{22}{5}$ as a mixed number.

Your solution

Example 4

Write $\frac{18}{6}$ as a whole number.

Solution

$\begin{array}{r} 3 \\ 6\overline{)\,18} \\ -18 \\ \hline 0 \end{array} \qquad \frac{18}{6} = 3$

Note: The remainder is zero.

You Try It 4

Write $\frac{28}{7}$ as a whole number.

Your solution

Example 5

Write $21\frac{3}{4}$ as an improper fraction.

Solution

$21\frac{3}{4} = \frac{84 + 3}{4} = \frac{87}{4}$

You Try It 5

Write $14\frac{5}{8}$ as an improper fraction.

Your solution

Solutions on p. S4

2.2 Exercises

Objective A

Identify the fraction as a proper fraction, an improper fraction, or a mixed number.

1. $\dfrac{12}{7}$

2. $5\dfrac{2}{11}$

3. $\dfrac{29}{40}$

4. $\dfrac{19}{13}$

Express the shaded portion of the circle as a fraction.

5.

6.

7.

8.

Express the shaded portion of the circles as a mixed number.

9.

10.

11.

12.

13.

14.

Express the shaded portion of the circles as an improper fraction.

15.

16.

17.

18.

19.

20.

21. Shade $\dfrac{5}{6}$ of

22. Shade $\dfrac{3}{8}$ of

23. Shade $1\dfrac{2}{5}$ of

24. Shade $1\dfrac{3}{4}$ of

25. Shade $\frac{6}{5}$ of

26. Shade $\frac{7}{3}$ of

Objective B

Write the improper fraction as a mixed number or a whole number.

27. $\frac{11}{4}$ **28.** $\frac{16}{3}$ **29.** $\frac{20}{4}$ **30.** $\frac{18}{9}$ **31.** $\frac{9}{8}$ **32.** $\frac{13}{4}$

33. $\frac{23}{10}$ **34.** $\frac{29}{2}$ **35.** $\frac{48}{16}$ **36.** $\frac{51}{3}$ **37.** $\frac{8}{7}$ **38.** $\frac{16}{9}$

39. $\frac{7}{3}$ **40.** $\frac{9}{5}$ **41.** $\frac{16}{1}$ **42.** $\frac{23}{1}$ **43.** $\frac{17}{8}$ **44.** $\frac{31}{16}$

45. $\frac{12}{5}$ **46.** $\frac{19}{3}$ **47.** $\frac{9}{9}$ **48.** $\frac{40}{8}$ **49.** $\frac{72}{8}$ **50.** $\frac{3}{3}$

Write the mixed number as an improper fraction.

51. $2\frac{1}{3}$ **52.** $4\frac{2}{3}$ **53.** $6\frac{1}{2}$ **54.** $8\frac{2}{3}$ **55.** $6\frac{5}{6}$ **56.** $7\frac{3}{8}$

57. $9\frac{1}{4}$ **58.** $6\frac{1}{4}$ **59.** $10\frac{1}{2}$ **60.** $15\frac{1}{8}$ **61.** $8\frac{1}{9}$ **62.** $3\frac{5}{12}$

63. $5\frac{3}{11}$ **64.** $3\frac{7}{9}$ **65.** $2\frac{5}{8}$ **66.** $12\frac{2}{3}$ **67.** $1\frac{5}{8}$ **68.** $5\frac{3}{7}$

69. $11\frac{1}{9}$ **70.** $12\frac{3}{5}$ **71.** $3\frac{3}{8}$ **72.** $4\frac{5}{9}$ **73.** $6\frac{7}{13}$ **74.** $8\frac{5}{14}$

APPLYING THE CONCEPTS

75. Name three situations in which fractions are used. Provide an example of a fraction that is used in each situation.

2.3 Writing Equivalent Fractions

Objective A   **To find equivalent fractions by raising to higher terms**

Equal fractions with different denominators are called **equivalent fractions.**

$\frac{4}{6}$ is equivalent to $\frac{2}{3}$.

Remember that the Multiplication Property of One states that the product of a number and one is the number. This is true for fractions as well as whole numbers. This property can be used to write equivalent fractions.

$$\frac{2}{3} \times 1 = \frac{2}{3} \times \frac{1}{1} = \frac{2 \cdot 1}{3 \cdot 1} = \frac{2}{3}$$

$$\frac{2}{3} \times 1 = \frac{2}{3} \times \boxed{\frac{2}{2}} = \frac{2 \cdot 2}{3 \cdot 2} = \frac{4}{6} \qquad \frac{4}{6} \text{ is equivalent to } \frac{2}{3}.$$

$$\frac{2}{3} \times 1 = \frac{2}{3} \times \boxed{\frac{4}{4}} = \frac{2 \cdot 4}{3 \cdot 4} = \frac{8}{12} \qquad \frac{8}{12} \text{ is equivalent to } \frac{2}{3}.$$

$\frac{2}{3}$ was rewritten as the equivalent fractions $\frac{4}{6}$ and $\frac{8}{12}$.

⇒ Write a fraction that is equivalent to $\frac{5}{8}$ and has a denominator of 32.

$$32 \div 8 = 4$$

$$\frac{5}{8} = \frac{5 \cdot 4}{8 \cdot 4} = \frac{20}{32}$$

• **Divide the larger denominator by the smaller.**

• **Multiply the numerator and denominator of the given fraction by the quotient (4).**

$\frac{20}{32}$ is equivalent to $\frac{5}{8}$.

Example 1   Write $\frac{2}{3}$ as an equivalent fraction that has a denominator of 42.

Solution $42 \div 3 = 14$ $\frac{2}{3} = \frac{2 \cdot 14}{3 \cdot 14} = \frac{28}{42}$

$\frac{28}{42}$ is equivalent to $\frac{2}{3}$.

You Try It 1   Write $\frac{3}{5}$ as an equivalent fraction that has a denominator of 45.

Your solution

Example 2   Write 4 as a fraction that has a denominator of 12.

Solution   Write 4 as $\frac{4}{1}$.

$12 \div 1 = 12$ $4 = \frac{4 \cdot 12}{1 \cdot 12} = \frac{48}{12}$

$\frac{48}{12}$ is equivalent to 4.

You Try It 2   Write 6 as a fraction that has a denominator of 18.

Your solution

Solutions on p. S4

Objective B   To write a fraction in simplest form

A fraction is in **simplest form** when there are no common factors in the numerator and the denominator.

The fractions $\frac{4}{6}$ and $\frac{2}{3}$ are equivalent fractions.

$\frac{4}{6}$ has been written in simplest form as $\frac{2}{3}$.

The Multiplication Property of One can be used to write fractions in simplest form. Write the numerator and denominator of the given fraction as a product of factors. Write factors common to both the numerator and denominator as an improper fraction equivalent to 1.

$$\frac{4}{6} = \frac{2 \cdot 2}{2 \cdot 3} = \boxed{\frac{2}{2}} \cdot \frac{2}{3} = 1 \cdot \frac{2}{3} = \frac{2}{3}$$

The process of eliminating common factors is displayed with slashes through the common factors as shown at the right.

To write a fraction in simplest form, eliminate the common factors.

$$\frac{4}{6} = \frac{\cancel{2} \cdot 2}{\cancel{2} \cdot 3} = \frac{2}{3}$$

$$\frac{18}{30} = \frac{\cancel{2} \cdot \cancel{3} \cdot 3}{\cancel{2} \cdot \cancel{3} \cdot 5} = \frac{3}{5}$$

An improper fraction can be changed to a mixed number.

$$\frac{22}{6} = \frac{\cancel{2} \cdot 11}{\cancel{2} \cdot 3} = \frac{11}{3} = 3\frac{2}{3}$$

Example 3

Write $\frac{15}{40}$ in simplest form.

Solution $\dfrac{15}{40} = \dfrac{3 \cdot \cancel{5}}{2 \cdot 2 \cdot 2 \cdot \cancel{5}} = \dfrac{3}{8}$

You Try It 3

Write $\frac{16}{24}$ in simplest form.

Your solution

Example 4

Write $\frac{6}{42}$ in simplest form.

Solution $\dfrac{6}{42} = \dfrac{\cancel{2} \cdot \cancel{3}}{\cancel{2} \cdot \cancel{3} \cdot 7} = \dfrac{1}{7}$

You Try It 4

Write $\frac{8}{56}$ in simplest form.

Your solution

Example 5

Write $\frac{8}{9}$ in simplest form.

Solution $\dfrac{8}{9} = \dfrac{2 \cdot 2 \cdot 2}{3 \cdot 3} = \dfrac{8}{9}$

$\frac{8}{9}$ is already in simplest form because there are no common factors in the numerator and denominator.

You Try It 5

Write $\frac{15}{32}$ in simplest form.

Your solution

Example 6

Write $\frac{30}{12}$ in simplest form.

Solution $\dfrac{30}{12} = \dfrac{\cancel{2} \cdot \cancel{3} \cdot 5}{\cancel{2} \cdot 2 \cdot \cancel{3}} = \dfrac{5}{2} = 2\frac{1}{2}$

You Try It 6

Write $\frac{48}{36}$ in simplest form.

Your solution

Solutions on p. S4

2.3 Exercises

Objective A

Write an equivalent fraction with the given denominator.

1. $\dfrac{1}{2} = \dfrac{}{10}$

2. $\dfrac{1}{4} = \dfrac{}{16}$

3. $\dfrac{3}{16} = \dfrac{}{48}$

4. $\dfrac{5}{9} = \dfrac{}{81}$

5. $\dfrac{3}{8} = \dfrac{}{32}$

6. $\dfrac{7}{11} = \dfrac{}{33}$

7. $\dfrac{3}{17} = \dfrac{}{51}$

8. $\dfrac{7}{10} = \dfrac{}{90}$

9. $\dfrac{3}{4} = \dfrac{}{16}$

10. $\dfrac{5}{8} = \dfrac{}{32}$

11. $3 = \dfrac{}{9}$

12. $5 = \dfrac{}{25}$

13. $\dfrac{1}{3} = \dfrac{}{60}$

14. $\dfrac{1}{16} = \dfrac{}{48}$

15. $\dfrac{11}{15} = \dfrac{}{60}$

16. $\dfrac{3}{50} = \dfrac{}{300}$

17. $\dfrac{2}{3} = \dfrac{}{18}$

18. $\dfrac{5}{9} = \dfrac{}{36}$

19. $\dfrac{5}{7} = \dfrac{}{49}$

20. $\dfrac{7}{8} = \dfrac{}{32}$

21. $\dfrac{5}{9} = \dfrac{}{18}$

22. $\dfrac{11}{12} = \dfrac{}{36}$

23. $7 = \dfrac{}{3}$

24. $9 = \dfrac{}{4}$

25. $\dfrac{7}{9} = \dfrac{}{45}$

26. $\dfrac{5}{6} = \dfrac{}{42}$

27. $\dfrac{15}{16} = \dfrac{}{64}$

28. $\dfrac{11}{18} = \dfrac{}{54}$

29. $\dfrac{3}{14} = \dfrac{}{98}$

30. $\dfrac{5}{6} = \dfrac{}{144}$

31. $\dfrac{5}{8} = \dfrac{}{48}$

32. $\dfrac{7}{12} = \dfrac{}{96}$

33. $\dfrac{5}{14} = \dfrac{}{42}$

34. $\dfrac{2}{3} = \dfrac{}{42}$

35. $\dfrac{17}{24} = \dfrac{}{144}$

36. $\dfrac{5}{13} = \dfrac{}{169}$

37. $\dfrac{3}{8} = \dfrac{}{408}$

38. $\dfrac{9}{16} = \dfrac{}{272}$

39. $\dfrac{17}{40} = \dfrac{}{800}$

40. $\dfrac{9}{25} = \dfrac{}{1000}$

Objective B

Write the fraction in simplest form.

41. $\dfrac{4}{12}$

42. $\dfrac{8}{22}$

43. $\dfrac{22}{44}$

44. $\dfrac{2}{14}$

45. $\dfrac{2}{12}$

46. $\dfrac{50}{75}$ **47.** $\dfrac{40}{36}$ **48.** $\dfrac{12}{8}$ **49.** $\dfrac{0}{30}$ **50.** $\dfrac{10}{10}$

51. $\dfrac{9}{22}$ **52.** $\dfrac{14}{35}$ **53.** $\dfrac{75}{25}$ **54.** $\dfrac{8}{60}$ **55.** $\dfrac{16}{84}$

56. $\dfrac{20}{44}$ **57.** $\dfrac{12}{35}$ **58.** $\dfrac{8}{36}$ **59.** $\dfrac{28}{44}$ **60.** $\dfrac{12}{16}$

61. $\dfrac{16}{12}$ **62.** $\dfrac{24}{18}$ **63.** $\dfrac{24}{40}$ **64.** $\dfrac{44}{60}$ **65.** $\dfrac{8}{88}$

66. $\dfrac{9}{90}$ **67.** $\dfrac{144}{36}$ **68.** $\dfrac{140}{297}$ **69.** $\dfrac{48}{144}$ **70.** $\dfrac{32}{120}$

71. $\dfrac{60}{100}$ **72.** $\dfrac{33}{110}$ **73.** $\dfrac{36}{16}$ **74.** $\dfrac{80}{45}$ **75.** $\dfrac{32}{160}$

APPLYING THE CONCEPTS

76. Make a list of five different fractions that are equivalent to $\dfrac{2}{3}$.

77. Make a list of five different fractions that are equivalent to 3.

78. Show that $\dfrac{15}{24} = \dfrac{5}{8}$ by using a diagram.

79. **a.** What fraction of the states in the United States of America have names that begin with the letter M?
   b. What fraction of the states have names that begin and end with a vowel?

2.4 Addition of Fractions and Mixed Numbers

Objective A   **To add fractions with the same denominator**

Fractions with the same denominator are added by adding the numerators and placing the sum over the common denominator. After adding, write the sum in simplest form.

➡ Add: $\frac{2}{7} + \frac{4}{7}$

$$\begin{array}{r} \frac{2}{7} \\ + \frac{4}{7} \\ \hline \frac{6}{7} \end{array}$$

- Add the numerators and place the sum over the common denominator.

$$\frac{2}{7} + \frac{4}{7} = \frac{2+4}{7} = \frac{6}{7}$$

Example 1

Add: $\frac{5}{12} + \frac{11}{12}$

Solution

$$\begin{array}{r} \frac{5}{12} \\ + \frac{11}{12} \\ \hline \frac{16}{12} = \frac{4}{3} = 1\frac{1}{3} \end{array}$$

You Try It 1

Add: $\frac{3}{8} + \frac{7}{8}$

Your solution

Solution on p. S4

Objective B   **To add fractions with unlike denominators**

To add fractions with unlike denominators, first rewrite the fractions as equivalent fractions with a common denominator. The common denominator is the LCM of the denominators of the fractions.

➡ Find the total of $\frac{1}{2}$ and $\frac{1}{3}$.

The common denominator is the LCM of 2 and 3. The LCM = 6. The LCM of denominators is sometimes called the **least common denominator** (LCD).

Write equivalent fractions using the LCM.

$$\frac{1}{2} = \frac{3}{6}$$
$$+ \frac{1}{3} = \frac{2}{6}$$

Add the fractions.

$$\frac{1}{2} = \frac{3}{6}$$
$$+ \frac{1}{3} = \frac{2}{6}$$
$$\frac{5}{6}$$

Example 2

Find $\frac{7}{12}$ more than $\frac{3}{8}$.

Solution

$$\frac{3}{8} = \frac{9}{24}$$
$$+\frac{7}{12} = \frac{14}{24}$$
$$\frac{23}{24}$$

• The LCM of 8 and 12 is 24.

You Try It 2

Find the sum of $\frac{5}{12}$ and $\frac{9}{16}$.

Your solution

Example 3

Add: $\frac{5}{8} + \frac{7}{9}$

Solution

$$\frac{5}{8} = \frac{45}{72}$$
$$+\frac{7}{9} = \frac{56}{72}$$
$$\frac{101}{72} = 1\frac{29}{72}$$

• The LCM of 8 and 9 is 72.

You Try It 3

Add: $\frac{7}{8} + \frac{11}{15}$

Your solution

Example 4

Add: $\frac{2}{3} + \frac{3}{5} + \frac{5}{6}$

Solution

$$\frac{2}{3} = \frac{20}{30}$$
$$\frac{3}{5} = \frac{18}{30}$$
$$+\frac{5}{6} = \frac{25}{30}$$
$$\frac{63}{30} = 2\frac{3}{30} = 2\frac{1}{10}$$

• The LCM of 3, 5, and 6 is 30.

You Try It 4

Add: $\frac{3}{4} + \frac{4}{5} + \frac{5}{8}$

Your solution

Solutions on p. S5

Objective C   **To add whole numbers, mixed numbers, and fractions**

TAKE NOTE

The procedure at the right illustrates why $2 + \frac{2}{3} = 2\frac{2}{3}$. You do not need to show these steps when adding a whole number and a fraction. Here are two more examples:

$$7 + \frac{1}{5} = 7\frac{1}{5}$$
$$6 + \frac{3}{4} = 6\frac{3}{4}$$

The sum of a whole number and a fraction is a mixed number.

➡ Add: $2 + \frac{2}{3}$

$$\boxed{2} + \frac{2}{3} = \boxed{\frac{6}{3}} + \frac{2}{3} = \frac{8}{3} = 2\frac{2}{3}$$

To add a whole number and a mixed number, write the fraction and then add the whole numbers.

➡ Add: $7\frac{2}{5} + 4$

Write the fraction.

$$7\frac{2}{5}$$
$$+\,4$$
$$\overline{\frac{2}{5}}$$

Add the whole numbers.

$$7\frac{2}{5}$$
$$+\,4$$
$$\overline{11\frac{2}{5}}$$

To add two mixed numbers, add the fractional parts and then add the whole numbers. Remember to reduce the sum to simplest form.

➡ What is $6\frac{14}{15}$ added to $5\frac{4}{9}$?

The LCM of 9 and 15 is 45.

Add the fractional parts.

$$5\frac{4}{9} = 5\frac{20}{45}$$
$$+\ 6\frac{14}{15} = 6\frac{42}{45}$$
$$\overline{\phantom{+\ 6\frac{14}{15} = }\ \frac{62}{45}}$$

Add the whole numbers.

$$5\frac{4}{9} = 5\frac{20}{45}$$
$$+\ 6\frac{14}{15} = 6\frac{42}{45}$$
$$\overline{\phantom{+\ 6\frac{14}{15} = }\ 11\frac{62}{45} = 11 + 1\frac{17}{45} = 12\frac{17}{45}}$$

Example 5

Add: $5 + \frac{3}{8}$

Solution $5 + \frac{3}{8} = 5\frac{3}{8}$

You Try It 5

What is 7 added to $\frac{6}{11}$?

Your solution

Example 6

Find 17 increased by $3\frac{3}{8}$.

Solution
$$17$$
$$+\ 3\frac{3}{8}$$
$$\overline{20\frac{3}{8}}$$

You Try It 6

Find the sum of 29 and $17\frac{5}{12}$.

Your solution

Example 7

Add: $5\frac{2}{3} + 11\frac{5}{6} + 12\frac{7}{9}$

Solution
$$5\frac{2}{3} = 5\frac{12}{18} \qquad \bullet \ \text{LCM} = 18$$
$$11\frac{5}{6} = 11\frac{15}{18}$$
$$+\ 12\frac{7}{9} = 12\frac{14}{18}$$
$$\overline{28\frac{41}{18} = 30\frac{5}{18}}$$

You Try It 7

Add: $7\frac{4}{5} + 6\frac{7}{10} + 13\frac{11}{15}$

Your solution

Example 8

Add: $11\frac{5}{8} + 7\frac{5}{9} + 8\frac{7}{15}$

Solution
$$11\frac{5}{8} = 11\frac{225}{360} \qquad \bullet \ \text{LCM} = 360$$
$$7\frac{5}{9} = 7\frac{200}{360}$$
$$+\ 8\frac{7}{15} = 8\frac{168}{360}$$
$$\overline{26\frac{593}{360} = 27\frac{233}{360}}$$

You Try It 8

Add: $9\frac{3}{8} + 17\frac{7}{12} + 10\frac{14}{15}$

Your solution

Solutions on p. S5

Example 9

A rain gauge collected $2\frac{1}{3}$ inches of rain in October, $5\frac{1}{2}$ inches in November, and $3\frac{3}{8}$ inches in December. Find the total rainfall for the 3 months.

Strategy

To find the total rainfall for the 3 months, add the 3 amounts of rainfall $\left(2\frac{1}{3}, 5\frac{1}{2}, \text{ and } 3\frac{3}{8}\right)$.

Solution

$$2\frac{1}{3} = 2\frac{8}{24}$$
$$5\frac{1}{2} = 5\frac{12}{24}$$
$$+\ 3\frac{3}{8} = 3\frac{9}{24}$$
$$\overline{\phantom{+\ 3\frac{3}{8}=\ }10\frac{29}{24} = 11\frac{5}{24}}$$

The total rainfall for the 3 months was $11\frac{5}{24}$ inches.

You Try It 9

On Monday, you spent $4\frac{1}{2}$ hours in class, $3\frac{3}{4}$ hours studying, and $1\frac{1}{3}$ hours driving. Find the number of hours spent on these three activities.

Your strategy

Your solution

Example 10

Barbara Walsh worked 4 hours, $2\frac{1}{3}$ hours, and $5\frac{2}{3}$ hours this week at a part-time job. Barbara is paid $9 an hour. How much did she earn this week?

Strategy

To find how much Barbara earned:
- Find the total number of hours worked.
- Multiply the total number of hours worked by the hourly wage (9).

Solution

$$
\begin{array}{ll}
4 & \quad\quad 12 \\
2\frac{1}{3} & \quad\ \times\ 9 \\
+\ 5\frac{2}{3} & \quad\overline{108} \\
\overline{11\frac{3}{3} = 12 \text{ hours worked}}
\end{array}
$$

Barbara earned $108 this week.

You Try It 10

Jeff Sapone, a carpenter, worked $1\frac{2}{3}$ hours of overtime on Monday, $3\frac{1}{3}$ hours of overtime on Tuesday, and 2 hours of overtime on Wednesday. At an overtime hourly rate of $36, find Jeff's overtime pay for these 3 days.

Your strategy

Your solution

Solutions on p. S5

2.4 Exercises

Objective A

Add.

1. $\dfrac{2}{7} + \dfrac{1}{7}$

2. $\dfrac{3}{11} + \dfrac{5}{11}$

3. $\dfrac{1}{2} + \dfrac{1}{2}$

4. $\dfrac{1}{3} + \dfrac{2}{3}$

5. $\dfrac{8}{11} + \dfrac{7}{11}$

6. $\dfrac{9}{13} + \dfrac{7}{13}$

7. $\dfrac{8}{5} + \dfrac{9}{5}$

8. $\dfrac{5}{3} + \dfrac{7}{3}$

9. $\dfrac{3}{5} + \dfrac{8}{5} + \dfrac{3}{5}$

10. $\dfrac{3}{8} + \dfrac{5}{8} + \dfrac{7}{8}$

11. $\dfrac{3}{4} + \dfrac{1}{4} + \dfrac{5}{4} = \dfrac{9}{4} = 2\dfrac{1}{4}$

12. $\dfrac{2}{7} + \dfrac{4}{7} + \dfrac{5}{7}$

13. $\dfrac{3}{8} + \dfrac{7}{8} + \dfrac{1}{8}$ $\dfrac{11}{8} = 1\dfrac{3}{8}$

14. $\dfrac{5}{12} + \dfrac{7}{12} + \dfrac{1}{12}$

15. $\dfrac{4}{15} + \dfrac{7}{15} + \dfrac{11}{15} = \dfrac{22}{15} = 1\dfrac{7}{15}$

16. $\dfrac{3}{4} + \dfrac{3}{4} + \dfrac{1}{4}$

17. $\dfrac{3}{16} + \dfrac{5}{16} + \dfrac{7}{16} = \dfrac{15}{16}$

18. $\dfrac{5}{18} + \dfrac{11}{18} + \dfrac{17}{18}$

19. $\dfrac{3}{11} + \dfrac{5}{11} + \dfrac{7}{11} = \dfrac{15}{11} = 1\dfrac{4}{11}$

20. $\dfrac{5}{7} + \dfrac{4}{7} + \dfrac{5}{7}$

21. Find the sum of $\dfrac{4}{9}$ and $\dfrac{5}{9}$.

22. Find the sum of $\dfrac{5}{12}$, $\dfrac{1}{12}$, and $\dfrac{11}{12}$.

23. Find the total of $\dfrac{5}{8}$, $\dfrac{3}{8}$, and $\dfrac{7}{8}$.

24. Find the total of $\dfrac{4}{13}$, $\dfrac{7}{13}$, and $\dfrac{11}{13}$.

Objective B

Add.

25. $\dfrac{1}{2} + \dfrac{2}{3}$ $\dfrac{3}{6} + \dfrac{4}{6}$ $\dfrac{7}{6} = 1\dfrac{1}{6}$

26. $\dfrac{2}{3} + \dfrac{1}{4}$

27. $\dfrac{3}{14} + \dfrac{5}{7}$ $\dfrac{3}{14} + \dfrac{10}{14} =$ $\dfrac{13}{14}$

28. $\dfrac{3}{5} + \dfrac{7}{10}$

29. $\dfrac{8}{15} + \dfrac{7}{20}$ $\dfrac{32}{60} + \dfrac{21}{60}$ $\dfrac{53}{60}$

30. $\dfrac{1}{6} + \dfrac{7}{9}$

31. $\dfrac{3}{8} + \dfrac{9}{14}$ $\dfrac{21}{56} + \dfrac{36}{56}$ $\dfrac{57}{56} = 1\dfrac{1}{56}$

32. $\dfrac{5}{12} + \dfrac{5}{16}$

33. $\dfrac{3}{20} + \dfrac{7}{30}$ $\dfrac{9}{60} + \dfrac{14}{60} = \dfrac{23}{60}$

34. $\dfrac{5}{12} + \dfrac{7}{30}$

35. $\dfrac{2}{3} + \dfrac{6}{19}$ $\dfrac{38}{19} + \dfrac{7}{19} + \dfrac{9}{19}$ $\dfrac{?}{76}$

36. $\dfrac{1}{2} + \dfrac{3}{29}$

37. $\dfrac{1}{3} + \dfrac{5}{6} + \dfrac{7}{9}$ **38.** $\dfrac{2}{3} + \dfrac{5}{6} + \dfrac{7}{12}$ **39.** $\dfrac{5}{6} + \dfrac{1}{12} + \dfrac{5}{16}$ **40.** $\dfrac{2}{9} + \dfrac{7}{15} + \dfrac{4}{21}$

41. $\dfrac{2}{3} + \dfrac{1}{5} + \dfrac{7}{12}$ **42.** $\dfrac{3}{4} + \dfrac{4}{5} + \dfrac{7}{12}$ **43.** $\dfrac{1}{4} + \dfrac{4}{5} + \dfrac{5}{9}$ **44.** $\dfrac{2}{3} + \dfrac{3}{5} + \dfrac{7}{8}$

45. $\dfrac{5}{16} + \dfrac{11}{18} + \dfrac{17}{24}$ **46.** $\dfrac{3}{10} + \dfrac{14}{15} + \dfrac{9}{25}$ **47.** $\dfrac{2}{3} + \dfrac{5}{8} + \dfrac{7}{9}$ **48.** $\dfrac{1}{3} + \dfrac{2}{9} + \dfrac{7}{8}$

49. What is $\dfrac{3}{8}$ added to $\dfrac{3}{5}$?   **50.** What is $\dfrac{5}{9}$ added to $\dfrac{7}{12}$?

51. Find the sum of $\dfrac{3}{8}$, $\dfrac{5}{6}$, and $\dfrac{7}{12}$.   **52.** Find the sum of $\dfrac{11}{12}$, $\dfrac{13}{24}$, and $\dfrac{4}{15}$.

53. Find the total of $\dfrac{1}{2}$, $\dfrac{5}{8}$, and $\dfrac{7}{9}$.   **54.** Find the total of $\dfrac{5}{14}$, $\dfrac{3}{7}$, and $\dfrac{5}{21}$.

Objective C

Add.

55. $\begin{array}{r} 1\dfrac{1}{2} \\ + 2\dfrac{1}{6} \\ \hline \end{array}$ **56.** $\begin{array}{r} 2\dfrac{2}{5} \\ + 3\dfrac{3}{10} \\ \hline \end{array}$ **57.** $\begin{array}{r} 4\dfrac{1}{2} \\ + 5\dfrac{7}{12} \\ \hline \end{array}$ **58.** $\begin{array}{r} 3\dfrac{3}{8} \\ + 2\dfrac{5}{16} \\ \hline \end{array}$

59. $\begin{array}{r} 4 \\ + 5\dfrac{2}{7} \\ \hline \end{array}$ **60.** $\begin{array}{r} 6\dfrac{8}{9} \\ + 12 \\ \hline \end{array}$ **61.** $\begin{array}{r} 3\dfrac{5}{8} \\ + 2\dfrac{11}{20} \\ \hline \end{array}$ **62.** $\begin{array}{r} 4\dfrac{5}{12} \\ + 6\dfrac{11}{18} \\ \hline \end{array}$

63. $7\dfrac{5}{12} + 2\dfrac{9}{16}$ **64.** $9\dfrac{1}{2} + 3\dfrac{3}{11}$ **65.** $6\dfrac{1}{3} + 2\dfrac{3}{13}$ **66.** $8\dfrac{21}{40} + 6\dfrac{21}{32}$

67. $8\dfrac{29}{30} + 7\dfrac{11}{40}$ **68.** $17\dfrac{5}{16} + 3\dfrac{11}{24}$ **69.** $17\dfrac{3}{8} + 7\dfrac{7}{20}$ **70.** $14\dfrac{7}{12} + 29\dfrac{13}{21}$

71. $5\dfrac{7}{8} + 27\dfrac{5}{12}$

72. $7\dfrac{5}{6} + 3\dfrac{5}{9}$

73. $7\dfrac{5}{9} + 2\dfrac{7}{12}$

74. $3\dfrac{1}{2} + 2\dfrac{3}{4} + 1\dfrac{5}{6}$

75. $2\dfrac{1}{2} + 3\dfrac{2}{3} + 4\dfrac{1}{4}$

76. $3\dfrac{1}{3} + 7\dfrac{1}{5} + 2\dfrac{1}{7}$

77. $3\dfrac{1}{2} + 3\dfrac{1}{5} + 8\dfrac{1}{9}$

78. $6\dfrac{5}{9} + 6\dfrac{5}{12} + 2\dfrac{5}{18}$

79. $2\dfrac{3}{8} + 4\dfrac{7}{12} + 3\dfrac{5}{16}$

80. $2\dfrac{1}{8} + 4\dfrac{2}{9} + 5\dfrac{17}{18}$

81. $6\dfrac{5}{6} + 17\dfrac{2}{9} + 18\dfrac{5}{27}$

82. $4\dfrac{7}{20} + \dfrac{17}{80} + 25\dfrac{23}{60}$

83. Find the sum of $2\dfrac{4}{9}$ and $5\dfrac{7}{12}$.

84. Find $5\dfrac{5}{6}$ more than $3\dfrac{3}{8}$.

85. What is $4\dfrac{3}{4}$ added to $9\dfrac{1}{3}$?

86. What is $4\dfrac{8}{9}$ added to $9\dfrac{1}{6}$?

87. Find the total of $2\dfrac{2}{3}$, $4\dfrac{5}{8}$, and $2\dfrac{2}{9}$.

88. Find the total of $1\dfrac{5}{8}$, $3\dfrac{5}{6}$, and $7\dfrac{7}{24}$.

Objective D   *Application Problems*

89. Find the length of the shaft.

90. Find the length of the shaft.

91. A table 30 inches high has a top that is $1\dfrac{1}{8}$ inches thick. Find the total thickness of the table top after a $\dfrac{3}{16}$-inch veneer is applied.

92. Fred Thomson worked $2\dfrac{2}{3}$ hours of overtime on Monday, $1\dfrac{1}{4}$ hours on Wednesday, $1\dfrac{1}{3}$ hours on Friday, and $6\dfrac{3}{4}$ hours on Saturday.
   a. Find the total number of overtime hours worked during the week.
   b. At an overtime hourly wage of $22 per hour, how much overtime pay does Fred receive?

93. You are working a part-time job that pays $9 an hour. You worked 5, $3\frac{3}{4}$, $2\frac{1}{3}$, $1\frac{1}{4}$, and $7\frac{2}{3}$ hours during the last five days.

    a. Find the total number of hours you worked during the last five days.
    b. Find your total wages for the five days.

94. The course of a yachting race is in the shape of a triangle with sides that measure $4\frac{3}{10}$ miles, $3\frac{7}{10}$ miles, and $2\frac{1}{2}$ miles. Find the total length of the course.

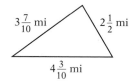

95. A flower garden in the yard of a colonial home is in the shape of a triangle as shown at the right. The wooden beams lining the edge of the garden need to be replaced. Find the total length of wood beams that must be purchased in order to replace the old beams.

96. Alan bought 200 shares of a utility stock for $26\frac{3}{16}$ per share. The monthly gains for the next three months were $1\frac{1}{2}$, $\frac{5}{16}$, and $2\frac{5}{8}$ per share. Find the value of one share of the stock at the end of the three months.

Where Americans Moved

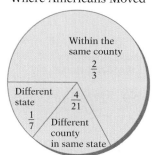

97. During a recent year, over 42 million Americans changed homes. The graph at the right shows what fractions of the people moved within the same county, moved to a different county in the same state, and moved to a different state. (Source: Census Bureau; *Geographical Mobility*) What fractional part of those who changed homes moved outside the county they had been living in?

APPLYING THE CONCEPTS

98. What is a unit fraction? Find the sum of the three largest unit fractions. Is there a smallest unit fraction? If so, write it down. If not, explain why.

99. Use a model to illustrate and explain the addition of fractions with unlike denominators.

100. A survey was conducted to determine people's favorite color from among blue, green, red, purple, and other. The surveyor claims that $\frac{1}{3}$ of the people responded blue, $\frac{1}{6}$ responded green, $\frac{1}{8}$ responded red, $\frac{1}{12}$ responded purple, and $\frac{2}{5}$ responded some other color. Is this possible? Explain your answer.

2.5 Subtraction of Fractions and Mixed Numbers

Objective A To subtract fractions with the same denominator

Fractions with the same denominator are subtracted by subtracting the numerators and placing the difference over the common denominator. After subtracting, write the fraction in simplest form.

➡ Subtract: $\frac{5}{7} - \frac{3}{7}$

$$\begin{array}{r} \frac{5}{7} \\ -\frac{3}{7} \\ \hline \frac{2}{7} \end{array}$$

• Subtract the numerators and place the difference over the common denominator.

$$\frac{5}{7} - \frac{3}{7} = \frac{5-3}{7} = \frac{2}{7}$$

Example 1

Find $\frac{17}{30}$ less $\frac{11}{30}$.

Solution

$$\begin{array}{r} \frac{17}{30} \\ -\frac{11}{30} \\ \hline \frac{6}{30} = \frac{1}{5} \end{array}$$

You Try It 1

Subtract: $\frac{16}{27} - \frac{7}{27}$

Your solution

Solution on p. S5

Objective B To subtract fractions with unlike denominators

To subtract fractions with unlike denominators, first rewrite the fractions as equivalent fractions with a common denominator. As with adding fractions, the common denominator is the LCM of the denominators of the fractions.

➡ Subtract: $\frac{5}{6} - \frac{1}{4}$

The common denominator is the LCM of 6 and 4. The LCM = 12.

Write equivalent fractions using the LCM.

$$\frac{5}{6} = \frac{10}{12}$$
$$-\frac{1}{4} = \frac{3}{12}$$

Subtract the fractions.

$$\begin{array}{r} \frac{5}{6} = \frac{10}{12} \\ -\frac{1}{4} = \frac{3}{12} \\ \hline \frac{7}{12} \end{array}$$

Example 2

Subtract: $\dfrac{11}{16} - \dfrac{5}{12}$

Solution

$\dfrac{11}{16} = \dfrac{33}{48}$ • LCM = 48

$-\dfrac{5}{12} = \dfrac{20}{48}$

$\phantom{-\dfrac{5}{12} = }\dfrac{13}{48}$

You Try It 2

Subtract: $\dfrac{13}{18} - \dfrac{7}{24}$

Your solution

Solution on p. S5

Objective C **To subtract whole numbers, mixed numbers, and fractions** [2]

To subtract mixed numbers without borrowing, subtract the fractional parts and then subtract the whole numbers.

➡ Subtract: $5\dfrac{5}{6} - 2\dfrac{3}{4}$

Subtract the fractional parts.

$5\dfrac{5}{6} = 5\dfrac{10}{12}$ • The LCM of 6 and 4 is 12.

$-2\dfrac{3}{4} = 2\dfrac{9}{12}$

$\phantom{-2\dfrac{3}{4} = }\dfrac{1}{12}$

Subtract the whole numbers.

$5\dfrac{5}{6} = 5\dfrac{10}{12}$

$-2\dfrac{3}{4} = 2\dfrac{9}{12}$

$\phantom{-2\dfrac{3}{4} = }3\dfrac{1}{12}$

Subtraction of mixed numbers sometimes involves borrowing.

➡ Subtract: $5 - 2\dfrac{5}{8}$

Borrow 1 from 5.

$5 = \overset{4}{\cancel{5}}\,1$

$-2\dfrac{5}{8} = 2\dfrac{5}{8}$

Write 1 as a fraction so that the fractions have the same denominators.

$5 = 4\dfrac{8}{8}$

$-2\dfrac{5}{8} = 2\dfrac{5}{8}$

Subtract the mixed numbers.

$5 = 4\dfrac{8}{8}$

$-2\dfrac{5}{8} = 2\dfrac{5}{8}$

$\phantom{-2\dfrac{5}{8} = }2\dfrac{3}{8}$

➡ Subtract: $7\dfrac{1}{6} - 2\dfrac{5}{8}$

Write equivalent fractions using the LCM.

$7\dfrac{1}{6} = 7\dfrac{4}{24}$

$-2\dfrac{5}{8} = 2\dfrac{15}{24}$

Borrow 1 from 7. Add the 1 to $\dfrac{4}{24}$. Write $1\dfrac{4}{24}$ as $\dfrac{28}{24}$.

$7\dfrac{1}{6} = \overset{6}{\cancel{7}}1\dfrac{4}{24} = 6\dfrac{28}{24}$

$-2\dfrac{5}{8} = 2\dfrac{15}{24} = 2\dfrac{15}{24}$

Subtract the mixed numbers.

$7\dfrac{1}{6} = 6\dfrac{28}{24}$

$-2\dfrac{5}{8} = 2\dfrac{15}{24}$

$\phantom{-2\dfrac{5}{8} = }4\dfrac{13}{24}$

Example 3

Subtract: $15\frac{7}{8} - 12\frac{2}{3}$

Solution

$$15\frac{7}{8} = 15\frac{21}{24} \quad \bullet \text{ LCM} = 24$$
$$-12\frac{2}{3} = 12\frac{16}{24}$$
$$\overline{\phantom{-12\frac{2}{3}=}3\frac{5}{24}}$$

You Try It 3

Subtract: $17\frac{5}{9} - 11\frac{5}{12}$

Your solution

Example 4

Subtract: $9 - 4\frac{3}{11}$

Solution

$$9 = 8\frac{11}{11} \quad \bullet \text{ LCM} = 11$$
$$-4\frac{3}{11} = 4\frac{3}{11}$$
$$\overline{\phantom{-4\frac{3}{11}=}4\frac{8}{11}}$$

You Try It 4

Subtract: $8 - 2\frac{4}{13}$

Your solution

Example 5

Find $11\frac{5}{12}$ decreased by $2\frac{11}{16}$.

Solution

$$11\frac{5}{12} = 11\frac{20}{48} = 10\frac{68}{48} \quad \bullet \text{ LCM} = 48$$
$$-2\frac{11}{16} = 2\frac{33}{48} = 2\frac{33}{48}$$
$$\overline{\phantom{-2\frac{11}{16}=2\frac{33}{48}=}8\frac{35}{48}}$$

You Try It 5

What is $21\frac{7}{9}$ minus $7\frac{11}{12}$?

Your solution

Solutions on p. S6

Objective D   **To solve application problems**

⇒ The outside diameter of a bushing is $3\frac{3}{8}$ inches and the wall thickness is $\frac{1}{4}$ inch. Find the inside diameter of the bushing.

$$\frac{1}{4} + \frac{1}{4} = \frac{2}{4} = \frac{1}{2}$$

$\bullet$ Add $\frac{1}{4}$ to $\frac{1}{4}$ to find the total thickness of the two walls.

$$3\frac{3}{8} = 3\frac{3}{8} = 2\frac{11}{8}$$
$$-\frac{1}{2} = \frac{4}{8} = \frac{4}{8}$$
$$\overline{\phantom{-\frac{1}{2}=\frac{4}{8}=}2\frac{7}{8}}$$

$\bullet$ Subtract the total thickness of the two walls to find the inside diameter.

The inside diameter of the bushing is $2\frac{7}{8}$ inches.

Example 6

A $2\frac{2}{3}$-inch piece is cut from a $6\frac{5}{8}$-inch board. How much of the board is left?

Strategy

To find the length remaining, subtract the length of the piece cut from the total length of the board.

$6\frac{5}{8}$ in.

Remaining Piece

$2\frac{2}{3}$ in.

Solution

$$6\frac{5}{8} = 6\frac{15}{24} = 5\frac{39}{24}$$
$$-\ 2\frac{2}{3} = 2\frac{16}{24} = 2\frac{16}{24}$$
$$\overline{\phantom{-2\frac{2}{3}=2\frac{16}{24}=}\ 3\frac{23}{24}}$$

$3\frac{23}{24}$ inches of the board are left.

Example 7

Two painters are staining a house. In 1 day one painter stained $\frac{1}{3}$ of the house, and the other stained $\frac{1}{4}$ of the house. How much of the job remains to be done?

Strategy

To find how much of the job remains:

• Find the total amount of the house already stained $\left(\frac{1}{3} + \frac{1}{4}\right)$.

• Subtract the amount already stained from 1, which represents the complete job.

Solution

$$\frac{1}{3} = \frac{4}{12} \qquad\qquad 1 = \frac{12}{12}$$
$$+\ \frac{1}{4} = \frac{3}{12} \qquad\qquad -\ \frac{7}{12} = \frac{7}{12}$$
$$\overline{\ \frac{7}{12}} \qquad\qquad \overline{\phantom{-\frac{7}{12}=}\ \frac{5}{12}}$$

$\frac{5}{12}$ of the house remains to be stained.

You Try It 6

A flight from New York to Los Angeles takes $5\frac{1}{2}$ hours. After the plane has been in the air for $2\frac{3}{4}$ hours, how much flight time remains?

Your strategy

Your solution

You Try It 7

A patient is put on a diet to lose 24 pounds in 3 months. The patient lost $7\frac{1}{2}$ pounds the first month and $5\frac{3}{4}$ pounds the second month. How much weight must be lost the third month to achieve the goal?

Your strategy

Your solution

Solutions on p. S6

2.5 Exercises

Objective A

Subtract.

1. $\dfrac{9}{17}$
    $-\dfrac{7}{17}$

2. $\dfrac{11}{15}$
    $-\dfrac{3}{15}$

3. $\dfrac{11}{12}$
    $-\dfrac{7}{12}$

4. $\dfrac{13}{15}$
    $-\dfrac{4}{15}$

5. $\dfrac{9}{20}$
    $-\dfrac{7}{20}$

6. $\dfrac{48}{55}$
    $-\dfrac{13}{55}$

7. $\dfrac{42}{65}$
    $-\dfrac{17}{65}$

8. $\dfrac{11}{24}$
    $-\dfrac{5}{24}$

9. $\dfrac{23}{30}$
    $-\dfrac{13}{30}$

10. $\dfrac{17}{42}$
    $-\dfrac{5}{42}$

11. What is $\dfrac{5}{14}$ less than $\dfrac{13}{14}$?

12. What is $\dfrac{7}{19}$ less than $\dfrac{17}{19}$?

13. Find the difference between $\dfrac{7}{8}$ and $\dfrac{5}{8}$.

14. Find the difference between $\dfrac{7}{12}$ and $\dfrac{5}{12}$.

15. What is $\dfrac{18}{23}$ minus $\dfrac{9}{23}$?

16. What is $\dfrac{7}{9}$ minus $\dfrac{3}{9}$?

17. Find $\dfrac{17}{24}$ decreased by $\dfrac{11}{24}$.

18. Find $\dfrac{19}{30}$ decreased by $\dfrac{11}{30}$.

Objective B

Subtract.

19. $\dfrac{2}{3}$
    $-\dfrac{1}{6}$

20. $\dfrac{7}{8}$
    $-\dfrac{5}{16}$

21. $\dfrac{5}{8}$
    $-\dfrac{2}{7}$

22. $\dfrac{5}{6}$
    $-\dfrac{3}{7}$

23. $\dfrac{5}{7}$
    $-\dfrac{3}{14}$

24. $\dfrac{5}{9}$
    $-\dfrac{7}{15}$

25. $\dfrac{8}{15}$
    $-\dfrac{7}{20}$

26. $\dfrac{7}{9}$
    $-\dfrac{1}{6}$

27. $\dfrac{9}{14}$
    $-\dfrac{3}{8}$

28. $\dfrac{5}{12}$
    $-\dfrac{5}{16}$

29. $\dfrac{46}{51}$
$-\dfrac{3}{17}$

30. $\dfrac{9}{16}$
$-\dfrac{17}{32}$

31. $\dfrac{21}{35}$
$-\dfrac{5}{14}$

32. $\dfrac{19}{40}$
$-\dfrac{3}{16}$

33. $\dfrac{29}{60}$
$-\dfrac{3}{40}$

34. What is $\dfrac{3}{5}$ less than $\dfrac{11}{12}$?

35. What is $\dfrac{5}{9}$ less than $\dfrac{11}{15}$?

36. Find the difference between $\dfrac{11}{24}$ and $\dfrac{7}{18}$.

37. Find the difference between $\dfrac{9}{14}$ and $\dfrac{5}{42}$.

38. Find $\dfrac{11}{12}$ decreased by $\dfrac{11}{15}$.

39. Find $\dfrac{17}{20}$ decreased by $\dfrac{7}{15}$.

40. What is $\dfrac{13}{20}$ minus $\dfrac{1}{6}$?

41. What is $\dfrac{5}{6}$ minus $\dfrac{7}{9}$?

Objective C

Subtract.

42. $5\dfrac{7}{12}$
$-2\dfrac{5}{12}$

43. $16\dfrac{11}{15}$
$-11\dfrac{8}{15}$

44. $72\dfrac{21}{23}$
$-16\dfrac{17}{23}$

45. $19\dfrac{16}{17}$
$-9\dfrac{7}{17}$

46. $6\dfrac{1}{3}$
-2

47. $5\dfrac{7}{8}$
-1

48. 10
$-6\dfrac{1}{3}$

49. 3
$-2\dfrac{5}{21}$

50. $6\dfrac{2}{5}$
$-4\dfrac{4}{5}$

51. $16\dfrac{3}{8}$
$-10\dfrac{7}{8}$

52. $25\dfrac{4}{9}$
$-16\dfrac{7}{9}$

53. $8\dfrac{3}{7}$
$-2\dfrac{6}{7}$

54. $16\dfrac{2}{5}$
$-8\dfrac{4}{9}$

55. $23\dfrac{7}{8}$
$-16\dfrac{2}{3}$

56. 6
$-4\dfrac{3}{5}$

57. $65\dfrac{8}{35}$
$-16\dfrac{11}{14}$

58. $82\dfrac{4}{33}$
$-16\dfrac{5}{22}$

59. $101\dfrac{2}{9}$
-16

60. $77\dfrac{5}{18}$
-61

61. 17
$-7\dfrac{8}{13}$

62. What is $5\frac{3}{8}$ less than $8\frac{1}{9}$?

63. What is $7\frac{3}{5}$ less than $23\frac{3}{20}$?

64. Find the difference between $9\frac{2}{7}$ and $3\frac{1}{4}$.

65. Find the difference between $12\frac{3}{8}$ and $7\frac{5}{12}$.

66. What is $10\frac{5}{9}$ minus $5\frac{11}{15}$?

67. Find $6\frac{1}{3}$ decreased by $3\frac{3}{5}$.

Objective D   *Application Problems*

68. Find the missing dimension.

$7\frac{7}{8}$ ft

?

$16\frac{2}{3}$ ft

69. Find the missing dimension.

$2\frac{7}{8}$ in.

?

$12\frac{3}{8}$ in.

70. In the Kentucky Derby the horses run $1\frac{1}{4}$ miles. In the Belmont Stakes they run $1\frac{1}{2}$ miles, and in the Preakness Stakes they run $1\frac{3}{16}$ miles. How much farther do the horses run in the Kentucky Derby than in the Preakness Stakes? How much farther do they run in the Belmont Stakes than in the Preakness Stakes?

71. In the running high jump in the 1948 Summer Olympic Games, Alice Coachman's distance was $66\frac{1}{8}$ inches. In the same event in the 1972 Summer Olympics, Urika Meyfarth jumped $75\frac{1}{2}$ inches, and in the 1996 Olympic Games, Stefka Kostadinova jumped $80\frac{3}{4}$ inches. Find the difference between Meyfarth's distance and Coachman's distance. Find the difference between Kostadinova's distance and Meyfarth's distance.

72. Two hikers plan a 3-day $27\frac{1}{2}$-mile backpack trip carrying a total of 80 pounds. The hikers plan to travel $7\frac{3}{8}$ miles the first day and $10\frac{1}{3}$ miles the second day.
a. How many miles do the hikers plan to travel the first two days?
b. How many miles will be left to travel on the third day?

$7\frac{3}{8}$ $10\frac{1}{3}$

Start

73. A 12-mile walkathon has three checkpoints. The first is $3\frac{3}{8}$ miles from the starting point. The second checkpoint is $4\frac{1}{3}$ miles from the first.

    a. How many miles is it from the starting point to the second checkpoint?
    b. How many miles is it from the second checkpoint to the finish line?

74. A patient with high blood pressure who weighs 225 pounds is put on a diet to lose 25 pounds in 3 months. The patient loses $8\frac{3}{4}$ pounds the first month and $11\frac{5}{8}$ pounds the second month. How much weight must be lost the third month for the goal to be achieved?

75. A wrestler is entered in the 172-pound weight class in the conference finals coming up in 3 weeks. The wrestler needs to lose $12\frac{3}{4}$ pounds. The wrestler loses $5\frac{1}{4}$ pounds the first week and $4\frac{1}{4}$ pounds the second week.

    a. Without doing the calculations, determine whether the wrestler can reach his weight class by losing less in the third week than was lost in the second week.
    b. How many pounds must be lost in the third week for the desired weight to be reached?

76. 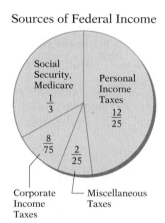 The figure at the right shows the sources of federal income. The category "Social Security, Medicare" includes unemployment and other retirement taxes. The category "Miscellaneous Taxes" includes excise, customs, estate, and gift taxes.

    a. What is the difference between the fraction of federal income derived from personal income taxes and the fraction derived from corporate income taxes?
    b. What is the difference between the fraction of federal income derived from the Social Security category and the fraction derived from miscellaneous taxes?

Sources of Federal Income

Social Security, Medicare $\frac{1}{3}$

Personal Income Taxes $\frac{12}{25}$

$\frac{8}{75}$ Corporate Income Taxes

$\frac{2}{25}$ Miscellaneous Taxes

APPLYING THE CONCEPTS

77. Fill in the square to produce a true statement: $5\frac{1}{3} - \boxed{} = 2\frac{1}{2}$.

78. Fill in the square to produce a true statement: $\boxed{} - 4\frac{1}{2} = 1\frac{5}{8}$.

79. Fill in the blank squares at the right so that the sum of the numbers is the same along any row, column, or diagonal. The resulting square is called a magic square.

		$\frac{3}{4}$
	$\frac{5}{8}$	
$\frac{1}{2}$		$\frac{7}{8}$

80. If $\frac{4}{15}$ of an electrician's income is spent for housing, what fraction of the electrician's income is not spent for housing?

2.6 Multiplication of Fractions and Mixed Numbers

Objective A   **To multiply fractions**

The product of two fractions is the product of the numerators over the product of the denominators.

➡ Multiply: $\dfrac{2}{3} \times \dfrac{4}{5}$

$$\frac{2}{3} \times \frac{4}{5} = \frac{2 \cdot 4}{3 \cdot 5} = \frac{8}{15}$$

- Multiply the numerators.
- Multiply the denominators.

The product $\dfrac{2}{3} \times \dfrac{4}{5}$ can be read "$\dfrac{2}{3}$ times $\dfrac{4}{5}$" or "$\dfrac{2}{3}$ of $\dfrac{4}{5}$."

Reading the times sign as "of" is useful in application problems.

$\dfrac{4}{5}$ of the bar is shaded.

Shade $\dfrac{2}{3}$ of the $\dfrac{4}{5}$ already shaded.

$\dfrac{8}{15}$ of the bar is then shaded light yellow.

$$\frac{2}{3} \text{ of } \frac{4}{5} = \frac{2}{3} \times \frac{4}{5} = \frac{8}{15}$$

After multiplying two fractions, write the product in simplest form.

➡ Multiply: $\dfrac{3}{4} \times \dfrac{14}{15}$

$$\frac{3}{4} \times \frac{14}{15} = \frac{3 \cdot 14}{4 \cdot 15}$$

- Multiply the numerators.
- Multiply the denominators.

$$= \frac{3 \cdot 2 \cdot 7}{2 \cdot 2 \cdot 3 \cdot 5}$$

- Write the prime factorization of each number.

$$= \frac{\overset{1}{\cancel{3}} \cdot \overset{1}{\cancel{2}} \cdot 7}{\underset{1}{\cancel{2}} \cdot 2 \cdot \underset{1}{\cancel{3}} \cdot 5} = \frac{7}{10}$$

- Eliminate the common factors. Then multiply the remaining factors in the numerator and denominator.

This example could also be worked by using the GCF.

$$\frac{3}{4} \times \frac{14}{15} = \frac{42}{60}$$

- Multiply the numerators.
- Multiply the denominators.

$$= \frac{6 \cdot 7}{6 \cdot 10}$$

- The GCF of 42 and 60 is 6. Factor 6 from 42 and 60.

$$= \frac{\overset{1}{\cancel{6}} \cdot 7}{\underset{1}{\cancel{6}} \cdot 10} = \frac{7}{10}$$

- Eliminate the GCF.

Example 1

Multiply $\frac{4}{15}$ and $\frac{5}{28}$.

Solution

$$\frac{4}{15} \times \frac{5}{28} = \frac{4 \cdot 5}{15 \cdot 28} = \frac{\overset{1}{2} \cdot \overset{1}{2} \cdot \overset{1}{5}}{3 \cdot \underset{1}{5} \cdot \underset{1}{2} \cdot \underset{1}{2} \cdot 7} = \frac{1}{21}$$

You Try It 1

Multiply $\frac{4}{21}$ and $\frac{7}{44}$.

Your solution

Example 2

Find the product of $\frac{9}{20}$ and $\frac{33}{35}$.

Solution

$$\frac{9}{20} \times \frac{33}{35} = \frac{9 \cdot 33}{20 \cdot 35} = \frac{3 \cdot 3 \cdot 3 \cdot 11}{2 \cdot 2 \cdot 5 \cdot 5 \cdot 7} = \frac{297}{700}$$

You Try It 2

Find the product of $\frac{2}{21}$ and $\frac{10}{33}$.

Your solution

Example 3

What is $\frac{14}{9}$ times $\frac{12}{7}$?

Solution

$$\frac{14}{9} \times \frac{12}{7} = \frac{14 \cdot 12}{9 \cdot 7} = \frac{2 \cdot \overset{1}{7} \cdot 2 \cdot 2 \cdot \overset{1}{3}}{3 \cdot \underset{1}{3} \cdot \underset{1}{7}} = \frac{8}{3} = 2\frac{2}{3}$$

You Try It 3

What is $\frac{16}{5}$ times $\frac{15}{24}$?

Your solution

Solutions on p. S6

Objective B **To multiply whole numbers, mixed numbers, and fractions**

To multiply a whole number by a fraction or mixed number, first write the whole number as a fraction with a denominator of 1.

➡ Multiply: $4 \times \frac{3}{7}$

$$4 \times \frac{3}{7} = \frac{4}{1} \times \frac{3}{7} = \frac{4 \cdot 3}{1 \cdot 7} = \frac{2 \cdot 2 \cdot 3}{7} = \frac{12}{7} = 1\frac{5}{7}$$

- Write 4 with a denominator of 1; then multiply the fractions.

When one or more of the factors in a product is a mixed number, write the mixed number as an improper fraction before multiplying.

➡ Multiply: $2\frac{1}{3} \times \frac{3}{14}$

$$2\frac{1}{3} \times \frac{3}{14} = \frac{7}{3} \times \frac{3}{14} = \frac{7 \cdot 3}{3 \cdot 14} = \frac{\overset{1}{7} \cdot \overset{1}{3}}{\underset{1}{3} \cdot 2 \cdot \underset{1}{7}} = \frac{1}{2}$$

- Write $2\frac{1}{3}$ as an improper fraction; then multiply the fractions.

Example 4

Multiply: $4\dfrac{5}{6} \times \dfrac{12}{13}$

Solution

$$4\frac{5}{6} \times \frac{12}{13} = \frac{29}{6} \times \frac{12}{13} = \frac{29 \cdot 12}{6 \cdot 13}$$

$$= \frac{29 \cdot \overset{1}{\cancel{2}} \cdot 2 \cdot \overset{1}{\cancel{3}}}{\underset{1}{\cancel{2}} \cdot \underset{1}{\cancel{3}} \cdot 13} = \frac{58}{13} = 4\frac{6}{13}$$

You Try It 4

Multiply: $5\dfrac{2}{5} \times \dfrac{5}{9}$

Your solution

Example 5

Find $5\dfrac{2}{3}$ times $4\dfrac{1}{2}$.

Solution

$$5\frac{2}{3} \times 4\frac{1}{2} = \frac{17}{3} \times \frac{9}{2} = \frac{17 \cdot 9}{3 \cdot 2}$$

$$= \frac{17 \cdot \overset{1}{\cancel{3}} \cdot 3}{\underset{1}{\cancel{3}} \cdot 2} = \frac{51}{2} = 25\frac{1}{2}$$

You Try It 5

Multiply: $3\dfrac{2}{5} \times 6\dfrac{1}{4}$

Your solution

Example 6

Multiply: $4\dfrac{2}{5} \times 7$

Solution

$$4\frac{2}{5} \times 7 = \frac{22}{5} \times \frac{7}{1} = \frac{22 \cdot 7}{5 \cdot 1}$$

$$= \frac{2 \cdot 11 \cdot 7}{5} = \frac{154}{5} = 30\frac{4}{5}$$

You Try It 6

Multiply: $3\dfrac{2}{7} \times 6$

Your solution

Solutions on p. S6

Objective C   **To solve application problems**

Length (ft)	Weight (lb/ft)
$6\frac{1}{2}$	$\frac{3}{8}$
$8\frac{5}{8}$	$1\frac{1}{4}$
$10\frac{3}{4}$	$2\frac{1}{2}$
$12\frac{7}{12}$	$4\frac{1}{3}$

The table at the left lists the length of steel rods and the weight per foot. The weight per foot is measured in pounds for each foot of rod and is abbreviated as lb/ft.

→ Find the weight of the steel bar that is $10\dfrac{3}{4}$ feet long.

Strategy
To find the weight of the steel bar, multiply its length by the weight per foot.

Solution $10\dfrac{3}{4} \times 2\dfrac{1}{2} = \dfrac{43}{4} \times \dfrac{5}{2} = \dfrac{43 \cdot 5}{4 \cdot 2} = \dfrac{215}{8} = 26\dfrac{7}{8}.$

The weight of the $10\frac{3}{4}$-foot rod is $26\frac{7}{8}$ pounds.

Example 7

An electrician earns $206 for each day worked. What are the electrician's earnings for working $4\frac{1}{2}$ days?

Strategy

To find the electrician's total earnings, multiply the daily earnings (206) by the number of days worked $\left(4\frac{1}{2}\right)$.

Solution $206 \times 4\frac{1}{2} = \frac{206}{1} \times \frac{9}{2}$

$$= \frac{206 \cdot 9}{1 \cdot 2}$$

$$= 927$$

The electrician's earnings are $927.

You Try It 7

Over the last 10 years, a house increased in value by $2\frac{1}{2}$ times. The price of the house 10 years ago was $70,000. What is the value of the house today?

Your strategy

Your solution

Example 8

The value of a small office building and the land on which it is built is $190,000. The value of the land is $\frac{1}{4}$ the total value. What is the dollar value of the building?

Strategy

To find the value of the building:

- Find the value of the land $\left(\frac{1}{4} \times 190,000\right)$.

- Subtract the value of the land from the total value (190,000).

Solution $\frac{1}{4} \times 190,000 = \frac{190,000}{4}$

$$= 47,500 \qquad \bullet \text{ Value of the land}$$

$190,000 - 47,500 = 142,500$

The value of the building is $142,500.

You Try It 8

A paint company bought a drying chamber and an air compressor for spray painting. The total cost of the two items was $60,000. The drying chamber's cost was $\frac{4}{5}$ of the total cost. What was the cost of the air compressor?

Your strategy

Your solution

Solutions on pp. S6–S7

2.6 Exercises

Objective A

Multiply.

1. $\frac{2}{3} \times \frac{7}{8}$ = $\frac{14}{24} = \frac{7}{12}$ **2.** $\frac{1}{2} \times \frac{2}{3}$ = $\frac{2}{6} = \frac{1}{3}$ **3.** $\frac{5}{16} \times \frac{7}{15}$ **4.** $\frac{3}{8} \times \frac{6}{7}$

5. $\frac{1}{6} \times \frac{1}{8}$ **6.** $\frac{2}{5} \times \frac{5}{6}$ **7.** $\frac{11}{12} \times \frac{6}{7}$ **8.** $\frac{11}{12} \times \frac{3}{5}$

9. $\frac{1}{6} \times \frac{6}{7}$ **10.** $\frac{3}{5} \times \frac{10}{11}$ **11.** $\frac{1}{5} \times \frac{5}{8}$ **12.** $\frac{6}{7} \times \frac{14}{15}$

13. $\frac{8}{9} \times \frac{27}{4}$ **14.** $\frac{3}{5} \times \frac{3}{10}$ **15.** $\frac{5}{6} \times \frac{1}{2}$ **16.** $\frac{3}{8} \times \frac{5}{12}$

17. $\frac{16}{9} \times \frac{27}{8}$ **18.** $\frac{5}{8} \times \frac{16}{15}$ **19.** $\frac{3}{2} \times \frac{4}{9}$ **20.** $\frac{5}{3} \times \frac{3}{7}$

21. $\frac{7}{8} \times \frac{3}{14}$ **22.** $\frac{2}{9} \times \frac{1}{5}$ **23.** $\frac{1}{10} \times \frac{3}{8}$ **24.** $\frac{5}{12} \times \frac{6}{7}$

25. $\frac{15}{8} \times \frac{16}{3}$ **26.** $\frac{5}{6} \times \frac{4}{15}$ **27.** $\frac{1}{2} \times \frac{2}{15}$ **28.** $\frac{3}{8} \times \frac{5}{16}$

29. $\frac{5}{7} \times \frac{14}{15}$ **30.** $\frac{3}{8} \times \frac{15}{41}$ **31.** $\frac{5}{12} \times \frac{42}{65}$ **32.** $\frac{16}{33} \times \frac{55}{72}$

33. $\frac{12}{5} \times \frac{5}{3}$ **34.** $\frac{17}{9} \times \frac{81}{17}$ **35.** $\frac{16}{85} \times \frac{125}{84}$ **36.** $\frac{19}{64} \times \frac{48}{95}$

37. Multiply $\frac{7}{12}$ and $\frac{15}{42}$.

38. Multiply $\frac{32}{9}$ and $\frac{3}{8}$.

39. Find the product of $\frac{5}{9}$ and $\frac{3}{20}$.

40. Find the product of $\frac{7}{3}$ and $\frac{15}{14}$.

41. What is $\frac{1}{2}$ times $\frac{8}{15}$?

42. What is $\frac{3}{8}$ times $\frac{12}{17}$?

Objective B

Multiply.

43. $4 \times \frac{3}{8}$ $= \frac{12}{8} = 1\frac{4}{8} = 1\frac{1}{2}$

44. $14 \times \frac{5}{7}$

45. $\frac{2}{3} \times 6$

46. $\frac{5}{12} \times 40$

47. $\frac{1}{3} \times 1\frac{1}{3}$

48. $\frac{2}{5} \times 2\frac{1}{2}$

49. $1\frac{7}{8} \times \frac{4}{15}$

50. $2\frac{1}{5} \times \frac{5}{22}$

51. $55 \times \frac{3}{10}$

52. $\frac{5}{14} \times 49$

53. $4 \times 2\frac{1}{2}$

54. $9 \times 3\frac{1}{3}$

55. $2\frac{1}{7} \times 3$

56. $5\frac{1}{4} \times 8$

57. $3\frac{2}{3} \times 5$

58. $4\frac{2}{9} \times 3$

59. $\frac{1}{2} \times 3\frac{3}{7}$

60. $\frac{3}{8} \times 4\frac{4}{5}$

61. $6\frac{1}{8} \times \frac{4}{7}$

62. $5\frac{1}{3} \times \frac{5}{16}$

63. $5\frac{1}{8} \times 5$

64. $6\frac{1}{9} \times 2$

65. $\frac{3}{8} \times 4\frac{1}{2}$

66. $\frac{5}{7} \times 2\frac{1}{3}$

67. $6 \times 2\frac{2}{3}$

68. $6\frac{1}{8} \times 0$

69. $1\frac{1}{3} \times 2\frac{1}{4}$

70. $2\frac{5}{8} \times \frac{3}{23}$

71. $2\frac{5}{8} \times 3\frac{2}{5}$

72. $5\frac{3}{16} \times 5\frac{1}{3}$

73. $3\frac{1}{7} \times 2\frac{1}{8}$

74. $16\frac{5}{8} \times 1\frac{1}{16}$

75. $2\frac{2}{5} \times 3\frac{1}{12}$

76. $2\frac{2}{3} \times \frac{3}{20}$

77. $5\frac{1}{5} \times 3\frac{1}{13}$

78. $3\frac{3}{4} \times 2\frac{3}{20}$

79. $10\frac{1}{4} \times 3\frac{1}{5}$

80. $12\frac{3}{5} \times 1\frac{3}{7}$

81. $5\frac{3}{7} \times 5\frac{1}{4}$

82. $6\frac{1}{2} \times 1\frac{3}{13}$

83. Multiply $2\frac{1}{2}$ and $3\frac{3}{5}$.

84. Multiply $4\frac{3}{8}$ and $3\frac{3}{5}$.

85. Find the product of $2\frac{1}{8}$ and $\frac{5}{17}$.

86. Find the product of $12\frac{2}{5}$ and $3\frac{7}{31}$.

87. What is $1\frac{3}{8}$ times $2\frac{1}{5}$?

88. What is $3\frac{1}{8}$ times $2\frac{4}{7}$?

Objective C   *Application Problems*

89. Salmon costs $8 per pound. Find the cost of $2\frac{3}{4}$ pounds of salmon.

90. Maria Rivera can walk $3\frac{1}{2}$ miles in 1 hour. At this rate, how far can Maria walk in $\frac{1}{3}$ hour?

91. A board that costs $6 is $9\frac{1}{4}$ feet long. One-third of the board is cut off.
   a. Without doing the calculation, is the piece being cut off at least 4 feet long?
   b. What is the length of the piece cut off?

92. The perimeter of a square is equal to four times the length of a side of the square. Find the perimeter of a square whose side measures $16\frac{3}{4}$ inches.

93. To find the area of a square, multiply the length of one side of the square times itself. What is the area of a square whose side measures $5\frac{1}{4}$ feet? The area of the square will be in square feet.

94. The area of a rectangle is equal to the product of the length of the rectangle times its width. Find the area of a rectangle that has a length of $4\frac{2}{5}$ miles and a width of $3\frac{3}{10}$ miles. The area will be in square miles.

95. A family budgets $\frac{2}{5}$ of its monthly income of $3200 per month for housing and utilities.
   a. What amount is budgeted for housing and utilities?
   b. What amount remains for purposes other than housing and utilities?

96. $\frac{5}{6}$ of a chemistry class of 36 has passing grades. $\frac{1}{5}$ of the students with passing grades received an A.
   a. How many students passed the chemistry course?
   b. How many students received A grades?

97. The parents of the Newton Junior High School Choir members are making robes for the choir. Each robe requires $2\frac{5}{8}$ yards of material at a cost of $8 per yard. Find the total cost of 24 choir robes.

The table at the right shows the length of steel rods and their weight per foot. Use this table for Exercises 98 to 100.

98. Find the weight of the $6\frac{1}{2}$-foot steel rod.

99. Find the weight of the $12\frac{7}{12}$-foot steel rod.

Length (ft)	Weight (lb/ft)
$6\frac{1}{2}$	$\frac{3}{8}$
$8\frac{5}{8}$	$1\frac{1}{4}$
$10\frac{3}{4}$	$2\frac{1}{2}$
$12\frac{7}{12}$	$4\frac{1}{3}$

100. Find the total weight of the $8\frac{5}{8}$-foot and the $10\frac{3}{4}$-foot steel rods.

101. The manager of a mutual fund has $\frac{1}{2}$ of the portfolio invested in bonds. Of the amount invested in bonds, $\frac{3}{8}$ is invested in corporate bonds. What fraction of the total portfolio is invested in corporate bonds?

APPLYING THE CONCEPTS

102. The product of 1 and a number is $\frac{1}{2}$. Find the number.

103. Our calendar is based on the solar year, which is $365\frac{1}{4}$ days. Use this fact to explain leap years.

104. Is the product of two positive fractions always greater than either one of the two numbers? If so, explain why. If not, give an example.

105. Which of the labeled points on the number line at the right could be the graph of the product of B and C?

106. Fill in the circles on the square at the right with the fractions $\frac{1}{6}$, $\frac{5}{18}$, $\frac{4}{9}$, $\frac{5}{9}$, $\frac{2}{3}$, $\frac{3}{4}$, $1\frac{1}{9}$, $1\frac{1}{2}$, and $2\frac{1}{4}$ so that the product of any row is equal to $\frac{5}{18}$.
(*Note:* There is more than one possible answer.)

Division of Fractions and Mixed Numbers

Objective A   **To divide fractions**

The **reciprocal** of a fraction is the fraction with the numerator and denominator interchanged.

The reciprocal of $\frac{2}{3}$ is $\frac{3}{2}$.

The process of interchanging the numerator and denominator is called **inverting** the fraction.

To find the reciprocal of a whole number, first write the whole number as a fraction with a denominator of 1; then find the reciprocal of that fraction.

The reciprocal of 5 is $\frac{1}{5}$. $\left(\text{Think } 5 = \frac{5}{1}.\right)$

Reciprocals are used to rewrite division problems as related multiplication problems. Look at the following two problems:

$$8 \div 2 = 4 \qquad\qquad 8 \times \frac{1}{2} = 4$$

8 divided by 2 is 4.   8 times the reciprocal of 2 is 4.

"Divided by" means the same as "times the reciprocal of." Thus "÷ 2" can be replaced with "$\times \frac{1}{2}$," and the answer will be the same. Fractions are divided by making this replacement.

➡ Divide: $\frac{2}{3} \div \frac{3}{4}$

$$\frac{2}{3} \div \frac{3}{4} = \frac{2}{3} \times \frac{4}{3} = \frac{2 \cdot 4}{3 \cdot 3} = \frac{2 \cdot 2 \cdot 2}{3 \cdot 3} = \frac{8}{9}$$

Example 1

Divide: $\frac{5}{8} \div \frac{4}{9}$

Solution $\frac{5}{8} \div \frac{4}{9} = \frac{5}{8} \times \frac{9}{4} = \frac{5 \cdot 9}{8 \cdot 4}$

$$= \frac{5 \cdot 3 \cdot 3}{2 \cdot 2 \cdot 2 \cdot 2 \cdot 2} = \frac{45}{32} = 1\frac{13}{32}$$

You Try It 1

Divide: $\frac{3}{7} \div \frac{2}{3}$

Your solution

Example 2

Divide: $\frac{3}{5} \div \frac{12}{25}$

Solution $\frac{3}{5} \div \frac{12}{25} = \frac{3}{5} \times \frac{25}{12} = \frac{3 \cdot 25}{5 \cdot 12}$

$$= \frac{\overset{1}{\cancel{3}} \cdot \overset{1}{\cancel{3}} \cdot 5}{\underset{1}{\cancel{3}} \cdot 2 \cdot 2 \cdot \underset{1}{\cancel{3}}} = \frac{5}{4} = 1\frac{1}{4}$$

You Try It 2

Divide: $\frac{3}{4} \div \frac{9}{10}$

Your solution

Solutions on p. S7

Objective B **To divide whole numbers, mixed numbers, and fractions**

To divide a fraction and a whole number, first write the whole number as a fraction with a denominator of 1.

➡ Divide: $\frac{3}{7} \div 5$

$$\frac{3}{7} \div \boxed{5} = \frac{3}{7} \div \boxed{\frac{5}{1}} = \frac{3}{7} \times \frac{1}{5} = \frac{3 \cdot 1}{7 \cdot 5} = \frac{3}{35}$$

• Write 5 with a denominator of 1. Then divide the fractions.

When a number in a quotient is a mixed number, write the mixed number as an improper fraction before dividing.

➡ Divide: $1\frac{13}{15} \div 4\frac{4}{5}$

Write the mixed numbers as improper fractions. Then divide the fractions.

$$1\frac{13}{15} \div 4\frac{4}{5} = \frac{28}{15} \div \frac{24}{5} = \frac{28}{15} \times \frac{5}{24} = \frac{28 \cdot 5}{15 \cdot 24} = \frac{\overset{1}{2} \cdot \overset{1}{2} \cdot 7 \cdot \overset{1}{5}}{3 \cdot \underset{1}{5} \cdot \underset{1}{2} \cdot \underset{1}{2} \cdot 2 \cdot 3} = \frac{7}{18}$$

Example 3

Divide $\frac{4}{9}$ by 5.

Solution

$$\frac{4}{9} \div 5 = \frac{4}{9} \div \frac{5}{1} = \frac{4}{9} \times \frac{1}{5}$$

$$= \frac{4 \cdot 1}{9 \cdot 5} = \frac{2 \cdot 2}{3 \cdot 3 \cdot 5} = \frac{4}{45}$$

You Try It 3

Divide $\frac{5}{7}$ by 6.

Your solution

Example 4

Find the quotient of $\frac{3}{8}$ and $2\frac{1}{10}$.

Solution

$$\frac{3}{8} \div 2\frac{1}{10} = \frac{3}{8} \div \frac{21}{10} = \frac{3}{8} \times \frac{10}{21}$$

$$= \frac{3 \cdot 10}{8 \cdot 21} = \frac{\overset{1}{3} \cdot \overset{1}{2} \cdot 5}{2 \cdot 2 \cdot 2 \cdot \underset{1}{3} \cdot 7} = \frac{5}{28}$$

You Try It 4

Find the quotient of $12\frac{3}{5}$ and 7.

Your solution

Example 5

Divide: $2\frac{3}{4} \div 1\frac{5}{7}$

Solution

$$2\frac{3}{4} \div 1\frac{5}{7} = \frac{11}{4} \div \frac{12}{7} = \frac{11}{4} \times \frac{7}{12} = \frac{11 \cdot 7}{4 \cdot 12}$$

$$= \frac{11 \cdot 7}{2 \cdot 2 \cdot 2 \cdot 2 \cdot 3} = \frac{77}{48} = 1\frac{29}{48}$$

You Try It 5

Divide: $3\frac{2}{3} \div 2\frac{2}{5}$

Your solution

Solutions on p. S7

Example 6

Divide: $1\frac{13}{15} \div 4\frac{1}{5}$

Solution

$$1\frac{13}{15} \div 4\frac{1}{5} = \frac{28}{15} \div \frac{21}{5} = \frac{28}{15} \times \frac{5}{21} = \frac{28 \cdot 5}{15 \cdot 21}$$

$$= \frac{2 \cdot 2 \cdot \overset{1}{\cancel{7}} \cdot \overset{1}{\cancel{5}}}{3 \cdot \underset{1}{\cancel{5}} \cdot 3 \cdot \underset{1}{\cancel{7}}} = \frac{4}{9}$$

You Try It 6

Divide: $2\frac{5}{6} \div 8\frac{1}{2}$

Your solution

Example 7

Divide: $4\frac{3}{8} \div 7$

Solution

$$4\frac{3}{8} \div 7 = \frac{35}{8} \div \frac{7}{1} = \frac{35}{8} \times \frac{1}{7}$$

$$= \frac{35 \cdot 1}{8 \cdot 7} = \frac{5 \cdot \overset{1}{\cancel{7}}}{2 \cdot 2 \cdot 2 \cdot \underset{1}{\cancel{7}}} = \frac{5}{8}$$

You Try It 7

Divide: $6\frac{2}{5} \div 4$

Your solution

Solutions on p. S7

Objective C   **To solve application problems**

Example 8

A car used $15\frac{1}{2}$ gallons of gasoline on a 310-mile trip. How many miles can this car travel on 1 gallon of gasoline?

Strategy

To find the number of miles, divide the number of miles traveled by the number of gallons of gasoline used.

Solution

$$310 \div 15\frac{1}{2} = \frac{310}{1} \div \frac{31}{2}$$

$$= \frac{310}{1} \times \frac{2}{31} = \frac{310 \cdot 2}{1 \cdot 31}$$

$$= \frac{2 \cdot 5 \cdot \overset{1}{\cancel{31}} \cdot 2}{1 \cdot \underset{1}{\cancel{31}}} = \frac{20}{1} = 20$$

The car travels 20 miles on 1 gallon of gasoline.

You Try It 8

Leon Dern purchased a $\frac{1}{2}$-ounce gold coin for $195. What would a similar coin weighing 1 ounce cost?

Your strategy

Your solution

Solution on p. S7

Example 9

A 12-foot board is cut into pieces $2\frac{1}{4}$ feet long for use as bookshelves. What is the length of the remaining piece after as many shelves as possible are cut?

Strategy

To find the length of the remaining piece:

- Divide the total length of the board (12) by the length of each shelf $\left(2\frac{1}{4}\right)$. This will give you the number of shelves cut, with a certain fraction of a shelf left over.
- Multiply the fractional part of the result in Step 1 by the length of one shelf to determine the length of the remaining piece.

Solution

$$12 \div 2\frac{1}{4} = \frac{12}{1} \div \frac{9}{4} = \frac{12}{1} \times \frac{4}{9}$$

$$= \frac{12 \cdot 4}{1 \cdot 9} = \frac{16}{3} = 5\frac{1}{3}$$

There are 5 pieces $2\frac{1}{4}$ feet long.

There is 1 piece that is $\frac{1}{3}$ of $2\frac{1}{4}$ feet long.

$$\frac{1}{3} \times 2\frac{1}{4} = \frac{1}{3} \times \frac{9}{4} = \frac{1 \cdot 9}{3 \cdot 4} = \frac{3}{4}$$

The length of the piece remaining is $\frac{3}{4}$ foot.

You Try It 9

A 16-foot board is cut into pieces $3\frac{1}{3}$ feet long for shelves for a bookcase. What is the length of the remaining piece after as many shelves as possible are cut?

Your strategy

Your solution

Solution on pp. S7–S8

2.7 Exercises

Objective A

Divide.

1. $\dfrac{1}{3} \div \dfrac{2}{5}$

2. $\dfrac{3}{7} \div \dfrac{3}{2}$

$\dfrac{3}{7} \times \dfrac{2}{3} = \dfrac{9}{21} + \dfrac{14}{11} =$

3. $\dfrac{3}{7} \div \dfrac{3}{7}$

4. $0 \div \dfrac{1}{2}$

5. $0 \div \dfrac{3}{4}$

6. $\dfrac{16}{33} \div \dfrac{4}{11}$

7. $\dfrac{5}{24} \div \dfrac{15}{36}$

8. $\dfrac{11}{15} \div \dfrac{1}{12}$

9. $\dfrac{15}{16} \div \dfrac{16}{39}$

10. $\dfrac{2}{15} \div \dfrac{3}{5}$

11. $\dfrac{8}{9} \div \dfrac{4}{5}$

12. $\dfrac{11}{15} \div \dfrac{5}{22}$

13. $\dfrac{1}{9} \div \dfrac{2}{3}$

14. $\dfrac{10}{21} \div \dfrac{5}{7}$

15. $\dfrac{2}{5} \div \dfrac{4}{7}$

16. $\dfrac{3}{8} \div \dfrac{5}{12}$

17. $\dfrac{1}{2} \div \dfrac{1}{4}$

18. $\dfrac{1}{3} \div \dfrac{1}{9}$

19. $\dfrac{1}{5} \div \dfrac{1}{10}$

20. $\dfrac{4}{15} \div \dfrac{2}{5}$

21. $\dfrac{7}{15} \div \dfrac{14}{5}$

22. $\dfrac{5}{8} \div \dfrac{15}{2}$

23. $\dfrac{14}{3} \div \dfrac{7}{9}$

24. $\dfrac{7}{4} \div \dfrac{9}{2}$

25. $\dfrac{5}{9} \div \dfrac{25}{3}$

26. $\dfrac{5}{16} \div \dfrac{3}{8}$

27. $\dfrac{2}{3} \div \dfrac{1}{3}$

28. $\dfrac{4}{9} \div \dfrac{1}{9}$

29. $\dfrac{5}{7} \div \dfrac{2}{7}$

30. $\dfrac{5}{6} \div \dfrac{1}{9}$

31. $\dfrac{2}{3} \div \dfrac{2}{9}$

32. $\dfrac{5}{12} \div \dfrac{5}{6}$

33. $4 \div \dfrac{2}{3}$

34. $\dfrac{2}{3} \div 4$

35. $\dfrac{3}{2} \div 3$

36. $3 \div \dfrac{3}{2}$

37. Divide $\frac{7}{8}$ by $\frac{3}{4}$.

38. Divide $\frac{7}{12}$ by $\frac{3}{4}$.

39. Find the quotient of $\frac{5}{7}$ and $\frac{3}{14}$.

40. Find the quotient of $\frac{6}{11}$ and $\frac{9}{32}$.

Objective B

Divide.

41. $\frac{5}{6} \div 25$

42. $22 \div \frac{3}{11}$

43. $6 \div 3\frac{1}{3}$

44. $5\frac{1}{2} \div 11$

45. $6\frac{1}{2} \div \frac{1}{2}$

46. $\frac{3}{8} \div 2\frac{1}{4}$

47. $\frac{5}{12} \div 4\frac{4}{5}$

48. $1\frac{1}{2} \div 1\frac{3}{8}$

49. $8\frac{1}{4} \div 2\frac{3}{4}$

50. $3\frac{5}{9} \div 32$

51. $4\frac{1}{5} \div 21$

52. $6\frac{8}{9} \div \frac{31}{36}$

53. $\frac{11}{12} \div 2\frac{1}{3}$

54. $\frac{7}{8} \div 3\frac{1}{4}$

55. $\frac{5}{16} \div 5\frac{3}{8}$

56. $\frac{9}{14} \div 3\frac{1}{7}$

57. $35 \div \frac{7}{24}$

58. $\frac{3}{8} \div 2\frac{3}{4}$

59. $\frac{11}{18} \div 2\frac{2}{9}$

60. $\frac{21}{40} \div 3\frac{3}{10}$

61. $2\frac{1}{16} \div 2\frac{1}{2}$

62. $7\frac{3}{5} \div 1\frac{7}{12}$

63. $1\frac{2}{3} \div \frac{3}{8}$

64. $16 \div \frac{2}{3}$

65. $1\frac{5}{8} \div 4$

66. $13\frac{3}{8} \div \frac{1}{4}$

67. $16 \div 1\frac{1}{2}$

68. $9 \div \frac{7}{8}$

69. $16\frac{5}{8} \div 1\frac{2}{3}$

70. $24\frac{4}{5} \div 2\frac{3}{5}$

71. $1\frac{1}{3} \div 5\frac{8}{9}$

72. $13\frac{2}{3} \div 0$

73. $82\dfrac{3}{5} \div 19\dfrac{1}{10}$

74. $45\dfrac{3}{5} \div 15$

75. $102 \div 1\dfrac{1}{2}$

76. $0 \div 3\dfrac{1}{2}$

77. $8\dfrac{2}{7} \div 1$

78. $6\dfrac{9}{16} \div 1\dfrac{3}{32}$

79. $8\dfrac{8}{9} \div 2\dfrac{13}{18}$

80. $10\dfrac{1}{5} \div 1\dfrac{7}{10}$

81. $7\dfrac{3}{8} \div 1\dfrac{27}{32}$

82. Divide $7\dfrac{7}{9}$ by $5\dfrac{5}{6}$.

83. Divide $2\dfrac{3}{4}$ by $1\dfrac{23}{32}$.

84. Find the quotient of $8\dfrac{1}{4}$ and $1\dfrac{5}{11}$.

85. Find the quotient of $\dfrac{14}{17}$ and $3\dfrac{1}{9}$.

Objective C   *Application Problems*

86. Individual cereal boxes contain $\dfrac{3}{4}$ ounce of cereal. How many boxes can be filled with 600 ounces of cereal?

87. A box of Post's Great Grains cereal costing $4 contains 16 ounces of cereal. How many $1\dfrac{1}{3}$-ounce portions can be served from this box?

88. A $\dfrac{5}{8}$-carat diamond was purchased for $1200. What would a similar diamond weighing 1 carat cost?

89. The Inverness Investor Group bought $8\dfrac{1}{3}$ acres of land for $200,000. What was the cost of each acre?

90. KU Energy stock is offered for $31\dfrac{5}{8}$ per share. How many shares can you buy for $1265?

91. A nut moves $\dfrac{5}{32}$ inch for each turn. Find the number of turns it will take for the nut to move $1\dfrac{7}{8}$ inches.

92. The Hammond Company purchased $9\dfrac{3}{4}$ acres for a housing project. One and one-half acres were set aside for a park.

  a. How many acres are available for housing?

  b. How many $\dfrac{1}{4}$-acre parcels of land can be sold after the land for the park is set aside?

93. A chef purchased a roast that weighed $10\frac{3}{4}$ pounds. After the fat was trimmed and the bone removed, the roast weighed $9\frac{1}{3}$ pounds.

  a. What was the total weight of the fat and bone?

  b. How many $\frac{1}{3}$-pound servings can be cut from the trimmed roast?

94. A 15-foot board is cut into pieces $3\frac{1}{2}$ feet long for a bookcase. What is the length of the piece remaining after as many shelves as possible have been cut?

95. A scale of $\frac{1}{2}$ inch to 1 foot is used to draw the plans for a house. The scale measurements for three walls are given in the table at the right. Complete the table to determine the actual wall lengths for the three walls a, b, and c.

Wall	Scale	Actual Wall Length
a	$6\frac{1}{4}$ in.	?
b	9 in.	?
c	$7\frac{7}{8}$ in.	?

APPLYING THE CONCEPTS

96. On a map, two cities are $4\frac{5}{8}$ inches apart. If $\frac{3}{8}$ inch on the map represents 60 miles, what is the number of miles between the two cities?

97. Is the quotient always less than the dividend in a division problem? Explain.

98. Fill in the box to make a true statement.

  a. $\frac{3}{4} \cdot \boxed{} = \frac{1}{2}$ **b.** $\frac{2}{3} \cdot \boxed{} = 1\frac{3}{4}$

99. A page of type in a certain textbook is $7\frac{1}{2}$ inches wide. If the page is divided into three equal columns, with $\frac{3}{8}$ inch between columns, how wide is each column?

100. A whole number is both multiplied and divided by the same proper fraction. Which is greater, the product or the quotient?

101. Fractions are multiplied by multiplying the numerators and multiplying the denominators. Consider dividing fractions in a similar manner. Divide the numerators and divide the denominators.

$$\frac{4}{15} \div \frac{2}{5} = \frac{4 \div 2}{15 \div 5} = \frac{2}{3}$$

This gives the same answer as the traditional method.

$$\frac{4}{15} \div \frac{2}{5} = \frac{4}{15} \cdot \frac{5}{2} = \frac{20}{30} = \frac{2}{3}$$

Try this method for the following division problems.

  a. $\frac{5}{6} \div \frac{1}{3}$ **b.** $\frac{4}{21} \div \frac{2}{7}$ **c.** $\frac{5}{9} \div \frac{5}{3}$ **d.** $\frac{15}{16} \div \frac{3}{8}$

102. Does the method of dividing fractions shown above always work? Give an example of when the traditional method would be a better choice.

2.8 Order, Exponents, and the Order of Operations Agreement

Objective A   **To identify the order relation between two fractions**

Recall that whole numbers can be graphed as points on the number line. Fractions can also be graphed as points on the number line.

The graph of $\frac{3}{4}$ on the number line

The number line can be used to determine the order relation between two fractions. A fraction that appears to the left of a given fraction is less than the given fraction. A fraction that appears to the right of a given fraction is greater than the given fraction.

$$\frac{1}{8} < \frac{3}{8} \qquad \frac{6}{8} > \frac{3}{8}$$

To find the order relation between two fractions with the same denominator, compare the numerators. The fraction that has the smaller numerator is the smaller fraction. When the denominators are different, begin by writing equivalent fractions with a common denominator; then compare numerators.

➡ Find the order relation between $\frac{11}{18}$ and $\frac{5}{8}$.

The LCM of 18 and 8 is 72.

$$\frac{11}{18} = \frac{44}{72} \leftarrow \text{Smaller numerator} \qquad \frac{11}{18} < \frac{5}{8} \quad \text{or} \quad \frac{5}{8} > \frac{11}{18}$$

$$\frac{5}{8} = \frac{45}{72} \leftarrow \text{Larger numerator}$$

Example 1   Place the correct symbol, $<$ or $>$, between the two numbers.

$$\frac{5}{12} \qquad \frac{7}{18}$$

Solution $\frac{5}{12} = \frac{15}{36} \qquad \frac{7}{18} = \frac{14}{36}$

$$\frac{5}{12} > \frac{7}{18}$$

You Try It 1   Place the correct symbol, $<$ or $>$, between the two numbers.

$$\frac{9}{14} \qquad \frac{13}{21}$$

Your solution

Solution on p. S8

Objective B   **To simplify expressions containing exponents**

Repeated multiplication of the same fraction can be written in two ways:

$$\frac{1}{2} \cdot \frac{1}{2} \cdot \frac{1}{2} \cdot \frac{1}{2} \qquad \text{or} \qquad \left(\frac{1}{2}\right)^4 \leftarrow \text{Exponent}$$

The exponent indicates how many times the fraction occurs as a factor in the multiplication. The expression $\left(\frac{1}{2}\right)^4$ is in exponential notation.

Example 2

Simplify: $\left(\dfrac{5}{6}\right)^3 \cdot \left(\dfrac{3}{5}\right)^2$

You Try It 2

Simplify: $\left(\dfrac{7}{11}\right)^2 \cdot \left(\dfrac{2}{7}\right)$

Solution

$\left(\dfrac{5}{6}\right)^3 \cdot \left(\dfrac{3}{5}\right)^2 = \left(\dfrac{5}{6} \cdot \dfrac{5}{6} \cdot \dfrac{5}{6}\right) \cdot \left(\dfrac{3}{5} \cdot \dfrac{3}{5}\right)$

$= \dfrac{\overset{1}{\cancel{5}} \cdot \overset{1}{\cancel{5}} \cdot 5 \cdot \overset{1}{\cancel{3}} \cdot \overset{1}{\cancel{3}}}{2 \cdot \underset{1}{\cancel{3}} \cdot 2 \cdot \underset{1}{\cancel{3}} \cdot 2 \cdot 3 \cdot \underset{1}{\cancel{5}} \cdot \underset{1}{\cancel{5}}} = \dfrac{5}{24}$

Your solution

Solution on p. S8

Objective C   **To use the Order of Operations Agreement to simplify expressions**

The Order of Operations Agreement is used for fractions as well as whole numbers.

> **Step 1** Do all operations inside parentheses.
> **Step 2** Simplify any number expressions containing exponents.
> **Step 3** Do multiplications and divisions as they occur from left to right.
> **Step 4** Do additions and subtractions as they occur from left to right.

➡ Simplify $\dfrac{14}{15} - \left(\dfrac{1}{2}\right)^2 \times \left(\dfrac{2}{3} + \dfrac{4}{5}\right)$.

$\dfrac{14}{15} - \left(\dfrac{1}{2}\right)^2 \times \underbrace{\left(\dfrac{2}{3} + \dfrac{4}{5}\right)}$ 1. Perform operations in parentheses.

$\dfrac{14}{15} - \underbrace{\left(\dfrac{1}{2}\right)^2} \times \dfrac{22}{15}$ 2. Simplify expressions with exponents.

$\dfrac{14}{15} - \underbrace{\dfrac{1}{4} \times \dfrac{22}{15}}$ 3. Do multiplications and divisions as they occur from left to right.

$\underbrace{\dfrac{14}{15} - \dfrac{11}{30}}$ 4. Do additions and subtractions as they occur from left to right.

$\dfrac{17}{30}$

One or more of the above steps may not be needed to simplify an expression. In that case, proceed to the next step in the Order of Operations Agreement.

Example 3

Simplify: $\left(\dfrac{3}{4}\right)^2 \div \left(\dfrac{3}{8} - \dfrac{1}{12}\right)$

You Try It 3

Simplify:

$\left(\dfrac{1}{13}\right)^2 \cdot \left(\dfrac{1}{4} + \dfrac{1}{6}\right) \div \dfrac{5}{13}$

Solution

$\left(\dfrac{3}{4}\right)^2 \div \left(\dfrac{3}{8} - \dfrac{1}{12}\right)$

$= \left(\dfrac{3}{4}\right)^2 \div \left(\dfrac{7}{24}\right) = \dfrac{9}{16} \div \dfrac{7}{24}$

$= \dfrac{9}{16} \cdot \dfrac{24}{7} = \dfrac{27}{14} = 1\dfrac{13}{14}$

Your solution

Solution on p. S8

2.8 Exercises

Objective A

Place the correct symbol, $<$ or $>$, between the two numbers.

1. $\dfrac{11}{40}$ $\dfrac{19}{40}$

2. $\dfrac{92}{103}$ $\dfrac{19}{103}$

3. $\dfrac{2}{3}$ $\dfrac{5}{7}$

4. $\dfrac{2}{5}$ $\dfrac{3}{8}$

5. $\dfrac{5}{8}$ $\dfrac{7}{12}$

6. $\dfrac{11}{16}$ $\dfrac{17}{24}$

7. $\dfrac{7}{9}$ $\dfrac{11}{12}$

8. $\dfrac{5}{12}$ $\dfrac{7}{15}$

9. $\dfrac{13}{14}$ $\dfrac{19}{21}$

10. $\dfrac{13}{18}$ $\dfrac{7}{12}$

11. $\dfrac{7}{24}$ $\dfrac{11}{30}$

12. $\dfrac{13}{36}$ $\dfrac{19}{48}$

Objective B

Simplify.

13. $\left(\dfrac{3}{8}\right)^2$

14. $\left(\dfrac{5}{12}\right)^2$

15. $\left(\dfrac{2}{9}\right)^3$

16. $\left(\dfrac{1}{2}\right)\cdot\left(\dfrac{2}{3}\right)^2$

17. $\left(\dfrac{2}{3}\right)\cdot\left(\dfrac{1}{2}\right)^4$

18. $\left(\dfrac{1}{3}\right)^2\cdot\left(\dfrac{3}{5}\right)^3$

19. $\left(\dfrac{2}{5}\right)^3\cdot\left(\dfrac{5}{7}\right)^2$

20. $\left(\dfrac{5}{9}\right)^3\cdot\left(\dfrac{18}{25}\right)^2$

21. $\left(\dfrac{1}{3}\right)^4\cdot\left(\dfrac{9}{11}\right)^2$

22. $\left(\dfrac{1}{2}\right)^6\cdot\left(\dfrac{32}{35}\right)^2$

23. $\left(\dfrac{2}{3}\right)^4\cdot\left(\dfrac{81}{100}\right)^2$

24. $\left(\dfrac{1}{6}\right)\cdot\left(\dfrac{6}{7}\right)^2\cdot\left(\dfrac{2}{3}\right)$

25. $\left(\dfrac{2}{7}\right)\cdot\left(\dfrac{7}{8}\right)^2\cdot\left(\dfrac{8}{9}\right)$

26. $3\cdot\left(\dfrac{3}{5}\right)^3\cdot\left(\dfrac{1}{3}\right)^2$

27. $4\cdot\left(\dfrac{3}{4}\right)^3\cdot\left(\dfrac{4}{7}\right)^2$

28. $11\cdot\left(\dfrac{3}{8}\right)^3\cdot\left(\dfrac{8}{11}\right)^2$

Objective C

Simplify.

29. $\dfrac{1}{2}-\dfrac{1}{3}+\dfrac{2}{3}$

30. $\dfrac{2}{5}+\dfrac{3}{10}-\dfrac{2}{3}$

31. $\dfrac{1}{3}\div\dfrac{1}{2}+\dfrac{3}{4}$

32. $\dfrac{4}{5}+\dfrac{3}{7}\cdot\dfrac{14}{15}$

33. $\left(\dfrac{3}{4}\right)^2 - \dfrac{5}{12}$

34. $\left(\dfrac{3}{5}\right)^3 - \dfrac{3}{25}$

35. $\dfrac{5}{6} \cdot \left(\dfrac{2}{3} - \dfrac{1}{6}\right) + \dfrac{7}{18}$

36. $\dfrac{3}{4} \cdot \left(\dfrac{11}{12} - \dfrac{7}{8}\right) + \dfrac{5}{16}$

37. $\dfrac{7}{12} - \left(\dfrac{2}{3}\right)^2 + \dfrac{5}{8}$

38. $\dfrac{11}{16} - \left(\dfrac{3}{4}\right)^2 + \dfrac{7}{12}$

39. $\dfrac{3}{4} \cdot \left(\dfrac{4}{9}\right)^2 + \dfrac{1}{2}$

40. $\dfrac{9}{10} \cdot \left(\dfrac{2}{3}\right)^3 + \dfrac{2}{3}$

41. $\left(\dfrac{1}{2} + \dfrac{3}{4}\right) \div \dfrac{5}{8}$

42. $\left(\dfrac{2}{3} + \dfrac{5}{6}\right) \div \dfrac{5}{9}$

43. $\dfrac{3}{8} \div \left(\dfrac{5}{12} + \dfrac{3}{8}\right)$

44. $\dfrac{7}{12} \div \left(\dfrac{2}{3} + \dfrac{5}{9}\right)$

45. $\left(\dfrac{3}{8}\right)^2 \div \left(\dfrac{3}{7} + \dfrac{3}{14}\right)$

46. $\left(\dfrac{5}{6}\right)^2 \div \left(\dfrac{5}{12} + \dfrac{2}{3}\right)$

47. $\dfrac{2}{5} \div \dfrac{3}{8} \cdot \dfrac{4}{5}$

APPLYING THE CONCEPTS

48. The table at the right shows the results of a survey that asked fast-food patrons their criteria for choosing where to go for fast food. For example, 3 out of every 25 people surveyed said that the speed of the service was most important.
a. According to the survey, do more people choose a fast-food restaurant on the basis of its location or the quality of the food?
b. Which criterion was cited by the most people?

Fast-Food Patrons' Top Criteria for Fast-Food Restaurants	
Food Quality	$\dfrac{1}{4}$
Location	$\dfrac{13}{50}$
Menu	$\dfrac{4}{25}$
Price	$\dfrac{2}{25}$
Speed	$\dfrac{3}{25}$
Other	$\dfrac{3}{100}$

Source: Maritz Marketing Research, Inc.

49. A farmer died and left 17 horses to be divided among 3 children. The first child was to receive $\dfrac{1}{2}$ of the horses, the second child $\dfrac{1}{3}$ of the horses, and the third child $\dfrac{1}{9}$ of the horses. The executor for the family's estate realized that 17 horses could not be divided by halves, thirds, or ninths and so added a neighbor's horse to the farmer's. With 18 horses, the executor gave 9 horses to the first child, 6 horses to the second child, and 2 horses to the third child. This accounted for the 17 horses, so the executor returned the borrowed horse to the neighbor. Explain why this worked.

50. $\dfrac{2}{3} < \dfrac{3}{4}$. Is $\dfrac{2+3}{3+4}$ less than $\dfrac{2}{3}$, greater than $\dfrac{2}{3}$, or between $\dfrac{2}{3}$ and $\dfrac{3}{4}$?

Focus on Problem Solving

Common Knowledge   An application problem may not provide all the information that is needed to solve the problem. Sometimes, however, the necessary information is common knowledge.

⇒ You are traveling by bus from Boston to New York. The trip is 4 hours long. If the bus leaves Boston at 10 A.M., what time should you arrive in New York?

What other information do you need to solve this problem?

You need to know that, using a 12-hour clock, the hours run

10 A.M.
11 A.M.
12 P.M.
1 P.M.
2 P.M.

Four hours after 10 A.M. is 2 P.M.

You should arrive in New York at 2 P.M.

⇒ You purchase a 34¢ stamp at the Post Office and hand the clerk a one-dollar bill. How much change do you receive?

What information do you need to solve this problem?

You need to know that there are 100¢ in one dollar.

Your change is 100¢ − 34¢.

100 − 34 = 66

You receive 66¢ in change.

What information do you need to know to solve each of the following problems?

1. You sell a dozen tickets to a fundraiser. Each ticket costs $10. How much money do you collect?

2. The weekly lab period for your science course is one hour and twenty minutes long. Find the length of the science lab period in minutes.

3. An employee's monthly salary is $3750. Find the employee's annual salary.

4. A survey revealed that eighth graders spend an average of 3 hours each day watching television. Find the total time an eighth grader spends watching TV each week.

5. You want to buy a carpet for a room that is 15 feet wide and 18 feet long. Find the amount of carpet that you need.

Projects and Group Activities

Music   In musical notation, notes are printed on a **staff**, which is a set of five horizontal lines and the spaces between them. The notes of a musical composition are grouped into **measures**, or **bars**. Vertical lines separate measures on a staff. The shape of a note indicates how long it should be held. The whole note has the longest time value of any note. Each time value is divided by 2 in order to find the next smallest time value.

The **time signature** is a fraction that appears at the beginning of a piece of music. The numerator of the fraction indicates the number of beats in a measure. The denominator indicates what kind of note receives 1 beat. For example, music written in $\frac{2}{4}$ time has 2 beats to a measure, and a quarter note receives 1 beat. One measure in $\frac{2}{4}$ time may have 1 half note, 2 quarter notes, 4 eighth notes, or any other combination of notes totaling 2 beats. Other common time signatures include $\frac{4}{4}$, $\frac{3}{4}$, and $\frac{6}{8}$.

1. Explain the meaning of the 6 and the 8 in the time signature $\frac{6}{8}$.

2. Give some possible combinations of notes in one measure of a piece written in $\frac{4}{4}$ time.

3. What does a dot at the right of a note indicate? What is the effect of a dot at the right of a half note? At the right of a quarter note? At the right of an eighth note?

4. Symbols called rests are used to indicate periods of silence in a piece of music. What symbols are used to indicate the different time values of rests?

5. Find some examples of musical compositions written in different time signatures. Use a few measures from each to show that the sum of the time values of the notes and rests in each measure equals the numerator of the time signature.

Construction   Suppose you are involved in building your own home. Design a stairway from the first floor of the house to the second floor. Some of the questions you will need to answer follow.

What is the distance from the floor of the first story to the floor of the second story?

Typically, what is the number of steps in a stairway?

What is a reasonable length for the run of each step?

What is the width of the wood being used to build the staircase?

In designing the stairway, remember that each riser should be the same height, that each run should be the same length, and that the width of the wood used for the steps will have to be incorporated in the calculation.

Fractions of Diagrams The diagram that follows has been broken up into nine areas separated by heavy lines. Eight of the areas have been labeled *A* through *H*. The ninth area is shaded. Determine which lettered areas would have to be shaded so that half of the entire diagram is shaded and half is not shaded. Write down the strategy that you or your group use to arrive at the solution. Compare your strategy with that of other individual students or groups.

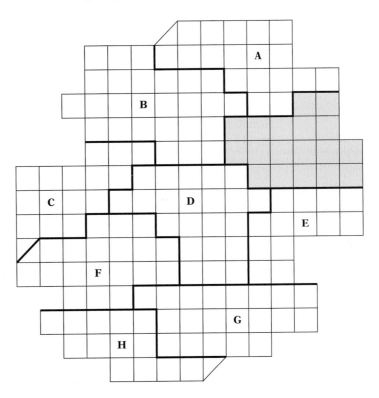

Using Patterns in Experimentation Show how to cut a pie into the greatest number of pieces with only five straight cuts of a knife. An illustration showing how five cuts can produce 13 pieces appears at the right. The correct answer, however, has more than thirteen pieces.

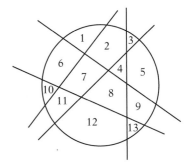

A reasonable question to ask is "How do I know when I have the maximum number of pieces?" To determine the answer, we suggest that you start with one cut, then two cuts, then three cuts, … . Try to discover a pattern for the greatest number of pieces that each number of cuts can produce.

Chapter Summary

Key Words The *least common multiple* (LCM) is the smallest common multiple of two or more numbers. [p. 63]

The *greatest common factor* (GCF) is the largest common factor of two or more numbers. [p. 64]

A *fraction* can represent the number of equal parts of a whole. [p. 67]

A *proper fraction* is a fraction less than 1. [p. 67]

A *mixed number* is a number greater than 1 with a whole-number part and a fractional part. [p. 67]

An *improper fraction* is a fraction greater than or equal to 1. [p. 67]

Equal fractions with different denominators are called *equivalent fractions*. [p. 71]

A fraction is in *simplest form* when there are no common factors in the numerator and the denominator. [p. 72]

The *reciprocal* of a fraction is the fraction with the numerator and denominator interchanged. The process of interchanging the numerator and denominator is called *inverting* the fraction. [p. 99]

Essential Rules

Addition of Fractions with the Same Denominator
To add fractions with the same denominator, add the numerators and place the sum over the common denominator. [p. 75]

Addition of Fractions with Unlike Denominators
To add fractions with unlike denominators, first rewrite the fractions as equivalent fractions with the same denominator. Then add the numerators and place the sum over the common denominator. [p. 75]

Subtraction of Fractions with the Same Denominator
To subtract fractions with the same denominator, subtract the numerators and place the difference over the common denominator. [p. 83]

Subtraction of Fractions with Unlike Denominators
To subtract fractions with unlike denominators, rewrite the fractions as equivalent fractions with the same denominator. Then subtract the numerators and place the difference over the common denominator. [p. 83]

Multiplication of Fractions
To multiply two fractions, multiply the numerators and place the product over the product of the denominators. [p. 91]

Division of Fractions
To divide two fractions, multiply by the reciprocal of the divisor. [p. 99]

Order Relations Between Fractions
To find the order relation between two fractions with the same denominator, compare the numerators. The fraction that has the smaller numerator is the smaller fraction. [p. 107]

Chapter Review

1. Write $\frac{30}{45}$ in simplest form.

2. Simplify: $\left(\frac{3}{4}\right)^3 \cdot \frac{20}{27}$

3. Express the shaded portion of the circles as an improper fraction.

4. Find the total of $\frac{2}{3}$, $\frac{5}{6}$, and $\frac{2}{9}$.

5. Place the correct symbol, $<$ or $>$, between the two numbers.

$$\frac{11}{18} \qquad \frac{17}{24}$$

6. Subtract:
$$18\frac{1}{6}$$
$$-\ 3\frac{5}{7}$$

7. Simplify: $\frac{2}{7}\left(\frac{5}{8} - \frac{1}{3}\right) \div \frac{3}{5}$

8. Multiply: $2\frac{1}{3} \times 3\frac{7}{8}$

9. Divide: $1\frac{1}{3} \div \frac{2}{3}$

10. Find $\frac{17}{24}$ decreased by $\frac{3}{16}$.

11. Divide: $8\frac{2}{3} \div 2\frac{3}{5}$

12. Find the GCF of 20 and 48.

13. Write an equivalent fraction with the given denominator.

$$\frac{2}{3} = \frac{}{36}$$

14. What is $\frac{15}{28}$ divided by $\frac{5}{7}$?

15. Write an equivalent fraction with the given denominator.

$$\frac{8}{11} = \frac{}{44}$$

16. Multiply: $2\frac{1}{4} \times 7\frac{1}{3}$

17. Find the LCM of 18 and 12.

18. Write $\frac{16}{44}$ in simplest form.

19. Add: $\dfrac{3}{8} + \dfrac{5}{8} + \dfrac{1}{8}$

20. Subtract:
$$\begin{array}{r} 16 \\ -\ 5\frac{7}{8} \\ \hline \end{array}$$

21. Add: $4\dfrac{4}{9} + 2\dfrac{1}{6} + 11\dfrac{17}{27}$

22. Find the GCF of 15 and 25.

23. Write $\dfrac{17}{5}$ as a mixed number.

24. Simplify: $\left(\dfrac{4}{5} - \dfrac{2}{3}\right)^2 \div \dfrac{4}{15}$

25. Add: $\dfrac{3}{8} + 1\dfrac{2}{3} + 3\dfrac{5}{6}$

26. Find the LCM of 18 and 27.

27. Subtract: $\dfrac{11}{18} - \dfrac{5}{18}$

28. Write $2\dfrac{5}{7}$ as an improper fraction.

29. Divide: $\dfrac{5}{6} \div \dfrac{5}{12}$

30. Multiply: $\dfrac{5}{12} \times \dfrac{4}{25}$

31. What is $\dfrac{11}{50}$ multiplied by $\dfrac{25}{44}$?

32. Express the shaded portion of the circles as a mixed number.

33. During three months of the rainy season, $5\dfrac{7}{8}$, $6\dfrac{2}{3}$, and $8\dfrac{3}{4}$ inches of rain fell. Find the total rainfall for the three months.

34. A home building contractor bought $4\dfrac{2}{3}$ acres for \$168,000. What was the cost of each acre?

35. A 15-mile race has three checkpoints. The first checkpoint is $4\dfrac{1}{2}$ miles from the starting point. The second checkpoint is $5\dfrac{3}{4}$ miles from the first checkpoint. How many miles is the second checkpoint from the finish line?

36. A compact car gets 36 miles on each gallon of gasoline. How many miles can the car travel on $6\dfrac{3}{4}$ gallons of gasoline?

Chapter Test

1. Multiply: $\dfrac{9}{11} \times \dfrac{44}{81}$

2. Find the GCF of 24 and 80.

3. Divide: $\dfrac{5}{9} \div \dfrac{7}{18}$

4. Simplify: $\left(\dfrac{3}{4}\right)^2 \div \left(\dfrac{2}{3} + \dfrac{5}{6}\right) - \dfrac{1}{12}$

5. Write $9\dfrac{4}{5}$ as an improper fraction.

6. What is $5\dfrac{2}{3}$ multiplied by $1\dfrac{7}{17}$?

7. Write $\dfrac{40}{64}$ in simplest form.

8. Place the correct symbol, $<$ or $>$, between the two numbers.
$\dfrac{3}{8} \quad \dfrac{5}{12}$

9. Simplify: $\left(\dfrac{1}{4}\right)^3 \div \left(\dfrac{1}{8}\right)^2 - \dfrac{1}{6}$

10. Find the LCM of 24 and 40.

11. Subtract: $\dfrac{17}{24} - \dfrac{11}{24}$

12. Write $\dfrac{18}{5}$ as a mixed number.

13. Find the quotient of $6\dfrac{2}{3}$ and $3\dfrac{1}{6}$.

14. Write an equivalent fraction with the given denominator.
$\dfrac{5}{8} = \dfrac{}{72}$

15. Add: $\dfrac{5}{6}$

$\dfrac{7}{9}$

$+\,\dfrac{1}{15}$

16. Subtract: $23\dfrac{1}{8}$

$-\ \ 9\dfrac{9}{44}$

17. What is $\dfrac{9}{16}$ minus $\dfrac{5}{12}$?

18. Simplify: $\left(\dfrac{2}{3}\right)^{4} \cdot \dfrac{27}{32}$

19. Add: $\dfrac{7}{12} + \dfrac{11}{12} + \dfrac{5}{12}$

20. What is $12\dfrac{5}{12}$ more than $9\dfrac{17}{20}$?

21. Express the shaded portion of the circles as an improper fraction.

22. An electrician earns $240 for each day worked. What is the total of the electrician's earnings for working $3\dfrac{1}{2}$ days?

23. Grant Miura bought $7\dfrac{1}{4}$ acres of land for a housing project. One and three-fourths acres were set aside for a park, and the remaining land was developed into $\dfrac{1}{2}$-acre lots. How many lots were available for sale?

24. Chris Aguilar bought 100 shares of a utility stock at $24\dfrac{1}{2}$ per share. The stock gained $5\dfrac{5}{8}$ during the first month of ownership and lost $2\dfrac{1}{4}$ during the second month. Find the value of one share of the utility stock at the end of the second month.

25. In 3 successive months, the rainfall measured $11\dfrac{1}{2}$ inches, $7\dfrac{5}{8}$ inches, and $2\dfrac{1}{3}$ inches. Find the total rainfall for the 3 months.

Cumulative Review

1. Round 290,496 to the nearest thousand.

2. Subtract:
    $$\begin{array}{r} 390{,}047 \\ -\ 98{,}769 \\ \hline \end{array}$$

3. Find the product of 926 and 79.

4. Divide: $57\overline{)30{,}792}$

5. Simplify: $4 \cdot (6 - 3) \div 6 - 1$

6. Find the prime factorization of 44.

7. Find the LCM of 30 and 42.

8. Find the GCF of 60 and 80.

9. Write $7\frac{2}{3}$ as an improper fraction.

10. Write $\frac{25}{4}$ as a mixed number.

11. Write an equivalent fraction with the given denominator.
    $$\frac{5}{16} = \frac{\ }{48}$$

12. Write $\frac{24}{60}$ in simplest form.

13. What is $\frac{9}{16}$ more than $\frac{7}{12}$?

14. Add:
    $$\begin{array}{r} 3\frac{7}{8} \\ 7\frac{5}{12} \\ +\ 2\frac{15}{16} \\ \hline \end{array}$$

15. Find $\frac{3}{8}$ less than $\frac{11}{12}$.

16. Subtract:
    $$\begin{array}{r} 5\frac{1}{6} \\ -\ 3\frac{7}{18} \\ \hline \end{array}$$

17. Multiply: $\dfrac{3}{8} \times \dfrac{14}{15}$

18. Multiply: $3\dfrac{1}{8} \times 2\dfrac{2}{5}$

19. Divide: $\dfrac{7}{16} \div \dfrac{5}{12}$

20. Find the quotient of $6\dfrac{1}{8}$ and $2\dfrac{1}{3}$.

21. Simplify: $\left(\dfrac{1}{2}\right)^3 \cdot \dfrac{8}{9}$

22. Simplify: $\left(\dfrac{1}{2} + \dfrac{1}{3}\right) \div \left(\dfrac{2}{5}\right)^2$

23. Molly O'Brien had $1359 in a checking account. During the week, Molly wrote checks for $128, $54, and $315. Find the amount in the checking account at the end of the week.

24. The tickets for a movie were $5 for an adult and $2 for a student. Find the total income from the sale of 87 adult tickets and 135 student tickets.

25. Find the total weight of three packages that weigh $1\dfrac{1}{2}$ pounds, $7\dfrac{7}{8}$ pounds, and $2\dfrac{2}{3}$ pounds.

26. A board $2\dfrac{5}{8}$ feet long is cut from a board $7\dfrac{1}{3}$ feet long. What is the length of the remaining piece?

27. A car travels 27 miles on each gallon of gasoline. How many miles can the car travel on $8\dfrac{1}{3}$ gallons of gasoline?

28. Jimmy Santos purchased $10\dfrac{1}{3}$ acres of land to build a housing development. Jimmy donated 2 acres for a park. How many $\dfrac{1}{3}$-acre parcels can be sold from the remaining land?

Chapter 3 Decimals

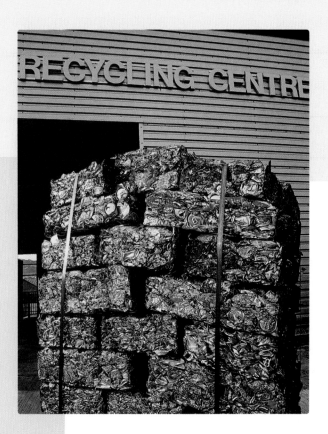

Objectives

Section 3.1

A To write decimals in standard form and in words

B To round a decimal to a given place value

Section 3.2

A To add decimals

B To solve application problems

Section 3.3

A To subtract decimals

B To solve application problems

Section 3.4

A To multiply decimals

B To solve application problems

Section 3.5

A To divide decimals

B To solve application problems

Section 3.6

A To convert fractions to decimals

B To convert decimals to fractions

C To identify the order relation between two decimals or between a decimal and a fraction

Recycling empty soda cans is easy and enables the tin to be reused, which saves our natural resources. Recycling centers offer money as an incentive for dropping off empty cans. The cans are weighed and, for each pound donated, a certain amount of money is offered. Both the weight and the payment are numbers calculated using decimals. **Exercises 122 and 123 on page 142** are examples of how a recycling center rewards the recycling of products.

 Need help? For on-line student resources, such as section quizzes, visit this textbook's web site at **math.college.hmco.com/students.**

1. Express the shaded portion of the rectangle as a fraction.

2. Round 36,852 to the nearest hundred.

3. Write 4791 in words.

4. Write six thousand eight hundred forty-two in standard form.

For Exercises 5 to 8, add, subtract, multiply or divide.

5. 37 + 8892 + 465

6. 2403 − 765

7. 844 × 91

8. 23)6412

Go Figure

Maria and Pedro are siblings. Pedro has as many brothers as sisters. Maria has twice as many brothers as sisters. How many children are in the family?

3.1   Introduction to Decimals

Objective A   **To write decimals in standard form and in words**

Point of Interest

Leonardo of Pisa (Fibonacci) anticipated the decimal system. He wrote $\dfrac{2}{10}\dfrac{7}{10}5$ to

mean $5 + \dfrac{7}{10} + \dfrac{2}{10 \cdot 10}$.

The smallest human bone is found in the middle ear and measures 0.135 inch in length. The number 0.135 is in **decimal notation.**

Note the relationship between fractions and numbers written in decimal notation.

Three tenths   Three hundredths   Three thousandths

$$\frac{3}{10} = 0.\underline{3}$$ $$\frac{3}{100} = 0.0\underline{3}$$ $$\frac{3}{1000} = 0.00\underline{3}$$

1 zero   1 decimal place   2 zeros   2 decimal places   3 zeros   3 decimal places

A number written in decimal notation has three parts.

351 . 7089

Whole-number part   Decimal point   Decimal part

A number written in decimal notation is often called simply a **decimal.** The position of a digit in a decimal determines the digit's place value.

In the decimal 351.7089, the position of the digit 9 determines that its place value is ten-thousandths.

When writing a decimal in words, write the decimal part as though it were a whole number. Then name the place value of the last digit.

0.6481   Six thousand four hundred eighty-one ten-thousandths

549.238   Five hundred forty-nine and two hundred thirty-eight thousandths
            (The decimal point is read as "and.")

To write a decimal in standard form, zeros may have to be inserted after the decimal point so that the last digit is in the given place-value position.

Five and thirty-eight <u>hundredths</u>

    8 is in the hundredths place.   5.3<u>8</u>

Nineteen and four <u>thousandths</u>

    4 is in the thousandths place.   19.00<u>4</u>
    Insert two zeros so that 4 is in
    the thousandths place.

Seventy-one <u>ten-thousandths</u>

    1 is in the ten-thousandths   0.007<u>1</u>
    place. Insert two zeros so that
    1 is in the ten-thousandths place.

Example 1	Write 307.4027 in words.	**You Try It 1**	Write 209.05838 in words.
Solution	Three hundred seven and four thousand twenty-seven ten-thousandths	**Your solution**	
Example 2	Write six hundred seven and seven hundred eight hundred-thousandths in standard form.	**You Try It 2**	Write forty-two thousand and two hundred seven millionths in standard form.
Solution	607.00708	**Your solution**	

Solutions on p. S8

Objective B   **To round a decimal to a given place value**

Rounding decimals is similar to rounding whole numbers, except that the digits to the right of the given place value are dropped instead of being replaced by zeros.

If the digit to the right of the given place value is less than 5, drop that digit and all digits to the right. If the digit to the right of the given place value is greater than or equal to 5, increase the number in the given place value by 1 and drop all digits to its right.

➡ Round 26.3799 to the nearest hundredth.

```
                   ┌── Given place value
         26.3799
                └── 9 > 5   Increase 7 by 1 and drop all digits to the right of 7.
```

26.3799 rounded to the nearest hundredth is 26.38.

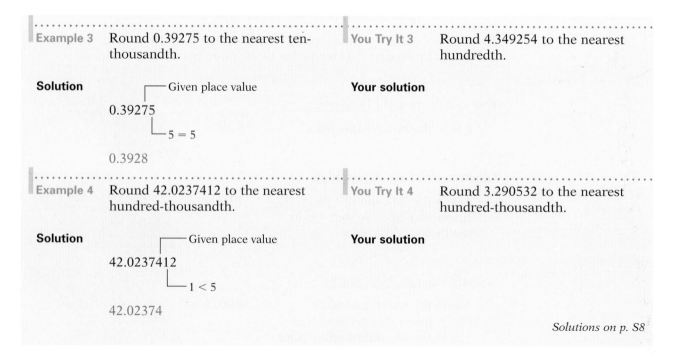

Example 3	Round 0.39275 to the nearest ten-thousandth.	**You Try It 3**	Round 4.349254 to the nearest hundredth.
Solution	┌── Given place value 0.39275 └── 5 = 5 0.3928	**Your solution**	
Example 4	Round 42.0237412 to the nearest hundred-thousandth.	**You Try It 4**	Round 3.290532 to the nearest hundred-thousandth.
Solution	┌── Given place value 42.0237412 └── 1 < 5 42.02374	**Your solution**	

Solutions on p. S8

3.1 Exercises

Objective A

Name the place value of the digit 4.

1. 36.8497   **2.** 0.473129   **3.** 107.35846   **4.** 2.03415

Write each decimal in words.

5. 0.27   **6.** 0.92   **7.** 1.005   **8.** 3.067

9. 36.4   **10.** 59.7   **11.** 0.00035   **12.** 0.00092

13. 10.007   **14.** 20.009   **15.** 52.00095   **16.** 64.00037

17. 0.0293   **18.** 0.0717   **19.** 6.324   **20.** 8.916

21. 276.3297   **22.** 418.3115   **23.** 216.0729   **24.** 976.0317

25. 4625.0379   **26.** 2986.0925   **27.** 1.00001   **28.** 3.00003

Write each decimal in standard form.

29. Seven hundred sixty-two thousandths   **30.** Two hundred ninety-five thousandths

31. Sixty-two millionths   **32.** Forty-one millionths

33. Eight and three hundred four ten-thousandths   **34.** Four and nine hundred seven ten-thousandths

35. Three hundred four and seven hundredths   **36.** Eight hundred ninety-six and four hundred seven thousandths

37. Three hundred sixty-two and forty-eight thousandths

38. Seven hundred eighty-four and eighty-four thousandths

39. Three thousand forty-eight and two thousand two ten-thousandths

40. Seven thousand sixty-one and nine thousand one ten-thousandths

Objective B

Round each decimal to the given place value.

41. 7.359   Tenths
7.4

42. 6.405   Tenths

43. 23.009   Tenths
23.0

44. 89.19204   Tenths

45. 22.68259   Hundredths
22.68

46. 16.30963   Hundredths

47. 7.072854   Thousandths
7.073

48. 1946.3745   Thousandths

49. 62.009435   Thousandths
62.009

50. 0.029876   Ten-thousandths

51. 0.012346   Ten-thousandths
0.0123

52. 1.702596   Nearest whole number

53. 2.079239   Hundred-thousandths
2.0792

54. 0.0102903   Millionths

55. 0.1009754   Millionths

APPLYING THE CONCEPTS

56. To what decimal place value are timed events in the Olympics recorded? Provide some specific examples of events and the winning times in each.

57. Provide an example of a situation in which a decimal is always rounded up, even if the digit to the right is less than 5. Provide an example of a situation in which a decimal is always rounded down, even if the digit to the right is 5 or greater than 5. (*Hint*: Think about situations in which money changes hands.)

58. Indicate which zeros of the number, if any, need not be entered on a calculator.
   a. 23.500   **b.** 0.000235   **c.** 300.0005   **d.** 0.004050

59. **a.** A decimal number was rounded to 6. Between what two numbers, to the nearest tenth, was the number?
   b. A decimal number was rounded to 10.2. Between what two numbers, to the nearest hundredth, was the number?

3.2 Addition of Decimals

To add decimals

To add decimals, write the numbers so that the decimal points are on a vertical line. Add as for whole numbers, and write the decimal point in the sum directly below the decimal points in the addends.

➡ Add: 0.237 + 4.9 + 27.32

Note that by placing the decimal points on a vertical line, we make sure that digits of the same place value are added.

Example 1   Find the sum of 42.3, 162.903, and 65.0729.

Solution
$$
\begin{array}{r}
\overset{1\,1\,1}{42.3} \\
162.903 \\
+\ \ 65.0729 \\
\hline
270.2759
\end{array}
$$

You Try It 1   Find the sum of 4.62, 27.9, and 0.62054.

Your solution

Example 2   Add: 0.83 + 7.942 + 15

Solution
$$
\begin{array}{r}
\overset{1\,1}{0.83} \\
7.942 \\
+\ 15. \\
\hline
23.772
\end{array}
$$

You Try It 2   Add: 6.05 + 12 + 0.374

Your solution

Solutions on pp. S8–S9

ESTIMATION

Estimating the Sum of Two or More Decimals

Estimate and then use your calculator to find 23.037 + 16.7892.

To estimate the sum of two or more numbers, round each number to the same place value. In this case, we will round to the nearest whole number. Then add. The estimated answer is 40.

$$
\begin{array}{rcr}
23.037 & \approx & 23 \\
+\ 16.7892 & \approx & +\ 17 \\
\hline
& & 40
\end{array}
$$

Now use your calculator to find the exact sum. The exact answer is 39.8262.

23.037 ⊞ 16.7892 ⊟ 39.8262

Objective B **To solve application problems**

A study determined the average amounts that boys and girls aged 12 to 16 are paid for the chores shown in the table at the right. Use the figures in the table for Example 3 and You Try It 3.

Chore	Girls	Boys
Clean room	$1.93	$2.61
Make bed	$1.30	$1.64
Do homework	$1.41	$2.86
Pick up or put away clothes	$1.35	$1.29
Clean bathroom	$1.73	$9.02
Do dishes	$1.65	$2.56
Dry or fold clothes	$1.52	$1.63
Load or empty dishwasher	$1.07	$2.06
Vacuum	$2.21	$2.76

Source: Copyright © 2000 *Time, Inc.* Reprinted by permission.

Example 3

On average, what would a girl be paid during a week in which she did each one of the chores listed in the table?

Strategy

To find the amount she would be paid, add the amounts in the column headed "Girls."

Solution

1.93 + 1.30 + 1.41 + 1.35 + 1.73
        + 1.65 + 1.52 + 1.07 + 2.21 = 14.17

For performing each one of the chores listed, a girl would be paid, on average, $14.17.

You Try It 3

On average, what would a boy be paid during a week in which he did each one of the chores listed in the table?

Your strategy

Your solution

Example 4

Dan Burhoe earned a salary of $138.50 for working 3 days this week as a food server. He also received $42.75, $35.80, and $59.25 in tips during the 3 days. Find his total income for the 3 days of work.

Strategy

To find the total income, add the tips (42.75, 35.80, and 59.25) to the salary (138.50).

Solution

138.50 + 42.75 + 35.80 + 59.25 = 276.30

Dan's total income for the 3 days of work was $276.30.

You Try It 4

Anita Khavari, an insurance executive, earns a salary of $875 every 4 weeks. During the past 4-week period, she received commissions of $985.80, $791.46, $829.75, and $635.42. Find her total income for the past 4-week period.

Your strategy

Your solution

Solutions on p. S9

3.2 Exercises

Objective A

Add.

1. 16.008 + 2.0385 + 132.06

2. 17.32 + 1.0579 + 16.5

3. 1.792 + 67 + 27.0526

4. 8.772 + 1.09 + 26.5027

5. 3.02 + 62.7 + 3.924

6. 9.06 + 4.976 + 59.6

7. 82.006 + 9.95 + 0.927

8. 0.826 + 8.76 + 79.005

9. 4.307 + 99.82 + 9.078

10.   0.3
      + 0.07

11.   0.29
      + 0.4

12.   1.007
      + 2.1

13.   7.3
      + 9.005

14.   4.9257
     27.05
     +   9.0063

15.   8.72
     99.073
     +   2.9736

16.   62.4
      9.827
     + 692.44

17.   8
     89.43
     +   7.0659

Estimate by rounding to the nearest whole number. Then use your calculator to add.

18.   342.42
     89.625
     + 176.2

19.   219.9
      0.872
     +   13.42

20.   823.9
     82.65
     +   46.923

21.   678.92
     97.6
     +   5.423

Objective B   *Application Problems*

22. A family has a mortgage of $814.72, a Visa bill of $216.40, and an electric bill of $87.32. Estimate by rounding the numbers to the nearest hundred dollars, and then calculate the exact amount of the three payments.

23. Find the length of the shaft.

24. Find the length of the shaft.

25. You have $2143.57 in your checking account. You make deposits of $210.98, $45.32, $1236.34, and $27.99. Find the amount in your checking account after you have made the deposits if no money has been withdrawn.

26. The perimeter of a triangle is the sum of the lengths of the three sides of the triangle. Find the perimeter of a triangle that has sides that measure 4.9 meters, 6.1 meters, and 7.5 meters.

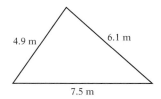

27. The world's population in 2050 is expected to be 8.9 billion people. It is projected that in that year, Asia's population will be 5.3 billion and Africa's population will be 1.8 billion. What are the combined populations of Asia and Africa expected to be in 2050? (Source: United Nations Population Division, World Population Prospects)

The table at the right shows the number of self-employed people, in millions, in the United States. They are listed by annual earnings. Use this table for Exercises 28–30.

Annual Earnings	Number of Self-Employed (in millions)
Less than $5000	5.1
$5000 – $24,999	3.9
$25,000 – $49,999	1.6
$50,000 or more	1.0

Source: U.S. Small Business Administration

28. How many self-employed people earn less than $50,000?

29. How many self-employed people earn $5000 or more?

30. How many people in the United States are self-employed?

APPLYING THE CONCEPTS

The table at the right gives the prices for selected products in a grocery store. Use this table for Exercises 31 and 32.

31. Does a customer with $10 have enough money to purchase raisin bran, bread, milk, and butter?

Product	Cost
Raisin bran	$3.29
Butter	$2.79
Bread	$1.49
Popcorn	$1.19
Potatoes	$2.49
Cola (6-pack)	$1.99
Mayonnaise	$2.99
Lunch meat	$3.39
Milk	$2.59
Toothpaste	$2.69

32. Name three items that would cost more than $7 but less than $8. (There is more than one answer.)

33. Can a piece of rope 4 feet long be wrapped around the box shown at the right?

1.4 ft
1.4 ft
1.4 ft

3.3 Subtraction of Decimals

Objective A   **To subtract decimals**

To subtract decimals, write the numbers so that the decimal points are on a vertical line. Subtract as for whole numbers, and write the decimal point in the difference directly below the decimal point in the subtrahend.

➡ Subtract 21.532 − 9.875 and check.

Placing the decimal points on a vertical line ensures that digits of the same place value are subtracted.

$$
\begin{array}{rl}
Check: & \text{Subtrahend} \\
& +\ \text{Difference} \\
\hline
& =\ \text{Minuend}
\end{array}
\qquad
\begin{array}{r}
{}^{1}{}^{1}\ {}^{1}{}^{1} \\
9.875 \\
+\ 11.657 \\
\hline
21.532
\end{array}
$$

➡ Subtract 4.3 − 1.7942 and check.

$$
\begin{array}{r}
{}^{3}\ {}^{12}\ {}^{9}\ {}^{9}\ {}^{10} \\
\cancel{4}.\cancel{3}\cancel{0}\cancel{0}\cancel{0} \\
-\ 1.7942 \\
\hline
2.5058
\end{array}
$$

If necessary, insert zeros in the minuend before subtracting.

$$
Check: \quad
\begin{array}{r}
{}^{1}\ {}^{1}{}^{1}\ {}^{1} \\
1.7942 \\
+\ 2.5058 \\
\hline
4.3000
\end{array}
$$

Example 1   Subtract 39.047 − 7.96 and check.

Solution
$$
\begin{array}{r}
{}^{8}\ {}^{9}\ {}^{14} \\
3\cancel{9}.\cancel{0}\cancel{4}7 \\
-\ 7.96 \\
\hline
31.087
\end{array}
\qquad
Check:
\begin{array}{r}
{}^{1}\ {}^{1} \\
7.96 \\
+\ 31.087 \\
\hline
39.047
\end{array}
$$

You Try It 1   Subtract 72.039 − 8.47 and check.

Your solution

Example 2   Find 9.23 less than 29 and check.

Solution
$$
\begin{array}{r}
{}^{1}\ {}^{18}\ {}^{9}\ {}^{10} \\
\cancel{2}\cancel{9}.\cancel{0}\cancel{0} \\
-\ 9.23 \\
\hline
19.77
\end{array}
\qquad
Check:
\begin{array}{r}
{}^{1}\ {}^{1}{}^{1} \\
9.23 \\
+\ 19.77 \\
\hline
29.00
\end{array}
$$

You Try It 2   Subtract 35 − 9.67 and check.

Your solution

Example 3   Subtract 1.2 − 0.8235 and check.

Solution
$$
\begin{array}{r}
{}^{0}\ {}^{11}\ {}^{9}\ {}^{9}\ {}^{10} \\
\cancel{1}.\cancel{2}\cancel{0}\cancel{0}\cancel{0} \\
-\ 0.8235 \\
\hline
0.3765
\end{array}
\qquad
Check:
\begin{array}{r}
{}^{1}\ {}^{1}{}^{1}{}^{1} \\
0.8235 \\
+\ 0.3765 \\
\hline
1.2000
\end{array}
$$

You Try It 3   Subtract 3.7 − 1.9715 and check.

Your solution

Solutions on p. S9

ESTIMATION

Estimating the Difference Between Two Decimals

Estimate and then use your calculator to find $820.2306 - 475.74815$.

To estimate the difference between two numbers, round each number to the same place value. In this case we will round to the nearest ten. Then subtract. The estimated answer is 340.

$$
\begin{array}{rcr}
820.2306 & \approx & 820 \\
- \ 475.74815 & \approx & - \ 480 \\
\hline
& & 340
\end{array}
$$

Now use your calculator to find the exact difference. The exact answer is 344.48245.

820.2306 $\boxed{-}$ 475.74815 $\boxed{=}$ 344.48245

Objective B   **To solve application problems**

Example 4

You bought a book for $15.87. How much change did you receive from a $20.00 bill?

Strategy

To find the amount of change, subtract the cost of the book (15.87) from 20.00.

Solution

$$
\begin{array}{r}
20.00 \\
- \ 15.87 \\
\hline
4.13
\end{array}
$$

You received $4.13 in change.

You Try It 4

Your breakfast cost $6.85. How much change did you receive from a $10.00 bill?

Your strategy

Your solution

Example 5

You had a balance of $87.93 on your student debit card. You then used the card, deducting $15.99 for a CD, $6.85 for lunch, and $8.50 for a ticket to the football game. What is your new student debit card balance?

Strategy

To find your new debit card balance:

- Add to find the total of the three deductions (15.99 + 6.85 + 8.50).
- Subtract the total of the three deductions from the old balance (87.93).

Solution

$$
\begin{array}{r}
15.99 \\
6.85 \\
+ \ 8.50 \\
\hline
31.34 \ \text{total of deductions}
\end{array}
\qquad
\begin{array}{r}
87.93 \\
- \ 31.34 \\
\hline
56.59
\end{array}
$$

Your new debit card balance is $56.59.

You Try It 5

You had a balance of $2472.69 in your checking account. You then wrote checks for $1025.60, $79.85, and $162.47. Find the new balance in your checking account.

Your strategy

Your solution

Solutions on p. S9

3.3 Exercises

Objective A

Subtract and check.

1. 24.037 − 18.41

2. 26.029 − 19.31

3. 123.07 − 9.4273

4. 214 − 7.143

5. 16.5 − 9.7902

6. 13.2 − 8.6205

7. 235.79 − 20.093

8. 463.27 − 40.095

9. 63.005 − 9.1274

10. 23.004 − 7.2175

11. 92 − 19.2909

12. 41.2405 − 25.2709

13.   0.32
     − 0.0058

14.   0.78
     − 0.0073

15.   3.005
     − 1.982

16.   6.007
     − 2.734

17.   352.16
     − 90.994

18.   872
     − 80.753

19.   724.32
     − 69.00
     _____
     655.32

20.   625.46
     − 77.509

21.   362.394
     − 19.4672

22.   421.385
     − 17.5293

23.   19
     − 10.372

24.   23.4
     − 0.921

Use the relationship between addition and subtraction to complete the statement.

25. _____ + 2.325 = 7.01

26. 5.392 + _____ = 8.07

27. _____ + 8.967 = 19.35

Estimate by rounding to the nearest whole number (ones place). Then use your calculator to subtract.

28.   93.079256
     − 66.09249

29.   3.7529
     − 1.00784

30.   76.53902
     − 45.73005

31.   9.07325
     − 1.924

32. The manager of the Edgewater Cafe takes a reading of the cash register tape each hour. At 1:00 P.M. the tape read $967.54. At 2:00 P.M. the tape read $1437.15. Find the amount of sales between 1:00 P.M. and 2:00 P.M.

33. Find the missing dimension.

34. Find the missing dimension.

35. You had a balance of $1029.74 in your checking account. You then wrote checks for $67.92, $43.10, and $496.34.
   a. Find the total amount of the checks written.
   b. Find the new balance in your checking account.

36. The price of gasoline is $1.42 per gallon after the price rose $.07 one month and $.12 the next month. Find the price of gasoline before these increases in price.

37. The average annual rainfall in Oakland, California, is 21.79 inches. Oakland's normal rainfall for July 1 to February 4 is 13.25 inches. From July 1 to February 4, 1998, Oakland, California, had 30.02 inches of rain. How many inches above normal was Oakland's rainfall from July 1 to February 4, 1998? (Source: National Weather Service)

38. Grace Herrera owned 357.448 shares of a mutual fund on January 1. On December 31 of the same year, she had 439.917 shares. What was the increase in the number of shares Grace owned during that year?

The table at the right shows the actual and projected number of households shopping online, in millions. Use the table for Exercises 39 and 40.

39. Calculate the growth in online shopping from 2000 to 2003.

40. Calculate the growth in online shopping from 1998 to 2003.

Year	Number of Households Shopping Online (in millions)
1998	8.7
1999	13.1
2000	17.7
2001	23.1
2002	30.3
2003	40.3

Source: Forrester Research Inc.

APPLYING THE CONCEPTS

41. Find the largest amount by which the estimate of the sum of two decimals rounded to **a.** tenths, **b.** hundredths, and **c.** thousandths places could differ from the exact sum.

3.4 Multiplication of Decimals

Objective A   To multiply decimals

Point of Interest

Benjamin Banneker (1731–1806) was the first African American to earn distinction as a mathematician and scientist. He was on the survey team that determined the boundaries of Washington, DC. The mathematics of surveying requires extensive use of decimals.

Decimals are multiplied as though they were whole numbers. Then the decimal point is placed in the product. Writing the decimals as fractions shows where to write the decimal point in the product.

$$0.\underline{3} \times 5 = \frac{3}{10} \times \frac{5}{1} = \frac{15}{10} = 1.\underline{5}$$

1 decimal place   1 decimal place

$$0.\underline{3} \times 0.\underline{5} = \frac{3}{10} \times \frac{5}{10} = \frac{15}{100} = 0.\underline{15}$$

1 decimal place   1 decimal place   2 decimal places

$$0.\underline{3} \times 0.\underline{05} = \frac{3}{10} \times \frac{5}{100} = \frac{15}{1000} = 0.\underline{015}$$

1 decimal place   2 decimal places   3 decimal places

To multiply decimals, multiply the numbers as with whole numbers. Write the decimal point in the product so that the number of decimal places in the product is the sum of the decimal places in the factors.

⇒ Multiply: 21.4×0.36

$$
\begin{array}{r}
21.4 \\
\times\ 0.36 \\
\hline
1284 \\
642 \\
\hline
7.704
\end{array}
$$

1 decimal place
2 decimal places

3 decimal places

⇒ Multiply: 0.037×0.08

$$
\begin{array}{r}
0.037 \\
\times\ \ 0.08 \\
\hline
0.00296
\end{array}
$$

3 decimal places
2 decimal places
5 decimal places

• **Two zeros must be inserted between the 2 and the decimal point so that there are 5 decimal places in the product.**

To multiply a decimal by a power of 10 (10, 100, 1000, . . .), move the decimal point to the right the same number of places as there are zeros in the power of 10.

$3.8925 \times 1\underline{0}$ $= 38.925$

     1 zero   1 decimal place

$3.8925 \times 1\underline{00}$ $= 389.25$

     2 zeros   2 decimal places

$3.8925 \times 1\underline{000}$ $= 3892.5$

     3 zeros   3 decimal places

$3.8925 \times 1\underline{0,000}$ $= 38,925.$

     4 zeros   4 decimal places

$3.8925 \times 1\underline{00,000}$ $= 389,250.$

     5 zeros   5 decimal places

• **Note that a zero must be inserted before the decimal point.**

Note that if the power of 10 is written in exponential notation, the exponent indicates how many places to move the decimal point.

$$3.8925 \times 10^1 = 38.925$$
1 decimal place

$$3.8925 \times 10^2 = 389.25$$
2 decimal places

$$3.8925 \times 10^3 = 3892.5$$
3 decimal places

$$3.8925 \times 10^4 = 38,925.$$
4 decimal places

$$3.8925 \times 10^5 = 389,250.$$
5 decimal places

Example 1   Multiply: 920×3.7

Solution

$$
\begin{array}{r}
920 \\
\times \quad 3.7 \\
\hline
644\ 0 \\
2760 \quad \\
\hline
3404.0
\end{array}
$$

You Try It 1   Multiply: 870×4.6

Your solution

Example 2   Find 0.00079 multiplied by 0.025.

Solution

$$
\begin{array}{r}
0.00079 \\
\times \quad 0.025 \\
\hline
395 \\
158 \quad \\
\hline
0.00001975
\end{array}
$$

You Try It 2   Find 0.000086 multiplied by 0.057.

Your solution

Example 3   Find the product of 3.69 and 2.07.

Solution

$$
\begin{array}{r}
3.69 \\
\times\ 2.07 \\
\hline
2583 \\
7380 \quad \\
\hline
7.6383
\end{array}
$$

You Try It 3   Find the product of 4.68 and 6.03.

Your solution

Example 4   Multiply: $42.07 \times 10,000$

Solution $42.07 \times 10,000 = 420,700$

You Try It 4   Multiply: 6.9×1000

Your solution

Example 5   Find 3.01 times 10^3.

Solution $3.01 \times 10^3 = 3010$

You Try It 5   Find 4.0273 times 10^2.

Your solution

Solutions on p. S9

ESTIMATION

Estimating the Product of Two Decimals

Estimate and then use your calculator to find 28.259 × 0.029.

To estimate a product, round each number so that there is one nonzero digit. Then multiply.

$$
\begin{array}{r}
28.259 \approx \quad 30 \\
\times\ 0.029 \approx \times\ 0.03 \\
\hline
0.90
\end{array}
$$

The estimated answer is 0.90.

Now use your calculator to find the exact product. The exact answer is 0.819511.

28.259 ☒ 0.029 ☐ 0.819511

Objective B   **To solve application problems**

The tables below list water rates and meter fees for a city. These tables are used for Example 6 and You Try It 6.

Water Charges	
Commercial	$1.39/1000 gal
Comm Restaurant	$1.39/1000 gal
Industrial	$1.39/1000 gal
Institutional	$1.39/1000 gal
Res—No Sewer	
Residential—SF	
>0 <200 gal per day	$1.15/1000 gal
>200 <1500 gal per day	$1.39/1000 gal
>1500 gal per day	$1.54/1000 gal

Meter Charges	
Meter	Meter Fee
5/8" & 3/4"	$13.50
1"	$21.80
1-1/2"	$42.50
2"	$67.20
3"	$133.70
4"	$208.20
6"	$415.10
8"	$663.70

Example 6

Find the total bill for an industrial water user with a 6-inch meter that used 152,000 gallons of water for July and August.

Strategy

To find the total cost of water:

- Find the cost of water by multiplying the cost per 1000 gallons (1.39) by the number of 1000-gallon units used.
- Add the cost of the water to the meter fee (415.10).

Solution

Cost of water = $\frac{152,000}{1000} \cdot 1.39 = 211.28$

Total cost = 211.28 + 415.10 = 626.38

The total cost is $626.38.

You Try It 6

Find the total bill for a commercial user that used 5000 gallons of water per day for July and August. The user has a 3-inch meter.

Your strategy

Your solution

Solution on pp. S9–S10

Example 7

It costs $.036 an hour to operate an electric motor. How much does it cost to operate the motor for 120 hours?

Strategy

To find the cost of running the motor for 120 hours, multiply the hourly cost (0.036) by the number of hours the motor is run (120).

Solution

```
   0.036
×    120
─────────
     720
    36
─────────
   4.320
```

The cost of running the motor for 120 hours is $4.32.

You Try It 7

The cost of electricity to run a freezer for 1 hour is $.035. This month the freezer has run for 210 hours. Find the total cost of running the freezer this month.

Your strategy

Your solution

Example 8

Jason Ng earns a salary of $360 for a 40-hour work week. This week he worked 12 hours of overtime at a rate of $13.50 for each hour of overtime worked. Find his total income for the week.

Strategy

To find Jason's total income for the week:

- Find the overtime pay by multiplying the hourly overtime rate (13.50) by the number of hours of overtime worked (12).
- Add the overtime pay to the weekly salary (360).

Solution

```
   13.50              360.00
×     12            + 162.00
────────            ─────────
   27 00              522.00
  135 0
────────
  162.00   Overtime pay
```

Jason's total income for this week is $522.00.

You Try It 8

You make a down payment of $175 on a stereo and agree to make payments of $37.18 a month for the next 18 months to repay the remaining balance. Find the total cost of the stereo.

Your strategy

Your solution

Solutions on p. S10

3.4 Exercises

Multiply.

1.   0.9
    × 0.4

.36

2.   0.7
    × 0.9

3.   0.5
    × 0.6

4.   0.3
    × 0.7

5.   0.5
    × 0.5

6.   0.7
    × 0.7

7.   0.9
    × 0.5

8.   0.2
    × 0.6

9.   7.7
    × 0.9

10.   3.4
    × 0.4

11.   9.2
    × 0.2

12.   2.6
    × 0.7

13.   7.2
    × 0.6

14.   6.8
    × 0.4

15.   7.4
    × 0.1

16.   3.8
    × 0.1

17.   7.9
    × 5

18.   9.3
    × 7

19.   0.68
    × 4

20.   0.83
    × 9

21.   0.67
    × 0.9

22.   0.84
    × 0.3

23.   0.16
    × 0.6

24.   0.47
    × 0.8

25.   2.5
    × 5.4

26.   3.9
    × 1.9

27.   8.4
    × 9.5

28.   7.6
    × 5.8

29.   0.83
    × 5.2

30.   0.24
    × 2.7

31.   0.46
    × 3.9

32.   0.78
    × 6.8

33.   0.2
    × 0.3

34.   0.3
    × 0.3

35.   0.24
    × 0.3

36.   0.17
    × 0.5

37.   1.47
    × 0.09

38.   6.37
    × 0.05

39.   8.92
    × 0.004

40.   6.75
    × 0.007

41. 0.49
     × 0.16

42. 0.38
     × 0.21

43. 7.6
     × 0.01

44. 5.1
     × 0.01

45. 8.62
     ×   4

46. 5.83
     ×   7

47. 64.5
     ×   9

48. 37.8
     ×   8

49. 2.19
     × 9.2

50. 1.25
     × 5.6

51. 1.85
     × 0.023

52. 37.8
     × 0.052

53. 0.478
     × 0.37

54. 0.526
     × 0.22

55. 48.3
     × 0.0041

56. 67.2
     × 0.0086

57. 2.437
     ×   6.1

58. 4.237
     × 0.54

59. 0.413
     × 0.0016

60. 0.517
     × 0.0029

61. 94.73
     × 0.57

62. 89.23
     × 0.62

63. 8.005
     × 0.067

64. 9.032
     × 0.019

65. 4.29×0.1

66. 6.78×0.1

67. 5.29×0.4

68. 6.78×0.5

69. 0.68×0.7

70. 0.56×0.9

71. 1.4×0.73

72. 6.3×0.37

73. 5.2×7.3

74. 7.4×2.9

75. 3.8×0.61

76. 7.2×0.72

77. 0.32×10

78. 6.93×10

79. 0.065×100

80. 0.039×100

81. 6.2856×1000

82. 3.2954×1000 **83.** 3.2×1000 **84.** $0.006 \times 10{,}000$ **85.** $3.57 \times 10{,}000$

86. 8.52×10^1 **87.** 0.63×10^1 **88.** 82.9×10^2

89. 0.039×10^2 **90.** 6.8×10^3 **91.** 4.9×10^4

92. 6.83×10^4 **93.** 0.067×10^2 **94.** 0.052×10^2

95. Find the product of 0.0035 and 3.45.   **96.** Find the product of 237 and 0.34.

97. Multiply 3.005 by 0.00392.   **98.** Multiply 20.34 by 1.008.

99. Multiply 1.348 by 0.23.   **100.** Multiply 0.000358 by 3.56.

101. Find the product of 23.67 and 0.0035.   **102.** Find the product of 0.00346 and 23.1.

103. Find the product of 5, 0.45, and 2.3.   **104.** Find the product of 0.03, 23, and 9.45.

 Estimate and then use your calculator to multiply.

105.
$$\begin{array}{r} 28.5 \\ \times\ \ 3.2 \\ \hline \end{array}$$

106.
$$\begin{array}{r} 86.3 \\ \times\ \ 4.4 \\ \hline \end{array}$$

107.
$$\begin{array}{r} 2.38 \\ \times\ 0.44 \\ \hline \end{array}$$

108.
$$\begin{array}{r} 9.82 \\ \times\ 0.77 \\ \hline \end{array}$$

109.
$$\begin{array}{r} 0.866 \\ \times\ \ \ 4.5 \\ \hline \end{array}$$

110.
$$\begin{array}{r} 0.239 \\ \times\ \ \ 8.2 \\ \hline \end{array}$$

111.
$$\begin{array}{r} 4.34 \\ \times\ 2.59 \\ \hline \end{array}$$

112.
$$\begin{array}{r} 6.87 \\ \times\ 9.98 \\ \hline \end{array}$$

113.
$$\begin{array}{r} 8.434 \\ \times\ 0.044 \\ \hline \end{array}$$

114.
$$\begin{array}{r} 7.037 \\ \times\ 0.094 \\ \hline \end{array}$$

115.
$$\begin{array}{r} 28.44 \\ \times\ \ 1.12 \\ \hline \end{array}$$

116.
$$\begin{array}{r} 86.57 \\ \times\ \ 7.33 \\ \hline \end{array}$$

117.
$$\begin{array}{r} 49.6854 \\ \times\ 39.0672 \\ \hline \end{array}$$

118.
$$\begin{array}{r} 2.00547 \\ \times\ \ \ \ 9.672 \\ \hline \end{array}$$

119.
$$\begin{array}{r} 0.00456 \\ \times\ 0.009542 \\ \hline \end{array}$$

120.
$$\begin{array}{r} 7.00637 \\ \times\ \ 0.0128 \\ \hline \end{array}$$

Objective B   *Application Problems*

121. An electric motor costing $315.45 has an operating cost of $.027 for 1 hour of operation. Find the cost to run the motor for 56 hours. Round to the nearest cent.

122. Four hundred empty soft drink cans weigh 18.75 pounds. A recycling center pays $.75 per pound for the cans. Find the amount received for the 400 cans. Round to the nearest cent.

123. A recycling center pays $.045 per pound for newspapers. Estimate the payment for recycling 520 pounds of newspapers. Find the actual amount received from recycling the newspapers.

124. The perimeter of a square is equal to four times the length of a side of the square. Find the perimeter of a square whose side measures 2.8 meters.

2.8 m

125. The area of a rectangle is equal to the product of the length of the rectangle times its width. Find the area of a rectangle that has a length of 6.75 feet and a width of 3.5 feet. The area will be in square feet.

3.5 ft

6.75 ft

126. You bought a car for $5000 down and made payments of $399.50 each month for 36 months.
a. Find the amount of the payments over the 36 months.
b. Find the total cost of the car.

127. A nurse earns a salary of $756 for a 40-hour week. This week the nurse worked 15 hours of overtime at a rate of $28.35 for each hour of overtime worked.
a. Find the nurse's overtime pay.
b. Find the nurse's total income for the week.

128. Bay Area Rental Cars charges $12 a day and $.12 per mile for renting a car. You rented a car for 3 days and drove 235 miles. Find the total cost of renting the car.

129. The graph at the right shows United States Postal Service rates for express mail. How much would it cost a company to mail 25 express mail packages, each weighing $\frac{1}{4}$ pound, post office to addressee?

130. A taxi costs $1.50 and $.20 for each $\frac{1}{8}$ mile driven. Find the cost of hiring a taxi to get from the airport to the hotel—a distance of 5.5 miles.

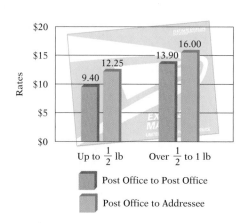

U.S. Postal Service Rates for Express Mail

The table at the right lists three pieces of steel required for a repair project. Use this table for Exercise 131.

Grade of Steel	Weight (pounds per foot)	Required Number of Feet	Cost per Pound
1	2.2	8	$1.20
2	3.4	6.5	$1.35
3	6.75	15.4	$1.94

131. a. Find the total cost of each of the grades of steel.
b. Find the total cost of the three pieces of steel.

A confectioner ships holiday packs of candy and nuts anywhere in the United States. At the right is a price list for nuts and candy, and below is a table of shipping charges to zones in the United States. For any fraction of a pound, use the next higher weight. Use these tables for Exercise 132. (16 oz = 1 lb)

Code	Description	Price
112	Almonds 16 oz	$4.75
116	Cashews 8 oz	$2.90
117	Cashews 16 oz	$5.50
130	Macadamias 7 oz	$5.25
131	Macadamias 16 oz	$9.95
149	Pecan halves 8 oz	$6.25
155	Mixed nuts 8 oz	$4.80
160	Cashew brittle 8 oz	$1.95
182	Pecan roll 8 oz	$3.70
199	Chocolate peanuts 8 oz	$1.90

Pounds	Zone 1	Zone 2	Zone 3	Zone 4
1–3	$6.55	$6.85	$7.25	$7.75
4–6	$7.10	$7.40	$7.80	$8.30
7–9	$7.50	$7.80	$8.20	$8.70
10–12	$7.90	$8.20	$8.60	$9.10

132. Find the cost of sending the following orders to the given mail zone.

a. Code	Quantity	b. Code	Quantity	c. Code	Quantity
116	2	112	1	117	3
130	1	117	4	131	1
149	3	131	2	155	2
182	4	160	3	160	4
Mail to zone 4.		182	5	182	1
		Mail to zone 3.		199	3
				Mail to zone 2.	

133. An emissions test for cars requires that of the total engine exhaust, less than 1 part per thousand $\left(\dfrac{1}{1000} = 0.001\right)$ be hydrocarbon emissions. Using this figure, determine which of the cars in the table below would fail the emissions test.

Car	Total Engine Exhaust	Hydrocarbon Emission
1	367,921	36
2	401,346	42
3	298,773	21
4	330,045	32
5	432,989	45

APPLYING THE CONCEPTS

Chris works at B & W Garage as an auto mechanic and has just completed an engine overhaul for a customer. To determine the cost of the repair job, Chris keeps a list of times worked and parts used. A parts list and a list of the times worked are shown below. Use these tables for Exercise 134.

Parts Used		Time Spent	
Item	Quantity	Day	Hours
Gasket set	1	Monday	7.0
Ring set	1	Tuesday	7.5
Valves	8	Wednesday	6.5
Wrist pins	8	Thursday	8.5
Valve springs	16	Friday	9.0
Rod bearings	8		
Main bearings	5		
Valve seals	16		
Timing chain	1		

Price List		
Item Number	Description	Unit Price
27345	Valve spring	$1.85
41257	Main bearing	$3.40
54678	Valve	$4.79
29753	Ring set	$33.98
45837	Gasket set	$48.99
23751	Timing chain	$42.95
23765	Fuel pump	$77.59
28632	Wrist pin	$2.71
34922	Rod bearing	$2.67
2871	Valve seal	$0.42

134. a. Organize a table of data showing the parts used, the unit price for each, and the price of the quantity used. *Hint:* Use the following headings for the table.

Quantity   *Item Number*   *Description*   *Unit Price*   *Total*

b. Add up the numbers in the "Total" column to find the total cost of the parts.
c. If the charge for labor is $36.75 per hour, compute the cost of labor.
d. What is the total cost for parts and labor?

135. Explain how the decimal point is placed when a number is multiplied by 10, 100, 1000, 10,000, etc.

136. Explain how the decimal point is placed in the product of two decimals.

137. Show how the decimal is placed in the product of 1.3×2.31 by first writing each number as a fraction and then multiplying. Now change back to decimal notation.

3.5 Division of Decimals

Objective A | **To divide decimals**

To divide decimals, move the decimal point in the divisor to the right to make the divisor a whole number. Move the decimal point in the dividend the same number of places to the right. Place the decimal point in the quotient directly over the decimal point in the dividend, and then divide as with whole numbers.

➡ Divide: $3.25\overline{)15.275}$

$$3.25.\overline{)15.27.5}$$

- Move the decimal point 2 places to the right in the divisor and then in the dividend. Place the decimal point in the quotient.

$$
\begin{array}{r}
4.7 \\
325.\overline{)\ 1527.5} \\
-1300 \\
\hline
227\ 5 \\
-227\ 5 \\
\hline
0
\end{array}
$$

- Divide as with whole numbers.

Moving the decimal point the same number of decimal places in the divisor and dividend does not change the value of the quotient, because this process is the same as multiplying the numerator and denominator of a fraction by the same number. In the example above,

$$3.25\overline{)15.275} = \frac{15.275}{3.25} = \frac{15.275 \times 100}{3.25 \times 100} = \frac{1527.5}{325} = 325\overline{)1527.5}$$

When dividing decimals, we usually round the quotient off to a specified place value, rather than writing the quotient with a remainder.

➡ Divide: $0.3\overline{)0.56}$
Round to the nearest hundredth.

$$
\begin{array}{r}
1.866 \approx 1.87 \\
0.3.\overline{)\ 0.5.600} \\
-\ 3 \\
\hline
2\ 6 \\
-2\ 4 \\
\hline
20 \\
-18 \\
\hline
20 \\
-18
\end{array}
$$

We must carry the division to the thousandths place to round the quotient to the nearest hundredth. Therefore, zeros must be inserted in the dividend so that the quotient has a digit in the thousandths place.

⇒ Divide 57.93 by 3.24. Round to the nearest thousandth.

$$17.8796 \approx 17.880$$

$$
\begin{array}{r}
3.24.\overline{)\,57.93.0000} \\
-32\ 4 \\
\hline
25\ 53 \\
-22\ 68 \\
\hline
2\ 85\ 0 \\
-2\ 59\ 2 \\
\hline
25\ 80 \\
-22\ 68 \\
\hline
3\ 120 \\
-2\ 916 \\
\hline
2040 \\
-1944 \\
\hline
\end{array}
$$

Zeros must be inserted in the dividend so that the quotient has a digit in the ten-thousandths place.

To divide a decimal by a power of 10 (10, 100, 1000, . . .), move the decimal point to the left the same number of places as there are zeros in the power of 10.

$34.65 \div 1\underline{0} = 3.465$

1 zero   1 decimal place

$34.65 \div 1\underline{00} = 0.3465$

2 zeros   2 decimal places

$34.65 \div 1\underline{000} = 0.03465$

3 zeros   3 decimal places

● Note that a zero must be inserted between the 3 and the decimal point.

$34.65 \div \underline{10,000} = 0.003465$

4 zeros   4 decimal places

● Note that two zeros must be inserted between the 3 and the decimal point.

If the power of 10 is written in exponential notation, the exponent indicates how many places to move the decimal point.

$34.65 \div 10^1 = 3.465$ 1 decimal place

$34.65 \div 10^2 = 0.3465$ 2 decimal places

$34.65 \div 10^3 = 0.03465$ 3 decimal places

$34.65 \div 10^4 = 0.003465$ 4 decimal places

Example 1   Divide: $0.1344 \div 0.032$

Solution

$$
\begin{array}{r}
4.2 \\
0.032.\overline{)\,0.134.4} \\
-128 \\
\hline
6\ 4 \\
-6\ 4 \\
\hline
0 \\
\end{array}
$$

You Try It 1   Divide: $0.1404 \div 0.052$

Your solution

Solution on p. S10

Example 2   Divide: $58.092 \div 82$
Round to the nearest
thousandth.

Solution $\underline{0.7084} \approx 0.708$
$82)\overline{58.0920}$
$-57\,4$
69
$\underline{-\,\,\,0}$
692
$\underline{-656}$
360
$\underline{-328}$

You Try It 2   Divide: $37.042 \div 76$
Round to the nearest
thousandth.

Your solution

Example 3   Divide: $420.9 \div 7.06$
Round to the nearest tenth.

Solution $\underline{59.61} \approx 59.6$
$7.06.)\overline{420.90.00}$
$\underline{-353\,0}$
$67\,90$
$\underline{-63\,54}$
$4\,36\,0$
$\underline{-4\,23\,6}$
$12\,40$
$\underline{-\,\,7\,06}$

You Try It 3   Divide: $370.2 \div 5.09$
Round to the nearest tenth.

Your solution

Example 4   Divide: $402.75 \div 1000$

Solution $402.75 \div 1000 = 0.40275$

You Try It 4   Divide: $309.21 \div 10,000$

Your solution

Example 5   What is 0.625 divided by 10^2?

Solution $0.625 \div 10^2 = 0.00625$

You Try It 5   What is 42.93 divided by 10^4?

Your solution

Solutions on p. S10

E S T I M A T I O N

Estimating the Quotient of Two Decimals

Estimate and then use your calculator to find $282.18 \div 0.48$.

To estimate a quotient, round each number so $282.18 \div 0.48 \approx$
that there is one nonzero digit. Then divide. $300 \div 0.5 = 600$

The estimated answer is 600.

Now use your calculator to find the $282.18 \boxed{\div} 0.48 \boxed{=} 587.875$
exact quotient. The exact answer is
587.875.

Objective B **To solve application problems**

The graph at the right shows average hourly earnings in the United States. Use this table for Example 6 and You Try It 6.

Average Hourly Earnings in the United States

Source: Statistical Abstract of the United States; Bureau of Labor Statistics

---

Example 6

How many times greater were the average hourly earnings in 2000 than in 1970? Round to the nearest whole number.

Strategy

To find how many times greater the average hourly earnings were, divide the 2000 average hourly earnings (14.38) by the 1970 average hourly earnings (3.23).

Solution

$14.38 \div 3.23 \approx 4$

The average hourly earnings in 2000 were about 4 times greater than in 1970.

You Try It 6

How many times greater were the average hourly earnings in 1990 than in 1970? Round to the nearest tenth.

Your strategy

Your solution

Example 7

A one-year subscription to a monthly magazine costs $45. The price of each issue at the newsstand is $4.90. How much would you save per issue by buying a year's subscription rather than buying each issue at the newsstand?

Strategy

To find the amount saved:

• Find the subscription price per issue by dividing the cost of the subscription (45) by the number of issues (12).
• Subtract the subscription price per issue from the newsstand price (4.90).

Solution

$45 \div 12 = 3.75$

$4.90 - 3.75 = 1.15$

The savings would be $1.15 per issue.

You Try It 7

A Nielsen survey of the number of people (in millions) who watch television during the week is given in the table below.

Mon.	*Tues.*	*Wed.*	*Thu.*	*Fri.*	*Sat.*	*Sun.*
91.9	89.8	90.6	93.9	78.0	77.1	87.7

Find the average number of people who watch television per day.

Your strategy

Your solution

Solutions on p. S10

3.5 Exercises

Objective A

Divide.

1. 3)2.46

2. 7)3.71

3. 0.8)3.84

4. 0.9)6.93

5. 0.7)62.3

6. 0.4)52.8

7. 0.4)24

8. 0.5)65

9. 0.7)59.01

10. 0.9)8.721

11. 0.5)16.15

12. 0.8)77.6

13. 0.7)3.542

14. 0.6)2.436

15. 6.3)8.19

16. 3.2)7.04

17. 3.6)0.396

18. 2.7)0.648

19. 6.9)26.22

20. 1.7)84.66

Divide. Round to the nearest tenth.

21. 55.62 ÷ 8.8

22. 25.43 ÷ 5.4

23. 5.427 ÷ 9.5

24. 1.837 ÷ 1.4

25. 18.4 ÷ 7.3

26. 52.9 ÷ 8.1

27. 0.183 ÷ 0.17

28. 0.381 ÷ 0.47

29. 6.924 ÷ 0.053

Divide. Round to the nearest hundredth.

30. 4.817 ÷ 16

31. 6.467 ÷ 8

32. 0.0418 ÷ 0.53

33. 0.0647 ÷ 0.72

34. 7 ÷ 0.55

35. 38.665 ÷ 0.95

36. 13.97 ÷ 25.4

37. 27.738 ÷ 60.8

38. 3.171 ÷ 45.6

Divide. Round to the nearest thousandth.

39. $1.028 \div 54$

40. $6.729 \div 27$

41. $0.0437 \div 0.5$

42. $75.469 \div 77.8$

43. $34.31 \div 95.3$

44. $0.2695 \div 2.67$

45. $0.4871 \div 4.72$

46. $0.1142 \div 17.2$

47. $0.2307 \div 26.7$

Divide. Round to the nearest whole number.

48. $16.5 \div 4$

49. $89.76 \div 90$

50. $1.94 \div 0.3$

51. $1.0478 \div 0.413$

52. $2.148 \div 0.519$

53. $0.79 \div 0.778$

54. $3.092 \div 0.075$

55. $392 \div 6.9$

56. $8.729 \div 0.075$

Divide.

57. $4.07 \div 10$

58. $0.039 \div 10$

59. $42.67 \div 10$

60. $389.7 \div 100$

61. $1.037 \div 100$

62. $237.835 \div 100$

63. $8.295 \div 1000$

64. $82,547 \div 1000$

65. $825.37 \div 1000$

66. $8.35 \div 10^1$

67. $0.32 \div 10^1$

68. $87.65 \div 10^1$

69. $23.627 \div 10^2$

70. $2.954 \div 10^2$

71. $0.0053 \div 10^2$

72. $289.32 \div 10^3$

73. $1.8932 \div 10^3$

74. $0.139 \div 10^3$

75. Divide 44.208 by 2.4.

76. Divide 0.04664 by 0.44.

77. Find the quotient of 723.15 and 45.

78. Find the quotient of 3.3463 and 3.07.

79. Divide 13.5 by 10^3.

80. Divide 0.045 by 10^5.

81. Find the quotient of 23.678 and 1000.

82. Find the quotient of 7.005 and 10,000.

83. What is 0.0056 divided by 0.05?

84. What is 123.8 divided by 0.02?

Estimate and then use your calculator to divide. Round your calculated answer to the nearest ten-thousandth.

85. $42.42 \div 3.8$

86. $69.8 \div 7.2$

87. $389 \div 0.44$

88. $642 \div 0.83$

89. $6.394 \div 3.5$

90. $8.429 \div 4.2$

91. $1.235 \div 0.021$

92. $7.456 \div 0.072$

93. $95.443 \div 1.32$

94. $423.0925 \div 4.0927$

95. $1.000523 \div 429.07$

96. $0.03629 \div 0.00054$

Objective B   *Application Problems*

97. Ramon, a high school football player, gained 162 yards on 26 carries in a high school football game. Find the average number of yards gained per carry. Round to the nearest hundredth.

98. A car with an odometer reading of 17,814.2 is filled with 9.4 gallons of gas. At an odometer reading of 18,130.4, the tank is empty and the car is filled with 12.4 gallons of gas. How many miles does the car travel on 1 gallon of gasoline?

99. A case of diet cola costs $6.79. If there are 24 cans in a case, find the cost per can. Round to the nearest cent.

100. Anne is building bookcases that are 3.4 feet long. How many complete shelves can be cut from a 12-foot board?

101. Earl is 52 years old and is buying $70,000 of life insurance for an annual premium of $703.80. If he pays each annual premium in 12 equal install-ments, how much is each monthly payment?

102. An oil company has issued 3,541,221,500 shares of stock. The company paid $6,090,990,120 in dividends. Find the dividend for each share of stock. Round to the nearest cent.

103. When the Massachusetts Turnpike opened, the toll for a passenger car that traveled the entire 136 miles of it was $5.60. Calculate the cost per mile. Round to the nearest cent.

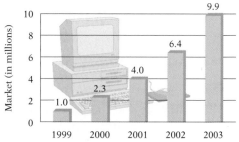

104. The graph at the right shows the expected growth of digital subscriber lines (DSL). Figures given are in millions. How many times greater was the DSL market expected to be in 2003 than in 2001? Round to the nearest tenth.

Digital Subscriber Lines (DSL) Market

Source: Copyright © 1999 *USA Today.* Reprinted with permission.

APPLYING THE CONCEPTS

105. Explain how the decimal point is moved when a number is divided by 10, 100, 1000, 10,000, etc.

106. A ballpoint pen priced at 50¢ was not selling. When the price was reduced to a different whole number of cents, the entire stock sold for $31.95. How many cents were charged per pen when the price was reduced? (*Hint*: There is more than one possible answer.)

107. Explain how baseball batting averages are determined. Then find Boston Red Sox shortstop Nomar Garciaparra's batting average with 190 hits out of 532 at bats. Round to the nearest thousandth.

108. Explain how the decimal point is placed in the quotient when a number is divided by a decimal.

For each of the problems below, insert $+$, $-$, $\times$, or $\div$ into the square so that the statement is true.

109. 3.45 ☐ 0.5 = 6.9   **110.** 3.46 ☐ 0.24 = 0.8304   **111.** 6.009 ☐ 4.68 = 1.329

112. 0.064 ☐ 1.6 = 0.1024   **113.** 9.876 ☐ 23.12 = 32.996   **114.** 3.0381 ☐ 1.23 = 2.47

Fill in the square to make a true statement.

115. 6.47 − ☐ = 1.253   **116.** 6.47 + ☐ = 9   **117.** 0.009 ÷ ☐ = 0.36

3.6 Comparing and Converting Fractions and Decimals

Objective A   **To convert fractions to decimals**

Every fraction can be written as a decimal. To write a fraction as a decimal, divide the numerator of the fraction by the denominator. The quotient can be rounded to the desired place value.

➡ Convert $\frac{3}{7}$ to a decimal.

$$\frac{0.42857}{7)3.00000}$$ $\frac{3}{7}$ rounded to the nearest hundredth is 0.43.

$\frac{3}{7}$ rounded to the nearest thousandth is 0.429.

$\frac{3}{7}$ rounded to the nearest ten-thousandth is 0.4286.

➡ Convert $3\frac{2}{9}$ to a decimal. Round to the nearest thousandth.

$$3\frac{2}{9} = \frac{29}{9} \qquad \frac{3.2222}{9)29.0000} \qquad 3\frac{2}{9} \text{ rounded to the nearest thousandth is } 3.222.$$

Example 1

Convert $\frac{3}{8}$ to a decimal.
Round to the nearest hundredth.

Solution $\frac{0.375}{8)3.000} \approx 0.38$

You Try It 1

Convert $\frac{9}{16}$ to a decimal.
Round to the nearest tenth.

Your solution

Example 2

Convert $2\frac{3}{4}$ to a decimal.
Round to the nearest tenth.

Solution $2\frac{3}{4} = \frac{11}{4} \qquad \frac{2.75}{4)11.00} \approx 2.8$

You Try It 2

Convert $4\frac{1}{6}$ to a decimal.
Round to the nearest hundredth.

Your solution

Solutions on p. S10

Objective B   **To convert decimals to fractions**

To convert a decimal to a fraction, remove the decimal point and place the decimal part over a denominator equal to the place value of the last digit in the decimal.

$$0.47 = \overset{\text{hundredths}}{\frac{47}{100}} \qquad\qquad 7.45 = 7\overset{\text{hundredths}}{\frac{45}{100}} = 7\frac{9}{20}$$

$$0.275 = \overset{\text{thousandths}}{\frac{275}{1000}} = \frac{11}{40} \qquad 0.16\frac{2}{3} = \overset{\text{hundredths}}{\frac{16\frac{2}{3}}{100}} = 16\frac{2}{3} \div 100 = \frac{50}{3} \times \frac{1}{100} = \frac{1}{6}$$

Example 3   Convert 0.82 and 4.75 to fractions.

Solution $0.82 = \dfrac{82}{100} = \dfrac{41}{50}$

$4.75 = 4\dfrac{75}{100} = 4\dfrac{3}{4}$

You Try It 3   Convert 0.56 and 5.35 to fractions.

Your solution

Example 4   Convert $0.15\dfrac{2}{3}$ to a fraction.

Solution $0.15\dfrac{2}{3} = \dfrac{15\dfrac{2}{3}}{100} = 15\dfrac{2}{3} \div 100$

$= \dfrac{47}{3} \times \dfrac{1}{100} = \dfrac{47}{300}$

You Try It 4   Convert $0.12\dfrac{7}{8}$ to a fraction.

Your solution

Solutions on pp. S10–S11

Objective C   **To identify the order relation between two decimals or between a decimal and a fraction**

Decimals, like whole numbers and fractions, can be graphed as points on the number line. The number line can be used to show the order of decimals. A decimal that appears to the right of a given number is greater than the given number. A decimal that appears to the left of a given number is less than the given number.

3.00   3.05   3.10   3.15   3.20   3.25   3.30   3.35   3.40

Note that 3, 3.0, and 3.00 represent the same number.

➡ Find the order relation between $\dfrac{3}{8}$ and 0.38.

$$\dfrac{3}{8} = 0.375 \qquad 0.38 = 0.380$$

$$0.375 < 0.380$$

$$\dfrac{3}{8} < 0.38$$

Example 5   Place the correct symbol, < or >, between the numbers.

$\dfrac{5}{16}$ 0.32

Solution $\dfrac{5}{16} \approx 0.313$

$0.313 < 0.32$

$\dfrac{5}{16} < 0.32$

You Try It 5   Place the correct symbol, < or >, between the numbers.

0.63 $\dfrac{5}{8}$

Your solution

Solution on p. S11

3.6 Exercises

Objective A

Convert the fraction to a decimal. Round to the nearest thousandth.

1. $\dfrac{5}{8}$ **2.** $\dfrac{7}{12}$ **3.** $\dfrac{2}{3}$ **4.** $\dfrac{5}{6}$ **5.** $\dfrac{1}{6}$ **6.** $\dfrac{7}{8}$

7. $\dfrac{5}{12}$ **8.** $\dfrac{9}{16}$ **9.** $\dfrac{7}{4}$ **10.** $\dfrac{5}{3}$ **11.** $1\dfrac{1}{2}$ **12.** $2\dfrac{1}{3}$

13. $\dfrac{16}{4}$ **14.** $\dfrac{36}{9}$ **15.** $\dfrac{3}{1000}$ **16.** $\dfrac{5}{10}$ **17.** $7\dfrac{2}{25}$ **18.** $16\dfrac{7}{9}$

19. $37\dfrac{1}{2}$ **20.** $\dfrac{5}{24}$ **21.** $\dfrac{4}{25}$ **22.** $3\dfrac{1}{3}$ **23.** $8\dfrac{2}{5}$ **24.** $5\dfrac{4}{9}$

Objective B

Convert the decimal to a fraction.

25. 0.8   **26.** 0.4   **27.** 0.32   **28.** 0.48   **29.** 0.125

30. 0.485   **31.** 1.25   **32.** 3.75   **33.** 16.9   **34.** 17.5

35. 8.4   **36.** 10.7   **37.** 8.437   **38.** 9.279   **39.** 2.25

40. 7.75   **41.** $0.15\dfrac{1}{3}$ **42.** $0.17\dfrac{2}{3}$ **43.** $0.87\dfrac{7}{8}$ **44.** $0.12\dfrac{5}{9}$

45. 7.38   **46.** 0.33   **47.** 0.57   **48.** $0.33\dfrac{1}{3}$ **49.** $0.66\dfrac{2}{3}$

Objective C

Place the correct symbol, < or >, between the numbers.

50. 0.15   0.5

51. 0.6   0.45

52. 6.65   6.56

53. 3.89   3.98

54. 2.504   2.054

55. 0.025   0.105

56. $\dfrac{3}{8}$ 0.365

57. $\dfrac{4}{5}$ 0.802

58. $\dfrac{2}{3}$ 0.65

59. 0.85 $\dfrac{7}{8}$

60. $\dfrac{5}{9}$ 0.55

61. $\dfrac{7}{12}$ 0.58

62. 0.62 $\dfrac{7}{15}$

63. $\dfrac{11}{12}$ 0.92

64. 0.161 $\dfrac{1}{7}$

65. 0.623   0.6023

66. 0.86   0.855

67. 0.87   0.087

68. 1.005   0.5

69. 0.033   0.3

APPLYING THE CONCEPTS

The graph at the right shows the breakdown by age group of Americans who are hearing impaired. Use this graph for Exercises 70 and 71.

70. Determine whether the number of hearing-impaired individuals under the age of 45 is more than or less than the number of hearing impaired individuals over the age of 64.

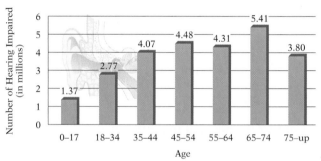

Breakdown by Age Group of Hearing-Impaired Americans

Source: American Speech-Language-Hearing Association

71. Is the number of hearing-impaired individuals who are aged 65–74 more than or less than 4 times the number of hearing-impaired individuals who are aged 0–17?

72. Is $\dfrac{7}{13}$ in decimal form a repeating decimal? Why or why not?

73. If a number is rounded to the nearest thousandth, is it always greater than if it were rounded to the nearest hundredth? Give examples to support your answer.

74. Explain how terminating, repeating, and nonrepeating decimals differ. Give an example of each kind of decimal.

Focus on Problem Solving

Relevant Information   Problems in mathematics or real life involve a question or a need and information or circumstances related to that need. Solving problems in the sciences usually involves a question, an observation, and measurements of some kind.

One of the challenges of problem solving in the sciences is to separate the information that is relevant to the problem from other information. Following is an example from the physical sciences in which some relevant information was omitted.

Hooke's Law states that the distance that a weight will stretch a spring is directly proportional to the weight on the spring. That is, $d = kF$, where d is the distance the spring is stretched and F is the force. In an experiment to verify this law, some physics students were continually getting inconsistent results. Finally, the instructor discovered that the heat produced when the lights were turned on was affecting the experiment. In this case, relevant information was omitted—namely, that the temperature of the spring can affect the distance it will stretch.

A lawyer drove 8 miles to the train station. After a 35-minute ride of 18 miles, the lawyer walked 10 minutes to the office. Find the total time it took the lawyer to get to work.

From this situation, answer the following before reading on.

a. What is asked for?

b. Is there enough information to answer the question?

c. Is information given that is not needed?

Here are the answers.

a. We want the total time for the lawyer to get to work.

b. No. We do not know the time it takes the lawyer to get to the train station.

c. Yes. Neither the distance to the train station nor the distance of the train ride is necessary to answer the question.

For each of the following problems, answer these questions:

a. What is asked for?

b. Is there enough information to answer the question?

c. Is information given that is not needed?

1. A customer bought 6 boxes of strawberries and paid with a $20 bill. What was the change?

2. A board is cut into two pieces. One piece is 3 feet longer than the other piece. What is the length of the original board?

3. A family rented a car for their vacation and drove 680 miles. The cost of the rental car was $21 per day with 150 free miles per day and $.15 for each mile driven above the number of free miles allowed. How many miles did the family drive per day?

4. An investor bought 8 acres of land for $80,000. One and one-half acres were set aside for a park, and the remainder was developed into one-half-acre lots. How many lots were available for sale?

5. You wrote checks of $43.67, $122.88, and $432.22 after making a deposit of $768.55. How much do you have left in your checking account?

Projects and Group Activities

Topographical Maps   Carpenters use fractions to measure the wood that is used to frame a house. For instance, a door entry may measure $42\frac{1}{2}$ inches. However, the grading contractor (the person who levels the lot on which the house is built) measures the height of the lot by using decimals. For instance, a certain place may be at a height of 554.2 feet. (The height is measured above sea level, so 554.2 means 554.2 feet above sea level.)

A surveyor provides the grading contractor with a grading plan that shows the elevation of each point of the lot. These plans are drawn so that the house can be sited in such a way that water will drain away from it. The diagram below is a **topographical map** for a lot. Along each closed curve, called a **contour curve,** the lot is at the same height above sea level. For instance, every point on the curve in red is 556.5 feet above sea level.

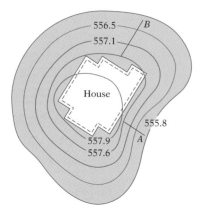

Using the map above, answer the following questions.

1. What is the elevation of the highest point on the lot?

2. What is the elevation of the lowest point on the lot?

3. What is the difference in elevation between the highest and lowest points on the lot?

4. Assuming that this map is drawn to scale, describe the significance of how steep the slope is along line *A* compared to that along line *B*.

5. Acquire a forest service contour map of a portion of a national forest and plan a cross-country hike, choosing a trail that avoids the steepest ascent.

International Math and Science Test Scores

The Third International Mathematics and Science Study (TIMSS) is responsible for comparing the performance of half a million students from 41 countries at five different grade levels on tests of their mathematics and science knowledge. The table below shows the performance of the fourth-graders from the 26 nations that participated at this level.

1. Calculate the average score in mathematics and the average score in science. (Find the sum of all the scores and then divide by the number of scores.) Round to the nearest tenth.

2. How far above or below each average are the U.S. scores?

3. Which countries scored above the United States in math? In science? Are they the same countries?

4. Discuss the following: The study reported that U.S. fourth-graders receive, on average, 4.2 hours of mathematics instruction per week; the international average is 3.9 hours. However, four of the nations that scored higher than the United States in mathematics spend less time per week on mathematics instruction.

5. One area of the mathematics testing in which the U.S. average was below the international average was measurement. On the international test, units cited in the questions were metric units. Discuss how these two facts might be related.

6. Where do you think the United States should rank internationally, given its resources and population?

7. What would you do to improve math and science education in the United States?

Nation	Average Score in Mathematics	Average Score in Science
Australia	546	562
Austria	559	565
Canada	532	549
Cyprus	502	475
Czech Republic	567	557
England	513	551
Greece	492	497
Hong Kong	587	533
Hungary	548	532
Iceland	474	505
Iran, Islamic Republic	429	416
Ireland	550	539
Israel	531	505
Japan	597	574
Korea	611	597
Kuwait	400	401
Latvia (LSS)	525	512
Netherlands	577	557
New Zealand	499	531
Norway	502	530
Portugal	475	480
Scotland	520	536
Singapore	625	547
Slovenia	552	546
Thailand	490	473
United States	545	565

Chapter Summary

Key Words

A number written in *decimal notation* has three parts: a whole-number part, a decimal point, and a decimal part. The *decimal part* of a number represents a number less than 1. [p. 123]

Essential Rules

To Write a Decimal in Words
Write the decimal part as though it were a whole number. Then name the place value of the last digit. [p. 123]

To Write a Decimal in Standard Form When It Is Written in Words
Write the whole-number part, replace the word "and" with a decimal point, and write the decimal part so that the last digit is in the given place-value position. [p. 123]

To Round a Decimal to a Given Place Value
If the digit to the right of the given place value is less than 5, drop that digit and all digits to the right. If the digit to the right of the given place value is greater than or equal to 5, increase the number in the given place value by 1, and drop all digits to its right. [p. 124]

To Add Decimals
Write the numbers so that the decimal points are on a vertical line. Add as you would with whole numbers. Then write the decimal point in the sum directly below the decimal points in the addends. [p. 127]

To Subtract Decimals
Write the numbers so that the decimal points are on a vertical line. Subtract as you would with whole numbers. Then write the decimal point in the difference directly below the decimal point in the subtrahend. [p. 131]

To Multiply Decimals
Multiply the numbers as you would whole numbers. Then write the decimal point in the product so that the number of decimal places in the product is the sum of the decimal places in the factors. [p. 135]

To Multiply a Decimal by a Power of 10
Move the decimal point to the right the same number of places as there are zeros in the power of 10. If the power of 10 is written in exponential notation, the exponent indicates how many places to move the decimal point. [pp. 135–136]

To Divide Decimals
Move the decimal point in the divisor to the right so that it is a whole number. Move the decimal point in the dividend the same number of places to the right. Place the decimal point in the quotient directly above the decimal point in the dividend. Then divide as you would with whole numbers. [p. 145]

To Divide a Decimal by a Power of 10
Move the decimal point to the left the same number of places as there are zeros in the power of 10. If the power of 10 is written in exponential notation, the exponent indicates how many places to move the decimal point. [p. 146]

To Convert a Fraction to a Decimal
Divide the numerator of the fraction by the denominator. Round the quotient to the desired place value. [p. 153]

To Convert a Decimal to a Fraction
Remove the decimal point and place the decimal part over a denominator equal to the place value of the last digit in the decimal. [p. 153]

To Find the Order Relation Between a Decimal and a Fraction
First rewrite the fraction as a decimal. Then compare the two decimals. [p. 154]

Chapter Review

1. Find the quotient of 3.6515 and 0.067.

2. Find the sum of 369.41, 88.3, 9.774, and 366.474.

3. Place the correct symbol, $<$ or $>$, between the two numbers.
   0.055   0.1

4. Write 22.0092 in words.

5. Round 0.05678235 to the nearest hundred-thousandth.

6. Convert $2\frac{1}{3}$ to a decimal. Round to the nearest hundredth.

7. Convert 0.375 to a fraction.

8. Add: $3.42 + 0.794 + 32.5$

9. Write thirty-four and twenty-five thousandths in standard form.

10. Place the correct symbol, $<$ or $>$, between the two numbers.
    $\frac{5}{8}$ 0.62

11. Convert $\frac{7}{9}$ to a decimal. Round to the nearest thousandth.

12. Convert $0.66\frac{2}{3}$ to a fraction.

13. Subtract: $27.31 - 4.4465$

14. Round 7.93704 to the nearest hundredth.

15. Find the product of 3.08 and 2.9.

16. Write 342.37 in words.

17. Write three and six thousand seven hundred fifty-three hundred-thousandths in standard form.

18. Multiply: 34.79
× 0.74

19. Divide: 0.053)‾0.349482‾

20. What is 7.796 decreased by 2.9175?

21. You had a balance of $895.68 in your checking account. You then wrote checks of $145.72 and $88.45. Find the new balance in your checking account.

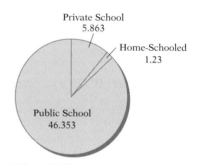

The graph at the right shows where American children in grades K–12 are being educated. Figures are in millions of children. Use this graph for Exercises 22 and 23.

22. Find the total number of American children in grades K–12.

Private School 5.863

Home-Schooled 1.23

Public School 46.353

Where Children in Grades K–12 Are Being Educated in America (in millions)

Source: U.S. Department of Education; Home School Legal Defense Association

23. How many more children are being educated in public school than in private school?

24. According to the American School Food Service Association, 1.9 million gallons of milk are served in school cafeterias every day. How many gallons of milk are served in school cafeterias during a 5-day school week?

8

6 — 6.2

Acreage (in millions)

4

2 — 1.5 — 1.2 — 0.97

0

Texas   Georgia   Mississippi   Arkansas

Acres Planted in Cotton

25. Almost 15 million acres are planted in cotton in the United States. The top four cotton states are shown in the graph at the right. How many times greater is the number of acres planted in Texas than the number planted in Mississippi? Round to the nearest tenth.

Source: Agriculture Department, *Cotton and Wool Outlook*, April 2000

Chapter Test

1. Place the correct symbol, $<$ or $>$, between the two numbers.
0.66   0.666

2. Subtract: 13.027
                  $-$ 8.94

3. Write 45.0302 in words.

4. Convert $\frac{9}{13}$ to a decimal. Round to the nearest thousandth.

5. Convert 0.825 to a fraction.

6. Round 0.07395 to the nearest ten-thousandth.

7. Find 0.0569 divided by 0.037. Round to the nearest thousandth.

8. Find 9.23674 less than 37.003.

9. Round 7.0954625 to the nearest thousandth.

10. Divide: $0.006\overline{)1.392}$

11. Add: 270.93
                  97.
                   1.976
             $+$ 88.675

12. Find the missing dimension.

4.86 in.

?

6.23 in.

13. Multiply: 1.37
$\times\, 0.004$

14. What is the total of 62.3, 4.007, and 189.65?

15. Write two hundred nine and seven thousand eighty-six hundred-thousandths in standard form.

16. A car was bought for $16,734.40, with a down payment of $2500. The balance was paid in 36 monthly payments. Find the amount of each monthly payment.

17. You received a salary of $727.50, a commission of $1909.64, and a bonus of $450. Find your total income.

18. A long-distance telephone call costs $.85 for the first 3 minutes and $.42 for each additional minute. Find the cost of a 12-minute long-distance telephone call.

The table at the right shows the average number of hours per week that students use a computer. Use this table for Exercises 19 and 20.

19. On average, how many hours per year does a 10th-grade student use a computer? Use a 52-week year.

Grade Level	Average Number of Hours of Computer Use per Week
Prekindergarten – kindergarten	3.9
1st – 3rd	4.9
4th – 6th	4.2
7th – 8th	6.9
9th – 12th	6.7

Source: Find/SVP American Learning Household Survey

20. On average, how many more hours per year does a 2nd-grade student use a computer than a 5th-grade student? Use a 52-week year.

Cumulative Review

1. Divide: $89\overline{)20{,}932}$

2. Simplify: $2^3 \cdot 4^2$

3. Simplify: $2^2 - (7 - 3) \div 2 + 1$

4. Find the LCM of 9, 12, and 24.

5. Write $\dfrac{22}{5}$ as a mixed number.

6. Write $4\dfrac{5}{8}$ as an improper fraction.

7. Write an equivalent fraction with the given denominator.
$$\frac{5}{12} = \frac{}{60}$$

8. Add: $\dfrac{3}{8} + \dfrac{5}{12} + \dfrac{9}{16}$

9. What is $5\dfrac{7}{12}$ increased by $3\dfrac{7}{18}$?

10. Subtract: $9\dfrac{5}{9} - 3\dfrac{11}{12}$

11. Multiply: $\dfrac{9}{16} \times \dfrac{4}{27}$

12. Find the product of $2\dfrac{1}{8}$ and $4\dfrac{5}{17}$.

13. Divide: $\dfrac{11}{12} \div \dfrac{3}{4}$

14. What is $2\dfrac{3}{8}$ divided by $2\dfrac{1}{2}$?

15. Simplify: $\left(\dfrac{2}{3}\right)^2 \cdot \left(\dfrac{3}{4}\right)^3$

16. Simplify: $\left(\dfrac{2}{3}\right)^2 - \left(\dfrac{2}{3} - \dfrac{1}{2}\right) + 2$

17. Write 65.0309 in words.

18. Add:
$$
\begin{aligned}
379.006 \\
27.523 \\
9.8707 \\
+\ 88.2994 \\
\hline
\end{aligned}
$$

19. What is 29.005 decreased by 7.9286?

20. Multiply: 9.074
$\times\ 6.09$

21. Divide: $8.09\overline{)17.42963}$.
Round to the nearest thousandth.

22. Convert $\frac{11}{15}$ to a decimal. Round to the nearest thousandth.

23. Convert $0.16\frac{2}{3}$ to a fraction.

24. Place the correct symbol, $<$ or $>$, between the two numbers.
$\frac{8}{9}$ 0.98

25. The graph at the right shows the number of vacation days per year that are legally mandated in several countries. How many more vacation days does Sweden mandate than Germany?

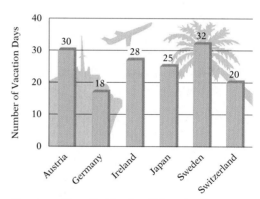

Number of Legally Mandated Vacation Days
Source: Economic Policy Institute; *World Almanac*

26. An investor purchased stock at $\$32\frac{1}{8}$ per share. During the first 2 months of ownership, the stock lost $\$\frac{3}{4}$ and then gained $\$1\frac{1}{2}$. Find the value of each share of stock at the end of the 2 months.

27. You have a checking account balance of $814.35. You then write checks for $42.98, $16.43, and $137.56. Find your checking account balance after you write the checks.

28. A machine lathe takes 0.017 inch from a brass bushing that is 1.412 inches thick. Find the resulting thickness of the bushing.

29. The state income tax on your business is $820 plus 0.08 times your profit. You made a profit of $64,860 last year. Find the amount of income tax you paid last year.

30. You bought a camera costing $210.96. The down payment was $20, and the balance is to be paid in 8 equal monthly payments. Find the monthly payment.

Chapter 4

Ratio and Proportion

What cargo might this freighter be carrying? Auto parts, clothing, watches, toys, agricultural goods? These are only a few of the products imported into the United States each year. When an American company imports goods from another country, U.S. dollars are exchanged for the currency of the other country in order to pay for the products. This exchange of currencies is discussed in **Exercise 33 on page 176.**

Need help? For on-line student resources, such as section quizzes, visit this textbook's web site at **math.college.hmco.com/students.**

1. Simplify: $\dfrac{8}{10}$

2. Simplify: $\dfrac{450}{650 + 250}$

3. Write as a decimal: $\dfrac{372}{15}$

4. Which is greater, 4×33 or 62×2?

5. Complete: $? \times 5 = 20$

Go Figure

Luis, Kim, Reggie, and Dave are standing in line. Dave is not first. Kim is between Luis and Reggie. Luis is between Dave and Kim. Give the order in which the men are standing.

4.1 Ratio

Objective A   **To write the ratio of two quantities in simplest form**   ◄ 4 ►

Quantities such as 3 feet, 12 cents, and 9 cars are number quantities written with **units**.

3 feet
12 cents   These are only some examples of units. Shirts, dollars, trees,
9 cars   miles, and gallons are further examples.
↑
units

A **ratio** is a comparison of two quantities that have the *same* units. This comparison can be written three different ways:

1. As a fraction
2. As two numbers separated by a colon (:)
3. As two numbers separated by the word *to*

The ratio of the lengths of two boards, one 8 feet long and the other 10 feet long, can be written as

1. $\dfrac{8 \text{ feet}}{10 \text{ feet}} = \dfrac{8}{10} = \dfrac{4}{5}$

2. $8 \text{ feet} : 10 \text{ feet} = 8:10 = 4:5$

3. $8 \text{ feet to } 10 \text{ feet} = 8 \text{ to } 10 = 4 \text{ to } 5$

A ratio is in **simplest form** when the two numbers do not have a common factor. Note that in a ratio, the units are not written.

This ratio means that the smaller board is $\dfrac{4}{5}$ the length of the longer board.

Example 1
Write the comparison $6 to $8 as a ratio in simplest form using a fraction, a colon, and the word *to*.

Solution $\dfrac{\$6}{\$8} = \dfrac{6}{8} = \dfrac{3}{4}$
$\$6:\$8 = 6:8 = 3:4$
$\$6 \text{ to } \$8 = 6 \text{ to } 8 = 3 \text{ to } 4$

You Try It 1
Write the comparison 20 pounds to 24 pounds as a ratio in simplest form using a fraction, a colon, and the word *to*.

Your solution

Example 2
Write the comparison 18 quarts to 6 quarts as a ratio in simplest form using a fraction, a colon, and the word *to*.

Solution $\dfrac{18 \text{ quarts}}{6 \text{ quarts}} = \dfrac{18}{6} = \dfrac{3}{1}$
$18 \text{ quarts} : 6 \text{ quarts} =$
$18:6 = 3:1$
$18 \text{ quarts to } 6 \text{ quarts} =$
$18 \text{ to } 6 = 3 \text{ to } 1$

You Try It 2
Write the comparison 64 miles to 8 miles as a ratio in simplest form using a fraction, a colon, and the word *to*.

Your solution

Solutions on p. S11

Objective B   **To solve application problems**

Use the table below for Example 3 and You Try It 3.

| Board Feet of Wood at a Lumber Store | | | |
Pine	Ash	Oak	Cedar
20,000	18,000	10,000	12,000

Example 3

Find, as a fraction in simplest form, the ratio of the number of board feet of pine to the number of board feet of oak.

Strategy

To find the ratio, write the ratio of board feet of pine (20,000) to board feet of oak (10,000) in simplest form.

Solution

$$\frac{20,000}{10,000} = \frac{2}{1}$$

The ratio is $\frac{2}{1}$.

You Try It 3

Find, as a fraction in simplest form, the ratio of the number of board feet of cedar to the number of board feet of ash.

Your strategy

Your solution

Example 4

The cost of building a patio cover was $250 for labor and $350 for materials. What, as a fraction in simplest form, is the ratio of the cost of materials to the total cost for labor and materials?

Strategy

To find the ratio, write the ratio of the cost of materials ($350) to the total cost ($250 + $350) in simplest form.

Solution

$$\frac{\$350}{\$250 + \$350} = \frac{350}{600} = \frac{7}{12}$$

The ratio is $\frac{7}{12}$.

You Try It 4

A company spends $20,000 a month for television advertising and $15,000 a month for radio advertising. What, as a fraction in simplest form, is the ratio of the cost of radio advertising to the total cost of radio and television advertising?

Your strategy

Your solution

Solutions on p. S11

4.1 Exercises

Objective A

Write the comparison as a ratio in simplest form using a fraction, a colon (:), and the word *to*.

1. 3 pints to 15 pints

2. 6 pounds to 8 pounds

3. $40 to $20

4. 10 feet to 2 feet

5. 3 miles to 8 miles

6. 2 hours to 3 hours

7. 37 hours to 24 hours

8. 29 inches to 12 inches

9. 6 minutes to 6 minutes

10. 8 days to 12 days

11. 35 cents to 50 cents

12. 28 inches to 36 inches

13. 30 minutes to 60 minutes

14. 25 cents to 100 cents

15. 32 ounces to 16 ounces

16. 12 quarts to 4 quarts

17. 3 cups to 4 cups

18. 6 years to 7 years

19. $5 to $3

20. 30 yards to 12 yards

21. 12 quarts to 18 quarts

22. 20 gallons to 28 gallons

23. 14 days to 7 days

24. 9 feet to 3 feet

Objective B   *Application Problems*

Write ratios in simplest form using a fraction.

Family Budget						
Housing	*Food*	*Transportation*	*Taxes*	*Utilities*	*Miscellaneous*	*Total*
$800	$400	$300	$350	$150	$400	$2400

25. Use the table to find the ratio of housing cost to total expenses.

26. Use the table to find the ratio of food cost to total expenses.

27. Use the table to find the ratio of utilities cost to food cost.

28. Use the table to find the ratio of transportation cost to housing cost.

29. National Collegiate Athletic Association (NCAA) statistics show that for every 154,000 high school seniors playing basketball, only 4000 will play college basketball as first-year students. Write the ratio of the number of first-year students playing college basketball to the number of high school seniors playing basketball.

30. NCAA statistics show that for every 2800 college seniors playing college basketball, only 50 will play as rookies in the National Basketball Association. Write the ratio of the number of National Basketball Association rookies to the number of college seniors playing basketball.

31. A transformer has 40 turns in the primary coil and 480 turns in the secondary coil. State the ratio of the number of turns in the primary coil to the number of turns in the secondary coil.

32. Rita Sterling bought a computer system for $2400. Five years later she sold the computer for $900. Find the ratio of the amount she received for the computer to the cost of the computer.

33. A house with an original value of $90,000 increased in value to $110,000 in 5 years.
    a. Find the increase in the value of the house.
    b. What is the ratio of the increase in value to the original value of the house?

34. The price of gasoline jumped from $1.35 to $1.62 in 1 year.
    a. What was the increase in the price per gallon?
    b. What is the ratio of the increase in price to the original price?

APPLYING THE CONCEPTS

A bank uses the ratio of a borrower's total monthly debts to the borrower's total monthly income to determine the maximum monthly payment for a potential homeowner. This ratio is called the debt–income ratio. Use the homeowner's income–debt table at the right for Exercises 35 and 36.

Income	Debts
$3500	$900
$250	$170
$140	$160
	$95

35. Compute the debt–income ratio for the potential homeowner.

36. If Central Trust Bank will make a loan to a customer whose debt–income ratio is less than $\frac{1}{3}$, will the potential homeowner qualify? Explain your answer.

37. To make a home loan, First National Bank requires a debt–income ratio that is less than $\frac{2}{5}$. Would the homeowner whose income–debt table is given at the right qualify for a loan using these standards? Explain.

Income		Debts	
Salary	$3400	Mortgage	$1800
Interest	$83	Property tax	$104
Rent	$650	Insurance	$35
Dividends	$34	Liabilities	$120
		Credit card	$234
		Car loan	$197

38. Is the value of a ratio always less than 1? Explain.

4.2 Rates

Objective A   **To write rates**

Crime	Every
Larceny	4 seconds
Burglary	14 seconds
Robbery	60 seconds
Rape	6 minutes
Murder	31 minutes

A **rate** is a comparison of two quantities that have *different* units. A rate is written as a fraction.

A distance runner ran 26 miles in 4 hours. The distance-to-time rate is written

$$\frac{26 \text{ miles}}{4 \text{ hours}} = \frac{13 \text{ miles}}{2 \text{ hours}}$$

A rate is in **simplest form** when the numbers that form the rate have no common factors. Note that the units are written as part of the rate.

Example 1   Write "6 roof supports for every 9 feet" as a rate in simplest form.

Solution $\frac{6 \text{ supports}}{9 \text{ feet}} = \frac{2 \text{ supports}}{3 \text{ feet}}$

You Try It 1   Write "15 pounds of fertilizer for 12 trees" as a rate in simplest form.

Your solution

Solution on p. S11

Objective B   **To write unit rates**

A **unit rate** is a rate in which the number in the denominator is 1.

$\frac{\$3.25}{1 \text{ pound}}$ or \$3.25/pound is read "\$3.25 per pound."

To find unit rates, divide the number in the numerator of the rate by the number in the denominator of the rate.

A car traveled 344 miles on 16 gallons of gasoline. To find the miles per gallon (unit rate), divide the numerator of the rate by the denominator of the rate.

$\frac{344 \text{ miles}}{16 \text{ gallons}}$ is the rate.

$16\overline{)344.0}$ 21.5 miles/gallon is the unit rate.
21.5

Example 2   Write "300 feet in 8 seconds" as a unit rate.

Solution $\frac{300 \text{ feet}}{8 \text{ seconds}}$ $8\overline{)300.0}$
37.5

37.5 feet/second

You Try It 2   Write "260 miles in 8 hours" as a unit rate.

Your solution

Solution on p. S11

Objective C   **To solve application problems**

→ The table at the right shows typical air fares for long routes. Find the cost per mile for the four routes in order to determine the most expensive route and the least expensive route on the basis of mileage flown.

Long Routes	Miles	Fare
New York–Los Angeles	2475	$683
San Francisco–Dallas	1464	$536
Denver–Pittsburgh	1302	$525
Minneapolis–Hartford	1050	$483

Strategy
To find the cost per mile, divide the fare by the miles flown for each route. Compare the costs per mile to determine the most expensive and least expensive routes per mile.

Solution   New York–Los Angeles $\quad \dfrac{683}{2475} \approx 0.28$

San Francisco–Dallas $\quad \dfrac{536}{1464} \approx 0.37$

Denver–Pittsburgh $\quad \dfrac{525}{1302} \approx 0.40$

Minneapolis–Hartford $\quad \dfrac{483}{1050} = 0.46$

$0.28 < 0.37 < 0.40 < 0.46$

The Minneapolis–Hartford route is the most expensive per mile, and the New York–Los Angeles route is the least expensive per mile.

Example 3

As an investor, Jung Ho purchased 100 shares of stock for $1500. One year later, Jung sold the 100 shares for $1800. What was his profit per share?

Strategy
To find Jung's profit per share:

• Find the total profit by subtracting the original cost (1500) from the selling price (1800).
• Find the profit per share (unit rate) by dividing the total profit by the number of shares of stock (100).

Solution
$1800 - 1500 = 300$

$300 \div 100 = 3$

Jung Ho's profit was $3/share.

You Try It 3

Erik Peltier, a jeweler, purchased 5 ounces of gold for $1625. Later, he sold the 5 ounces for $1720. What was Erik's profit per ounce?

Your strategy

Your solution

Solution on p. S11

4.2 Exercises

Objective A

Write each phrase as a rate in simplest form.

1. 3 pounds of meat for 4 people

2. 30 ounces in 24 glasses

3. $80 for 12 boards

4. 84 cents for 6 bars of soap

5. 300 miles on 15 gallons

6. 88 feet in 8 seconds

7. 20 children in 8 families

8. 48 leaves on 9 plants

9. 16 gallons in 2 hours

10. 25 ounces in 5 minutes

Objective B

Write each phrase as a unit rate.

11. 10 feet in 4 seconds

12. 816 miles in 6 days

13. $1300 earned in 4 weeks

14. $27,000 earned in 12 months

15. 1100 trees planted on 10 acres

16. 3750 words on 15 pages

17. $32.97 earned in 7 hours

18. $315.70 earned in 22 hours

19. 628.8 miles in 12 hours

20. 388.8 miles in 8 hours

21. 344.4 miles on 12.3 gallons of gasoline

22. 409.4 miles on 11.5 gallons of gasoline

23. $349.80 for 212 pounds

24. $11.05 for 3.4 pounds

Objective C   *Application Problems*

25. An automobile was driven 326.6 miles on 11.5 gallons of gas. Find the number of miles driven per gallon of gas.

26. You drive 246.6 miles in 4.5 hours. Find the number of miles you drove per hour.

27. The Saturn-5 rocket uses 534,000 gallons of fuel in 2.5 minutes. How much fuel does the rocket use per minute?

28. Assume that Regency Computer produced 5000 zip disks for $26,536.32. Of the disks made, 122 did not meet company standards.
  a. How many disks did meet company standards?
  b. What was the cost per disk for those disks that met company standards?

29. The Pierre family purchased a 250-pound side of beef for $365.75 and had it packaged. During the packaging, 75 pounds of beef were discarded as waste.
  a. How many pounds of beef were packaged?
  b. What was the cost per pound for the packaged beef?

30. The Bear Valley Fruit Stand purchased 250 boxes of strawberries for $162.50. All the strawberries were sold for $312.50. What was the profit per box of strawberries?

31. During filming, an IMAX camera uses 65-mm film at a rate of 5.6 feet per second.
  a. At what rate per minute does the camera go through film?
  b. At what rate does it use a 500-foot roll of 65-mm film? Round to the nearest second.

32. The table at the right shows the population and area of three countries. The population density of a country is the number of people per square mile.
  a. Which country has the least population density?
  b. How many more people per square mile are there in India than in the United States? Round to the nearest whole number.

Country	Population	Area (in square miles)
Australia	18,322,000	2,968,000
India	936,546,000	1,269,000
United States	267,000,000	3,619,000

Another application of rates is in the area of international trade. Suppose a company in Canada purchases a shipment of sneakers from an American company. The Canadian company must exchange the Canadian dollar for U.S. dollars in order to pay for the order. The number of Canadian dollars that are equivalent to one U.S. dollar is called the **exchange rate.**

33. The table at the right shows the exchange rates per U.S. dollar for three foreign countries and for the Euro at the time of this writing.
  a. How many euros would be paid for an order of American computer hardware costing $38,000?
  b. Calculate the cost, in Japanese yen, of an American car costing $24,000.

Exchange Rates per U.S. Dollar	
Australian Dollar	1.741
Canadian Dollar	1.477
Japanese Yen	106.500
The Euro	1.118

APPLYING THE CONCEPTS

34. You have a choice of receiving a wage of $34,000 per year, $2840 per month, $650 per week, or $18 per hour. Which pay choice would you take? Assume a 40-hour week with 52 weeks per year.

35. The price–earnings ratio of a company's stock is one measure used by stock market analysts to assess the financial well-being of the company. Explain the meaning of the price–earnings ratio.

4.3 Proportions

Objective A   **To determine whether a proportion is true**

Point of Interest

Proportions were studied by the earliest mathematicians. Clay tablets uncovered by archaeologists show evidence of proportions in Egyptian and Babylonian cultures dating from 1800 B.C.

A **proportion** is an expression of the equality of two ratios or rates.

$$\frac{50 \text{ miles}}{4 \text{ gallons}} = \frac{25 \text{ miles}}{2 \text{ gallons}}$$

Note that the units of the numerators are the same and the units of the denominators are the same.

$$\frac{3}{6} = \frac{1}{2}$$

This is the equality of two ratios.

A proportion is **true** if the fractions are equal when written in lowest terms.

In any true proportion, the "cross products" are equal.

➡ Is $\frac{2}{3} = \frac{8}{12}$ a true proportion?

$$\frac{2}{3} \diagdown\diagup \frac{8}{12} \quad \begin{array}{l} 3 \times 8 = 24 \\ 2 \times 12 = 24 \end{array}$$

The cross products *are* equal.

$\frac{2}{3} = \frac{8}{12}$ is a true proportion.

A proportion is **not true** if the fractions are not equal when reduced to lowest terms.

If the cross products are not equal, then the proportion is not true.

➡ Is $\frac{4}{5} = \frac{8}{9}$ a true proportion?

$$\frac{4}{5} \diagdown\diagup \frac{8}{9} \quad \begin{array}{l} 5 \times 8 = 40 \\ 4 \times 9 = 36 \end{array}$$

The cross products *are not* equal.

$\frac{4}{5} = \frac{8}{9}$ is not a true proportion.

Example 1

Is $\frac{5}{8} = \frac{10}{16}$ a true proportion?

Solution

$$\frac{5}{8} \diagdown\diagup \frac{10}{16} \quad \begin{array}{l} 8 \times 10 = 80 \\ 5 \times 16 = 80 \end{array}$$

The cross products are equal.
The proportion is true.

You Try It 1

Is $\frac{6}{10} = \frac{9}{15}$ a true proportion?

Your solution

Example 2

Is $\frac{62 \text{ miles}}{4 \text{ gallons}} = \frac{33 \text{ miles}}{2 \text{ gallons}}$ a true proportion?

Solution

$$\frac{62}{4} \diagdown\diagup \frac{33}{2} \quad \begin{array}{l} 4 \times 33 = 132 \\ 62 \times 2 = 124 \end{array}$$

The cross products are not equal.
The proportion is not true.

You Try It 2

Is $\frac{\$32}{6 \text{ hours}} = \frac{\$90}{8 \text{ hours}}$ a true proportion?

Your solution

Solutions on p. S11

Objective B **To solve proportions**

Sometimes one of the numbers in a proportion is unknown. In this case, it is necessary to *solve* the proportion.

To **solve** a proportion, find a number to replace the unknown so that the proportion is true.

➡ Solve: $\dfrac{9}{6} = \dfrac{3}{n}$

$$\dfrac{9}{6} = \dfrac{3}{n}$$

$9 \times n = 6 \times 3$ • Find the cross products.

$9 \times n = 18$

$n = 18 \div 9$ • Think of $9 \times n = 18$ as $9\overline{)18}$.

$n = 2$

Check:

$$\dfrac{9}{6} \bowtie \dfrac{3}{2} \quad \begin{array}{l} \to 6 \times 3 = 18 \\ \to 9 \times 2 = 18 \end{array}$$

Example 3

Solve $\dfrac{n}{12} = \dfrac{25}{60}$ and check.

Solution

$n \times 60 = 12 \times 25$
$n \times 60 = 300$
$n = 300 \div 60$
$n = 5$

Check:

$$\dfrac{5}{12} \bowtie \dfrac{25}{60} \quad \begin{array}{l} \to 12 \times 25 = 300 \\ \to 5 \times 60 = 300 \end{array}$$

You Try It 3

Solve $\dfrac{n}{14} = \dfrac{3}{7}$ and check.

Your solution

Example 4

Solve $\dfrac{4}{9} = \dfrac{n}{16}$. Round to the nearest tenth.

Solution

$4 \times 16 = 9 \times n$
$64 = 9 \times n$
$64 \div 9 = n$
$7.1 \approx n$

Note: A rounded answer is an approximation. Therefore, the answer to a check will not be exact.

You Try It 4

Solve $\dfrac{5}{7} = \dfrac{n}{20}$. Round to the nearest tenth.

Your solution

Solutions on pp. S11–S12

Example 5

Solve $\frac{28}{52} = \frac{7}{n}$ and check.

Solution

$28 \times n = 52 \times 7$
$28 \times n = 364$
$\quad\quad n = 364 \div 28$
$\quad\quad n = 13$

Check:

$\frac{28}{52} \,\rangle\!\!\langle\, \frac{7}{13}$ → $52 \times 7 = 364$
→ $28 \times 13 = 364$

You Try It 5

Solve $\frac{15}{20} = \frac{12}{n}$ and check.

Your solution

Example 6

Solve $\frac{15}{n} = \frac{8}{3}$. Round to the nearest hundredth.

Solution

$15 \times 3 = n \times 8$
$\quad\quad 45 = n \times 8$
$45 \div 8 = n$
$\quad 5.63 \approx n$

You Try It 6

Solve $\frac{12}{n} = \frac{7}{4}$. Round to the nearest hundredth.

Your solution

Example 7

Solve $\frac{n}{9} = \frac{3}{1}$ and check.

Solution

$n \times 1 = 9 \times 3$
$n \times 1 = 27$
$\quad\quad n = 27 \div 1$
$\quad\quad n = 27$

Check:

$\frac{27}{9} \,\rangle\!\!\langle\, \frac{3}{1}$ → $9 \times 3 = 27$
→ $27 \times 1 = 27$

You Try It 7

Solve $\frac{n}{12} = \frac{4}{1}$ and check.

Your solution

Example 8

Solve $\frac{5}{9} = \frac{15}{n}$ and check.

Solution

$5 \times n = 9 \times 15$
$5 \times n = 135$
$\quad\quad n = 135 \div 5$
$\quad\quad n = 27$

Check:

$\frac{5}{9} \,\rangle\!\!\langle\, \frac{15}{27}$ → $9 \times 15 = 135$
→ $5 \times 27 = 135$

You Try It 8

Solve $\frac{3}{8} = \frac{12}{n}$ and check.

Your solution

Solutions on p. S12

Objective C   **To solve application problems**

. . . .

Example 9

A mason determines that 9 cement blocks are required for a retaining wall 2 feet long. At this rate, how many cement blocks are required for a retaining wall that is 24 feet long?

Strategy

To find the number of cement blocks for a retaining wall 24 feet long, write and solve a proportion using n to represent the number of blocks required.

Solution

$$\frac{9 \text{ cement blocks}}{2 \text{ feet}} = \frac{n \text{ cement blocks}}{24 \text{ feet}}$$

$$9 \times 24 = 2 \times n$$
$$216 = 2 \times n$$
$$216 \div 2 = n$$
$$108 = n$$

A 24-foot retaining wall requires 108 cement blocks.

You Try It 9

Twenty-four jars can be packed in 6 identical boxes. At this rate, how many jars can be packed in 15 boxes?

Your strategy

Your solution

. . . .

Example 10

The dosage of a certain medication is 2 ounces for every 50 pounds of body weight. How many ounces of this medication are required for a person who weighs 175 pounds?

Strategy

To find the number of ounces of medication for a person weighing 175 pounds, write and solve a proportion using n to represent the number of ounces of medication for a 175-pound person.

Solution

$$\frac{2 \text{ ounces}}{50 \text{ pounds}} = \frac{n \text{ ounces}}{175 \text{ pounds}}$$

$$2 \times 175 = 50 \times n$$
$$350 = 50 \times n$$
$$350 \div 50 = n$$
$$7 = n$$

A 175-pound person requires 7 ounces of medication.

You Try It 10

Three tablespoons of a liquid plant fertilizer are to be added to every 4 gallons of water. How many tablespoons of fertilizer are required for 10 gallons of water?

Your strategy

Your solution

Solutions on p. S12

4.3 Exercises

Objective A

Determine whether the proportion is true or not true.

1. $\dfrac{4}{8} = \dfrac{10}{20}$

2. $\dfrac{39}{48} = \dfrac{13}{16}$

3. $\dfrac{7}{8} = \dfrac{11}{12}$

4. $\dfrac{15}{7} = \dfrac{17}{8}$

5. $\dfrac{27}{8} = \dfrac{9}{4}$

6. $\dfrac{3}{18} = \dfrac{4}{19}$

7. $\dfrac{45}{135} = \dfrac{3}{9}$

8. $\dfrac{3}{4} = \dfrac{54}{72}$

9. $\dfrac{16}{3} = \dfrac{48}{9}$

10. $\dfrac{15}{5} = \dfrac{3}{1}$

11. $\dfrac{7}{40} = \dfrac{7}{8}$

12. $\dfrac{9}{7} = \dfrac{6}{5}$

13. $\dfrac{50 \text{ miles}}{2 \text{ gallons}} = \dfrac{25 \text{ miles}}{1 \text{ gallon}}$

14. $\dfrac{16 \text{ feet}}{10 \text{ seconds}} = \dfrac{24 \text{ feet}}{15 \text{ seconds}}$

15. $\dfrac{6 \text{ minutes}}{5 \text{ cents}} = \dfrac{30 \text{ minutes}}{25 \text{ cents}}$

16. $\dfrac{16 \text{ pounds}}{12 \text{ days}} = \dfrac{20 \text{ pounds}}{14 \text{ days}}$

17. $\dfrac{\$15}{4 \text{ pounds}} = \dfrac{\$45}{12 \text{ pounds}}$

18. $\dfrac{270 \text{ trees}}{6 \text{ acres}} = \dfrac{90 \text{ trees}}{2 \text{ acres}}$

19. $\dfrac{300 \text{ feet}}{4 \text{ rolls}} = \dfrac{450 \text{ feet}}{7 \text{ rolls}}$

20. $\dfrac{1 \text{ gallon}}{4 \text{ quarts}} = \dfrac{7 \text{ gallons}}{28 \text{ quarts}}$

21. $\dfrac{\$65}{5 \text{ days}} = \dfrac{\$26}{2 \text{ days}}$

22. $\dfrac{80 \text{ miles}}{2 \text{ hours}} = \dfrac{110 \text{ miles}}{3 \text{ hours}}$

23. $\dfrac{7 \text{ tiles}}{4 \text{ feet}} = \dfrac{42 \text{ tiles}}{20 \text{ feet}}$

24. $\dfrac{15 \text{ feet}}{3 \text{ yards}} = \dfrac{90 \text{ feet}}{18 \text{ yards}}$

Objective B

Solve. Round to the nearest hundredth.

25. $\dfrac{n}{4} = \dfrac{6}{8}$

26. $\dfrac{n}{7} = \dfrac{9}{21}$

27. $\dfrac{12}{18} = \dfrac{n}{9}$

28. $\dfrac{7}{21} = \dfrac{35}{n}$

29. $\dfrac{6}{n} = \dfrac{24}{36}$

30. $\dfrac{3}{n} = \dfrac{15}{10}$

31. $\dfrac{n}{45} = \dfrac{17}{135}$

32. $\dfrac{9}{4} = \dfrac{18}{n}$

33. $\dfrac{n}{6} = \dfrac{2}{3}$

34. $\dfrac{5}{12} = \dfrac{n}{144}$

35. $\dfrac{n}{5} = \dfrac{7}{8}$

36. $\dfrac{4}{n} = \dfrac{9}{5}$

37. $\dfrac{n}{11} = \dfrac{32}{4}$

38. $\dfrac{3}{4} = \dfrac{8}{n}$

39. $\dfrac{5}{12} = \dfrac{n}{8}$

40. $\dfrac{36}{20} = \dfrac{12}{n}$

41. $\dfrac{n}{15} = \dfrac{21}{12}$

42. $\dfrac{40}{n} = \dfrac{15}{8}$

43. $\dfrac{32}{n} = \dfrac{1}{3}$

44. $\dfrac{5}{8} = \dfrac{42}{n}$

45. $\dfrac{18}{11} = \dfrac{16}{n}$

46. $\dfrac{25}{4} = \dfrac{n}{12}$

47. $\dfrac{28}{8} = \dfrac{12}{n}$

48. $\dfrac{n}{30} = \dfrac{65}{120}$

49. $\dfrac{0.3}{5.6} = \dfrac{n}{25}$

50. $\dfrac{1.3}{16} = \dfrac{n}{30}$

51. $\dfrac{0.7}{9.8} = \dfrac{3.6}{n}$

52. $\dfrac{1.9}{7} = \dfrac{13}{n}$

Objective C   *Application Problems*

Solve. Round to the nearest hundredth.

53. A 6-ounce package of Puffed Wheat contains 600 calories. How many calories are in a 0.5-ounce serving of the cereal?

54. A car travels 70.5 miles on 3 gallons of gas. Find the distance that the car can travel on 14 gallons of gas.

55. Ron Stokes uses 2 pounds of fertilizer for every 100 square feet of lawn for landscape maintenance. At this rate, how many pounds of fertilizer did he use on a lawn that measures 3500 square feet?

56. A nursery prepares a liquid plant food by adding 1 gallon of water for each 2 ounces of plant food. At this rate, how many gallons of water are required for 25 ounces of plant food?

57. A manufacturer of baseball equipment makes 4 aluminum bats for every 15 bats made from wood. On a day when 100 aluminum bats are made, how many wooden bats are produced?

58. A brick wall 20 feet in length contains 1040 bricks. At the same rate, how many bricks would it take to build a wall 48 feet in length?

59. The scale on the map at the right is "1.25 inches equals 10 miles." Find the distance between Carlsbad and Del Mar, which are 2 inches apart on the map.

60. The scale on the plans for a new house is "1 inch equals 3 feet." Find the length and width of a room that measures 5 inches by 8 inches on the drawing.

61. The dosage for a medication is $\frac{1}{3}$ ounce for every 40 pounds of body weight. At this rate, how many ounces of medication should a physician prescribe for a patient who weighs 150 pounds?

62. A bank requires a monthly payment of $33.45 on a $2500 loan. At the same rate, find the monthly payment on a $10,000 loan.

63. A pre-election survey showed that 2 out of every 3 eligible voters would cast ballots in the county election. At this rate, how many people in a county of 240,000 eligible voters would vote in the election?

64. A paint manufacturer suggests using 1 gallon of paint for every 400 square feet of a wall. At this rate, how many gallons of paint would be required for a room that has 1400 square feet of wall?

65. A 60-year-old male can obtain $10,000 of life insurance for $35.35 per month. At this rate, what is the monthly cost of $50,000 of life insurance?

66. Suppose a computer chip manufacturer knows from experience that in an average production run of 2000 circuit boards, 60 will be defective. How many defective circuit boards can be expected from a run of 25,000 circuit boards?

67. You own 240 shares of a computer stock. The company declares a stock split of 5 shares for every 3 owned. How many shares of stock will you own after the stock split?

68. The director of data processing at a college estimates that the ratio of student time to administrative time on a certain computer is 3:2. During a month in which the computer was used 200 hours for administration, how many hours was it used by students?

69. The ratio of weight on the moon to weight on Earth is 1:6. If a bowling ball weighs 16 pounds on Earth, what would it weigh on the moon?

70. When engineers designed a new car, they first built a model of the car. The ratio of the size of a part on the model to the actual size of the part is 2:5. If a door is 1.3 feet long on the model, what is the length of the door on the car?

71. Carlos Capasso owns 50 shares of Texas Utilities that pay dividends of $153. At this rate, what dividend would Carlos receive after buying 300 additional shares of Texas Utilities?

APPLYING THE CONCEPTS

The table at the right shows how each dollar of projected spending by the federal government for a recent year was distributed. The costs for Social Security, health, income security, net interest, and the category education, veterans, and transportation are fixed expenditures. Defense and "Other" represent the only discretionary spending by the government. The projected budget for the year was $1977.6 billion.

How Your Federal Tax Dollar Is Spent	
Social Security	23 cents
Health	21 cents
Income security	14 cents
Net interest	11 cents
Edu., Vets, Transp.	10 cents
Defense	16 cents
Other	5 cents

Source: Tax Foundation, Office of Management and Budget

72. a. Is at least one-half of federal spending discretionary spending?
   b. Using a colon, write the ratio of the expenditures for "health" to the discretionary spending.
   c. Find the amount of the budget to be spent on fixed expenditures.
   d. Find the amount of the budget to be spent on Social Security.

73. A survey of voters in a city claimed that 2 people of every 5 who voted cast a ballot in favor of city amendment A and that 3 people of every 4 who voted cast a ballot against amendment A. Is this possible? Explain your answer.

74. Write a word problem that requires solving a proportion to find the answer.

Focus on Problem Solving

Looking for a Pattern

A very useful problem-solving strategy is looking for a pattern.

Problem   A legend says that a peasant invented the game of chess and gave it to a very rich king as a present. The king so enjoyed the game that he gave the peasant the choice of anything in the kingdom. The peasant's request was simple: "Place one grain of wheat on the first square, 2 grains on the second square, 4 grains on the third square, 8 on the fourth square, and continue doubling the number of grains until the last square of the chessboard is reached." How many grains of wheat must the king give the peasant?

Solution   A chessboard consists of 64 squares. To find the total number of grains of wheat on the 64 squares, we begin by looking at the amount of wheat on the first few squares.

Square 1	Square 2	Square 3	Square 4	Square 5	Square 6	Square 7	Square 8
1	2	4	8	16	32	64	128
1	3	7	15	31	63	127	255

The bottom row of numbers represents the sum of the number of grains of wheat up to and including that square. For instance, the number of grains of wheat on the first 7 squares is $1 + 2 + 4 + 8 + 16 + 32 + 64 = 127$.

One pattern to observe is that the number of grains of wheat on a square can be expressed as a power of 2.

The number of grains on square $n = 2^{n-1}$.

For example, the number of grains on square $7 = 2^{7-1} = 2^6 = 64$.

A second pattern of interest is that **the number *below* a square** (the total number of grains up to and including that square) **is one less than the number of grains of wheat *on the next* square.** For example, the number *below* square 7 is one less than the number *on* square 8 ($128 - 1 = 127$). From this observation, the number of grains of wheat on the first 8 squares is the number on square 8 (128) plus one less than the number on square 8 (127): The total number of grains of wheat on the first 8 squares is $128 + 127 = 255$.

From this observation,

$$\begin{array}{c}\text{Number of grains of}\\ \text{wheat on the chessboard}\end{array} = \begin{array}{c}\text{number of grains}\\ \text{on square 64}\end{array} + \begin{array}{c}\text{one less than the number}\\ \text{of grains on square 64}\end{array}$$

$$= 2^{64-1} + (2^{64-1} - 1)$$

$$= 2^{63} + 2^{63} - 1 \approx 18{,}000{,}000{,}000{,}000{,}000{,}000$$

To give you an idea of the magnitude of this number, this is more wheat than has been produced in the world since chess was invented.

The same king decided to have a banquet in the long banquet room of the palace to celebrate the invention of chess. The king had 50 square tables, and each table could seat only one person on each side. The king pushed the tables together to form one long banquet table. How many people could sit at this table? *Hint:* Try constructing a pattern by using 2 tables, 3 tables, and 4 tables.

Projects and Group Activities

The Golden Ratio   There are certain designs that have been repeated over and over in both art and architecture. One of these involves the **golden rectangle.**

A golden rectangle is drawn at the right. Begin with a square that measures, say, 2 inches on a side. Now measure the distance from *A* to *B*. Place this length along the bottom of the square, starting at *A*. The resulting rectangle is a golden rectangle.

The **golden ratio** is the ratio of the length of the golden rectangle to its width. If you have drawn the rectangle following the procedure above, you will find that the golden ratio is approximately 1.6.

The golden ratio appears in many different situations. Some historians claim that some of the great pyramids of Egypt are based on the golden ratio. The drawing at the right shows the Pyramid of Gizeh, which dates from approximately 2600 B.C. The ratio of the height to a side of the base is approximately 1.6.

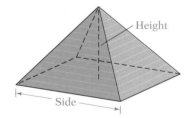

1. The canvas of the Mona Lisa painted by Leonardo da Vinci is a golden rectangle. However, there are other instances of the golden rectangle in the painting itself. Do some research on this painting and write a few paragraphs summarizing your findings.

2. What do 3 × 5 and 5 × 8 index cards have to do with the golden rectangle?

3. What does the United Nations Building in New York City have to do with the golden rectangle?

4. When was the Parthenon in Athens, Greece, built? What does the front of that building have to do with the golden rectangle?

Drawing the Floor Plans for a Building

The drawing at the left is a sketch of the floor plan for a cabin at a resort in the mountains of Utah. The measurements are missing. Assume that you are the architect and will finish the drawing. You will have to decide the size of the rooms and put in the measurements to scale.

Design a cabin that you would like to own. Select a scale and draw all the rooms to scale.

If you are interested in architecture, visit an architect who is using **CAD** (computer-aided design) to create a floor plan. Computer technology has revolutionized the field of architectural design.

Small Business and Gross Income

Do some research on your local newspaper. Answer each of the following questions.

1. What is the newspaper's circulation?
2. How often is an edition of the newspaper published?
3. What is the cost per issue when it is purchased at a newsstand?
4. Is there a special rate for subscribers to the newspaper?
5. What is the cost of a subscription?
6. What is the length of time for which a subscription is paid?
7. How many of the newspaper's readers are subscribers?
8. Use the answers to the above questions to determine the number of copies sold at newsstands.
9. Use the figures that you have gathered to determine the total annual income, or **gross income,** derived from sales of the newspaper.

The U.S. House of Representatives   The framers of the Constitution decided to use a ratio to determine the number of representatives from each state. It was determined that each state would have one representative for every 30,000 citizens, with a minimum of one representative. Congress has changed this ratio over the years, and we now have 435 representatives.

Find the number of representatives from your state. Determine the ratio of citizens to representatives. Also do this for the most populous state and for the least populous state.

 You might consider getting information on the number of representatives for each state and the populations of different states via the Internet.

Chapter Summary

Key Words   Quantities such as 8 feet and 60 miles are number quantities written with *units*. [p. 169]

A *ratio* is a comparison of two quantities that have the same units. [p. 169]

A ratio is in *simplest form* when the two numbers that make up the ratio have no common factors. [p. 169]

A *rate* is a comparison of two quantities that have different units. [p. 173]

A rate is in *simplest form* when the numbers that make up the rate have no common factors. [p. 173]

A *unit rate* is a rate in which the number in the denominator is 1. [p. 173]

A *proportion* is an expression of the equality of two ratios or rates. [p. 177]

A proportion is true if the fractions are equal when written in lowest terms. In any true proportion, the "cross products" are equal. [p. 177]

Essential Rules   **Ways to Express a Ratio**
A ratio can be written in three different ways:

1. As a fraction
2. As two numbers separated by a colon (:)
3. As two numbers separated by the word *to* [p. 169]

To Find Unit Rates
To find unit rates, divide the number in the numerator of the rate by the number in the denominator of the rate. [p. 173]

To Solve a Proportion
One of the numbers in a proportion may be unknown. To solve a proportion, find a number to replace the unknown so that the proportion is true. [p. 178]

Chapter Review

1. Determine whether the proportion is true or not true.
$$\frac{2}{9} = \frac{10}{45}$$

2. Write the comparison 32 dollars to 80 dollars as a ratio in simplest form using a fraction, a colon (:), and the word *to*.

3. Write "250 miles in 4 hours" as a unit rate.

4. Determine whether the proportion is true or not true.
$$\frac{8}{15} = \frac{32}{60}$$

5. Solve the proportion.
$$\frac{16}{n} = \frac{4}{17}$$

6. Write "$300 earned in 40 hours" as a unit rate.

7. Write "$8.75 for 5 pounds" as a unit rate.

8. Write the comparison 8 feet to 28 feet as a ratio in simplest form using a fraction, a colon (:), and the word *to*.

9. Solve the proportion.
$$\frac{n}{8} = \frac{9}{2}$$

10. Solve the proportion. Round to the nearest hundredth.
$$\frac{18}{35} = \frac{10}{n}$$

11. Write the comparison 6 inches to 15 inches as a ratio in simplest form using a fraction, a colon (:), and the word *to*.

12. Determine whether the proportion is true or not true.
$$\frac{3}{8} = \frac{10}{24}$$

13. Write "$15 in 4 hours" as a rate in simplest form.

14. Write "326.4 miles on 12 gallons" as a unit rate.

15. Write the comparison 12 days to 12 days as a ratio in simplest form using a fraction, a colon (:), and the word *to*.

16. Determine whether the proportion is true or not true.
$$\frac{5}{7} = \frac{25}{35}$$

17. Solve the proportion. Round to the nearest hundredth.

$$\frac{24}{11} = \frac{n}{30}$$

18. Write "100 miles in 3 hours" as a rate in simplest form.

19. In 5 years, the price of a calculator went from $40 to $24. What is the ratio, as a fraction in simplest form, of the decrease in price to the original price?

20. The property tax on a $45,000 home is $900. At the same rate, what is the property tax on a home valued at $120,000?

21. The high temperature during a 24-hour period was 84 degrees, and the low temperature was 42 degrees. Write the ratio, as a fraction in simplest form, of the high temperature to the low temperature for the 24-hour period.

22. The total cost of manufacturing 1000 radios was $36,600. Of the radios made, 24 did not pass inspection. Find the cost per radio that the manufacturer must now assume in pricing the radios. That is, what is the cost per radio of the radios that *did* pass inspection?

23. A brick wall 40 feet in length contains 448 concrete blocks. At the same rate, how many blocks would it take to build a wall that is 120 feet in length?

24. A retail computer store spends $30,000 a year on TV advertising and $12,000 on newspaper advertising. Find the ratio, as a fraction in simplest form, of TV advertising to newspaper advertising.

25. A 15-pound turkey costs $10.20. What is the cost per pound?

26. Mahesh drove 198.8 miles in 3.5 hours. Find the average number of miles he drove per hour.

27. An insurance policy costs $3.87 for every $1000 of insurance. At this rate, what is the cost of $50,000 of insurance?

28. Pascal Hollis purchased 80 shares of stock for $3580. What was the cost per share?

29. Monique uses 1.5 pounds of fertilizer for every 200 square feet of lawn. How many pounds of fertilizer will she have to use on a lawn that measures 3000 square feet?

30. A house had an original value of $80,000, but its value increased to $120,000 in 2 years. Find the ratio, as a fraction in simplest form, of the increase to the original value.

Chapter Test

1. Write "$22,036.80 earned in 12 months" as a unit rate.

2. Write the comparison 40 miles to 240 miles as a ratio in simplest form using a fraction, a colon (:), and the word *to*.

3. Write "18 supports for every 8 feet" as a rate in simplest form.

4. Determine whether the proportion is true or not true.
$$\frac{40}{125} = \frac{5}{25}$$

5. Write the comparison 12 days to 8 days as a ratio in simplest form using a fraction, a colon (:), and the word *to*.

6. Solve the proportion.
$$\frac{5}{12} = \frac{60}{n}$$

7. Write "256.2 miles on 8.4 gallons of gas" as a unit rate.

8. Write the comparison 27 dollars to 81 dollars as a ratio in simplest form using a fraction, a colon (:), and the word *to*.

9. Determine whether the proportion is true or not true.
$$\frac{5}{14} = \frac{25}{70}$$

10. Solve the proportion.
$$\frac{n}{18} = \frac{9}{4}$$

11. Write "$81 for 12 boards" as a rate in simplest form.

12. Write the comparison 18 feet to 30 feet as a ratio in simplest form using a fraction, a colon (:), and the word *to*.

13. Fifty shares of a utility stock pay a dividend of $62.50. At the same rate, find the dividend paid on 500 shares of the utility stock.

14. The average summer temperature in a California desert is 112 degrees. In a city 100 miles away, the average summer temperature is 86 degrees. Write the ratio, as a fraction in simplest form, of the average city temperature to the average desert temperature.

15. A plane travels 2421 miles in 4.5 hours. Find the plane's speed in miles per hour.

16. A research scientist estimates that the human body contains 88 pounds of water for every 100 pounds of body weight. At this rate, estimate the number of pounds of water in a college student who weighs 150 pounds.

17. If 40 feet of lumber costs $69.20, what is the per-foot cost of the lumber?

18. The dosage of a certain medication is $\frac{1}{4}$ ounce for every 50 pounds of body weight. How many ounces of this medication are required for a person who weighs 175 pounds? Write the answer as a decimal.

19. An automobile sales company spends $25,000 each month for television advertising and $40,000 each month for radio advertising. Write, as a fraction in simplest form, the ratio of the cost of radio advertising to the total cost of advertising.

20. The property tax on a house valued at $175,000 is $1500. At the same rate, find the property tax on a house valued at $140,000.

Cumulative Review

1. Subtract: 20,095
   − 10,937

2. Write $2 \cdot 2 \cdot 2 \cdot 2 \cdot 3 \cdot 3 \cdot 3$ in exponential notation.

3. Simplify: $4 - (5 - 2)^2 \div 3 + 2$

4. Find the prime factorization of 160.

5. Find the LCM of 9, 12, and 18.

6. Find the GCF of 28 and 42.

7. Write $\frac{40}{64}$ in simplest form.

8. Find $4\frac{7}{15}$ more than $3\frac{5}{6}$.

9. What is $4\frac{5}{9}$ less than $10\frac{1}{6}$?

10. Multiply: $\frac{11}{12} \times 3\frac{1}{11}$

11. Find the quotient of $3\frac{1}{3}$ and $\frac{5}{7}$.

12. Simplify: $\left(\frac{2}{5} + \frac{3}{4}\right) \div \frac{3}{2}$

13. Write 4.0709 in words.

14. Round 2.09762 to the nearest hundredth.

15. Divide: $8.09\overline{)16.0976}$
    Round to the nearest thousandth.

16. Convert $0.06\frac{2}{3}$ to a fraction.

17. Write the comparison 25 miles to 200 miles as a ratio in simplest form using a fraction.

18. Write "87 cents for 6 pencils" as a rate in simplest form.

19. Write "250.5 miles on 7.5 gallons of gas" as a unit rate.

20. Solve $\dfrac{40}{n} = \dfrac{160}{17}$.

21. A car traveled 457.6 miles in 8 hours. Find the car's speed in miles per hour.

22. Solve the proportion.

$$\frac{12}{5} = \frac{n}{15}$$

23. You had $1024 in your checking account. You then wrote checks for $192 and $88. What is your new checking account balance?

24. Malek Khatri buys a tractor for $22,760. A down payment of $5000 is required. The balance remaining is paid in 48 equal monthly installments. What is the monthly payment?

25. Yuko is assigned to read a book containing 175 pages. She reads $\dfrac{2}{5}$ of the book during Thanksgiving vacation. How many pages of the assignment remain to be read?

26. A building contractor bought $2\dfrac{1}{3}$ acres of land for $84,000. What was the cost of each acre?

27. Benjamin Eli bought a shirt for $22.79 and a tie for $9.59. He used a $50 bill to pay for the purchases. Find the amount of change.

28. If you earn an annual salary of $41,619, what is your monthly salary?

29. A soil conservationist estimates that a river bank is eroding at the rate of 3 inches every 6 months. At this rate, how many inches will be eroded in 50 months?

30. The dosage of a certain medication is $\dfrac{1}{2}$ ounce for every 50 pounds of body weight. How many ounces of this medication are required for a person who weighs 160 pounds? Write the answer as a decimal.

Chapter 5 Percents

Objectives

Section 5.1
A To write a percent as a fraction or a decimal
B To write a fraction or a decimal as a percent

Section 5.2
A To find the amount when the percent and the base are given
B To solve application problems

Section 5.3
A To find the percent when the base and amount are given
B To solve application problems

Section 5.4
A To find the base when the percent and amount are given
B To solve application problems

Section 5.5
A To solve percent problems using proportions
B To solve application problems

The number of children in this kindergarten class represents a percent of the population of a school district. Local populations and estimates of growth or decline are used to predict future job markets in that area. For example, population growth may indicate a need for more teachers in the local schools. The demography **project on page 219** describes different segments of the U.S. population as percents of the total population.

Need help? For on-line student resources, such as section quizzes, visit this textbook's web site at **math.college.hmco.com/students.**

For Exercises 1 to 6, multiply or divide.

1. $19 \times \dfrac{1}{100}$

2. 23×0.01

3. 0.47×100

4. $0.06 \times 47{,}500$

5. $60 \div 0.015$

6. $8 \div \dfrac{1}{4}$

7. Multiply $\dfrac{5}{8} \times 100$. Write the answer as a decimal.

8. Write $\dfrac{200}{3}$ as a mixed number.

9. Divide $28 \div 16$. Write the answer as a decimal.

Go Figure

A palindrome is a whole number that remains unchanged when its digits are written in reverse order. For example, 818 is a palindrome.

a. Find the smallest three-digit multiple of 6 that is a palindrome.

b. Many nonpalindrome numbers can be converted into palindromes: reverse the digits of the number and add the result to the original; continue until a palindrome is achieved. Using this process, what is the palindrome created from 874?

5.1 Introduction to Percents

Objective A   **To write a percent as a fraction or a decimal**

Percent means "parts of 100." In the figure at the right, there are 100 parts. Because 13 of the 100 parts are shaded, 13% of the figure is shaded.

In most applied problems involving percents, it is necessary either to rewrite a percent as a fraction or a decimal or to rewrite a fraction or a decimal as a percent.

To write a percent as a fraction, remove the percent sign and multiply by $\frac{1}{100}$.

$$13\% = 13 \times \frac{1}{100} = \frac{13}{100}$$

To rewrite a percent as a decimal, remove the percent sign and multiply by 0.01.

$$13\% \quad = \quad 13 \times 0.01 \quad = \quad 0.13$$

Move the decimal point two places to the left. Then remove the percent sign.

Example 1   Write 120% as a fraction and as a decimal.

Solution
$$120\% = 120 \times \frac{1}{100} = \frac{120}{100}$$
$$= 1\frac{1}{5}$$
$$120\% = 120 \times 0.01 = 1.2$$

Note that percents larger than 100 are greater than 1.

You Try It 1   Write 125% as a fraction and as a decimal.

Your solution

Example 2

Write $16\frac{2}{3}\%$ as a fraction.

Solution
$$16\frac{2}{3}\% = 16\frac{2}{3} \times \frac{1}{100}$$
$$= \frac{50}{3} \times \frac{1}{100} = \frac{50}{300} = \frac{1}{6}$$

You Try It 2

Write $33\frac{1}{3}\%$ as a fraction.

Your solution

Example 3   Write 0.5% as a decimal.

Solution $0.5\% = 0.5 \times 0.01 = 0.005$

You Try It 3   Write 0.25% as a decimal.

Your solution

Solutions on p. S12

Objective B   **To write a fraction or a decimal as a percent**

A fraction or a decimal can be written as a percent by multiplying by 100%.

➡ Write $\frac{3}{8}$ as a percent.

$$\frac{3}{8} = \frac{3}{8} \times 100\% = \frac{3}{8} \times \frac{100}{1}\% = \frac{300}{8}\% = 37\frac{1}{2}\% \text{ or } 37.5\%$$

➡ Write 0.37 as a percent.

$$0.37 \quad = \quad 0.37 \times 100\% \quad = \quad 37\%$$

Move the decimal point two places to the right. Then write the percent sign.

Example 4   Write 0.015 as a percent.

Solution   0.015 = 0.015 × 100%
            = 1.5%

You Try It 4   Write 0.048 as a percent.

Your solution

Example 5   Write 2.15 as a percent.

Solution   2.15 = 2.15 × 100% = 215%

You Try It 5   Write 3.67 as a percent.

Your solution

Example 6

Write $0.33\frac{1}{3}$ as a percent.

Solution $0.33\frac{1}{3} = 0.33\frac{1}{3} \times 100\%$

$= 33\frac{1}{3}\%$

You Try It 6

Write $0.62\frac{1}{2}$ as a percent.

Your solution

Example 7

Write $\frac{2}{3}$ as a percent.
Write the remainder in fractional form.

Solution $\frac{2}{3} = \frac{2}{3} \times 100\% = \frac{200}{3}\%$

$= 66\frac{2}{3}\%$

You Try It 7

Write $\frac{5}{6}$ as a percent.
Write the remainder in fractional form.

Your solution

Example 8

Write $2\frac{2}{7}$ as a percent.
Round to the nearest tenth.

Solution $2\frac{2}{7} = \frac{16}{7} = \frac{16}{7} \times 100\%$

$= \frac{1600}{7}\% \approx 228.6\%$

You Try It 8

Write $1\frac{4}{9}$ as a percent.
Round to the nearest tenth.

Your solution

Solutions on pp. S12–S13

5.1 Exercises

Objective A

Write as a fraction and as a decimal.

1. 25%   **2.** 40%   **3.** 130%   **4.** 150%

5. 100%   **6.** 87%   **7.** 73%   **8.** 45%

9. 383%   **10.** 425%   **11.** 70%   **12.** 55%

13. 88%   **14.** 64%   **15.** 32%   **16.** 18%

Write as a fraction.

17. $66\frac{2}{3}\%$ **18.** $12\frac{1}{2}\%$ **19.** $83\frac{1}{3}\%$ **20.** $3\frac{1}{8}\%$ **21.** $11\frac{1}{9}\%$ **22.** $\frac{3}{8}\%$

23. $45\frac{5}{11}\%$ **24.** $15\frac{3}{8}\%$ **25.** $4\frac{2}{7}\%$ **26.** $5\frac{3}{4}\%$ **27.** $6\frac{2}{3}\%$ **28.** $8\frac{2}{3}\%$

Write as a decimal.

29. 6.5%   **30.** 9.4%   **31.** 12.3%   **32.** 16.7%   **33.** 0.55%

34. 0.45%   **35.** 8.25%   **36.** 6.75%   **37.** 5.05%   **38.** 3.08%

39. 2%   **40.** 7%   **41.** 80.4%   **42.** 36.2%   **43.** 4.9%

Objective B

Write as a percent.

44. 0.16   **45.** 0.73   **46.** 0.05   **47.** 0.01   **48.** 1.07   **49.** 2.94

50. 0.004   **51.** 0.006   **52.** 1.012   **53.** 3.106   **54.** 0.8   **55.** 0.7

Write as a percent. Round to the nearest tenth.

56. $\dfrac{27}{50}$ **57.** $\dfrac{37}{100}$ **58.** $\dfrac{1}{3}$ **59.** $\dfrac{2}{5}$

60. $\dfrac{5}{8}$ **61.** $\dfrac{1}{8}$ **62.** $\dfrac{1}{6}$ **63.** $1\dfrac{1}{2}$

64. $\dfrac{7}{40}$ **65.** $1\dfrac{2}{3}$ **66.** $1\dfrac{7}{9}$ **67.** $\dfrac{7}{8}$

Write as a percent. Write the remainder in fractional form.

68. $\dfrac{15}{50}$ **69.** $\dfrac{12}{25}$ **70.** $\dfrac{7}{30}$ **71.** $\dfrac{1}{3}$

72. $2\dfrac{3}{8}$ **73.** $1\dfrac{2}{3}$ **74.** $2\dfrac{1}{6}$ **75.** $\dfrac{7}{8}$

76. Write the part of the square that is shaded as a fraction, as a decimal, and as a percent. Write the part of the square that is not shaded as a fraction, as a decimal, and as a percent.

APPLYING THE CONCEPTS

77. Determine whether the statement is true or false. If the statement is false, give an example to show that the statement is false.
a. Multiplying a number by a percent always decreases the number.
b. Dividing by a percent always increases the number.
c. The word *percent* means "per hundred."

78. Explain in your own words how to change a percent to a decimal and how to change a decimal to a percent.

79. A sale on computers advertised $\dfrac{1}{3}$ off the regular price. What percent of the regular price does this represent?

80. A suit was priced at 50% off the regular price. What fraction of the regular price does this represent?

81. If $\dfrac{2}{5}$ of the population voted in an election, what percent of the population did not vote?

82. How can you recognize a fraction that represents a number that is less than 1%?

5.2 Percent Equations: Part I

Objective A

To find the amount when the percent and the base are given

A real estate broker receives a payment that is 4% of an $85,000 sale. To find the amount the broker receives requires answering the question "4% of $85,000 is what?"

This sentence can be written using mathematical symbols and then solved for the unknown number.

4%	of	$85,000	is	what?
↓	↓	↓	↓	↓

$$\boxed{\begin{matrix}\text{percent} \\ 4\%\end{matrix}} \times \boxed{\begin{matrix}\text{base} \\ 85{,}000\end{matrix}} = \boxed{\begin{matrix}\text{amount} \\ n\end{matrix}}$$

of is written as × (times)
is is written as = (equals)
what is written as n (the unknown number)

$$0.04 \quad \times \quad 85{,}000 \quad = \quad n$$
$$3400 \quad = \quad n$$

Note that the percent is written as a decimal.

The broker receives a payment of $3400.

The solution was found by solving the basic percent equation for amount.

The Basic Percent Equation

$$\boxed{\text{Percent}} \times \boxed{\text{base}} = \boxed{\text{amount}}$$

In most cases, the percent is written as a decimal before the basic percent equation is solved. However, some percents are more easily written as a fraction than as a decimal. For example,

$$33\tfrac{1}{3}\% = \tfrac{1}{3} \qquad 66\tfrac{2}{3}\% = \tfrac{2}{3} \qquad 16\tfrac{2}{3}\% = \tfrac{1}{6} \qquad 83\tfrac{1}{3}\% = \tfrac{5}{6}$$

Example 1   Find 5.7% of 160.

Solution $n = 0.057 \times 160$
$n = 9.12$

Note that the words "what is" are missing from the problem but are implied by the word "Find."

You Try It 1   Find 6.3% of 150.

Your solution

Example 2

What is $33\tfrac{1}{3}\%$ of 90?

Solution $n = \tfrac{1}{3} \times 90$
$n = 30$

You Try It 2

What is $16\tfrac{2}{3}\%$ of 66?

Your solution

Solutions on p. S13

Objective B   **To solve application problems**

Solving percent problems requires identifying the three elements of the basic percent equation. Recall that these three parts are the *percent,* the *base,* and the *amount.* Usually the base follows the phrase "percent of."

During a recent year, charitable contributions totaled $190 billion. The table at the right shows where this money was donated. Use these data for Example 3 and You Try It 3.

Recipients	Percent of Total Charitable Contributions
Religion	43%
Education	14%
Health	9%
Other	34%

Source: Giving USA 2000/AAFRC Trust for Philanthropy

Example 3

How much money was donated to religious organizations?

Strategy

To find the amount donated, write and solve the basic percent equation using n to represent the amount. The percent is 43%. The base is $190 billion.

Solution $43\% \times 190 = n$
$0.43 \times 190 = n$
$81.7 = n$

$87.1 billion was donated to religious organizations.

You Try It 3

Determine how much money was donated to educational organizations.

Your strategy

Your solution

Example 4

A quality control inspector found that 1.2% of 2500 telephones inspected were defective. How many telephones inspected were not defective?

Strategy

To find the number of nondefective phones:

• Find the number of defective phones. Write and solve the basic percent equation using n to represent the number of defective phones (amount). The percent is 1.2% and the base is 2500.
• Subtract the number of defective phones from the number of phones inspected (2500).

Solution $1.2\% \times 2500 = n$
$0.012 \times 2500 = n$
$30 = n$ defective phones

$2500 - 30 = 2470$

2470 telephones were not defective.

You Try It 4

An electrician's hourly wage was $23.50 before an 8% raise. What is the new hourly wage?

Your strategy

Your solution

Solutions on p. S13

5.2 Exercises

Objective A

Solve.

1. 8% of 100 is what?

2. 16% of 50 is what?

3. 27% of 40 is what?

4. 52% of 95 is what?

5. 0.05% of 150 is what?

6. 0.075% of 625 is what?

7. 125% of 64 is what?

8. 210% of 12 is what?

9. Find 10.7% of 485.

10. Find 12.8% of 625.

11. What is 0.25% of 3000?

12. What is 0.06% of 250?

13. 80% of 16.25 is what?

14. 26% of 19.5 is what?

15. What is $1\frac{1}{2}$% of 250?

16. What is $5\frac{3}{4}$% of 65?

17. $16\frac{2}{3}$% of 120 is what?

18. $83\frac{1}{3}$% of 246 is what?

19. What is $33\frac{1}{3}$% of 630?

20. What is $66\frac{2}{3}$% of 891?

21. Which is larger: 5% of 95, or 75% of 6?

22. Which is larger: 112% of 5, or 0.45% of 800?

23. Which is smaller: 79% of 16, or 20% of 65?

24. Which is smaller: 15% of 80, or 95% of 15?

25. Which is smaller: 2% of 1500, or 72% of 40?

26. Which is larger: 22% of 120, or 84% of 32?

27. Find 31.294% of 82,460.

28. Find 123.94% of 275,976.

Objective B   *Application Problems*

29. Approximately 30% of the 44 million people in the United States who do not have health insurance are between the ages of 18 and 24. (Source: Census Bureau) About how many people in the United States aged 18 to 24 do not have health insurance?

30. During January of 2000, approximately 28.4% of the total auto industry sales of 1,212,321 vehicles were General Motors sales. (Source: Autodata) Find the approximate number of vehicles sold by General Motors in January of 2000.

During the first quarter of 2000, approximately 11 million personal computers were sold in the United States. The table at the right shows the leading PC makers and their percents of those sales. Use these data for Exercises 31 and 32.

PC Makers	Percent of Sales
Dell	17%
Compaq	16%
Hewlett-Packard	13%
Gateway	9%

GartnerGroup's Dataquest, April 2000

31. How many of the 11 million personal computers were sold by Dell?

32. How many more personal computers were sold by Compaq than by Gateway? Give the answer in thousands.

33. A sales tax of 6% of the cost of a car was added to the purchase price of $19,500.
a. How much was the sales tax?
b. What is the total cost of the car, including sales tax?

34. During the packaging process for oranges, spoiled oranges are discarded by an inspector. In one day an inspector found that 4.8% of the 20,000 pounds of oranges inspected were spoiled.
a. How many pounds of oranges were spoiled?
b. How many pounds of oranges were not spoiled?

35. In 1995, Americans ate an average of 236 eggs during the year. In 2000, that number increased by approximately 9%. (Source: USDA Economic Research Service)
a. Find the increase in the number of eggs eaten annually from 1995 to 2000. Round to the nearest whole number.
b. Americans ate an average of how many eggs during 2000?

36. Funtimes Amusement Park has 550 employees and must hire an additional 22% for the vacation season. What is the total number of employees needed for the vacation season?

APPLYING THE CONCEPTS

The two circle graphs at the right show how surveyed employees actually spend their time and the way they would prefer to spend their time. Assuming that employees have 112 hours a week of time that is not spent sleeping, answer Exercises 37 to 39. Round to the nearest tenth of an hour.

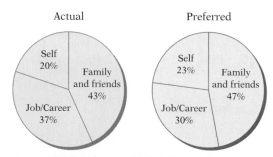

Source: WSJ Supplement, Work & Family; from *Families and Work Institute*

37. What is the actual number of hours per week that employees spend with family and friends?

38. What is the number of hours per week that employees would prefer to spend on job/career?

39. What is the difference between the number of hours an employee preferred to spend on self and the actual amount of time the employee spent on self?

5.3 Percent Equations: Part II

Objective A   **To find the percent when the base and amount are given**

A recent promotional game at a grocery store listed the probability of winning a prize as "1 chance in 2." A percent can be used to describe the chance of winning. This requires answering the question "What percent of 2 is 1?"

The chance of winning can be found by solving the basic percent equation for *percent*.

What percent of 2 is 1?

$$\text{percent } n \times \text{base } 2 = \text{amount } 1$$

$$n \times 2 = 1$$
$$n = 1 \div 2$$
$$n = 0.5$$
$$n = 50\%$$

• The solution must be written as a percent to answer the question.

There is a 50% chance of winning a prize.

Example 1   What percent of 40 is 30?

Solution
$$n \times 40 = 30$$
$$n = 30 \div 40$$
$$n = 0.75$$
$$n = 75\%$$

You Try It 1   What percent of 32 is 16?

Your solution

Example 2   What percent of 12 is 27?

Solution
$$n \times 12 = 27$$
$$n = 27 \div 12$$
$$n = 2.25$$
$$n = 225\%$$

You Try It 2   What percent of 15 is 48?

Your solution

Example 3   25 is what percent of 75?

Solution
$$25 = n \times 75$$
$$25 \div 75 = n$$
$$\frac{1}{3} = n$$
$$33\frac{1}{3}\% = n$$

You Try It 3   30 is what percent of 45?

Your solution

Solutions on p. S13

Objective B   **To solve application problems**

To solve percent problems, remember that it is necessary to identify the percent, base, and amount. Usually the base follows the phrase "percent of."

Example 4

The monthly house payment for the Kaminski family is $787.50. What percent of the Kaminskis' monthly income of $3750 is the house payment?

Strategy

To find what percent of the income the house payment is, write and solve the basic percent equation using n to represent the percent. The base is $3750 and the amount is $787.50.

Solution

$n \times 3750 = 787.50$
$n = 787.50 \div 3750$
$n = 0.21 = 21\%$

The house payment is 21% of the monthly income.

Example 5

On one Monday night, 31.39 million of the approximately 40.76 million households watching television were not watching David Letterman. What percent of these households were watching David Letterman? Round to the nearest percent.

Strategy

To find the percent of households watching David Letterman:

• Subtract to find the number of households that were watching David Letterman (40.76 million − 31.39 million).

• Write and solve the basic percent equation using n to represent the percent. The base is 40.76, and the amount is the number of households watching David Letterman.

Solution

40.76 million − 31.39 million = 9.37 million

9.37 million households watching David Letterman

$n \times 40.76 = 9.37$
$n = 9.37 \div 40.76$
$n \approx 0.23$

Approximately 23% of the households were watching David Letterman.

You Try It 4

Tomo Nagata had an income of $33,500 and paid $5025 in income tax. What percent of the income is the income tax?

Your strategy

Your solution

You Try It 5

According to the U.S. Department of Defense, of the 518,921 enlisted personnel in the U.S. Army in 1950, 512,370 people were men. What percent of the enlisted personnel in the U.S. Army in 1950 were women? Round to the nearest tenth of a percent.

Your strategy

Your solution

Solutions on p. S13

5.3 Exercises

Objective A

Solve.

1. What percent of 75 is 24?

2. What percent of 80 is 20?

3. 15 is what percent of 90?

4. 24 is what percent of 60?

5. What percent of 12 is 24?

6. What percent of 6 is 9?

7. What percent of 16 is 6?

8. What percent of 24 is 18?

9. 18 is what percent of 100?

10. 54 is what percent of 100?

11. 5 is what percent of 2000?

12. 8 is what percent of 2500?

13. What percent of 6 is 1.2?

14. What percent of 2.4 is 0.6?

15. 16.4 is what percent of 4.1?

16. 5.3 is what percent of 50?

17. 1 is what percent of 40?

18. 0.3 is what percent of 20?

19. What percent of 48 is 18?

20. What percent of 11 is 88?

21. What percent of 2800 is 7?

22. What percent of 400 is 12?

23. 4.2 is what percent of 175?

24. 41.79 is what percent of 99.5?

25. What percent of 86.5 is 8.304?

26. What percent of 1282.5 is 2.565?

Objective B   *Application Problems*

27. Seven in ten couples disagree about financial issues. (Source: Yankelovich Partners for Lutheran Brotherhood) What percent of couples disagree about financial matters?

28. In a survey, 1236 adults nationwide were asked, "What irks you most about the actions of other motorists?" The response "tailgaters" was given by 293 people. (Source: Reuters/Zogby) What percent of those surveyed were most irked by tailgaters? Round to the nearest tenth of a percent.

29. According to the U.S. Department of Agriculture, of the 63 billion pounds of vegetables produced in the United States in one year, 16 billion pounds were wasted. What percent of vegetables produced were wasted? Round to the nearest tenth of a percent.

30. In a recent year, 2400 billion cigarettes were produced. China National Tobacco Corporation produced the largest number, approximately 590.4 billion. (Source: World Tobacco File 1998–2000) What percent of the total cigarette production did China National Tobacco Corporation produce?

31. The typical American household spends $1355 a year on energy utilities. Of this amount, approximately $81.30 is spent on lighting. (Source: Department of Energy, Owens Corning) What percent of the total amount spent on energy utilities is spent on lighting?

32. To receive a license to sell insurance, an insurance account executive must answer correctly 70% of the 250 questions on a test. Nicholas Mosley answered 177 questions correctly. Did he pass the test?

33. On average, an employee in the United States spends 3.1 hours per workday at a computer. (Source: John H. Heldrich Center for Workforce Development, Rutgers University; Center for Survey Research and Analysis, University of Connecticut; *Work Trends*, February 10, 2000) What percent of the workday is not spent at a computer? Use an 8-hour workday.

34. In a test of the breaking strength of concrete slabs for freeway construction, 3 of the 200 slabs tested did not meet safety requirements. What percent of the slabs did meet safety requirements?

APPLYING THE CONCEPTS

The graph at the right shows several categories of average lifetime costs of dog ownership. Use this graph for Exercises 35 and 36. Round answers to the nearest tenth of a percent.

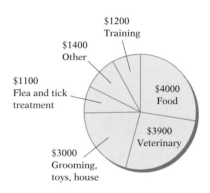

$1200 Training
$1400 Other
$1100 Flea and tick treatment
$4000 Food
$3900 Veterinary
$3000 Grooming, toys, house

35. What percent of the total amount is spent on food?

36. What percent of the total is spent on veterinary care?

37. Public utility companies will provide consumers with an analysis of their energy bills. For one customer, it was determined that $76 of a total bill of $134 was spent for home heating. What percent (to the nearest tenth) of the total bill was for home heating?

38. The Fun in the Sun organization claims to have taken a survey of 350 people, asking them to give their favorite outdoor temperature for hiking. The results are given in the table at the right. Explain why these results are not possible.

Favorite Temperature	Percent
Greater than 90	5%
80–89	28%
70–79	35%
60–69	32%
Below 60	13%

5.4 Percent Equations: Part III

Objective A   To find the base when the percent and amount are given

In 1780, the population of Virginia was approximately 538,000. This was 19% of the total population of the United States at that time. To find the total population at that time, you must answer the question "19% of what number is 538,000?"

The population of the United States in 1780 can be found by solving the basic percent equation for the base.

19%   of what   is   538,000?
↓   ↓   ↓

| percent 19% | base n | amount 538,000 |

$$0.19 \times n = 538,000$$
$$n = 538,000 \div 0.19$$
$$n \approx 2,832,000$$

The population of the United States in 1780 was approximately 2,832,000.

Example 1   18% of what is 900?

Solution
$$0.18 \times n = 900$$
$$n = 900 \div 0.18$$
$$n = 5000$$

You Try It 1   86% of what is 215?

Your solution

Example 2   30 is 1.5% of what?

Solution
$$30 = 0.015 \times n$$
$$30 \div 0.015 = n$$
$$2000 = n$$

You Try It 2   15 is 2.5% of what?

Your solution

Example 3
$$33\frac{1}{3}\% \text{ of what is 7?}$$

Solution
$$\frac{1}{3} \times n = 7$$

• Note that the percent is written as a fraction.

$$n = 7 \div \frac{1}{3}$$
$$n = 21$$

You Try It 3
$$16\frac{2}{3}\% \text{ of what is 5?}$$

Your solution

Solutions on pp. S13–S14

Objective B   To solve application problems

To solve percent problems, it is necessary to identify the percent, base, and amount. Usually the base follows the phrase "percent of."

Example 4

A business office bought a used copy machine for $900, which was 75% of the original cost. What was the original cost of the copier?

Strategy

To find the original cost of the copier, write and solve the basic percent equation, using n to represent the original cost (base). The percent is 75% and the amount is $900.

Solution

$75\% \times n = 900$
$0.75 \times n = 900$
$n = 900 \div 0.75$
$n = 1200$

The original cost of the copier was $1200.

You Try It 4

A used car has a value of $10,458, which is 42% of the car's original value. What was the car's original value?

Your strategy

Your solution

Example 5

A carpenter's wage this year is $26.40 per hour, which is 110% of last year's wage. What was the increase in the hourly wage over last year?

Strategy

To find the increase in the hourly wage over last year:

- Find last year's wage. Write and solve the basic percent equation, using n to represent last year's wage (base). The percent is 110% and the amount is $26.40.
- Subtract last year's wage from this year's wage (26.40).

Solution

$110\% \times n = 26.40$
$1.10 \times n = 26.40$
$n = 26.40 \div 1.10$
$n = 24.00$ • **Last year's wage**

$26.40 - 24.00 = 2.40$

The increase in the hourly wage was $2.40.

You Try It 5

Chang's Sporting Goods has a tennis racket on sale for $89.60, which is 80% of the original price. What is the difference between the original price and the sale price?

Your strategy

Your solution

Solutions on p. S14

5.4 Exercises

Objective A

Solve. Round to the nearest hundredth.

1. 12% of what is 9?

2. 38% of what is 171?

3. 8 is 16% of what?

4. 54 is 90% of what?

5. 10 is 10% of what?

6. 37 is 37% of what?

7. 30% of what is 25.5?

8. 25% of what is 21.5?

9. 2.5% of what is 30?

10. 10.4% of what is 52?

11. 125% of what is 24?

12. 180% of what is 21.6?

13. 18 is 240% of what?

14. 24 is 320% of what?

15. 4.8 is 15% of what?

16. 87.5 is 50% of what?

17. 25.6 is 12.8% of what?

18. 45.014 is 63.4% of what?

19. 0.7% of what is 0.56?

20. 0.25% of what is 1?

21. 30% of what is 2.7?

22. 78% of what is 3.9?

23. 84 is $16\frac{2}{3}$% of what?

24. 120 is $33\frac{1}{3}$% of what?

25. $66\frac{2}{3}$% of what is 72?

26. $83\frac{1}{3}$% of what is 13.5?

Objective B   *Application Problems*

27. Of the travelers who, during a recent year, allowed their children to miss school to go along on a trip, approximately 1.738 million allowed their children to miss school for more than a week. This represented 11% of the travelers who allowed their children to miss school. (Source: Travel Industry Association) About how many travelers allowed their children to miss school to go along on a trip?

28. In the United States today, 23.1% of the women and 27.5% of the men have attained a bachelor's or graduate degree. (Source: Census Bureau) How many women in the United States have earned a bachelor's or graduate degree?

29. The average cost of attending a four-year public college in 1999–2000 was $3356. This was 103.4% of the cost for the 1998–1999 school year. (Source: College Board) What was the average cost of attending a four-year public college in 1998–1999? Round to the nearest dollar.

30. The Internal Revenue Service says that the average deduction for medical expenses for taxpayers in the $40,000–$50,000 bracket is $4500. This is 26% of the medical expenses claimed by taxpayers in the over-$200,000 bracket. How much is the average deduction for medical expenses claimed by taxpayers in the over-$200,000 bracket? Round to the nearest thousand.

31. During a quality control test, Micronics found that 24 computer boards were defective. This amount was 0.8% of the computer boards tested.
a. How many computer boards were tested?
b. How many computer boards tested were not defective?

32. Of the calls a directory assistance operator received, 441 were requests for telephone numbers listed in the current directory. This accounted for 98% of the calls for assistance that the operator received.
a. How many calls did the operator receive?
b. How many telephone numbers requested were not listed in the current directory?

33. In the New Hampshire presidential primary election on February 1, 2000, Al Gore received 76,897 votes. This represented approximately 52% of the Democratic votes cast. (Source: www.state.nh.us) How many of the Democratic votes were not for Al Gore?

APPLYING THE CONCEPTS

The table at the right contains nutrition information about a breakfast cereal. Solve Exercises 34 and 35 using information from this table.

34. The recommended daily amount of thiamin for an adult is 1.5 milligrams. Find the amount of thiamin in one serving of this cereal with skim milk.

35. The amount of copper in one serving of cereal with skim milk is 0.08 milligram. Find the recommended daily amount of copper for an adult.

36. Increase a number by 10%. Now decrease the number by 10%. Is the result the original number? Explain.

37. When a company goes bankrupt, another company may offer to purchase the assets of the company for "25 cents on the dollar." What percent of the value of the company is the buyer paying?

NUTRITION INFORMATION

SERVING SIZE: 1.4 OZ WHEAT FLAKES WITH 0.4 OZ. RAISINS: 39.4 g. ABOUT 1/2 CUP
SERVINGS PER PACKAGE: 14

	CEREAL & RAISINS	WITH 1/2 CUP VITAMINS A & D SKIM MILK

PERCENTAGE OF U.S. RECOMMENDED DAILY ALLOWANCES (U.S. RDA)

	CEREAL & RAISINS	WITH 1/2 CUP SKIM MILK
PROTEIN	4	15
VITAMIN A	15	20
VITAMIN C	**	2
THIAMIN	25	30
RIBOFLAVIN	25	35
NIACIN	25	35
CALCIUM	**	15
IRON	100	100
VITAMIN D	10	25
VITAMIN B₆	25	25
FOLIC ACID	25	25
VITAMIN B₁₂	25	30
PHOSPHOROUS	10	15
MAGNESIUM	10	20
ZINC	25	30
COPPER	2	4

* 2% MILK SUPPLIES AN ADDITIONAL 20 CALORIES. 2 g FAT, AND 10 mg CHOLESTEROL.
** CONTAINS LESS THAN 2% OF THE U.S. RDA OF THIS NUTRIENT

Percent Problems: Proportion Method

Objective A   **To solve percent problems using proportions**

Problems that can be solved using the basic percent equation can also be solved using proportions.

The proportion method is based on writing two ratios. One ratio is the percent ratio, written as $\frac{\text{percent}}{100}$. The second ratio is the amount-to-base ratio, written as $\frac{\text{amount}}{\text{base}}$. These two ratios form the proportion

$$\frac{\textbf{Percent}}{\textbf{100}} = \frac{\textbf{amount}}{\textbf{base}}$$

To use the proportion method, first identify the percent, the amount, and the base (the base usually follows the phrase "percent of").

What is 23% of 45?	What percent of 25 is 4?	12 is 60% of what number?
$\frac{23}{100} = \frac{n}{45}$	$\frac{n}{100} = \frac{4}{25}$	$\frac{60}{100} = \frac{12}{n}$
$23 \times 45 = 100 \times n$	$n \times 25 = 100 \times 4$	$60 \times n = 100 \times 12$
$1035 = 100 \times n$	$n \times 25 = 400$	$60 \times n = 1200$
$1035 \div 100 = n$	$n = 400 \div 25$	$n = 1200 \div 60$
$10.35 = n$	$n = 16\%$	$n = 20$

Example 1   15% of what is 7? Round to the nearest hundredth.

Solution
$$\frac{15}{100} = \frac{7}{n}$$
$$15 \times n = 100 \times 7$$
$$15 \times n = 700$$
$$n = 700 \div 15$$
$$n \approx 46.67$$

You Try It 1   26% of what is 22? Round to the nearest hundredth.

Your solution

Example 2   30% of 63 is what?

Solution
$$\frac{30}{100} = \frac{n}{63}$$
$$30 \times 63 = 100 \times n$$
$$1890 = 100 \times n$$
$$1890 \div 100 = n$$
$$18.90 = n$$

You Try It 2   16% of 132 is what?

Your solution

Solutions on p. S14

> *Objective B*   **To solve application problems**   « 5 »

. . . .

Example 3

An antiques dealer found that 86% of the 250 items that were sold during one month sold for under $1000. How many items sold for under $1000?

Strategy

To find the number of items that sold for under $1000, write and solve a proportion, using n to represent the number of items sold (amount) for less than $1000. The percent is 86% and the base is 250.

Solution

$$\frac{86}{100} = \frac{n}{250}$$
$$86 \times 250 = 100 \times n$$
$$21{,}500 = 100 \times n$$
$$21{,}500 \div 100 = n$$
$$215 = n$$

215 items sold for under $1000.

You Try It 3

Last year it snowed 64% of the 150 days of the ski season at a resort. How many days did it snow?

Your strategy

Your solution

. . . .

Example 4

In a test of the strength of nylon rope, 5 pieces of the 25 pieces tested did not meet the test standards. What percent of the nylon ropes tested did meet the standards?

Strategy

To find the percent of ropes tested that met the standards:

- Find the number of ropes that met the test standards (25 − 5).
- Write and solve a proportion, using n to represent the percent of ropes that met the test standards. The base is 25. The amount is the number of ropes that met the standards.

Solution

$25 - 5 = 20$ ropes met test standards

$$\frac{n}{100} = \frac{20}{25}$$
$$n \times 25 = 100 \times 20$$
$$n \times 25 = 2000$$
$$n = 2000 \div 25$$
$$n = 80\%$$

80% of the ropes tested did meet the test standards.

You Try It 4

Five ballpoint pens in a box of 200 were found to be defective. What percent of the pens were not defective?

Your strategy

Your solution

Solutions on p. S14

5.5 Exercises

Objective A

Solve.

1. 26% of 250 is what?

2. What is 18% of 150?

3. 37 is what percent of 148?

4. What percent of 150 is 33?

5. 68% of what is 51?

6. 126 is 84% of what?

7. What percent of 344 is 43?

8. 750 is what percent of 50?

9. 82 is 20.5% of what?

10. 2.4% of what is 21?

11. What is 6.5% of 300?

12. 96% of 75 is what?

13. 7.4 is what percent of 50?

14. What percent of 1500 is 693?

15. 50.5% of 124 is what?

16. What is 87.4% of 255?

17. 120% of what is 6?

18. 14 is 175% of what?

19. What is 250% of 18?

20. 325% of 4.4 is what?

21. 33 is 220% of what?

22. 160% of what is 40?

Objective B   *Application Problems*

23. A charitable organization spent $2940 for administrative expenses. This amount is 12% of the money it collected. What is the total amount of money that the organization collected?

24. A manufacturer of an anti-inflammatory drug claims that the drug will be effective for 6 hours. An independent testing service determined that the drug was effective for only 80% of the length of time claimed by the manufacturer. Find the length of time the drug will be effective as determined by the testing service.

25. The land area of North America is approximately 9,400,000 square miles. This represents approximately 16% of the total land area of the world. What is the approximate total land area of the world?

26. The Rincon Fire Department received 24 false alarms out of a total of 200 alarms received. What percent of the alarms received were false alarms?

27. 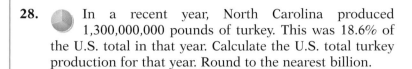 According to the National Education Association, elementary and secondary school revenues in 2000 totaled $335 billion. Use the figure at the right to determine how much of that amount came from the federal government.

State revenues 50%

Federal revenues 7%

Local/other revenues 43%

Elementary and Secondary School Revenues in 2000

Source: U.S. Census Bureau

28. In a recent year, North Carolina produced 1,300,000,000 pounds of turkey. This was 18.6% of the U.S. total in that year. Calculate the U.S. total turkey production for that year. Round to the nearest billion.

29. During one year, approximately 2,240,000 ounces of gold went into the manufacturing of electronic equipment in the United States. This is 16% of all the gold mined in the United States that year. How many ounces of gold were mined in the United States that year?

30. The table at the right shows the predicted increase in population from 2000 to 2040 for each of four counties in the Central Valley of California.
  a. What percent of the 2000 population of Sacramento County is the increase in population?
  b. What percent of the 2000 population of Kern County is the increase in population? Round to the nearest tenth of a percent.

County	2000 Population	Projected Increase
Sacramento	1,200,000	900,000
Kern	651,700	948,300
Fresno	794,200	705,800
San Joaquin	562,000	737,400

Source: California Department of Finance

APPLYING THE CONCEPTS

31. The graph at the right shows the causes of death for all police officers killed in the line of duty during a recent year. What percent of the deaths were due to traffic accidents? Round to the nearest tenth of a percent.

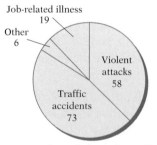

Job-related illness 19

Other 6

Violent attacks 58

Traffic accidents 73

Causes of Death for Police Officers Killed in the Line of Duty

Source: International Union of Police Associations

32. A certain solution of iodine is used to study the function of the thyroid gland. Approximately every 8 days, the potency of the solution is reduced by 50% of the amount present at that time. By what percent has the potency decreased after 24 days?

33. A certain solution of iron is used to study the formation of blood cells. Approximately every 15 hours, the potency of the solution is reduced by 50% of the amount present at that time. By what percent has the potency decreased after 60 hours?

Focus on Problem Solving

Using a Calculator as a Problem-Solving Tool

A calculator is an important tool for problem solving. Here are a few problems to solve with a calculator. You may need to research some of the questions to find information you do not know.

1. Choose any single-digit positive number. Multiply the number by 1507 and 7373. What is the answer? Choose another positive single-digit number and again multiply by 1507 and 7373. What is the answer? What pattern do you see? Why does this work?

2. The gross domestic product in 2000 was $9,963,100,000,000. Is this more or less than the amount of money that would be placed on the last square of a standard checkerboard if 1 cent were placed on the first square, 2 cents were placed on the second square, 4 cents were placed on the third square, 8 cents were placed on the fourth square, and so on until the 64th square was reached?

3. Which of the reciprocals of the first 16 natural numbers have a terminating-decimal representation and which have a repeating-decimal representation?

4. What is the largest natural number n for which $4^n > 1 \cdot 2 \cdot 3 \cdot 4 \cdot 5 \cdot \cdots \cdot n$?

5. If $1000 bills are stacked one on top of another, is the height of $1 billion less than or greater than the height of the Washington Monument?

6. What is the value of $1 + \cfrac{1}{1 + \cfrac{1}{1 + \cfrac{1}{1 + \cfrac{1}{1 + 1}}}}$?

7. Calculate 15^2, 35^2, 65^2, and 85^2. Study the results. Make a conjecture about a relationship between a number ending in 5 and its square. Use your conjecture to find 75^2 and 95^2. Does your conjecture work for 125^2?

8. Find the sum of the first 1000 natural numbers. (*Hint:* You could just start adding $1 + 2 + 3 + \cdots$, but even if you performed one operation every 3 seconds, it would take you an hour to find the sum. Instead, try pairing the numbers and then adding the pairs. Pair 1 and 1000, 2 and 999, 3 and 998, and so on. What is the sum of each pair? How many pairs are there? Use this information to answer the original question.)

9. To qualify for a home loan, a bank requires that the monthly mortgage payment be less than 25% of a borrower's monthly take-home income. A laboratory technician has deductions for taxes, insurance, and retirement that amount to 25% of the technician's monthly gross income. What minimum monthly income must this technician earn to receive a bank loan that has a mortgage payment of $1200 per month?

Projects and Group Activities

Consumer Price Index   The Consumer Price Index (CPI) is a percent that is written without the percent sign. For instance, a CPI of 160.1 means 160.1%. This number means that an item that cost $100 between 1982 and 1984 (the base years) would cost $160.10 today. Determining the cost is an application of the basic percent equation.

$$\text{Percent} \times \text{base} = \text{amount}$$
$$\text{CPI} \times \text{cost in base year} = \text{cost today}$$
$$1.601 \times 100 = 160.1 \qquad \bullet \; 160.1\% = 1.601$$

 The table below gives the CPI for various products in June of 2000. If you have Internet access, you can obtain current data for the items below, as well as other items not on this list, by visiting the web site of the Bureau of Labor Statistics.

Product	CPI
All items	172.3
Food and beverages	167.9
Housing	169.4
Clothes	128.3
Transportation	155.7
Medical care	260.5
Entertainment	103.4
Education	101.5

1. Of the items listed, are there any items that in 2000 cost more than twice as much as they cost during the base year? If so, which items?

2. Of the items listed, are there any items that in 2000 cost more than one-and-one-half times as much as they cost during the base years but less than twice as much as they cost during the base years? If so, which items?

3. If the cost for textbooks for one semester was $120 in the base years, how much did similar textbooks cost in 2000? Use the "Education" category.

4. If a new car cost $20,000 in 2000, what would a comparable new car have cost during the base years? Use the "Transportation" category.

5. If a movie ticket cost $8 in 2000, what would a comparable movie ticket have cost during the base years? Use the "Entertainment" category.

6. The base year for the CPI was 1967 before the change to 1982–1984. If 1967 were still used as the base year, the CPI for all items in 2000 (not just those listed above) would be 516.1.
   a. Using the base year of 1967, explain the meaning of a CPI of 516.1.
   b. Using the base year of 1967 and a CPI of 516.1, if textbooks cost $75 for one semester in 1967, how much did similar textbooks cost in 2000?
   c. Using the base year of 1967 and a CPI of 516.1, if a family's food budget in 2000 is $800 per month, what would a comparable family budget have been in 1967?

Demography   **Demography** is the statistical study of human populations. Many groups are interested in the size of certain segments of the population and in projections of population growth. For example, public school administrators want estimates of the number of school-age children who will be living in their districts 10 years from now.

The U.S. government provides estimates of what the population in the United States will be in the future. You can find these projections at the Census Bureau web site at www.census.gov. The table below contains data from that web site.

	Age:	Under 5	5–17	18–24	25–34	35–44	45–54	55–64	65 and Older
2010	Male	10,272	26,639	15,388	19,286	19,449	21,706	16,973	16,966
	Female	9,827	25,363	14,774	19,565	19,993	22,455	18,457	22,749
2050	Male	13,748	35,227	18,734	25,034	24,550	22,340	21,127	36,289
	Female	13,165	33,608	18,069	25,425	25,039	23,106	22,517	45,710

Population Projections for the United States, by Age and Sex (in thousands)

For these exercises, round all percents to the nearest tenth of a percent.

1. Which of the age groups listed are of interest to public school officials?

2. Which age group is of most interest to retirement home administrators?

3. Which age groups are of concern to accountants determining future benefits to be paid out by the Social Security Administration?

4. The population of which age group is of interest to manufacturers of disposable diapers?

5. The population of which age group is of primary concern to college and university admissions officers?

6. In which age groups do the males outnumber the females? In which do the females outnumber the males?

7. What percent of the projected population aged 65 and older in the year 2010 is female? Does this percent decrease in the projected population in 2050? If so, by how much?

8. Find the difference between the percent of the population that will be under 18 in 2010 and the percent that will be under 18 in 2050.

9. Assume that the work force consists of people aged 25 to 64. Find the decrease in the percents of the population for that group from 2010 to 2050.

10. Find the increase in the percents of the population for those aged 65 and over from 2010 to 2050.

11. Why are the sizes of the work force and the population aged 65 and over of special concern to the Social Security Administration?

12. Describe any patterns you see in the table.

13. Calculate a statistic based on the data in the table and explain why it would be of interest to an institution (such as a school system) or a manufacturer of consumer goods (such as baby food).

Health The American College of Sports Medicine (ACSM) recommends that you know how to determine your target heart rate in order to get the full benefit of exercise. Your **target heart rate** is the rate at which your heart should beat during any aerobic exercise such as running, cycling, fast walking, or participating in an aerobics class. According to the ACSM, you should reach your target rate and then maintain it for 20 minutes or more to achieve cardiovascular fitness. The intensity level varies for different individuals. A sedentary person might begin at the 60% level and gradually work up to 70%, whereas athletes and very fit individuals might work at the 85% level. The ACSM suggests that you calculate both 50% and 85% of your maximum heart rate. This will give you the low and high ends of the range within which your heart rate should stay.

To calculate your target heart rate:

	Example
Subtract your age from 220. This is your maximum heart rate.	$220 - 20 = 200$
Multiply your maximum heart rate by 50%. This is the low end of your range.	$200(0.50) = 100$
Divide the low end by 6. This is your low 10-second heart rate.	$100 \div 6 \approx 17$
Multiply your maximum heart rate by 85%. This is the high end of your range.	$200(0.85) = 170$
Divide the high end by 6. This is your high 10-second heart rate.	$170 \div 6 \approx 28$

1. Why are the low end and high end divided by 6 in order to determine the low and high 10-second heart rates?

2. Calculate your target heart rate, both the low and high end of your range.

3. What is the percent increase in your 10-second heart rate from the low end to the high end?

Chapter Summary

Key Words *Percent* means "parts of 100." [p. 197]

Essential Rules **To Write a Percent as a Fraction**
To write a percent as a fraction, remove the percent sign and multiply by $\frac{1}{100}$. [p. 197]

To Write a Percent as a Decimal
To write a percent as a decimal, remove the percent sign and multiply by 0.01. [p. 197]

To Write a Decimal as a Percent
To write a decimal as a percent, multiply by 100%. [p. 198]

To Write a Fraction as a Percent
To write a fraction as a percent, multiply by 100%. [p. 198]

Basic Percent Equation   Percent $\times$ base = amount [p. 201]

Proportion Method to Solve Percent Equations $\dfrac{\text{Percent}}{100} = \dfrac{\text{amount}}{\text{base}}$ [p. 213]

Chapter Review

1. What is 30% of 200?

2. 16 is what percent of 80?

3. Write $1\frac{3}{4}$ as a percent.

4. 20% of what is 15?

5. Write 12% as a fraction.

6. Find 22% of 88.

7. What percent of 20 is 30?

8. $16\frac{2}{3}$% of what is 84?

9. Write 42% as a decimal.

10. What is 7.5% of 72?

11. $66\frac{2}{3}$% of what is 105?

12. Write 7.6% as a decimal.

13. Find 125% of 62.

14. Write $16\frac{2}{3}$% as a fraction.

15. Use the proportion method to find what percent of 25 is 40.

16. 20% of what number is 15? Use the proportion method.

17. Write 0.38 as a percent.

18. 78% of what is 8.5? Round to the nearest tenth.

19. What percent of 30 is 2.2? Round to the nearest tenth of a percent.

20. What percent of 15 is 92? Round to the nearest tenth of a percent.

21. Trent missed 9 out of 60 questions on a history exam. What percent of the questions did he answer correctly? Use the proportion method.

22. A company used 7.5% of its $60,000 advertising budget for TV advertising. How much of the advertising budget was spent for TV advertising?

23. In a two-year period, the population of Houston, Texas, increased from 1,600,000 to 1,700,000. What percent increase does this represent?

24. Joshua purchased a video camera for $980 and paid a sales tax of 6.25% of the cost. What was the total cost of the video camera?

25. In a survey of 350 women and 420 men, 275 of the women and 300 of the men reported that they wore sunscreen often. To the nearest tenth of a percent, what percent of the women wore sunscreen often?

26. It is estimated that the world's population will be 9,100,000,000 by the year 2050. This is 149% of the population in 2000. (Source: U.S. Bureau of the Census) What was the world's population in 2000? Round to the nearest hundred million.

27. A computer system can be purchased for $1800. This is 60% of what the computer cost 4 years ago. What was the cost of the computer 4 years ago? Use the proportion method.

28. In one basketball game, Michael Jordan scored 33 points. This was 30% of all the points scored by the team. How many points were scored by the team?

Chapter Test

1. Write 97.3% as a decimal.

2. Write $83\frac{1}{3}\%$ as a fraction.

3. Write 0.3 as a percent.

4. Write 1.63 as a percent.

5. Write $\frac{3}{2}$ as a percent.

6. Write $\frac{2}{3}$ as a percent.

7. What is 77% of 65?

8. 47.2% of 130 is what?

9. Which is larger:
7% of 120, or 76% of 13?

10. Which is smaller:
13% of 200, or 212% of 12?

11. A fast-food company uses 6% of its $75,000 budget for advertising. What amount of the budget is spent on advertising?

12. During the packaging process for vegetables, spoiled vegetables are discarded by an inspector. In one day an inspector found that 6.4% of the 1250 pounds of vegetables were spoiled. How many pounds of vegetables were not spoiled?

The table at the right contains nutrition information about a breakfast cereal. Solve Exercises 13 and 14 with information taken from this table.

13. The recommended amount of potassium per day for an adult is 3000 milligrams (mg). What percent, to the nearest tenth of a percent, of the daily recommended amount of potassium is provided by one serving of this cereal with skim milk?

14. The daily recommended number of calories for a 190-pound man is 2200 calories. What percent, to the nearest tenth of a percent, of the daily recommended number of calories is provided by one serving of this cereal with 2% milk?

NUTRITION INFORMATION

SERVING SIZE: 1.4 OZ WHEAT FLAKES WITH 0.4 OZ. RAISINS: 39.4 g. ABOUT 1/2 CUP
SERVINGS PER PACKAGE: 14

	CEREAL & RAISINS	WITH 1/2 CUP VITAMINS A & D SKIM MILK
CALORIES	120	180
PROTEIN, g	3	7
CARBOHYDRATE, g	28	34
FAT, TOTAL, g	1	1*
UNSATURATED, g	1	
SATURATED, g	0	
CHOLESTEROL, mg	0	0*
SODIUM, mg	125	190
POTASSIUM, mg	240	440

* 2% MILK SUPPLIES AN ADDITIONAL 20 CALORIES.
2 g FAT, AND 10 mg CHOLESTEROL.
** CONTAINS LESS THAN 2% OF THE U.S. RDA OF THIS NUTRIENT

15. The Urban Center Department Store has 125 permanent employees and must hire an additional 20 temporary employees for the holiday season. What percent of the permanent employees is the number hired as temporary employees for the holiday season?

16. Conchita missed 7 out of 80 questions on a math exam. What percent of the questions did she answer correctly? (Round to the nearest tenth of a percent.)

17. 12 is 15% of what?

18. 42.5 is 150% of what? Round to the nearest tenth.

19. A manufacturer of transistors found 384 defective transistors during a quality control study. This amount was 1.2% of the transistors tested. Find the number of transistors tested.

20. A new house was bought for $95,000. Five years later the house sold for $152,000. The increase was what percent of the original price?

21. 123 is 86% of what number? Use the proportion method. Round to the nearest tenth.

22. What percent of 12 is 120? Use the proportion method.

23. A secretary receives a wage of $12.88 per hour. This amount is 112% of last year's wage. What is the dollar increase in the hourly wage over last year? Use the proportion method.

24. A city has a population of 71,500. Ten years ago the population was 32,500. The population now is what percent of what the population was 10 years ago? Use the proportion method.

25. The annual license fee on a car is 1.4% of the value of the car. If the license fee during a year is $91.00, what is the value of the car?

Cumulative Review

1. Simplify: $18 \div (7 - 4)^2 + 2$.

2. Find the LCM of 16, 24, and 30.

3. Find the sum of $2\frac{1}{3}$, $3\frac{1}{2}$, and $4\frac{5}{8}$.

4. Subtract: $27\frac{5}{12} - 14\frac{9}{16}$

5. Multiply: $7\frac{1}{3} \times 1\frac{5}{7}$

6. What is $\frac{14}{27}$ divided by $1\frac{7}{9}$?

7. Simplify: $\left(\frac{3}{4}\right)^3 \cdot \left(\frac{8}{9}\right)^2$

8. Simplify: $\left(\frac{2}{3}\right)^2 - \left(\frac{3}{8} - \frac{1}{3}\right) \div \frac{1}{2}$

9. Round 3.07973 to the nearest hundredth.

10. Subtract: 3.0902
                 $-$ 1.9706

11. Divide: $0.032\overline{)1.097}$
    Round to the nearest ten-thousandth.

12. Convert $3\frac{5}{8}$ to a decimal.

13. Convert 1.75 to a fraction.

14. Place the correct symbol, $<$ or $>$, between the two numbers.
    $\frac{3}{8}$ 0.87

15. Solve the proportion $\frac{3}{8} = \frac{20}{n}$. Round to the nearest tenth.

16. Write "$76.80 earned in 8 hours" as a unit rate.

17. Write $18\frac{1}{3}\%$ as a fraction.

18. Write $\frac{5}{6}$ as a percent.

19. 16.3% of 120 is what? Round to the nearest hundredth.

20. 24 is what percent of 18?

21. 12.4 is 125% of what?

22. What percent of 35 is 120? Round to the nearest tenth.

23. Sergio has an income of $740 per week. One-fifth of his income is deducted for income tax payments. Find his take-home pay.

24. Eunice bought a used car for $8353, with a down payment of $1000. The balance was paid in 36 equal monthly payments. Find the monthly payment.

25. The gasoline tax is $0.19 a gallon. Find the number of gallons of gasoline used during a month in which $79.80 was paid in gasoline taxes.

26. The real estate tax on a $72,000 home is $1440. At the same rate, find the real estate tax on a home valued at $150,000.

27. Ken purchased a stereo set for $490 and paid $29.40 in sales tax. What percent of the purchase price was the sales tax?

28. A survey of 300 people showed that 165 people favored a certain candidate for mayor. What percent of the people surveyed did not favor this candidate?

29. According to the Cabletelevision Advertising Bureau, cable households watch television 36.5% of the time. On average, how many hours per week do cable households spend watching TV? Round to the nearest tenth.

30. The Environmental Protection Agency found that 990 out of 5500 children tested had levels of lead in their blood exceeding federal guidelines. What percent of the children tested had levels of lead in the blood that exceeded federal standards?

6 Applications for Business and Consumers

Objectives

Your favorite cereal comes in two sizes, 12 ounces and 18 ounces. Would you save money in the long run by purchasing the larger size, even though it costs more than the smaller size? Calculating and comparing the unit price of a product enables you to make the more economical purchase, as shown in **Exercises 13 to 24 on pages 231 and 232**.

Need help? For on-line student resources, such as section quizzes, visit this textbook's web site at **math.college.hmco.com/students**.

1. $3.75 \div 5$

2. 3.47×15

3. $874.50 - 369.99$

4. $0.065 \times 150{,}000$

5. $1500 \times 0.06 \times 0.5$

6. $1372.47 + 36.91 + 5.00 + 2.86$

7. Divide $10 \div 3$. Round to the nearest hundredth.

8. Divide $345 \div 570$. Round to the nearest thousandth.

9. Place the correct symbol, $<$ or $>$, between the two numbers.

0.379   0.397

Go Figure

A store manager priced a pair of earrings in dollars and cents such that when 4% sales tax was added, the result was a whole number of dollars and 0 cents. Find the smallest possible number of dollars the items sold for, including the sales tax.

6.1 Applications to Purchasing

Objective A   To find unit cost

Frequently, stores promote items for purchase by advertising, say, 2 Red Baron Bake to Rise Pizzas for $10.50 or 5 cans of StarKist tuna for $4.25.

The **unit cost** is the cost of *one* Red Baron Pizza or of *one* can of StarKist tuna. To find the unit cost, divide the total cost by the number of units.

2 pizzas for $10.50   5 cans for $4.25
10.50 ÷ 2 = 5.25   4.25 ÷ 5 = 0.85
$5.25 is the cost of one pizza. $.85 is the cost of one can.
Unit cost: $5.25 per pizza   Unit cost: $.85 per can

Example 1

Find the unit cost. Round to the nearest tenth of a cent.
a. 3 gallons of mint chip ice cream for $17
b. 4 ounces of Crest toothpaste for $2.29

Strategy

To find the unit cost, divide the total cost by the number of units.

Solution
a. 17 ÷ 3 ≈ 5.667
      $5.667 per gallon
b. 2.29 ÷ 4 = 0.5725
      $.573 per ounce

You Try It 1

Find the unit cost. Round to the nearest tenth of a cent.
a. 8 size-AA Energizer batteries for $7.67
b. 15 ounces of Revlon shampoo for $2.29

Your strategy

Your solution

Solution on p. S15

Objective B   To find the most economical purchase

Comparison shoppers often find the most economical buy by comparing unit costs.

One store is selling 6 twelve-ounce cans of ginger ale for $2.99, and a second store is selling 24 twelve-ounce cans of ginger ale for $11.79. To find the better buy, compare the unit costs.

2.99 ÷ 6 ≈ 0.498   11.79 ÷ 24 ≈ 0.491
Unit cost: $.498 per can   Unit cost: $.491 per can

Because $.491 < $.498, the better buy is 24 cans for $11.79.

Example 2

Find the more economical purchase: 5 pounds of nails for $4.80, or 4 pounds of nails for $3.78.

Strategy
To find the more economical purchase, compare the unit costs.

Solution
$4.80 \div 5 = 0.96$
$3.78 \div 4 = 0.945$
$\$.945 < \$.96$

The more economical purchase is 4 pounds for $3.78.

You Try It 2

Find the more economical purchase: 6 cans of fruit for $5.70, or 4 cans of fruit for $3.96.

Your strategy

Your solution

Solution on p. S15

Objective C **To find total cost**

An installer of floor tile found the unit cost of identical floor tiles at three stores.

Store 1	Store 2	Store 3
$1.22 per tile	$1.18 per tile	$1.28 per tile

By comparing the unit costs, the installer determined that store 2 would provide the most economical purchase.

The installer also uses the unit cost to find the total cost of purchasing 300 floor tiles at store 2. The **total cost** is found by multiplying the unit cost by the number of units purchased.

Unit cost	×	number of units	=	total cost
1.18	×	300	=	354

The total cost is $354.

Example 3

Clear redwood lumber costs $5.43 per foot. How much would 25 feet of clear redwood cost?

Strategy
To find the total cost, multiply the unit cost (5.43) by the number of units (25).

Solution

Unit cost	×	number of units	=	total cost
5.43	×	25	=	135.75

The total cost is $135.75.

You Try It 3

Pine saplings cost $9.96 each. How much would 7 pine saplings cost?

Your strategy

Your solution

Solution on p. S15

6.1 Exercises

Application Problems

Find the unit cost. Round to the nearest tenth of a cent.

1. Heinz B·B·Q sauce, 18 ounces for $.99

2. Birds-eye maple, 6 feet for $18.75

3. Diamond walnuts, $2.99 for 8 ounces

4. A&W root beer, 6 cans for $2.99

5. Ibuprofen, 50 tablets for $3.99

6. Visine eye drops, 0.5 ounce for $3.89

7. Adjustable wood clamps, 2 for $13.95

8. Corn, 6 ears for $1.85

9. Cheerios cereal, 15 ounces for $2.99

10. Doritos Cool Ranch chips, 14.5 ounces for $2.99

11. Sheet metal screws, 8 for $.95

12. Folgers coffee, 11.5 ounces for $2.79

Application Problems

Suppose your local supermarket offers the following products at the given prices. Find the more economical purchase.

13. Sutter Home pasta sauce, 25.5 ounces for $3.29, or Muir Glen Organic pasta sauce, 26 ounces for $3.79

14. Kraft mayonnaise, 40 ounces for $2.98, or Springfield mayonnaise, 32 ounces for $2.39

15. Ortega salsa, 20 ounces for $3.29 or 12 ounces for $1.99

16. L'Oreal shampoo, 13 ounces for $3.99, or Cortexx shampoo, 12 ounces for $3.69

17. Golden Sun vitamin E, 200 tablets for $7.39 or 400 tablets for $12.99

18. Ultra Mr. Clean, 20 ounces for $2.67, or Ultra Spic and Span, 14 ounces for $2.19

19. 16 ounces of Kraft cheddar cheese, $4.37, or 9 ounces of Land to Lake cheddar cheese, $2.29

20. Bertolli olive oil, 34 ounces for $9.49, or Pompeian olive oil, 8 ounces for $2.39

21. Maxwell House coffee, 4 ounces for $3.99, or Sanka coffee, 2 ounces for $2.39

22. Wagner's vanilla extract, $3.29 for 1.5 ounces, or Durkee vanilla extract, 1 ounce for $2.74

23. Purina Cat Chow, $4.19 for 56 ounces, or Friskies Chef's Blend, $3.37 for 50.4 ounces

24. Kleenex tissues, $1.73 for 250 tissues, or Puffs tissues, $1.23 for 175 tissues

Objective C   *Application Problems*

25. If sliced bacon costs $4.59 per pound, find the total cost of 3 pounds.

26. Used red brick costs $.98 per brick. Find the total cost of 75 bricks.

27. Kiwi fruit cost $.23 each. Find the total cost of 8 kiwi.

28. Boneless chicken filets cost $4.69 per pound. Find the cost of 3.6 pounds. Round to the nearest cent.

29. Herbal tea costs $.98 per ounce. Find the total cost of 6.5 ounces.

30. If Stella Swiss Lorraine cheese costs $5.99 per pound, find the total cost of 0.65 pound. Round to the nearest cent.

31. Red Delicious apples cost $1.29 per pound. Find the total cost of 2.1 pounds. Round to the nearest cent.

32. Choice rib eye steak costs $4.49 per pound. Find the total cost of 2.8 pounds. Round to the nearest cent.

33. If Godiva chocolate costs $7.95 per pound, find the total cost of $\frac{3}{4}$ pound. Round to the nearest cent.

34. Color photocopying costs $.89 per page. Find the total cost for photocopying 120 pages.

APPLYING THE CONCEPTS

35. Explain in your own words the meaning of unit pricing.

36. What is the UPC (Universal Product Code) and how is it used?

6.2 Percent Increase and Percent Decrease

Objective A   **To find percent increase**

Percent increase is used to show how much a quantity has increased over its original value. The statements "food prices increased by 2.3% last year" and "city council members received a 4% pay increase" are examples of percent increase.

According to the Energy Information Administration, the number of alternative-fuel vehicles increased from approximately 277,000 to 352,000 in four years. Find the percent increase in alternative-fuel vehicles. Round to the nearest percent.

$$\boxed{\text{New value}} - \boxed{\text{original value}} = \boxed{\text{amount of increase}}$$

$$352{,}000 - 277{,}000 = 75{,}000$$

Now solve the basic percent equation for percent.

$$\text{Percent} \times \text{base} = \text{amount}$$

$$\boxed{\text{Percent increase}} \times \boxed{\text{original value}} = \boxed{\text{amount of increase}}$$

$$n \times 277{,}000 = 75{,}000$$
$$n = 75{,}000 \div 277{,}000$$
$$n \approx 0.27$$

The number of alternative-fuel vehicles increased by approximately 27%.

Example 1

The average wholesale price of coffee increased from $2 per pound to $3 per pound in one year. What was the percent increase in the price of 1 pound of coffee?

Strategy

To find the percent increase:

- Find the amount of the increase.
- Solve the basic percent equation for *percent*.

Solution

$$\boxed{\text{New value}} - \boxed{\text{original value}} = \boxed{\text{amount of increase}}$$

$$3 - 2 = 1$$

Percent × base = amount
$$n \times 2 = 1$$
$$n = 1 \div 2$$
$$n = 0.5 = 50\%$$

The percent increase was 50%.

You Try It 1

The average price of gasoline rose from $1.46 to $1.83 in 5 months. What was the percent increase in the price of gasoline? Round to the nearest percent.

Your strategy

Your solution

Solution on p. S15

Example 2

Chris Carley was earning $9.50 an hour as a nursing assistant before receiving a 10% increase in pay. What is Chris's new hourly pay?

Strategy

To find the new hourly wage:

- Solve the basic percent equation for *amount*.
- Add the amount of the increase to the original wage.

Solution

Percent × base = amount

    0.10 × 9.50 =   *n*

            0.95 = *n*

The amount of the increase was $.95.

9.50 + 0.95 = 10.45

The new hourly wage is $10.45.

You Try It 2

Yolanda Liyama was making a wage of $12.50 an hour as a baker before receiving a 14% increase in hourly pay. What is Yolanda's new hourly wage?

Your strategy

Your solution

Solution on p. S15

Objective B   **To apply percent increase to business—markup**

Some of the expenses involved in operating a business are salaries, rent, equipment, and utilities. To pay these expenses and earn a profit, a business must sell a product at a higher price than it paid for the product.

Cost is the price a business pays for a product, and **selling price** is the price at which a business sells a product to a customer. The difference between selling price and cost is called **markup.**

| Selling price | − | cost | = | markup |

or

| Cost | + | markup | = | selling price |

Markup is frequently expressed as a percent of a product's cost. This percent is called the **markup rate.**

| Markup rate | × | cost | = | markup |

Point of Interest

According to *Managing a Small Business*, from Liraz Publishing Company, goods in a store are often marked up 50% to 100% of the cost. This allows a business to make a profit of 5% to 10%.

➡ Suppose Bicycles Galore purchases an AMP Research B4 bicycle for $2119.20 and sells it for $2649. What markup rate does Bicycles Galore use?

| Selling price | − | cost | = | markup |

    2649.00   − 2119.20 =   529.80   • First find the markup.

    Percent   × base   =   amount   • Then solve the basic

| Markup rate | × | cost | = | markup | percent equation for

      n   × 2119.20 =   529.80   *percent.*

                    n = 529.80 ÷ 2119.20 = 0.25

The markup rate is 25%.

Example 3

The manager of a sporting goods store determines that a markup rate of 36% is necessary to make a profit. What is the markup on a pair of skis that costs the store $225?

Strategy

To find the markup, solve the basic percent equation for *amount*.

Solution

Percent × base = amount

| Markup rate | × | cost | = | markup |

$$0.36 \times 225 = n$$
$$81 = n$$

The markup is $81.

You Try It 3

A bookstore manager determines that a markup rate of 20% is necessary to make a profit. What is the markup on a book that costs the bookstore $8?

Your strategy

Your solution

Example 4

A plant nursery bought a yellow twig dogwood for $4.50 and used a markup rate of 46%. What is the selling price?

Strategy

To find the selling price:

- Find the markup by solving the basic percent equation for *amount*.
- Add the markup to the cost.

Solution

Percent × base = amount

| Markup rate | × | cost | = | markup |

$$0.46 \times 4.50 = n$$
$$2.07 = n$$

| Cost | + | markup | = | selling price |

$$4.50 + 2.07 = 6.57$$

The selling price is $6.57.

You Try It 4

A clothing store bought a suit for $72 and used a markup rate of 55%. What is the selling price?

Your strategy

Your solution

Solutions on p. S15

Objective C **To find percent decrease**

Percent decrease is used to show how much a quantity has decreased from its original value. The statements "the number of family farms decreased by 2% last year" and "there has been a 50% decrease in the cost of a Pentium chip" are examples of percent decrease.

During a two-year period, the value of U.S. agricultural products exported decreased from approximately $60.6 billion to $52.0 billion. Find the percent decrease in the value of agricultural exports. Round to the nearest tenth of a percent.

$$\boxed{\begin{array}{c}\text{Original}\\\text{value}\end{array}} - \boxed{\begin{array}{c}\text{new}\\\text{value}\end{array}} = \boxed{\begin{array}{c}\text{amount of}\\\text{decrease}\end{array}}$$

$$60.6 \quad - \quad 52.0 \quad = \quad 8.6$$

Now solve the basic percent equation for percent.

$$\text{Percent} \times \text{base} = \text{amount}$$

$$\boxed{\begin{array}{c}\text{Percent}\\\text{decrease}\end{array}} \times \boxed{\begin{array}{c}\text{original}\\\text{value}\end{array}} = \boxed{\begin{array}{c}\text{amount of}\\\text{decrease}\end{array}}$$

$$\begin{array}{rcl} n \times 60.6 &=& 8.6 \\ n &=& 8.6 \div 60.6 \\ n &\approx& 0.142 \end{array}$$

The value of agricultural exports decreased approximately 14.2%.

Point of Interest

A *bear market* is one in which the total value of all stocks decreases. During 1972 and 1973, the stock market experienced a 55% decrease. This means that a $10,000 investment would have decreased to $4500 during that period.

....

Example 5

During an 8-year period, Baltimore, Maryland's population decreased from approximately 736,000 to 646,000. Find the percent decrease in Baltimore's population. Round to the nearest tenth of a percent.

Strategy

To find the percent decrease:

• Find the amount of decrease.
• Solve the basic percent equation for *percent*.

Solution

$$\boxed{\begin{array}{c}\text{Original}\\\text{value}\end{array}} - \boxed{\begin{array}{c}\text{new}\\\text{value}\end{array}} = \boxed{\begin{array}{c}\text{amount of}\\\text{decrease}\end{array}}$$

$$736,000 \quad - \quad 646,000 \quad = \quad 90,000$$

$$\begin{array}{rcl} \text{Percent} \times \text{base} &=& \text{amount} \\ n \times 736,000 &=& 90,000 \\ n &=& 90,000 \div 736,000 \\ n &\approx& 0.122 \end{array}$$

Baltimore's population decreased approximately 12.2%.

You Try It 5

During an 8-year period, Norfolk, Virginia's population decreased from approximately 261,000 to 215,000. Find the percent decrease in Norfolk's population. Round to the nearest tenth of a percent.

Your strategy

Your solution

Solution on p. S15

Example 6

The total sales for December for a stationery store were $16,000. For January, total sales showed an 8% decrease from December's sales. What were the total sales for January?

Strategy

To find the total sales for January:

• Find the amount of decrease by solving the basic percent equation for *amount*.
• Subtract the amount of decrease from the December sales.

Solution

Percent × base = amount
$0.08 \times 16{,}000 = n$
$1280 = n$

The decrease in sales was $1280.

$16{,}000 - 1280 = 14{,}720$

The total sales for January were $14,720.

You Try It 6

Fog decreased the normal 5-mile visibility at an airport by 40%. What was the visibility in the fog?

Your strategy

Your solution

Solution on p. S15

Objective D   **To apply percent decrease to business—discount**

To promote sales, a store may reduce the regular price of some of its products temporarily. The reduced price is called the **sale price.** The difference between the regular price and the sale price is called the **discount.**

| Regular price | − | sale price | = | discount |

or

| Regular price | − | discount | = | sale price |

Discount is frequently stated as a percent of a product's regular price. This percent is called the **discount rate.**

| Discount rate | × | regular price | = | discount |

Example 7

A GE 25-inch stereo television that regularly sells for $299 is on sale for $250. Find the discount rate. Round to the nearest tenth of a percent.

Strategy

To find the discount rate:

• Find the discount.
• Solve the basic percent equation for *percent*.

Solution

Regular price	−	sale price	=	discount
299	−	250	=	49

Percent	×	base	=	amount
Discount rate	×	regular price	=	discount
n	×	299	=	49

$$n = 49 \div 299$$
$$n \approx 0.1639$$

The discount rate is 16.4%.

You Try It 7

A white azalea that regularly sells for $12.50 is on sale for $10.99. Find the discount rate. Round to the nearest tenth of a percent.

Your strategy

Your solution

Example 8

A 20-horsepower lawn mower is on sale for 25% off the regular price of $1125. Find the sale price.

Strategy

To find the sale price:

• Find the discount by solving the basic percent equation for *amount*.
• Subtract to find the sale price.

Solution

Percent	×	base	=	amount
Discount rate	×	regular price	=	discount
0.25	×	1125	=	n

$$281.25 = n$$

Regular price	−	discount	=	sale price
1125	−	281.25	=	843.75

The sale price is $843.75.

You Try It 8

A hardware store is selling a Newport security door for 15% off the regular price of $110. Find the sale price.

Your strategy

Your solution

Solutions on p. S16

6.2 Exercises

Objective A   *Application Problems*

Solve. Round percents to the nearest tenth of a percent.

1. The figure at the right shows the actual and projected enrollments in grades K–12 in the United States. What percent increase is expected from 1988 to 2008?

2. An automobile manufacturer increased the average mileage on a car from 17.5 miles per gallon to 18.2 miles per gallon. Find the percent increase in mileage.

3. In the 1990s, the number of Target stores increased from 420 stores to 914 stores. (Source: Target) What was the percent increase in the number of Target stores in the 1990s?

4. The graph at the right shows the number of women attending four-year colleges. Find the percent increase in the number of women at four-year colleges from 1983 to 2003.

5. In 1924, the number of events in the Winter Olympics was 14. The 2002 Winter Olympics in Salt Lake City included 78 medal events. (Source: David Wallenchinsky's *The Complete Book of the Winter Olympics*) Find the percent increase in the number of events in the Winter Olympics from 1924 to 2002.

6. The table at the right shows the estimated number of millionaire households in the United States. Find the percent increase in the estimated number of households containing millionaires from 1975 to 2005.

7. The population of Boise City, Idaho, increased 24.3% over an 8-year period. The population initially was approximately 127,000. (Source: The Census Bureau) Find the population of Boise City 8 years later.

8. During one year, the number of people subscribing to direct broadcasting satellite systems increased 87%. If the number of subscribers at the beginning of the year was 2.3 million, how many subscribers were there at the end of the year?

9. From 1998 to 2000, the amount of time the average Internet user spent online each week rose 72.7%. If 4.4 hours were spent online in 1998, how many hours per week were spent online in 2000? Round to the nearest tenth. (Source: *Time*, February 21, 2000)

Elementary and Secondary
School Enrollments

Source: National Center for
Education

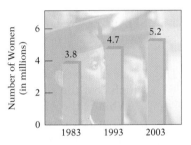

Women Attending Four-Year Colleges

Source: Copyright © 2000 *USA Today*.
Reprinted with permission.

Year	Number of Households Containing Millionaires
1975	350,000
1997	3,500,000
2005	5,600,000

Source: Affluent Market Institute

Objective B   *Application Problems*

10. A window air conditioner cost AirRite Air Conditioning Systems $285. Find the markup on the air conditioner if the markup rate is 25% of the cost.

11. The owner of Kerr Electronics purchased 300 Craig portable CD players at a cost of $85 each. If the owner uses a markup rate of 42%, what is the markup on each of the Craig CD players?

12. The manager of Brass Antiques has determined that a markup rate of 38% is necessary for a profit to be made. What is the markup on a brass door-knob that costs $45?

13. Computer Inc. uses a markup of $975 on a computer system that costs $3250. What is the markup rate on this system?

14. Saizon Pen & Office Supply uses a markup of $12 on a calculator that costs $20. What markup rate does this amount represent?

15. Giant Photo Service uses a markup rate of 48% on its Model ZA cameras, which cost the shop $162.
   a. What is the markup?
   b. What is the selling price?

16. The Circle R golf pro shop uses a markup rate of 45% on a set of Tour Pro golf clubs that costs the shop $210.
   a. What is the markup?
   b. What is the selling price?

17. Harvest Time Produce Inc. uses a 55% markup rate and pays $2 for a box of strawberries.
   a. What is the markup?
   b. What is the selling price of a box of strawberries?

18. Resner Builders' Hardware uses a markup rate of 42% for a table saw that costs $160. What is the selling price of the table saw?

19. Brad Burt's Magic Shop uses a markup rate of 48%. What is the selling price of a telescoping sword that costs $50?

Objective C   *Application Problems*

20. A new bridge reduced the normal 45-minute travel time between two cities by 18 minutes. What percent decrease does this represent?

21. By installing energy-saving equipment, the Pala Rey Youth Camp reduced its normal $800-per-month utility bill by $320. What percent decrease does this amount represent?

22. It is estimated that the value of a new car is reduced 30% after 1 year of ownership. Using this estimate, find how much value a $18,200 new car loses after 1 year.

23. A department store employs 1200 people during the holiday. At the end of the holiday season, the store reduces the number of employees by 45%. What is the decrease in the number of employees?

24. Because of a decrease in demand for super-8 video cameras, Kit's Cameras reduced the orders for these models from 20 per month to 8 per month.
 a. What is the amount of the decrease?
 b. What percent decrease does this amount represent?

25. A new computer system reduced the time for printing the payroll from 52 minutes to 39 minutes.
 a. What is the amount of the decrease?
 b. What percent decrease does this amount represent?

26. Juanita's average monthly expense for gasoline was $76. After joining a car pool, she was able to reduce the expense by 20%.
 a. What was the amount of the decrease?
 b. What is the average monthly gasoline bill now?

27. An oil company paid a dividend of $1.60 per share. After a reorganization, the company reduced the dividend by 37.5%.
 a. What was the amount of the decrease?
 b. What is the new dividend?

28. Because of an improved traffic pattern at a sports stadium, the average amount of time a fan waits to park decreased from 3.5 minutes to 2.8 minutes. What percent decrease does this amount represent?

29. One configuration of the Boeing 777-300 has a seating capacity of 394. This is 26 more than that of the corresponding 777-200 jet. What is the percent decrease in capacity from the 777-300 to the 777-200 model?

30. In 2000, the Pentagon revised its account of the number of Americans killed in the Korean War from 54,246 to 36,940. (Source: *Time,* June 12, 2000) What is the percent decrease in the reported number of military personnel killed in the Korean War?

Objective D   *Application Problems*

31. The Austin College Bookstore is giving a discount of $8 on calculators that normally sell for $24. What is the discount rate?

32. A discount clothing store is selling a $72 sport jacket for $24 off the regular price. What is the discount rate?

33. A disk player that regularly sells for $340 is selling for 20% off the regular price. What is the discount?

34. Dacor Appliances is selling its $450 washing machine for 15% off the regular price. What is the discount?

35. An electric grill that regularly sells for $140 is selling for $42 off the regular price. What is the discount rate?

36. Quick Service Gas Station has its regularly priced $45 tune-up on sale for 16% off the regular price.
  a. What is the discount?
  b. What is the sale price?

37. Tomatoes that regularly sell for $.85 per pound are on sale for 20% off the regular price.
  a. What is the discount?
  b. What is the sale price?

38. An outdoor supply store has its regularly priced $160 sleeping bags on sale for $120.
  a. What is the discount?
  b. What is the discount rate?

39. Standard Brands paint that regularly sells for $20 per gallon is on sale for $16 per gallon.
  a. What is the discount?
  b. What is the discount rate?

40. The graph at the right shows the number of active-duty U.S. military personnel in 1990 and in 2000. Which branch of the military had the greatest percent decrease in personnel from 1990 to 2000? What was the percent decrease for this branch of the service?

Number of Active-Duty U.S. Military Personnel

Source: Fiscal 2000 Annual Report to the President and Congress by the Secretary of Defense

APPLYING THE CONCEPTS

41. A welder earning $12 per hour is given a 10% raise. To find the new wage, we can multiply $12 by 0.10 and add the product to $12. Can the new wage be found by multiplying $12 by 1.10? Try both methods and compare your answers.

42. Grocers, florists, bakers, and other businesses must consider spoilage when deciding the markup of a product. For instance, suppose a florist purchased 200 roses at a cost of $.86 per rose. The florist wants a markup rate of 50% of the total cost of all the roses and expects 7% of the roses to wilt and therefore not be salable. Find the selling price per rose by answering each of the following questions.
  a. What is the florist's total cost for the 200 roses?
  b. Find the total selling price without spoilage.
  c. Find the number of roses the florist expects to sell. *Hint:* The number of roses the florist expects to sell is

    % of salable roses × number of roses purchased

  d. To find the selling price per rose, divide the total selling price without spoilage by the number of roses the florist expects to sell. Round to the nearest cent.

43. A promotional sale at a department store offers 25% off the sale price. The sale price is 25% off the regular price. Is this the same as a sale that offers 50% off the regular price? If not, which sale gives the better price? Explain your answer.

6.3 Interest

Objective A   **To calculate simple interest**

When you deposit money in a bank—for example, in a savings account—you are permitting the bank to use your money. The bank may use the deposited money to lend customers the money to buy cars or make renovations on their homes. The bank pays you for the privilege of using your money. The amount paid to you is called **interest.** If you are the one borrowing money from the bank, the amount you pay for the privilege of using that money is also called interest.

> **TAKE NOTE**
>
> If you deposit $1000 in a savings account paying 5% interest, the $1000 is the principal and 5% is the interest rate.

The original amount deposited or borrowed is called the **principal.** The amount of interest paid is usually given as a percent of the principal. The percent used to determine the amount of interest is the **interest rate.**

Interest paid on the original principal is called **simple interest.** To calculate simple interest, multiply the principal by the interest rate per period by the number of time periods. In this objective, we are working with annual interest rates, so the time periods are years. The simple interest formula for an annual interest rate is given below.

> **Simple Interest Formula for Annual Interest Rates**
>
> Principal × annual interest rate × time (in years) = interest

Interest rates are generally given as percents. Before performing calculations involving an interest rate, write the interest rate as a decimal.

➡ Calculate the simple interest due on a 2-year loan of $1500 that has an annual interest rate of 7.5%.

Principal	×	annual interest rate	×	time (in years)	=	interest
1500	×	0.075	×	2	=	225

The simple interest due is $225.

When we borrow money, the total amount to be repaid to the lender is the sum of the principal and the interest. This amount is called the **maturity value** of the loan.

> **Maturity Value Formula for Simple Interest Loans**
>
> Principal + interest = maturity value

In the example above, the simple interest due on the loan of $1500 was $225. The maturity value of the loan is therefore $1500 + $225 = $1725.

➡ Calculate the maturity value of a simple interest, 8-month loan of $8000 if the annual interest rate is 9.75%.

First find the interest due on the loan.

Principal	×	annual interest rate	×	time (in years)	=	interest
8000	×	0.0975	×	$\frac{8}{12}$	=	520

Find the maturity value.

Principal + interest = maturity value
8000 + 520 = 8520

The maturity value of the loan is $8520.

The monthly payment on a loan can be calculated by dividing the maturity value by the length of the loan in months.

> **Monthly Payment on a Simple Interest Loan**
> Maturity value ÷ length of the loan in months = monthly payment

In the example above, the maturity value of the loan is $8520. To find the monthly payment on the 8-month loan, divide 8520 by 8.

Maturity value	÷	length of the loan in months	=	monthly payment
8520	÷	8	=	1065

The monthly payment on the loan is $1065.

Example 1

Kamal borrowed $500 from a savings and loan association for 180 days at an annual interest rate of 7%. What is the simple interest due on the loan?

Strategy

To find the simple interest due, multiply the principal (500) times the annual interest rate (7% = 0.07) times the time, in years (180 days = $\frac{180}{365}$ year).

Solution

Principal	×	annual interest rate	×	time (in years)	=	interest
500	×	0.07	×	$\frac{180}{365}$	≈	17.26

The simple interest due is $17.26.

You Try It 1

A company borrowed $15,000 from a bank for 18 months at an annual interest rate of 8%. What is the simple interest due on the loan?

Your strategy

Your solution

Solution on p. S16

Example 2

Calculate the maturity value of a simple interest, nine-month loan of $4000 if the annual interest rate is 8.75%.

Strategy

To find the maturity value:

- Use the simple interest formula to find the simple interest due.
- Find the maturity value by adding the principal and the interest.

Solution

Principal	$\times$	annual interest rate	$\times$	time (in years)	$=$	interest

$$4000 \quad \times \; 0.0875 \; \times \quad \frac{9}{12} \quad = \quad 262.5$$

Principal + interest = maturity value
  4000   + 262.50 =   4262.50

The maturity value is $4262.50.

You Try It 2

Calculate the maturity value of a simple interest, 90-day loan of $3800. The annual interest rate is 6%.

Your strategy

Your solution

Example 3

The simple interest due on a 3-month loan of $1400 is $26.25. Find the monthly payment on the loan.

Strategy

To find the monthly payment:

- Find the maturity value by adding the principal and the interest.
- Divide the maturity value by the length of the loan in months (3).

Solution

Principal + interest = maturity value
  1400   + 26.25 =   1426.25

Maturity value ÷ length of the loan = payment
  1426.25   ÷   3   ≈ 475.42

The monthly payment is $475.42.

You Try It 3

The simple interest due on a 1-year loan of $1900 is $152. Find the monthly payment on the loan.

Your strategy

Your solution

Solutions on p. S16

Objective B   **To calculate finance charges on a credit card bill**

When a customer uses a credit card to make a purchase, the customer is actually receiving a loan. Therefore, there is frequently an added cost to the consumer who purchases on credit. This may be in the form of an annual fee and interest charges on purchases. The interest charges on purchases are called **finance charges.**

The finance charge on a credit card bill is calculated using the simple interest formula. In the last objective, the interest rates were annual interest rates. However, credit card companies generally issue *monthly* bills and express interest rates on credit card purchases as *monthly* interest rates. Therefore, when using the simple interest formula to calculate finance charges on credit card purchases, use a monthly interest rate and express the time in months.

Note: In the simple interest formula, the time must be expressed in the same period as the rate. For an *annual* interest rate, the time must be expressed in years. For a *monthly* interest rate, the time must be in months.

Example 4	**You Try It 4**
A credit card company charges a customer 1.5% per month on the unpaid balance of charges on the credit card. What is the finance charge in a month when the customer has an unpaid balance of $254?	The credit card that Francesca uses charges her 1.6% per month on her unpaid balance. Find the finance charge when her unpaid balance one month is $1250.

Strategy
To find the finance charge, multiply the principal, or unpaid balance (254), times the monthly interest rate (1.5%) times the number of months (1).

Your strategy

Solution

$$254 \times 0.015 \times 1 = 3.81$$

The simple interest due is $3.81.

Your solution

Solution on p. S16

Objective C **To calculate compound interest**

Usually, the interest paid on money deposited or borrowed is compound interest. **Compound interest** is computed not only on the original principal but also on interest already earned. Here is an illustration.

Suppose $1000 is invested for 3 years at an annual interest rate of 9% compounded annually. Because this is an *annual* interest rate, we will calculate the interest earned each year.

During the first year, the interest earned is calculated as follows:

Principal	×	annual interest rate	×	time (in years)	=	interest
1000	×	0.09	×	1	=	90

At the end of the first year, the total amount in the account is

$$1000 + 90 = 1090$$

During the second year, the interest earned is calculated on the amount in the account at the end of the first year.

Principal	×	annual interest rate	×	time (in years)	=	interest
1090	×	0.09	×	1	=	98.10

Note that the interest earned during the second year ($98.10) is greater than the interest earned during the first year ($90). This is because the interest earned during the first year was added to the original principal, and the interest for the second year was calculated using this sum. If the account earned simple interest, the interest earned would be the same every year ($90).

At the end of the second year, the total amount in the account is the sum of the amount in the account at the end of the first year and the interest earned during the second year.

$$1090 + 98.10 = 1188.10$$

The interest earned during the third year is calculated using the amount in the account at the end of the second year ($1188.10).

Principal	×	annual interest rate	×	time (in years)	=	interest
1188.10	×	0.09	×	1	≈	106.93

The amount in the account at the end of the third year is

$$1188.10 + 106.93 = 1295.03$$

To find the interest earned for the three years, subtract the original principal from the new principal.

New principal	−	original principal	=	interest earned
1295.03	−	1000	=	295.03

Note that the compound interest earned is $295.03. The simple interest earned on the investment would have been only $1000 × 0.09 × 3 = $270.

In this example, the interest was compounded annually. However, compound interest can be compounded

annually (once a year)
semiannually (twice a year)
quarterly (four times a year)
monthly (twelve times a year)
daily (365 times a year)

The more frequent the compounding periods, the more interest the account earns. For example, if, in the above example, the interest had been compounded quarterly rather than annually, the interest earned would have been greater.

Calculating compound interest can be very tedious, so there are tables that can be used to simplify these calculations. A portion of a Compound Interest Table is given in the Appendix.

➡ What is the value after 5 years of $1000 invested at 7% annual interest compounded quarterly?

To find the interest earned, multiply the original principal (1000) by the factor found in the Compound Interest Table. To find the factor, first find the table headed "Compounded Quarterly" in the Compound Interest Table in the Appendix. Then look at the number where the 7% column and the 5-year row meet.

	4%	*5%*	*6%*	*7%*	*8%*	*9%*	*10%*
				Compounded Quarterly			
1 year	1.04060	1.05094	1.06136	1.07186	1.08243	1.09308	1.10381
5 years	1.22019	1.28204	1.34686	**1.41478**	1.48595	1.56051	1.63862
10 years	1.48886	1.64362	1.81402	2.00160	2.20804	2.43519	2.68506
15 years	1.81670	2.10718	2.44322	2.83182	3.28103	3.80013	4.39979
20 years	2.21672	2.70148	3.29066	4.00639	4.87544	5.93015	7.20957

The factor is 1.41478.

$$1000 \times 1.41478 = 1414.78$$

The value of the investment after 5 years is $1414.78.

Example 5

An investment of $650 pays 8% annual interest compounded semiannually. What is the interest earned in 5 years?

Strategy
To find the interest earned:

- Find the new principal by multiplying the original principal (650) by the factor found in the Compound Interest Table (1.48024).
- Subtract the original principal from the new principal.

Solution
$650 \times 1.48024 \approx 962.16$

The new principal is $962.16.

$962.16 - 650 = 312.16$

The interest earned is $312.16.

You Try It 5

An investment of $1000 pays 6% annual interest compounded quarterly. What is the interest earned in 20 years?

Your strategy

Your solution

Solution on p. S16

6.3 Exercises

Application Problems

1. A 2-year student loan of $10,000 is made at an annual simple interest rate of 10.25%. The simple interest on the loan is $2050. Identify **a.** the principal, **b.** the interest, **c.** the interest rate, and **d.** the time period of the loan.

2. A contractor obtained a 9-month loan for $80,000 at an annual simple interest rate of 9.75%. The simple interest on the loan is $5850. Identify **a.** the principal, **b.** the interest, **c.** the interest rate, and **d.** the time period of the loan.

3. Find the simple interest Jacob Zucker owes on a 2-year student loan of $8000 at an annual interest rate of 9%.

4. Find the simple interest Kara Tanamachi owes on a $1\frac{1}{2}$-year loan of $1500 at an annual interest rate of 7.5%.

5. To finance the purchase of 15 new cars, the Tropical Car Rental Agency borrowed $100,000 for 9 months at an annual interest rate of 9%. What is the simple interest due on the loan?

6. A home builder obtained a preconstruction loan of $50,000 for 8 months at an annual interest rate of 9.5%. What is the simple interest due on the loan?

7. A bank lent Gloria Masters $20,000 at an annual interest rate of 8.8%. The period of the loan was 9 months. Find the simple interest due on the loan.

8. Eugene Madison obtained an 8-month loan of $4500 at an annual interest rate of 10.2%. Find the simple interest Eugene owes on the loan.

9. Shannon O'Hara borrowed $5000 for 90 days at an annual simple interest rate of 7.5%. Find the simple interest due on the loan.

10. Jon McCloud borrowed $8500 for 180 days at an annual simple interest rate of 9.25%. Find the simple interest due on the loan.

11. Jorge Elizondo took out a 75-day loan of $7500 at an annual interest rate of 9.5%. Find the simple interest due on the loan.

12. Kristi Yang borrowed $15,000. The term of the loan was 90 days, and the annual simple interest rate was 7.4%. Find the simple interest due on the loan.

13. The simple interest due on a 4-month loan of $4800 is $320. What is the maturity value of the loan?

14. The simple interest due on a 60-day loan of $6500 is $80.14. Find the maturity value of the loan.

15. An auto parts dealer borrowed $150,000 at a 9.5% annual simple interest rate for 1 year. Find the maturity value of the loan.

16. A corporate executive took out a $25,000 loan at an 8.2% annual simple interest rate for 1 year. Find the maturity value of the loan.

17. William Carey borrowed $12,500 for 8 months at an annual simple interest rate of 9.5%. Find the total amount due on the loan.

18. You arrange for a 9-month bank loan of $9000 at an annual simple interest rate of 8.5%. Find the total amount you must repay to the bank.

19. Capital City Bank approves a home-improvement loan application for $14,000 at an annual simple interest rate of 10.25% for 270 days. What is the maturity value of the loan?

20. A credit union lends a member $5000 for college tuition. The loan is made for 18 months at an annual simple interest rate of 6.9%. What is the maturity value of this loan?

21. Action Machining Company purchased a robot-controlled lathe for $225,000 and financed the full amount at 8% annual simple interest for 4 years. The simple interest on the loan is $72,000. Find the monthly payment.

22. For the purchase of an entertainment center, a $1900 loan is obtained for 2 years at an annual simple interest rate of 9.4%. The simple interest due on the loan is $357.20. What is the monthly payment on the loan?

23. To attract new customers, Heller Ford is offering car loans at an annual simple interest rate of 4.5%.
   a. Find the interest charged to a customer who finances a car loan of $12,000 for 2 years.
   b. Find the monthly payment.

24. Cimarron Homes Inc. purchased a small plane for $57,000 and financed the full amount for 5 years at an annual simple interest rate of 9%.
   a. Find the interest due on the loan.
   b. Find the monthly payment.

25. Dennis Pappas decided to build onto his present home instead of buying a new, larger house. He borrowed $42,000 for $3\frac{1}{2}$ years at an annual simple interest rate of 9.5%. Find the monthly payment.

26. Rosalinda Johnson took out a 6-month, $12,000 loan. The annual simple interest rate on the loan was 8.5%. Find the monthly payment.

Objective B   *Application Problems*

27. A credit card company charges a customer 1.25% per month on the unpaid balance of charges on the credit card. What is the finance charge in a month when the customer has an unpaid balance of $118.72?

28. The credit card that Dee Brown uses charges her 1.75% per month on her unpaid balance. Find the finance charge when her unpaid balance one month is $391.64.

29. What is the finance charge on an unpaid balance of $12,368.92 on a credit card that charges 1.5% per month on any unpaid balance?

30. Suppose you have an unpaid balance of $995.04 on a credit card that charges 1.2% per month on any unpaid balance. What finance charge do you owe the company?

31. A credit card customer has an unpaid balance of $1438.20. What is the difference between monthly finance charges of 1.15% per month on the unpaid balance and monthly finance charges of 1.85% per month?

32. One credit card company charges 1.25% per month on any unpaid balance, and a second company charges 1.75%. What is the difference between the finance charges that these two companies assess on an unpaid balance of $687.45?

Objective C   *Application Problems*

33. North Island Federal Credit Union pays 4% annual interest, compounded daily, on time savings deposits. Find the value of $750 deposited in this account after 1 year.

34. Tanya invested $2500 in a tax-sheltered annuity that pays 8% annual interest compounded daily. Find the value of her investment after 20 years.

35. Sal Travato invested $3000 in a corporate retirement account that pays 6% annual interest compounded semiannually. Find the value of his investment after 15 years.

36. To replace equipment, a farmer invested $20,000 in an account that pays 7% annual interest compounded monthly. What is the value of the investment after 5 years?

37. Green River Lodge invests $75,000 in a trust account that pays 8% interest compounded quarterly.
   a. What will the value of the investment be in 5 years?
   b. How much interest will be earned in the 5 years?

38. To save for retirement, a couple deposited $3000 in an account that pays 7% annual interest compounded daily.
   a. What will the value of the investment be in 10 years?
   b. How much interest will be earned in the 10 years?

39. To save for a child's education, the Petersens deposited $2500 into an account that pays 6% annual interest compounded daily. Find the amount of interest earned on this account over a 20-year period.

40. How much interest is earned in 2 years on $4000 deposited in an account that pays 6% interest, compounded quarterly?

APPLYING THE CONCEPTS

41. The Mission Valley Credit Union charges its customers an interest rate of 2% per month on money that is transferred into an account that is overdrawn. Find the interest owed to the credit union for 1 month when $800 is transferred into an overdrawn account.

42. Suppose you have a savings account that earns interest at the rate of 6% per year compounded monthly. On January 1, you open this acount with a deposit of $100.
   a. On February 1, you deposit an additional $100 into the account. What is the value of the account after the deposit?
   b. On March 1, you deposit an additional $100 into the account. What is the value of the account after the deposit?
   Note: This type of savings plan, wherein equal amounts ($100) are saved at equal time intervals (every month), is called an annuity.

43. At 4 P.M. on July 31, you open a savings account that pays 5% annual interest and you deposit $500 in the account. Your deposit is credited as of August 1. At the beginning of September, you receive a statement from the bank that shows that during the month of August, you received $2.12 in interest. The interest has been added to your account, bringing the total on deposit to $502.12. At the beginning of October, you receive a statement from the bank that shows that during the month of September, you received $2.06 in interest on the $502.12 on deposit. Explain why you received less interest during the second month when there was more money on deposit.

6.4 Real Estate Expenses

Objective A   **To calculate the initial expenses of buying a home**

One of the largest investments most people ever make is the purchase of a home. The major initial expense in the purchase is the down payment. The amount of the down payment is normally a percent of the purchase price. This percent varies among banks, but it usually ranges from 5% to 25%.

The **mortgage** is the amount that is borrowed to buy real estate. The mortgage amount is the difference between the purchase price and the down payment.

➡ A home is purchased for $140,000, and a down payment of $21,000 is made. Find the mortgage.

Purchase price	−	down payment	=	mortgage
140,000	−	21,000	=	119,000

The mortgage is $119,000.

TAKE NOTE
Because *points* means percent, a loan origination fee of
$2\frac{1}{2}$ points = $2\frac{1}{2}$% =
2.5% = 0.025.

Another initial expense in buying a home is the **loan origination fee,** which is a fee that the bank charges for processing the mortgage papers. The loan origination fee is usually a percent of the mortgage and is expressed in **points,** which is the term banks use to mean percent. For example, "5 points" means "5 percent."

Points	×	mortgage	=	loan origination fee

....

Example 1

A house is purchased for $125,000, and a down payment, which is 20% of the purchase price, is made. Find the mortgage.

Strategy
To find the mortgage:

- Find the down payment by solving the basic percent equation for *amount*.
- Subtract the down payment from the purchase price.

Solution

Percent × base = amount

Percent	×	purchase price	=	down payment
0.20	×	125,000	=	n
		25,000	=	n

Purchase price	−	down payment	=	mortgage
125,000	−	25,000	=	100,000

The mortgage is $100,000.

You Try It 1

An office building is purchased for $216,000, and a down payment, which is 25% of the purchase price, is made. Find the mortgage.

Your strategy

Your solution

Solution on p. S16

Example 2

A home is purchased with a mortgage of $65,000. The buyer pays a loan origination fee of $3\frac{1}{2}$ points. How much is the loan origination fee?

Strategy

To find the loan origination fee, solve the basic percent equation for *amount*.

Solution

Percent $\times$ base $=$ amount

| Points | $\times$ | mortgage | $=$ | fee |

$0.035 \quad \times \quad 65,000 \quad = \quad n$

$2275 = n$

The loan origination fee is $2275.

You Try It 2

The mortgage on a real estate investment is $80,000. The buyer paid a loan origination fee of $4\frac{1}{2}$ points. How much was the loan origination fee?

Your strategy

Your solution

Solution on p. S16

Objective B **To calculate ongoing expenses of owning a home**

Point of Interest

The number-one response of adults asked what they would spend money on first if they suddenly became wealthy (for example, by winning the lottery) was a house; 31% gave this response. (Source: Yankelovich Partners for Lutheran Brotherhood)

Besides the initial expenses of buying a house, there are continuing monthly expenses involved in owning a home. The monthly mortgage payment, utilities, insurance, and taxes are some of these ongoing expenses. Of these expenses, the largest one is normally the monthly mortgage payment.

For a fixed-rate mortgage, the monthly mortgage payment remains the same throughout the life of the loan. The calculation of the monthly mortgage payment is based on the amount of the loan, the interest rate on the loan, and the number of years required to pay back the loan. Calculating the monthly mortgage payment is fairly difficult, so tables such as the one in the Appendix are used to simplify these calculations.

⇒ Find the monthly mortgage payment on a 30-year $60,000 mortgage at an interest rate of 9%. Use the Monthly Payment Table in the Appendix.

$60,000 \times \underline{0.0080462} \approx 482.77$

$\uparrow$

From the table

The monthly mortgage payment is $482.77.

The monthly mortgage payment includes the payment of both principal and interest on the mortgage. The interest charged during any one month is charged on the unpaid balance of the loan. Therefore, during the early years of the mortgage, when the unpaid balance is high, most of the monthly mortgage payment is interest charged on the loan. During the last few years of a mortgage, when the unpaid balance is low, most of the monthly mortgage payment goes toward paying off the loan.

Point of Interest

Home buyers rated the following characteristics "extremely important" in their purchase decision.

Natural, open space: 77%
Walking and biking paths: 74%
Gardens with native plants: 56%
Clustered retail stores: 55%
Wilderness area: 52%
Outdoor pool: 52%
Community recreation center: 52%
Interesting little parks: 50%

(Source: American Lives, Inc. and Intercommunications, Inc.)

⇒ Find the interest paid on a mortgage during a month when the monthly mortgage payment is $186.26 and $58.08 of that amount goes toward paying off the principal.

| Monthly mortgage payment | − | principal | = | interest |

186.26 − 58.08 = 128.18

The interest paid on the mortgage is $128.18.

Property tax is another ongoing expense of owning a house. Property tax is normally an annual expense that may be paid on a monthly basis. The monthly property tax, which is determined by dividing the annual property tax by 12, is usually added to the monthly mortgage payment.

⇒ A homeowner must pay $534 in property tax annually. Find the property tax that must be added each month to the homeowner's monthly mortgage payment.

$534 \div 12 = 44.5$

Each month, $44.50 must be added to the monthly mortgage payment for property tax.

Example 3

Serge purchased some land for $120,000 and made a down payment of $25,000. The savings and loan association charges an annual interest rate of 8% on Serge's 25-year mortgage. Find the monthly mortgage payment.

Strategy

To find the monthly mortgage payment:

• Subtract the down payment from the purchase price to find the mortgage.
• Multiply the mortgage by the factor found in the Monthly Payment Table in the Appendix.

Solution

| Purchase price | − | down payment | = | mortgage |

120,000 − 25,000 = 95,000

$95,000 \times 0.0077182 \approx 733.23$
↑
From the table

The monthly mortgage payment is $733.23.

You Try It 3

A new condominium project is selling townhouses for $75,000. A down payment of $15,000 is required, and a 20-year mortgage at an annual interest rate of 9% is available. Find the monthly mortgage payment.

Your strategy

Your solution

Solution on p. S17

Example 4

A home has a mortgage of $134,000 for 25 years at an annual interest rate of 7%. During a month when $375.88 of the monthly mortgage payment is principal, how much of the payment is interest?

Strategy

To find the interest:

- Multiply the mortgage by the factor found in the Monthly Payment Table in the Appendix to find the monthly mortgage payment.
- Subtract the principal from the monthly mortgage payment.

Solution

$134,000 \times 0.0070678 \approx 947.09$

↑	↑
From the table	Monthly mortgage payment

$$\boxed{\begin{array}{c}\text{Monthly}\\\text{mortgage}\\\text{payment}\end{array}} - \boxed{\text{principal}} = \boxed{\text{interest}}$$

$947.09 \quad - \quad 375.88 \quad = \quad 571.21$

$571.21 of the payment is interest on the mortgage.

You Try It 4

An office building has a mortgage of $125,000 for 25 years at an annual interest rate of 9%. During a month when $492.65 of the monthly mortgage payment is principal, how much of the payment is interest?

Your strategy

Your solution

Example 5

The monthly mortgage payment for a home is $598.75. The annual property tax is $900. Find the total monthly payment for the mortgage and property tax.

Strategy

To find the monthly payment:

- Divide the annual property tax by 12 to find the monthly property tax.
- Add the monthly property tax to the monthly mortgage payment.

Solution

$900 \div 12 = 75$ monthly property tax

$598.75 + 75 = 673.75$

The total monthly payment is $673.75.

You Try It 5

The monthly mortgage payment for a home is $415.20. The annual property tax is $744. Find the total monthly payment for the mortgage and property tax.

Your strategy

Your solution

Solutions on p. S17

6.4 Exercises

Objective A   *Application Problems*

1. A condominium at Mt. Baldy Ski Resort was purchased for $97,000, and a down payment of $14,550 was made. Find the mortgage.

2. An insurance business was purchased for $173,000, and a down payment of $34,600 was made. Find the mortgage.

3. A building lot was purchased for $25,000. The lender requires a down payment of 30% of the purchase price. Find the down payment.

4. Ian Goldman purchased a new home for $88,500. The lender requires a down payment of 20% of the purchase price. Find the down payment.

5. Brian Stedman made a down payment of 25% of the $850,000 purchase price of an apartment building. How much was the down payment?

6. A clothing store was purchased for $125,000, and a down payment that was 25% of the purchase price was made. How much was the down payment?

7. A loan of $150,000 is obtained to purchase a home. The loan origination fee is $2\frac{1}{2}$ points. Find the amount of the loan origination fee.

8. Security Savings & Loan requires a borrower to pay $3\frac{1}{2}$ points for a loan. Find the amount of the loan origination fee for a loan of $90,000.

9. Baja Construction Inc. is selling homes for $150,000. A down payment of 5% is required.
   a. Find the down payment.
   b. Find the mortgage.

10. A cattle rancher purchased some land for $240,000. The bank requires a down payment of 15% of the purchase price.
    a. Find the down payment.
    b. Find the mortgage.

11. Vivian Tom purchased a home for $210,000. Find the mortgage if the down payment Vivian made is 10% of the purchase price.

12. A mortgage lender requires a down payment of 5% of the $80,000 purchase price of a condominium. How much is the mortgage?

Objective B   *Application Problems*

Solve. Use the Monthly Payment Table in the Appendix. Round to the nearest cent.

13. An investor obtained a loan of $150,000 to buy a car wash business. The monthly mortgage payment was based on 25 years at 8%. Find the monthly mortgage payment.

14. A beautician obtained a 20-year mortgage of $90,000 to expand the business. The credit union charges an annual interest rate of 9%. Find the monthly mortgage payment.

15. A couple interested in buying a home determines that they can afford a monthly mortgage payment of $800. Can they afford to buy a home with a 30-year $110,000 mortgage at 8% interest?

16. A lawyer is considering purchasing a new office building with a 15-year $400,000 mortgage at 8% interest. The lawyer can afford a monthly mortgage payment of $4000. Can the lawyer afford the monthly mortgage payment on the new office building?

17. The county tax assessor has determined that the annual property tax on a $125,000 house is $1348.20. Find the monthly property tax.

18. The annual property tax on a $155,000 home is $1992. Find the monthly property tax.

19. Abacus Imports Inc. has a warehouse with a 25-year mortgage of $200,000 at an annual interest rate of 9%.
   a. Find the monthly mortgage payment.
   b. During a month when $941.72 of the monthly mortgage payment is principal, how much of the payment is interest?

20. A vacation home has a mortgage of $135,000 for 30 years at an annual interest rate of 7%.
   a. Find the monthly mortgage payment.
   b. During a month when $392.47 of the monthly mortgage payment is principal, how much of the payment is interest?

21. The annual mortgage payment on a duplex is $10,844.40. The owner must pay an annual property tax of $948. Find the total monthly payment for the mortgage and property tax.

22. The monthly mortgage payment on a home is $716.40, and the homeowner pays an annual property tax of $792. Find the total monthly payment for the mortgage and property tax.

23. Maria Hernandez purchased a home for $210,000 and made a down payment of $15,000. The balance was financed for 15 years at an annual interest rate of 9%. Find the monthly mortgage payment.

24. A customer of a savings and loan purchased a $185,000 home and made a down payment of $20,000. The savings and loan charges its customers an annual interest rate of 7% for 30 years for a home mortgage. Find the monthly mortgage payment.

APPLYING THE CONCEPTS

25. A couple considering a mortgage of $100,000 have a choice of loans. One loan is an 8% loan for 20 years, and the other loan is at 8% for 30 years. Find the amount of interest that the couple can save by choosing the 20-year loan.

26. Find out what an adjustable-rate mortgage is. What is the difference between this type of loan and a fixed-rate mortgage? List some of the advantages and disadvantages of each.

6.5 Car Expenses

Objective A   **To calculate the initial expenses of buying a car**

The initial expenses in the purchase of a car usually include the down payment, the license fees, and the sales tax. The down payment may be very small or as much as 25% or 30% of the purchase price of the car, depending on the lending institution. License fees and sales tax are regulated by each state, so these expenses vary from state to state.

Example 1

A car is purchased for $18,500, and the lender requires a down payment of 15% of the purchase price. Find the amount financed.

Strategy

To find the amount financed:

- Find the down payment by solving the basic percent equation for *amount*.
- Subtract the down payment from the purchase price.

Solution

Percent $\times$ base $=$ amount

$$\boxed{\text{Percent}} \times \boxed{\begin{array}{c}\text{purchase}\\\text{price}\end{array}} = \boxed{\begin{array}{c}\text{down}\\\text{payment}\end{array}}$$

$$0.15 \times 18{,}500 = n$$
$$2775 = n$$

$$18{,}500 - 2775 = 15{,}725$$

The amount financed is $15,725.

You Try It 1

A down payment of 20% of the $19,200 purchase price of a new car is made. Find the amount financed.

Your strategy

Your solution

Example 2

A sales clerk purchases a car for $16,500 and pays a sales tax that is 5% of the purchase price. How much is the sales tax?

Strategy

To find the sales tax, solve the basic percent equation for *amount*.

Solution

Percent $\times$ base $=$ amount

$$\boxed{\text{Percent}} \times \boxed{\begin{array}{c}\text{purchase}\\\text{price}\end{array}} = \boxed{\begin{array}{c}\text{sales}\\\text{tax}\end{array}}$$

$$0.05 \times 16{,}500 = n$$
$$825 = n$$

The sales tax is $825.

You Try It 2

A car is purchased for $17,350. The car license fee is 1.5% of the purchase price. How much is the license fee?

Your strategy

Your solution

Solutions on p. S17

Objective B   **To calculate ongoing expenses of owning a car**

Besides the initial expenses of buying a car, there are continuing expenses involved in owning a car. These ongoing expenses include car insurance, gas and oil, general maintenance, and the monthly car payment. The monthly car payment is calculated in the same manner as monthly mortgage payments on a home loan. A monthly payment table, such as the one in the Appendix, is used to simplify the calculation of monthly car payments.

Example 3

At a cost of $.33 per mile, how much does it cost to operate a car during a year in which the car is driven 15,000 miles?

Strategy

To find the cost, multiply the cost per mile (0.33) by the number of miles driven (15,000).

Solution

$15,000 \times 0.33 = 4950$

The cost is $4950.

You Try It 3

At a cost of $.31 per mile, how much does it cost to operate a car during a year in which the car is driven 23,000 miles?

Your strategy

Your solution

Example 4

During one month, your total gasoline bill was $84 and the car was driven 1200 miles. What was the cost per mile for gasoline?

Strategy

To find the cost per mile, divide the cost for gasoline (84) by the number of miles driven (1200).

Solution

$84 \div 1200 = 0.07$

The cost per mile was $.07.

You Try It 4

In a year in which your total car insurance bill was $360 and the car was driven 15,000 miles, what was the cost per mile for car insurance?

Your strategy

Your solution

Example 5

A car is purchased for $18,500 with a down payment of $3700. The balance is financed for 3 years at an annual interest rate of 9%. Find the monthly car payment.

Strategy

To find the monthly payment:

• Subtract the down payment from the purchase price to find the amount financed.
• Multiply the amount financed by the factor found in the Monthly Payment Table in the Appendix.

Solution

$18,500 - 3700 = 14,800$
$14,800 \times 0.0317997 \approx 470.64$

The monthly payment is $470.64.

You Try It 5

A truck is purchased for $25,900 with a down payment of $6475. The balance is financed for 4 years at an annual interest rate of 8%. Find the monthly car payment.

Your strategy

Your solution

Solutions on p. S17

6.5 Exercises

Objective A   *Application Problems*

1. Amanda has saved $780 to make a down payment on a used minivan that costs $7100. The car dealer requires a down payment of 12% of the purchase price. Has she saved enough money to make the down payment?

2. A sedan was purchased for $23,500. A down payment of 15% of the purchase price was required. How much was the down payment?

3. A drapery installer bought a van to carry drapery samples. The purchase price of the van was $26,500, and a 4.5% sales tax was paid. How much was the sales tax?

4. A & L Lumber Company purchased a delivery truck for $28,500. A sales tax of 4% of the purchase price was paid. Find the sales tax.

5. A license fee of 2% of the purchase price is paid on a pickup truck costing $22,500. Find the license fee for the truck.

6. Your state charges a license fee of 1.5% on the purchase price of a car. How much is the license fee for a car that costs $16,998?

7. An electrician bought a $32,000 flatbed truck. A state license fee of $275 and a sales tax of 3.5% of the purchase price are required.
   a. Find the sales tax.
   b. Find the total cost of the sales tax and the license fee.

8. A physical therapist bought a used car for $9375 and made a down payment of $1875. The sales tax is 5% of the purchase price.
   a. Find the sales tax.
   b. Find the total cost of the sales tax and the down payment.

9. Martin bought a motorcycle for $16,200 and made a down payment of 25% of the purchase price.
   a. Find the down payment.
   b. Find the amount financed.

10. A carpenter bought a utility van for $24,900 and made a down payment of 15% of the purchase price.
    a. Find the down payment.
    b. Find the amount financed.

11. An author bought a sports car for $35,000 and made a down payment of 20% of the purchase price. Find the amount financed.

12. Tania purchased a used car for $13,500 and made a down payment of 25% of the cost. Find the amount financed.

Objective B   *Application Problems*

Solve. Use the Monthly Payment Table in the Appendix. Round to the nearest cent.

13. A rancher financed $14,000 for the purchase of a truck through a credit union at 9% interest for 4 years. Find the monthly truck payment.

14. A car loan of $8000 is financed for 3 years at an annual interest rate of 10%. Find the monthly car payment.

15. An estimate of the cost of owning a compact car is $.32 per mile. Using this estimate, find how much it costs to operate a car during a year in which the car is driven 16,000 miles.

16. An estimate of the cost of care and maintenance of automobile tires is $.015 per mile. Using this estimate, find how much it costs for care and maintenance of tires during a year in which the car is driven 14,000 miles.

17. A family spent $1600 on gas, oil, and car insurance during a period in which the car was driven 14,000 miles. Find the cost per mile for gas, oil, and car insurance.

18. Last year you spent $2100 for gasoline for your car. The car was driven 15,000 miles. What was your cost per mile for gasoline?

19. Elena's monthly car payment is $143.50. During a month in which $68.75 of the monthly payment is principal, how much of the payment is interest?

20. The cost for a pizza delivery truck for the year included $2868 in truck payments, $2400 for gasoline, and $675 for insurance. Find the total cost for truck payments, gasoline, and insurance for the year.

21. The city of Colton purchased a fire truck for $82,000 and made a down payment of $5400. The balance is financed for 5 years at an annual interest rate of 9%.
   a. Find the amount financed.
   b. Find the monthly truck payment.

22. A used car is purchased for $4995, and a down payment of $995 is made. The balance is financed for 3 years at an annual interest rate of 8%.
   a. Find the amount financed.
   b. Find the monthly car payment.

23. An artist purchased a new car costing $27,500 and made a down payment of $5500. The balance is financed for 3 years at an annual interest rate of 10%. Find the monthly car payment.

24. A camper is purchased for $19,500, and a down payment of $2500 is made. The balance is financed for 4 years at an annual interest rate of 9%. Find the monthly payment.

APPLYING THE CONCEPTS

25. One bank offers a 4-year car loan at an annual interest rate of 10% plus a loan application fee of $45. A second bank offers 4-year car loans at an annual interest rate of 11% but charges no loan application fee. If you need to borrow $5800 to purchase a car, which of the two bank loans has the lesser loan costs? Assume you keep the car for 4 years.

26. How much interest is paid on a 5-year car loan of $9000 if the interest rate is 9%?

6.6 Wages

Objective A   **To calculate commissions, total hourly wages, and salaries**

Commissions, hourly wage, and salary are three ways to receive payment for doing work.

Commissions are usually paid to salespersons and are calculated as a percent of total sales.

➡ As a real estate broker, Emma Smith receives a commission of 4.5% of the selling price of a house. Find the commission she earned for selling a home for $175,000.

To find the commission Emma earned, solve the basic percent equation for *amount*.

Percent   ×   base   =   amount

| Commission rate | × | total sales | = | commission |

0.045   ×   175,000   =   7875

The commission is $7875.

An employee who receives an **hourly wage** is paid a certain amount for each hour worked.

➡ A plumber receives an hourly wage of $18.25. Find the plumber's total wages for working 37 hours.

To find the plumber's total wages, multiply the hourly wage by the number of hours worked.

| Hourly wage | × | number of hours worked | = | total wages |

18.25   ×   37   =   675.25

The plumber's total wages for working 37 hours are $675.25.

An employee who is paid a **salary** receives payment based on a weekly, biweekly (every other week), monthly, or annual time schedule. Unlike the employee who receives an hourly wage, the salaried worker does not receive additional pay for working more than the regularly scheduled workday.

➡ Ravi Basar is a computer operator who receives a weekly salary of $695. Find his salary for 1 month (4 weeks).

To find Ravi's salary for 1 month, multiply the salary per pay period by the number of pay periods.

| Salary per pay period | × | number of pay periods | = | total salary |

695   ×   4   =   2780

Ravi's total salary for 1 month is $2780.

Example 1

A pharmacist's hourly wage is $28. On Saturday, the pharmacist earns time and a half (1.5 times the regular hourly wage). How much does the pharmacist earn for working 6 hours on Saturday?

Strategy

To find the pharmacist's earnings:

• Find the hourly wage for working on Saturday by multiplying the hourly wage by 1.5.
• Multiply the hourly wage by the number of hours worked.

Solution

$28 \times 1.5 = 42 \qquad 42 \times 6 = 252$

The pharmacist earns $252.

You Try It 1

A construction worker, whose hourly wage is $18.50, earns double time (2 times the regular hourly wage) for working overtime. Find the worker's wages for working 8 hours of overtime.

Your strategy

Your solution

Example 2

An efficiency expert received a contract for $3000. The consultant spent 75 hours on the project. Find the consultant's hourly wage.

Strategy

To find the hourly wage, divide the total earnings by the number of hours worked.

Solution

$3000 \div 75 = 40$

The hourly wage was $40.

You Try It 2

A contractor for a bridge project receives an annual salary of $48,228. What is the contractor's salary per month?

Your strategy

Your solution

Example 3

Dani Greene earns $28,500 per year plus a 5.5% commission on sales over $100,000. During one year, Dani sold $150,000 worth of computers. Find Dani's total earnings for the year.

Strategy

To find the total earnings:

• Find the sales over $100,000.
• Multiply the commission rate by sales over $100,000.
• Add the commission to the annual pay.

Solution

$150,000 - 100,000 = 50,000$
$50,000 \times 0.055 = 2750 \qquad$ Commission
$28,500 + 2750 = 31,250$

Dani earned $31,250.

You Try It 3

An insurance agent earns $27,000 per year plus a 9.5% commission on sales over $50,000. During one year, the agent's sales totaled $175,000. Find the agent's total earnings for the year.

Your strategy

Your solution

Solutions on pp. S17–S18

6.6 Exercises

Objective A   *Application Problems*

1. Lewis works in a clothing store and earns $9.50 per hour. How much does he earn in a 40-hour week?

2. Sasha pays a gardener an hourly wage of $11. How much does she pay the gardener for working 25 hours?

3. A real estate agent receives a 3% commission for selling a house. Find the commission that the agent earned for selling a house for $131,000.

4. Ron Caruso works as an insurance agent and receives a commission of 40% of the first year's premium. Find Ron's commission for selling a life insurance policy with a first-year premium of $1050.

5. A stockbroker receives a commission of 1.5% of the price of stock that is bought or sold. Find the commission on 100 shares of stock that were bought for $5600.

6. The owner of the Carousel Art Gallery receives a commission of 20% on paintings that are sold on consignment. Find the commission on a painting that sold for $22,500.

7. Keisha Brown receives an annual salary of $38,928 as a teacher of Italian. How much does Keisha receive each month?

8. An apprentice plumber receives an annual salary of $27,900. How much does the plumber receive per month?

9. An electrician's hourly wage is $25.80. For working overtime, the electrician earns double time. What is the electrician's hourly wage for working overtime?

10. Carlos receives a commission of 12% of his weekly sales as a sales representative for a medical supply company. Find the commission he earned during a week in which sales were $4500.

11. A golf pro receives a commission of 25% for selling a golf set. Find the commission the pro earned for selling a golf set costing $450.

12. Steven receives $3.75 per square yard to install carpet. How much does he receive for installing 160 square yards of carpet?

13. A typist charges $2.75 per page for typing technical material. How much does the typist earn for typing a 225-page book?

14. A nuclear chemist received $15,000 in consulting fees while working on a nuclear power plant. The chemist worked 120 hours on the project. Find the consultant's hourly wage.

15. Maxine received $3400 for working on a project as a computer consultant for 40 hours. Find her hourly wage.

16. Gil Stratton's hourly wage is $10.78. For working overtime, he receives double time.
a. What is Gil's hourly wage for working overtime?
b. How much does he earn for working 16 hours of overtime?

17. Mark is a lathe operator and receives an hourly wage of $15.90. When working on Saturday, he receives time and a half.
a. What is Mark's hourly wage on Saturday?
b. How much does he earn for working 8 hours on Saturday?

18. A stock clerk at a supermarket earns $8.20 an hour. For working the night shift, the clerk's wage increases by 15%.
a. What is the increase in hourly pay for working the night shift?
b. What is the clerk's hourly wage for working the night shift?

19. A nurse earns $16.50 an hour. For working the night shift, the nurse receives a 10% increase in pay.
a. What is the increase in hourly pay for working the night shift?
b. What is the hourly pay for working the night shift?

20. Tony's hourly wage as a service station attendant is $9.40. For working the night shift, his wage is increased 25%. What is Tony's hourly wage for working the night shift?

21. Nicole Tobin, a salesperson, receives a salary of $250 per week plus a commission of 15% on all sales over $1500. Find her earnings during a week in which sales totaled $3000.

APPLYING THE CONCEPTS

The table at the right shows the top five and the bottom five starting salaries for recent college graduates. Use this table for Exercises 22 to 25.

22. What was the amount of increase in the starting salary for journalism majors from the previous year? Round to the nearest dollar.

23. What was the starting salary in the previous year for an industrial engineer? Round to the nearest dollar.

24. Between electrical engineers and computer scientists, which received the larger amount of increase in starting salary from the previous year?

25. Between telecommunications majors and liberal arts majors, which received the smaller amount of increase in starting salary from the previous year?

Estimated Starting Salaries		
Bachelor's Degree	Estimated Starting Salary	% Change from Previous Year
Top 5		
Chemical engineering	$42,758	4.3%
Mechanical engineering	$39,852	4.5%
Electrical engineering	$38,811	4.0%
Industrial engineering	$37,732	4.0%
Computer science	$36,964	4.5%
Bottom 5		
Liberal arts	$24,102	3.5%
Natural resources	$22,950	3.5%
Home economics	$22,916	3.5%
Telecommunications	$22,447	4.0%
Journalism	$22,102	4.0%

Source: Michigan State University

Bank Statements

To calculate checkbook balances

Point of Interest

There are a number of computer programs that serve as "electronic" checkbooks. With these programs, you can pay your bills by using a computer to write the check and then transmit the check over telephone lines using a modem.

A checking account can be opened at most banks and savings and loan associations by depositing an amount of money in the bank. A checkbook contains checks and deposit slips and a checkbook register in which to record checks written and amounts deposited in the checking account. Each time a check is written, the amount of the check is subtracted from the amount in the account. When a deposit is made, the amount deposited is added to the amount in the account.

A portion of a checkbook register is shown below. The account holder had a balance of $587.93 before writing two checks, one for $286.87 and the other for $202.38, and making one deposit of $345.00.

		RECORD ALL CHARGES OR CREDITS THAT AFFECT YOUR ACCOUNT					BALANCE	
NUMBER	DATE	DESCRIPTION OF TRANSACTION	PAYMENT/DEBIT (−)	√ T	FEE (IF ANY) (−)	DEPOSIT/CREDIT (+)	$ 587	93
108	8/4	*Plumber*	$ 286 | 87		$	$	301	06
109	8/10	*Car Payment*	202 | 38				98	68
	8/14	*Deposit*				345 | 00	443	68

To find the current checking account balance, subtract the amount of each check from the previous balance. Then add the amount of the deposit.

The current checking account balance is $443.68.

Example 1 A mail carrier had a checking account balance of $485.93 before writing two checks, one for $18.98 and another for $35.72, and making a deposit of $250. Find the current checking account balance.

You Try It 1 A cement mason had a checking account balance of $302.46 before writing a check for $20.59 and making two deposits, one in the amount of $176.86 and another in the amount of $94.73. Find the current checking account balance.

Strategy To find the current balance:
- Subtract the amount of each check from the old balance.
- Add the amount of the deposit.

Your strategy

Solution

```
  485.93
−  18.98   first check
  466.95
−  35.72   second check
  431.23
+ 250.00   deposit
  681.23
```

The current checking account balance is $681.23.

Your solution

Solution on p. S18

Objective B **To balance a checkbook**

Each month a bank statement is sent to the account holder. The bank statement shows the checks that the bank has paid, the deposits received, and the current bank balance.

A bank statement and checkbook register are shown on the next page.

Balancing the checkbook, or determining whether the checking account balance is accurate, requires a number of steps.

1. In the checkbook register, put a check mark (√) by each check paid by the bank and by each deposit recorded by the bank.

		RECORD ALL CHARGES OR CREDITS THAT AFFECT YOUR ACCOUNT							BALANCE	
NUMBER	DATE	DESCRIPTION OF TRANSACTION	PAYMENT/DEBIT (−)		√ T	FEE (IF ANY) (−)	DEPOSIT/CREDIT (+)		$ 840	27
263	5/20	Dentist	$ 75	00	√	$	$		765	27
264	5/22	Post Office	33	61	√				731	66
265	5/22	Gas Company	67	14					664	52
	5/29	Deposit			√		192	00	856	52
266	5/29	Pharmacy	38	95	√				817	57
267	5/30	Telephone	63	85					753	72
268	6/2	Groceries	73	19	√				680	53
	6/3	Deposit			√		215	00	895	53
269	6/7	Insurance	103	00	√				792	53
	6/10	Deposit					225	00	1017	53
270	6/15	Photo Shop	16	63	√				1000	90
271	6/18	Newspaper	27	00					973	90

CHECKING ACCOUNT Monthly Statement			Account Number: 924-297-8
Date	Transaction	Amount	Balance
5/20	OPENING BALANCE		840.27
5/21	CHECK	75.00	765.27
5/23	CHECK	33.61	731.66
5/29	DEPOSIT	192.00	923.66
6/1	CHECK	38.95	884.71
6/1	INTEREST	4.47	889.18
6/3	CHECK	73.19	815.99
6/3	DEPOSIT	215.00	1030.99
6/9	CHECK	103.00	927.99
6/16	CHECK	16.63	911.36
6/20	SERVICE CHARGE	3.00	908.36
6/20	CLOSING BALANCE		908.36

2. Add to the current checkbook balance all checks that have been written but have not yet been paid by the bank and any interest paid on the account.

3. Subtract any service charges and any deposits not yet recorded by the bank. This is the checkbook balance.

4. Compare the balance with the bank balance listed on the bank statement. If the two numbers are equal, the bank statement and the checkbook "balance."

Current checkbook balance:	973.90
Checks: 265	67.14
267	63.85
271	27.00
Interest:	+ 4.47
	1136.36
Service charge:	− 3.00
	1133.36
Deposit:	− 225.00
Checkbook balance:	908.36

Closing bank balance from bank statement   Checkbook balance

$908.36   = $908.36

The bank statement and checkbook balance.

RECORD ALL CHARGES OR CREDITS THAT AFFECT YOUR ACCOUNT								BALANCE	
NUMBER	DATE	DESCRIPTION OF TRANSACTION	PAYMENT/DEBIT (−)	√ T	FEE (IF ANY) (−)	DEPOSIT/CREDIT (+)	$	1620	42
413	3/2	Car Payment	$ 232 | 15	√	$	$		1388	27
414	3/2	Utility	67 | 14	√				1321	13
415	3/5	Restaurant	78 | 14					1242	99
	3/8	Deposit		√		1842 | 66		3085	65
416	3/10	House Payment	672 | 14	√				2413	51
417	3/14	Insurance	177 | 10					2236	41

CHECKING ACCOUNT Monthly Statement			Account Number: 924-297-8
Date	Transaction	Amount	Balance
3/1	OPENING BALANCE		1620.42
3/4	CHECK	232.15	1388.27
3/5	CHECK	67.14	1321.13
3/8	DEPOSIT	1842.66	3163.79
3/10	INTEREST	6.77	3170.56
3/12	CHECK	672.14	2498.42
3/25	SERVICE CHARGE	2.00	2496.42
3/30	CLOSING BALANCE		2496.42

Balance the bank statement shown above.

1. In the checkbook register, put a check mark (√) by each check paid by the bank and by each deposit recorded by the bank.

2. Add to the current checkbook balance all checks that have been written but have not yet been paid by the bank and any interest paid on the account.

3. Subtract any service charges and any deposits not yet recorded by the bank. This is the checkbook balance.

4. Compare the balance with the bank balance listed on the bank statement. If the two numbers are equal, the bank statement and the checkbook "balance."

Current checkbook
balance: 2236.41

Checks: 415   78.14

         417   177.10

Interest: +   6.77

                                 2498.42

Service charge: −   2.00

Checkbook balance: 2496.42

Closing bank balance   Checkbook
from bank statement   balance

     $2496.42   = $2496.42

The bank statement and checkbook balance.

Example 2   Balance the bank statement shown below.

RECORD ALL CHARGES OR CREDITS THAT AFFECT YOUR ACCOUNT

NUMBER	DATE	DESCRIPTION OF TRANSACTION	PAYMENT/DEBIT (–)	√ T	FEE (IF ANY) (–)	DEPOSIT/CREDIT (+)	BALANCE $ 412 64
345	1/14	Phone Bill	$ 54 75	√	$	$	357 89
346	1/19	News Shop	18 98	√			338 91
347	1/23	Theatre Tickets	95 00				243 91
	1/31	Deposit		√		947 00	1190 91
348	2/5	Cash	250 00	√			940 91
349	2/12	Rent	840 00				100 91

CHECKING ACCOUNT Monthly Statement   Account Number: 924-297-8

Date	Transaction	Amount	Balance
1/10	OPENING BALANCE		412.64
1/18	CHECK	54.75	357.89
1/23	CHECK	18.98	338.91
1/31	DEPOSIT	947.00	1285.91
2/1	INTEREST	4.52	1290.43
2/10	CHECK	250.00	1040.43
2/10	CLOSING BALANCE		1040.43

Solution

Current checkbook balance:	100.91
Checks: 347	95.00
349	840.00
Interest:	+ 4.52
	1040.43
Service charge:	− 0.00
	1040.43
Deposit:	− 0.00
Checkbook balance:	1040.43

Closing bank balance from bank statement: $1040.43

Checkbook balance: $1040.43

The bank statement and the checkbook balance.

You Try It 2

Balance the bank statement shown below.

		RECORD ALL CHARGES OR CREDITS THAT AFFECT YOUR ACCOUNT							BALANCE	
NUMBER	DATE	DESCRIPTION OF TRANSACTION	PAYMENT/DEBIT (–)		√ T	FEE (IF ANY) (–)	DEPOSIT/CREDIT (+)		$ 903	17
	2/15	Deposit	$			$	$ 523	84	1427	01
234	2/20	Mortgage	773	21					653	80
235	2/27	Cash	200	00					453	80
	3/1	Deposit					523	84	977	64
236	3/12	Insurance	275	50					702	14
237	3/12	Telephone	78	73					623	41

CHECKING ACCOUNT Monthly Statement   Account Number: 314-271-4

Date	Transaction	Amount	Balance
2/14	OPENING BALANCE		903.17
2/15	DEPOSIT	523.84	1427.01
2/21	CHECK	773.21	653.80
2/28	CHECK	200.00	453.80
3/1	INTEREST	2.11	455.91
3/14	CHECK	275.50	180.41
3/14	CLOSING BALANCE		180.41

Your solution

Solution on p. S18

6.7 Exercises

Objective A   *Application Problems*

1. You had a checking account balance of $342.51 before making a deposit of $143.81. What is your new checking account balance?

2. Carmen had a checking account balance of $493.26 before writing a check for $48.39. What is the current checking account balance?

3. A real estate firm had a balance of $2431.76 in its rental-property checking account. What is the balance in this account after a check for $1209.29 has been written?

4. The business checking account for R and R Tires showed a balance of $1536.97. What is the balance in this account after a deposit of $439.21 has been made?

5. A nutritionist had a checking account balance of $1204.63 before writing one check for $119.27 and another check for $260.09. Find the current checkbook balance.

6. Sam had a checking account balance of $3046.93 before writing a check for $1027.33 and making a deposit of $150.00. Find the current checkbook balance.

7. The business checking account for Rachael's Dry Cleaning had a balance of $3476.85 before a deposit of $1048.53 was made. The store manager then wrote checks, one for $848.37 and another for $676.19. Find the current checkbook balance.

8. Joel had a checking account balance of $427.38 before a deposit of $127.29 was made. Joel then wrote two checks, one for $43.52 and one for $249.78. Find the current checkbook balance.

9. A carpenter had a checkbook balance of $404.96 before making a deposit of $350 and writing a check for $71.29. Is there enough money in the account to purchase a refrigerator for $675?

10. A taxi driver had a checkbook balance of $149.85 before making a deposit of $245 and writing a check for $387.68. Is there enough money in the account for the bank to pay the check?

11. A sporting goods store has the opportunity to buy downhill skis and cross-country skis at a manufacturer's closeout sale. The downhill skis will cost $3500, and the cross-country skis will cost $2050. There is currently $5625.42 in the sporting goods store's checking account. Is there enough money in the account to make both purchases by check?

12. A lathe operator's current checkbook balance is $1143.42. The operator wants to purchase a utility trailer for $525 and a used piano for $650. Is there enough money in the account to make the two purchases?

Objective B   *Application Problems*

13. Balance the checkbook.

RECORD ALL CHARGES OR CREDITS THAT AFFECT YOUR ACCOUNT

NUMBER	DATE	DESCRIPTION OF TRANSACTION	PAYMENT/DEBIT (−)		√ T	FEE (IF ANY) (−)	DEPOSIT/CREDIT (+)		BALANCE $	
									2466	79
223	3/2	Groceries	$ 167	32		$	$		2299	47
	3/5	Deposit					960	70	3260	17
224	3/5	Rent	860	00					2400	17
225	3/7	Gas & Electric	142	35					2257	82
226	3/7	Cash	300	00					1957	82
227	3/7	Insurance	218	44					1739	38
228	3/7	Credit Card	419	32					1320	06
229	3/12	Dentist	92	00					1228	06
230	3/13	Drug Store	47	03					1181	03
	3/19	Deposit					960	70	2141	73
231	3/22	Car Payment	241	35					1900	38
232	3/25	Cash	300	00					1600	38
233	3/25	Oil Company	166	40					1433	98
234	3/28	Plumber	155	73					1278	25
235	3/29	Department Store	288	39					989	86

CHECKING ACCOUNT Monthly Statement   Account Number: 122-345-1

Date	Transaction	Amount	Balance
3/1	OPENING BALANCE		2466.79
3/5	DEPOSIT	960.70	3427.49
3/7	CHECK	167.32	3260.17
3/8	CHECK	860.00	2400.17
3/8	CHECK	300.00	2100.17
3/9	CHECK	142.35	1957.82
3/12	CHECK	218.44	1739.38
3/14	CHECK	92.00	1647.38
3/18	CHECK	47.03	1600.35
3/19	DEPOSIT	960.70	2561.05
3/25	CHECK	241.35	2319.70
3/27	CHECK	300.00	2019.70
3/29	CHECK	155.73	1863.97
3/30	INTEREST	13.22	1877.19
4/1	CLOSING BALANCE		1877.19

14. Balance the checkbook.

		RECORD ALL CHARGES OR CREDITS THAT AFFECT YOUR ACCOUNT							BALANCE	
NUMBER	DATE	DESCRIPTION OF TRANSACTION	PAYMENT/DEBIT (–)		√ T	FEE (IF ANY) (–)	DEPOSIT/CREDIT (+)		$ 1219	43
	5/1	Deposit	$			$	$ 619	14	1838	57
515	5/2	Electric Bill	42	35					1796	22
516	5/2	Groceries	95	14					1701	08
517	5/4	Insurance	122	17					1578	91
518	5/5	Theatre Tickets	84	50					1494	41
	5/8	Deposit					619	14	2113	55
519	5/10	Telephone	37	39					2076	16
520	5/12	Newspaper	22	50					2053	66
	5/15	Interest					7	82	2061	48
	5/15	Deposit					619	14	2680	62
521	5/20	Computer Store	172	90					2507	72
522	5/21	Credit Card	313	44					2194	28
523	5/22	Eye Exam	82	00					2112	28
524	5/24	Groceries	107	14					2005	14
525	5/24	Deposit					619	14	2624	28
526	5/25	Oil Company	144	16					2480	12
527	5/30	Car Payment	288	62					2191	50
528	5/30	Mortgage Payment	877	42					1314	08

CHECKING ACCOUNT Monthly Statement		Account Number: 122-345-1	
Date	Transaction	Amount	Balance
5/1	OPENING BALANCE		1219.43
5/1	DEPOSIT	619.14	1838.57
5/3	CHECK	95.14	1743.43
5/4	CHECK	42.35	1701.08
5/6	CHECK	84.50	1616.58
5/8	CHECK	122.17	1494.41
5/8	DEPOSIT	619.14	2113.55
5/15	INTEREST	7.82	2121.37
5/15	CHECK	37.39	2083.98
5/15	DEPOSIT	619.14	2703.12
5/23	CHECK	82.00	2621.12
5/23	CHECK	172.90	2448.22
5/24	CHECK	107.14	2341.08
5/24	DEPOSIT	619.14	2960.22
5/30	CHECK	288.62	2671.60
6/1	CLOSING BALANCE		2671.60

15. Balance the checkbook.

RECORD ALL CHARGES OR CREDITS THAT AFFECT YOUR ACCOUNT

NUMBER	DATE	DESCRIPTION OF TRANSACTION	PAYMENT/DEBIT (−)		√ T	FEE (IF ANY) (−)	DEPOSIT/CREDIT (+)		BALANCE $ 2035	18
218	7/2	Mortgage	$ 984	60		$	$		1050	58
219	7/4	Telephone	63	36					987	22
220	7/7	Cash	200	00					787	22
	7/12	Deposit					792	60	1579	82
221	7/15	Insurance	292	30					1287	52
222	7/18	Investment	500	00					787	52
223	7/20	Credit Card	414	83					372	69
	7/26	Deposit					792	60	1165	29
224	7/27	Department Store	113	37					1051	92

CHECKING ACCOUNT Monthly Statement

Account Number: 122-345-1

Date	Transaction	Amount	Balance
7/1	OPENING BALANCE		2035.18
7/1	INTEREST	5.15	2040.33
7/4	CHECK	984.60	1055.73
7/6	CHECK	63.36	992.37
7/12	DEPOSIT	792.60	1784.97
7/20	CHECK	292.30	1492.67
7/24	CHECK	500.00	992.67
7/26	DEPOSIT	792.60	1785.27
7/28	CHECK	200.00	1585.27
7/30	CLOSING BALANCE		1585.27

APPLYING THE CONCEPTS

16. When a check is written, the amount is _____ from the balance.

17. When a deposit is made, the amount is _____ to the balance.

18. In checking the bank statement, _____ to the checkbook balance all checks that have been written but not processed.

19. In checking the bank balance, _____ any service charge and any deposits not yet recorded.

20. Define the words *credit* and *debit* as they apply to checkbooks.

Focus on Problem Solving

Counterexamples

An example that is given to show that a statement is not true is called a **counterexample.** For instance, suppose someone makes the statement "All colors are red." A counterexample to that statement would be to show someone the color blue or some other color.

If a statement is *always* true, there are no counterexamples. The statement "All even numbers are divisible by 2" is always true. It is not possible to give an example of an even number that is not divisible by 2.

In mathematics, statements that are always true are called *theorems,* and mathematicians are always searching for theorems. Sometimes a conjecture by a mathematician appears to be a theorem. That is, the statement appears to be always true, but later on someone finds a counterexample.

One example of this occurred when the French mathematician Pierre de Fermat (1601–1665) conjectured that $2^{(2^n)} + 1$ is always a prime number for any natural number n. For instance, when $n = 3$, we have $2^{(2^3)} + 1 = 2^8 + 1 = 257$, and 257 is a prime number. However, in 1732 Leonhard Euler (1707–1783) showed that when $n = 5$, $2^{(2^5)} + 1 = 4{,}294{,}967{,}297$ and that $4{,}294{,}967{,}297 = 641 \cdot 6{,}700{,}417$—without a calculator! Because $4{,}294{,}967{,}297$ was the product of two numbers (other than itself and 1), it is not a prime number. This counterexample showed that Fermat's conjecture is not a theorem.

For Exercises 1 and 5, find at least one counterexample.

1. All composite numbers are divisible by 2.

2. All prime numbers are odd numbers.

3. The square of any number is always bigger than the number.

4. The reciprocal of a number is always less than 1.

5. A number ending in 9 is always larger than a number ending in 3.

When a problem is posed, it may not be known whether the problem statement is true or false. For instance, Christian Goldbach (1690–1764) stated that every even number greater than 2 can be written as the sum of two prime numbers. For example,

$$12 = 5 + 7 \qquad 32 = 3 + 29$$

Although this problem is approximately 250 years old, mathematicians have not been able to prove it is a theorem, nor have they been able to find a counterexample.

For Exercises 6 to 9, answer true if the statement is always true. If there is an instance in which the statement is false, give a counterexample.

6. The sum of two positive numbers is always larger than either of the two numbers.

7. The product of two positive numbers is always larger than either of the two numbers.

8. Percents always represent a number less than or equal to 1.

9. It is never possible to divide by zero.

TAKE NOTE

Recall that a prime number is a natural number greater than 1 that can be divided by only itself and 1. For instance, 17 is a prime number. 12 is not a prime number because 12 is divisible by numbers other than 1 and 12—for example, 4.

Projects and Group Activities

Buying a Car Suppose a student has an after-school job to earn money to buy and maintain a car. We will make assumptions about the monthly costs in several categories in order to determine how many hours per week the student must work to support the car. Assume the student earns $8.50 per hour.

1. Monthly payment

   Assume that the car cost $8500 with a down payment of $1020. The remainder is financed for 3 years at an annual simple interest rate of 9%.

   Monthly payment = _____

2. Insurance

   Assume that insurance costs $1500 per year.

   Monthly insurance payment = _____

3. Gasoline

   Assume that the student travels 750 miles per month, that the car travels 25 miles per gallon of gasoline, and that gasoline costs $1.50 per gallon.

   Number of gallons of gasoline purchased per month = _____

   Monthly cost for gasoline = _____

4. Miscellaneous

   Assume $.33 per mile for upkeep.

   Monthly expense for upkeep = _____

5. Total monthly expenses for the monthly payment, insurance, gasoline, and miscellaneous = _____

6. To find the number of hours per month that the student must work to finance the car, divide the total monthly expenses by the hourly rate.

   Number of hours per month = _____

7. To find the number of hours per week that the student must work, divide the number of hours per month by 4.

   Number of hours per week = _____

   The student has to work almost _____ hours per week to pay the monthly car expenses.

If you own a car, make out your own expense record. If you do not own a car, make assumptions about the kind of car that you would like to purchase, and calculate the total monthly expenses that you would have. An insurance company will give you rates on different kinds of insurance. An automobile club can give you approximations of miscellaneous expenses.

Credit Card Charges

All credit card companies assess a finance charge when a credit card balance is not paid within a certain number of days of the *billing date* (the date the credit card bill is sent). There may also be an annual fee.

The table below shows the charges for five banks in September of 2000. (If you have Internet access, you can get current rates by visiting **http://www.bankrate.com**.)

Bank	Annual Interest Rate	Annual Fee
Pulaski Bank & Trust Co.	8.75%	$35
Wachovia Bank Card Services	9.50%	$88
USAA Savings Bank	10.50%	$45
Metropolitan National Bank	10.98%	$25
First Internet Bank of Indiana	12.00%	$0

The amount of the monthly finance charge is usually based on the average daily balance. For instance, say your credit card was issued by the USAA Savings Bank. If your average daily balance for one month was $375.89, then the monthly finance charge is calculated, using the simple interest formula, as follows:

TAKE NOTE
The monthly interest rate is the annual interest rate divided by 12.

$$\boxed{\text{Interest}} = \boxed{\text{principal}} \times \boxed{\text{monthly interest rate}} \times \boxed{\text{time (in months)}}$$

$$= 375.89 \times \frac{0.1050}{12} \times 1$$

$$\approx 3.29$$

The finance charge is $3.29.

1. Suppose you have an average daily balance of $421.65 for 12 months. With which bank will your *annual* charges (the finance charges and the annual fee) be least?

2. Suppose you have an average daily balance of $1175.91 for 12 months. With which bank will your *annual* charges (the finance charges and the annual fee) be least?

Chapter Summary

Key Words

The *unit cost* is the cost of one item. [p. 229]

Percent increase is used to show how much a quantity has increased over its original value. [p. 233]

Cost is the price a business pays for a product. [p. 234]

Selling price is the price at which a business sells a product to a customer. [p. 234]

Markup is the difference between selling price and cost. [p. 234]

Markup rate is the markup expressed as a percent of a product's cost. [p. 234]

Percent decrease is used to show how much a quantity has decreased from its original value. [p. 236]

Sale price is the price after a reduction from the regular price. [p. 237]

Discount is the difference between the regular price and the sale price. [p. 237]

Discount rate is the discount as a percent of a product's regular price. [p. 237]

Interest is the amount of money paid for the privilege of using someone else's money. [p. 243]

Principal is the amount of money originally deposited or borrowed. [p. 243]

The percent used to determine the amount of interest is the *interest rate*. [p. 243]

Interest computed on the original principal is called *simple interest*. [p. 243]

Compound interest is computed not only on the original principal but also on interest already earned. [p. 246]

A *mortgage* is an amount that is borrowed to buy real estate. [p. 253]

The *loan origination fee* is usually a percent of the mortgage and is expressed in *points*. [p. 253]

Commissions are usually paid to salespersons and are calculated as a percent of total sales. [p. 263]

An employee who receives an *hourly wage* is paid a certain amount for each hour worked. [p. 263]

An employee who is paid a *salary* receives payment based on a weekly, biweekly, monthly, or annual time schedule. [p. 263]

To *balance a checkbook* means to determine whether the checkbook balance is accurate. [p. 268]

Essential Rules	**To Find Unit Cost**	To find the unit cost, divide the total cost by the number of units. [p. 229]
	To Find Total Cost	To find the total cost, multiply the unit cost by the number of units. [p. 230]
	Basic Markup Equations	Selling price = cost + markup Markup = markup rate × cost [p. 234]
	Basic Discount Equations	Sale price = regular price − discount Discount = discount rate × regular price [p. 237]
	Annual Simple Interest Formula	Principal × $\dfrac{\text{annual}}{\text{interest rate}}$ × $\dfrac{\text{time}}{\text{in years}}$ = interest [p. 243]
	Maturity Value Formula for Simple Interest	Principal + interest = maturity value [p. 243]
	Monthly Payment on a Simple Interest Loan	Maturity value ÷ $\dfrac{\text{length of the}}{\text{loan in months}}$ = monthly payment [p. 244]

Chapter Review

1. A 20-ounce box of cereal costs $3.90. Find the unit cost.

2. An account executive had car expenses of $1025.58 for insurance, $605.82 for gas, $37.92 for oil, and $188.27 for maintenance during a year in which 11,320 miles were driven. Find the cost per mile for these four items taken as a group. Round to the nearest tenth of a cent.

3. An oil stock was bought for $42.375 per share. Six months later, the stock was selling for $55.25 per share. Find the percent increase in the price of the stock for the 6 months. Round to the nearest tenth of a percent.

4. A sporting goods store uses a markup rate of 40%. What is the markup on a ski suit that costs the store $180?

5. A contractor borrowed $100,000 from a credit union for 9 months at an annual interest rate of 9%. What is the simple interest due on the loan?

6. A computer programmer invested $25,000 in a retirement account that pays 6% interest, compounded daily. What is the value of the investment in 10 years? Use the Compound Interest Table in the Appendix. Round to the nearest cent.

7. Last year an oil company had earnings of $4.12 per share. This year the earnings are $4.73 per share. What is the percent increase in earnings per share? Round to the nearest percent.

8. The monthly mortgage payment for a condominium is $523.67. The owner must pay an annual property tax of $658.32. Find the total monthly payment for the mortgage and property tax.

9. A used pickup truck is purchased for $14,450. A down payment of 8% is made, and the remaining cost is financed for 4 years at an annual interest rate of 9%. Find the monthly payment. Use the Monthly Payment Table in the Appendix. Round to the nearest cent.

10. A fast-food restaurant invested $50,000 in an account that pays 7% annual interest compounded quarterly. What is the value of the investment in 1 year? Use the Compound Interest Table in the Appendix.

11. Paula Mason purchased a home for $125,000. The lender requires a down payment of 15%. Find the amount of the down payment.

12. A plumber bought a truck for $18,500. A state license of $315 and a sales tax of 6.25% of the purchase price are required. Find the total cost of the sales tax and the license fee.

13. Techno-Center uses a markup rate of 35% on all computer systems. Find the selling price of a computer system that costs the store $1540.

14. Mien pays a monthly car payment of $122.78. During a month in which $25.45 is principal, how much of the payment is interest?

15. The manager of the retail store at a ski resort receives a commission of 3% on all sales at the alpine shop. Find the total commission received during a month in which the shop had $108,000 in sales.

16. A suit that regularly costs $235 is on sale for 40% off the regular price. Find the sale price.

17. Luke had a checking account balance of $1568.45 before writing checks for $123.76, $756.45, and $88.77. He then deposited a check for $344.21. Find Luke's current checkbook balance.

18. Pros' Sporting Goods borrowed $30,000 at an annual interest rate of 8% for 6 months. Find the maturity value of the loan.

19. A credit union requires a borrower to pay $2\frac{1}{2}$ points for a loan. Find the origination fee for a $75,000 loan.

20. Sixteen ounces of mouthwash cost $3.49. A 33-ounce container of the same brand of mouthwash costs $6.99. Which is the better buy?

21. The Sweeneys bought a home for $156,000. The family made a 10% down payment and financed the remainder with a 30-year loan at an annual interest rate of 7%. Find the monthly mortgage payment. Use the Monthly Payment Table in the Appendix. Round to the nearest cent.

22. Richard Valdez receives $12.60 per hour for working 40 hours a week and time and a half for working over 40 hours. Find his total income during a week in which he worked 48 hours.

23. The business checking account of a donut shop showed a balance of $9567.44 before checks of $1023.55, $345.44, and $23.67 were written and checks of $555.89 and $135.91 were deposited. Find the current checkbook balance.

24. The simple interest due on a 4-month loan of $55,000 is $1375. Find the monthly payment on the loan.

25. A credit card company charges a customer 1.25% per month on the unpaid balance of charges on the card. What is the finance charge in a month in which the customer has an unpaid balance of $576?

Chapter Test

1. Twenty feet of lumber cost $138.40. What is the cost per foot?

2. Which is the more economical purchase: 3 pounds of tomatoes for $7.49 or 5 pounds of tomatoes for $12.59?

3. Red snapper costs $4.15 per pound. Find the cost of $3\frac{1}{2}$ pounds. Round to the nearest cent.

4. An exercise bicycle increased in price from $415 to $498. Find the percent increase in the cost of the exercise bicycle.

5. A department store uses a 40% markup rate. Find the selling price of a compact disc player that the store purchased for $215.

6. The price of gold dropped from $390 per ounce to $360 per ounce. What percent decrease does this amount represent? Round to the nearest tenth of a percent.

7. The price of a video camera dropped from $1120 to $896. What percent decrease does this price drop represent?

8. A corner hutch with a regular price of $299 is on sale for 30% off the regular price. Find the sale price.

9. A box of stationery that regularly sells for $9.50 is on sale for $5.70. Find the discount rate.

10. A construction company borrowed $75,000 for 4 months at an annual interest rate of 8%. Find the simple interest due on the loan.

11. Craig Allen borrowed $25,000 at an annual interest rate of 9.2% for 9 months. Find the maturity value of the loan.

12. A credit card company charges a customer 1.2% per month on the unpaid balance of charges on the card. What is the finance charge in a month in which the customer has an unpaid balance of $374.95?

13. Jorge, who is self-employed, placed $30,000 in an account that pays 6% annual interest compounded quarterly. How much interest was earned in 10 years? Use the Compound Interest Table in the Appendix.

14. A savings and loan institution is offering mortgage loans that have a loan origination fee of $2\frac{1}{2}$ points. Find the loan origination fee when a home is purchased with a loan of $134,000.

15. A new housing development offers homes with a mortgage of $222,000 for 25 years at an annual interest rate of 8%. Find the monthly mortgage payment. Use the Monthly Payment Table in the Appendix.

16. A Chevrolet was purchased for $23,750, and a 20% down payment was made. Find the amount financed.

17. A rancher purchased an SUV for $23,714 and made a down payment of 15% of the cost. The balance was financed for 4 years at an annual interest rate of 7%. Find the monthly truck payment. Use the Monthly Payment Table in the Appendix.

18. Shaney receives an hourly wage of $18.40 an hour as an emergency room nurse. When called in at night, she receives time and a half. How much does Shaney earn in a week when she works 30 hours at normal rates and 15 hours during the night?

19. The business checking account for a pottery store had a balance of $7349.44 before checks for $1349.67 and $344.12 were written. The store manager then made a deposit of $956.60. Find the current checkbook balance.

20. Balance the checkbook shown.

		RECORD ALL CHARGES OR CREDITS THAT AFFECT YOUR ACCOUNT							BALANCE	
NUMBER	DATE	DESCRIPTION OF TRANSACTION	PAYMENT/DEBIT (−)		√ T	FEE (IF ANY) (−)	DEPOSIT/CREDIT (+)		$ 1422	13
843	8/1	House Payment	$ 713	72		$	$		708	41
	8/4	Deposit					852	60	1561	01
844	8/5	Loan Payment	162	40					1398	61
845	8/6	Groceries	166	44					1232	17
846	8/10	Car Payment	322	37					909	80
	8/15	Deposit					852	60	1762	40
847	8/16	Credit Card	413	45					1348	95
848	8/18	Pharmacy	92	14					1256	81
849	8/22	Utilities	72	30					1184	51
850	8/28	Telephone	78	20					1106	31

CHECKING ACCOUNT Monthly Statement			Account Number: 122-345-1	
Date	Transaction		Amount	Balance
8/1	OPENING BALANCE			1422.13
8/3	CHECK		713.72	708.41
8/4	DEPOSIT		852.60	1561.01
8/8	CHECK		166.44	1394.57
8/8	CHECK		162.40	1232.17
8/15	DEPOSIT		852.60	2084.77
8/23	CHECK		72.30	2012.47
8/24	CHECK		92.14	1920.33
9/1	CLOSING BALANCE			1920.33

Cumulative Review

1. Simplify: $12 - (10 - 8)^2 \div 2 + 3$

2. Add: $3\frac{1}{3} + 4\frac{1}{8} + 1\frac{1}{12}$

3. Find the difference between $12\frac{3}{16}$ and $9\frac{5}{12}$.

4. Find the product of $5\frac{5}{8}$ and $1\frac{9}{15}$.

5. Divide: $3\frac{1}{2} \div 1\frac{3}{4}$

6. Simplify $\left(\frac{3}{4}\right)^2 \div \left(\frac{3}{8} - \frac{1}{4}\right) + \frac{1}{2}$.

7. Divide: $0.059\overline{)3.0792}$
Round to the nearest tenth.

8. Convert $\frac{17}{12}$ to a decimal. Round to the nearest thousandth.

9. Write "$410 in 8 hours" as a unit rate.

10. Solve the proportion $\frac{5}{n} = \frac{16}{35}$.
Round to the nearest hundredth.

11. Write $\frac{5}{8}$ as a percent.

12. Find 6.5% of 420.

13. Write 18.2% as a decimal.

14. What percent of 20 is 8.4?

15. 30 is 12% of what?

16. 65 is 42% of what? Round to the nearest hundredth.

17. A series of late-summer storms produced rainfall of $3\frac{3}{4}$, $8\frac{1}{2}$, and $1\frac{2}{3}$ inches during a 3-week period. Find the total rainfall during the 3 weeks.

18. The Homer family pays $\frac{1}{5}$ of its total monthly income for taxes. The family has a total monthly income of $4850. Find the amount of their monthly income that the Homers pay in taxes.

19. In 5 years, the cost of a scientific calculator went from $75 to $30. What is the ratio of the decrease in price to the original price?

20. A compact car was driven 417.5 miles on 12.5 gallons of gasoline. Find the number of miles driven per gallon of gasoline.

21. A 14-pound turkey costs $12.96. Find the unit cost. Round to the nearest cent.

22. Eighty shares of a stock paid a dividend of $112. At the same rate, find the dividend on 200 shares of the stock.

23. A video camera that regularly sells for $900 is on sale for 20% off the regular price. What is the sale price?

24. A department store bought a portable disc player for $85 and used a markup rate of 40%. Find the selling price of the disc player.

25. Sook Kim, an elementary school teacher, received an increase in salary from $2800 per month to $3024 per month. Find the percent increase in her salary.

26. A contractor borrowed $120,000 for 6 months at an annual interest rate of 10%. How much simple interest is due on the loan?

27. A red Ford Mustang was purchased for $26,900, and a down payment of $2000 was made. The balance is financed for 3 years at an annual interest rate of 9%. Find the monthly payment. Use the Monthly Payment Table in the Appendix. Round to the nearest cent.

28. A family had a checking account balance of $1846.78. A check of $568.30 was deposited into the account, and checks of $123.98 and $47.33 were written. Find the new checking account balance.

29. During one year, Anna Gonzalez spent $840 on gasoline and oil, $520 on insurance, $185 on tires, and $432 on repairs. Find the cost per mile to drive the car 10,000 miles during the year. Round to the nearest cent.

30. A house has a mortgage of $72,000 for 20 years at an annual interest rate of 11%. Find the monthly mortgage payment. Use the Monthly Payment Table in the Appendix. Round to the nearest cent.

Chapter 7

Statistics and Probability

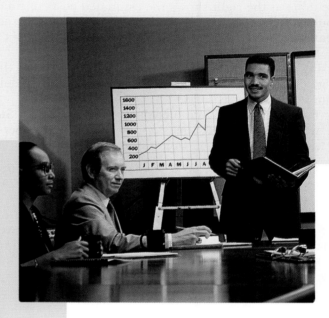

Objectives

Section 7.1

A To read a pictograph
B To read a circle graph

Section 7.2

A To read a bar graph
B To read a broken-line graph

Section 7.3

A To read a histogram
B To read a frequency polygon

Section 7.4

A To find the mean, median, and mode of a distribution
B To draw a box-and-whiskers plot

Section 7.5

A To calculate the probability of simple events

When business people use graphs during presentations, the information they present is often more easily understood and evaluated. Data displays, such as circle graphs and bar graphs, are used to assist us in analyzing a lot of information at one time because they condense all data into a single figure. Line graphs, like the one shown here and in the **examples on page 298,** provide the advantage of illustrating the trend of data over time.

Need help? For on-line student resources, such as section quizzes, visit this textbook's web site at **math.college.hmco.com/students.**

Prep Test

1. Bill-related mail accounted for 49 billion of the 102 billion pieces of first-class mail handled by the U.S. Postal Service during a recent year. (Source: USPS) What percent of the pieces of first-class handled by the U.S. Postal Service was bill-related mail? Round to the nearest tenth of a percent.

2. The table at the right shows the estimated costs of funding an education at a public college. Between which two enrollment years is the increase in cost greatest? What is the increase between these two years?

Enrollment Year	Cost of Public College
2005	$70,206
2006	$74,418
2007	$78,883
2008	$83,616
2009	$88,633
2010	$93,951

Source: The College Board's Annual Survey of Colleges

3. During the 1924 Summer Olympics in Paris, France, the United States won 45 gold medals, 27 silver medals, and 27 bronze medals. (Source: The Ultimate Book of Sports Lists)
   a. Find the ratio of gold medals won by the United States to silver medals won by the United States during the 1924 Summer Olympics. Write the ratio as a fraction in simplest form.

   b. Find the ratio of silver medals won by the United States to bronze medals won by the United States during the 1924 Summer Olympics. Write the ratio using a colon.

4. The table below shows the number of television viewers, in millions, who watch pay-cable channels, such as HBO and Showtime, each night of the week. (Source: Neilsen Media Research analyzed by Initiative Media North America)

Mon	Tue	Wed	Thu	Fri	Sat	Sun
3.9	4.5	4.2	3.9	5.2	7.1	5.5

   a. Arrange the numbers in the table from smallest to largest.
   b. Find the average number of viewers per night.

5. Approximately 90,000 women serve in the U.S. military. Five percent of these women serve in the Marine Corps. (Source: U.S. Department of Defense)
   a. Approximately how many women are in the Marine Corps?
   b. What fractional amount of women in the military are in the Marine Corps?

Go Figure

I have 2 brothers and 1 sister. My father's parents have 10 grandchildren. My mother's parents have 11 grandchildren. If no divorces or remarriages occurred, how many first cousins do I have?

7.1 Pictographs and Circle Graphs

Objective A   **To read a pictograph**

Statistics is the branch of mathematics concerned with **data,** or numerical information. **Graphs** are displays that provide a pictorial representation of data. The advantage of graphs is that they present information in a way that is easily read. The disadvantage of graphs is that they can be misleading. (See "Projects and Group Activities" at the end of this chapter.)

A **pictograph** uses symbols to represent information. The pictograph in Figure 1 represents the net worth of America's richest billionaires. Each symbol represents ten billion dollars.

Net Worth (in tens of billions of dollars)

Bill Gates

Larry Ellison

Warren Buffet

Paul Allen

S. Robson Walton

Figure 1 Net worth of America's richest billionaires
Source: Forbes

From the pictograph, we can determine that Bill Gates has the greatest net worth. Paul Allen's net worth is $10 billion more than S. Robson Walton's net worth.

The pictograph in Figure 2 represents the responses of 600 young Americans when asked what they would like to have with them on a desert island. "Books" was the response of what percent of the respondents?

Strategy
Use the basic percent equation. The base is 600 (the total number of responses), and the amount is 90 (the number responding "Books").

Music
Parents
Computer
Books
TV

= 30 responses

Figure 2 What 600 young Americans want on a desert island
Source: Time, June 30, 1997

Solution

Percent × base = amount

$$n \times 600 = 90$$

$$n = 90 \div 600$$

$$n = 0.15$$

15% of the respondents wanted books on a desert island.

The pictograph in Figure 3 shows the number of new cellular phones purchased in a particular city during a 4-month period.

The ratio of the number of cellular phones purchased in March to the number purchased in January is

$$\frac{3000}{4500} = \frac{2}{3}$$

January
February
March
April

= 1000 cellular phones

Figure 3 Monthly cellular phone purchases

Example 1

Use Figure 3 to find the total number of cellular phones purchased during the 4-month period.

Strategy

To find the total number of cellular phones purchased in the 4-month period:

- Read the pictograph to determine the number of cellular phones purchased each month.
- Add the four numbers.

Solution

Purchases for January: 4500
Purchases for February: 3500
Purchases for March: 3000
Purchases for April: 1500

Total purchases for the 4-month period:

```
   4,500
   3,500
   3,000
+  1,500
  _____
  12,500
```

There were 12,500 cellular phones purchased in the 4-month period.

You Try It 1

According to Figure 3, the number of cellular phones purchased in March represents what percent of the total number of cellular phones purchased in that city during that 4-month period?

Your strategy

Your solution

Solution on pp. S18–S19

Objective B   **To read a circle graph**

Point of Interest

Fossil fuels include coal, natural gas, and petroleum. Renewable energy includes hydroelectric power, solar energy, wood burning, and wind energy.

A **circle graph** represents data by the size of the sectors.

The circle graph in Figure 4 shows the consumption of energy sources in the United States during a recent year. The complete circle graph represents the total amount of energy consumed, 94.4 quadrillion Btu. Each sector of the circle represents the consumption of energy from a different source.

To find the percent of the total energy consumed that originated from nuclear power, solve the basic percent equation for percent. The base is 94.4 quadrillion Btu, and the amount is 7.2 quadrillion Btu.

Percent × base = amount

$$n \times 94.4 = 7.2$$
$$n = 7.2 \div 94.4$$
$$n \approx 0.076$$

To the nearest tenth of a percent, 7.6% of the energy consumed originated from nuclear power.

Figure 4 Annual energy consumption in quadrillion Btu in the United States

Source: The World Almanac and Book of Facts 2000

The circle graph in Figure 5 shows what percents of the $17.8 billion that the top 25 companies in the United States spent for national advertising in one year went to the various advertising media. The complete circle represents 100% of all the money spent by these companies. Each sector of the graph represents the percent of the total spent for a particular medium.

⟹ According to Figure 5, how much money was spent for magazine advertising? Round to the nearest hundred million dollars.

Strategy
Use the basic percent equation. The base is $17.8 billion, and the percent is 16%.

Solution
Percent × base = amount

$$0.16 \times 17.8 = n$$
$$2.848 = n$$

2.848 billion = 2,848,000,000

To the nearest hundred million, the amount spent for magazine advertising was $2,800,000,000.

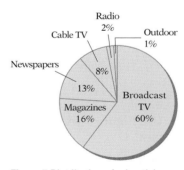

Figure 5 Distribution of advertising dollars for 25 companies

Source: Interep research

The circle graph in Figure 6 shows typical annual expenses of owning, operating, and financing a new car. Use this figure for Example 2 and You Try It 2.

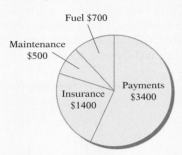

Figure 6 Annual expenses of $6000 for owning, operating, and financing a car
Source: Based on data from IntelliChoice

The circle graph in Figure 7 shows the distribution of an employee's gross monthly income. Use this figure for Example 3 and You Try It 3.

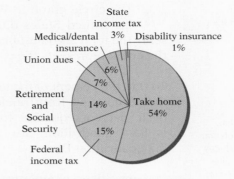

Figure 7 Distribution of gross monthly income of $2900

Example 2

Use Figure 6 to find the ratio of the annual insurance expense to the total annual cost of the car.

Strategy

To find the ratio:

• Locate the annual insurance expense in the circle graph.
• Write in simplest form the ratio of the annual insurance expense to the total annual cost of operating the car.

Solution

Annual insurance expense: $1400

$$\frac{1400}{6000} = \frac{7}{30}$$

The ratio is $\frac{7}{30}$.

You Try It 2

Use Figure 6 to find the ratio of the annual cost of fuel to the annual cost of maintenance.

Your strategy

Your solution

Example 3

Use Figure 7 to find the employee's take-home pay.

Strategy

To find the take-home pay:

• Locate the percent of the distribution that is take-home pay.
• Solve the basic percent equation for amount.

Solution

Take-home pay: 54%

Percent × base = amount
0.54 × 2900 = n
1566 = n

The employee's take-home pay is $1566.

You Try It 3

Use Figure 7 to find the amount paid for medical/dental insurance.

Your strategy

Your solution

Solutions on p. S19

7.1 Exercises

Objective A

The pictograph in Figure 8 shows the approximate gross revenues in the United States from four Walt Disney animated movies.

1. Find the total gross revenues from the four movies.

2. Find the ratio of the gross revenue of *Beauty and the Beast* to the gross revenue of *The Hunchback of Notre Dame*.

3. Find the percent of the total gross revenue that was earned by *The Lion King*.

Figure 8 Gross revenues of four Walt Disney animated movies

Source: **www.worldboxoffice.com**

The pictograph in Figure 9 is based on a survey of adults who were asked whether they agreed with each statement.

4. Find the ratio of the number of people who agree that space exploration impacts daily life to the number of people who agree that space will be colonized in their lifetime.

5. How many more people agreed that humanity should explore planets than agreed that space exploration impacts daily life?

6. Is the number of people who said they would travel in space more than twice the number of people who agreed that space would be colonized in their lifetime?

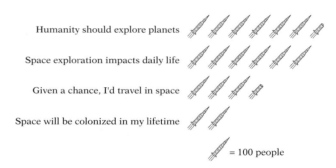

Figure 9 Number of adults who agree with the statement

Source: Opinion Research for Space Day Partners

The pictograph in Figure 10 is based on a survey of children aged 7 through 12. The percent of children's responses to the survey are shown. Assume that 500 children were surveyed.

7. Find the number of children who said they hid vegetables under a napkin.

8. What is the difference between the number of children who fed vegetables to the dog and the number who dropped them on the floor?

9. Were the responses given in the graph the only responses given by the children? Explain your answer.

Figure 10 How children try to hide vegetables

Source: Strategic Consulting and Research for Del Monte

An accounting major recorded the number of units required in each discipline to graduate with a degree in accounting. The results are shown in the circle graph in Figure 11.

10. How many units are required to graduate with a degree in accounting?

11. What is the ratio of the number of units in finance to the number of units in accounting?

12. What percent of the units required to graduate are taken in accounting? Round to the nearest tenth of a percent.

13. What percent of the units required to graduate are taken in mathematics? Round to the nearest tenth of a percent.

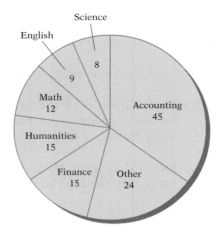

Figure 11 Number of units required to graduate with an accounting degree

The circle graph in Figure 12 shows the result of a survey in which people were asked, "What bothers you most about movie theaters?"

14. **a.** What complaint was mentioned the most often?
b. What complaint was mentioned the least often?

15. How many people were surveyed?

16. What is the ratio of the number of people responding "Dirty floors" to the number responding "High ticket prices"?

17. What percent of the respondents said that people talking bothered them most?

Figure 12 Distribution of responses in a survey

The circle graph in Figure 13 shows the breakdown of the approximately $3,100,000,000 that Americans spent on home video game equipment in one year.

18. Find the amount of money spent on TV game machines.

19. Find the amount of money spent on portable game machines.

20. What fractional amount of the total money spent was spent on accessories?

Figure 13 Percents of $3,100,000,000 spent annually on home video games

Source: The NPD Group, Toy Manufacturers of America

21. Is the amount spent for TV game machines more than three times the amount spent for portable game machines?

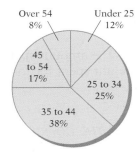 The circle graph in Figure 14 shows a breakdown, according to age, of the homeless in America.

22. What age group represents the largest segment of the homeless population?

23. Is the number of homeless who are aged 25 to 34 more or less than twice the number who are under the age of 25?

24. What percent of the homeless population is under the age of 35?

Figure 14 Ages of the homeless in America

Source: The Department of Housing and Urban Development

25. On average, how many of every 100,000 homeless people in America are over the age of 54?

The circle graph in Figure 15 shows the land area of each of the seven continents in square miles.

26. Find the total land area of the seven continents.

27. How much larger is North America than South America?

28. What percent of the total land area is the land area of Asia? Round to the nearest tenth of a percent.

29. What percent of the total land area is the land area of Australia? Round to the nearest tenth of a percent.

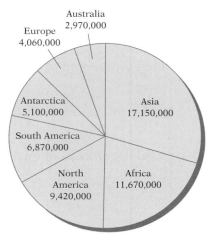

Figure 15 Land area of the seven continents (in square miles)

The circle graph in Figure 16 shows the average percent distribution of a family's income of $34,000 per year.

30. How much of the family's income was spent on food?

31. How much of the family's income was spent on entertainment?

32. How much more of the family's income was spent on medical needs than was spent on clothing?

33. Was the amount spent for housing more than twice the amount spent for food?

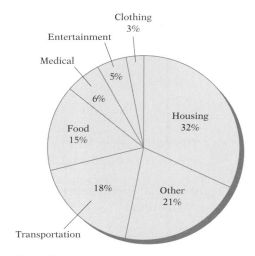

Figure 16 Average percent distribution of a family's income

Source: Consumer Expenditure Survey for the Bureau of Labor Statistics

APPLYING THE CONCEPTS

34. a. What are the advantages of presenting data in the form of a pictograph?
b. What are the disadvantages?

35. The circle graph at the right shows a couple's expenditures last month. Write two observations about this couple's expenses.

7.2 Bar Graphs and Broken-Line Graphs

Objective A To read a bar graph

Point of Interest

The first bar graph appeared in 1786 in the book *The Commercial and Political Atlas*. The author, William Playfair (1759–1823), was a pioneer in the use of graphical displays.

A **bar graph** represents data by the height of the bars. The bar graph in Figure 17 shows temperature data recorded for Cincinnati, Ohio, for the months March through November. For each month, the height of the bar indicates the normal daily high temperature during that month. The jagged line near the bottom of the graph indicates that the vertical scale is missing the numbers between 0 and 50.

The daily high temperature in September was 78°F. Because the bar for July is the tallest, the daily high temperature was highest in July.

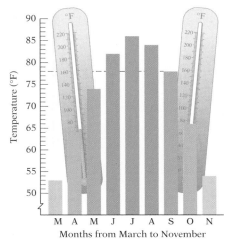

Figure 17 Daily high temperatures in Cincinnati, Ohio

Source: U.S. Weather Bureau

TAKE NOTE

The bar for athletic females is halfway between the marks for 50 and 60. Therefore, we estimate that the lung capacity is halfway between these two numbers, at 55.

A **double-bar graph** is used to display data for purposes of comparison. The double-bar graph in Figure 18 shows the lung capacity of inactive, versus that of athletic, 45-year-olds.

The lung capacity of an athletic female is 55 milliliters of oxygen per kilogram of body weight per minute.

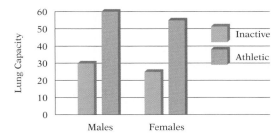

Figure 18 Lung capacity (in milliliters of oxygen per kilogram of body weight per minute)

Example 1

What is the ratio of the lung capacity of an inactive male to that of an athletic male?

Strategy

To write the ratio:

- Read the graph to find the lung capacity of an inactive male and of an athletic male.
- Write the ratio in simplest form.

Solution

Lung capacity of inactive male: 30
Lung capacity of athletic male: 60

$$\frac{30}{60} = \frac{1}{2}$$

The ratio is $\frac{1}{2}$.

You Try It 1

What is the ratio of the lung capacity of an inactive female to that of an athletic female?

Your strategy

Your solution

Solution on p. S19

Objective B   **To read a broken-line graph**

A **broken-line graph** represents data by the position of the lines. It is used to show trends.

The broken-line graph in Figure 19 shows the effect of inflation on the value of a $100,000 life insurance policy. The height of each dot indicates the value of the policy.

After 10 years, the purchasing power of the $100,000 has decreased to approximately $60,000.

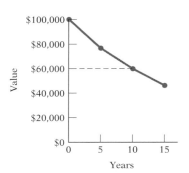

Figure 19 Effect of inflation on the value of a $100,000 life insurance policy

Two broken-line graphs are often shown in the same figure for comparison. Figure 20 shows the net incomes of two software companies, Math Associates and MatheMentors, before their merger.

Several things can be determined from the graph:

The net income for Math Associates in 2004 was $12 million.

The net income for MatheMentors declined from 2000 to 2001.

The net income for Math Associates increased for each year shown.

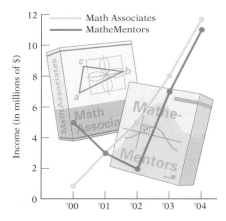

Figure 20 Net incomes of Math Associates and MatheMentors

Example 2

Use Figure 20 to approximate the difference between the net income of Math Associates and that of MatheMentors in 2002.

Strategy

To write the difference:

• Read the line graph to determine the net income of Math Associates and that of MatheMentors in 2002.
• Subtract to find the difference.

Solution

Net income for Math Associates: $5 million
Net income for MatheMentors: $2 million

$5 - 2 = 3$

The difference between the net incomes in 2002 was $3 million.

You Try It 2

Use Figure 20 to determine between which two years the net income of Math Associates increased the most.

Your strategy

Your solution

Solution on p. S19

7.2 Exercises

The bar graph in Figure 21 shows the regions in which all the passenger cars were produced during a recent year.

1. How many passenger cars were produced worldwide?

2. What is the difference between the number of passenger cars produced in Western Europe and the number produced in North America?

3. What percent of the passenger cars were produced in Asia? Round to the nearest percent.

Figure 21 Number of passenger cars produced (in a recent year)
Source: Reprinted with permission.

The double-bar graph in Figure 22 shows the fuel efficiency of four vehicles, as rated by the Environmental Protection Agency. They are among the most fuel-efficient 2000 model-year cars for city and highway mileage.

4. Is the fuel efficiency of the Chevrolet Metro greater on the highway or in city driving?

5. Approximately how many more miles per gallon does the Honda Insight get while traveling on the highway than in the city?

6. Estimate the difference between the fuel efficiency of the Honda Insight in city driving and the fuel efficiency of the VW Beetle in city driving.

Figure 22 Fuel efficiency ratings
Source: Environmental Protection Agency

The double-bar graph in Figure 23 shows maximum salaries for police officers in selected cities and the corresponding maximum salaries for officers in the suburbs of that city.

7. Estimate the difference between the maximum salaries of police officers in the suburbs of New York City and in the city of New York.

8. Is there a city for which the maximum salary of a police officer in the city is greater than the salary in the suburbs?

9. For which city is the difference between the maximum salary in the suburbs and that in the city the greatest?

10. Of the cities shown on the graph, which city has the lowest maximum salary for police officers in the suburbs?

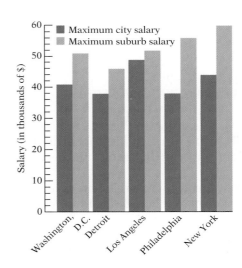

Figure 23 Maximum salaries of police officers in the city and the suburbs
Source: USA Today

Objective B

 The broken-line graph in Figure 24 shows the average monthly snowfall during ski season around Aspen, Colorado.

11. What is the average snowfall during January?

12. During which month is the snowfall the greatest?

13. What is the total average snowfall during March and April?

14. Find the ratio of the average snowfall in November to the average snowfall in December.

Figure 24 Average snowfall in Aspen, Colorado
Source: Weather America, by Alfred Garwood

 The double-broken-line graph in Figure 25 shows the number of calories per day that should be consumed by women and men in various age groups.

15. What is the difference between the number of calories recommended for men and the number recommended for women 19–22 years of age?

16. People of what age and gender have the lowest recommended number of calories?

17. Find the ratio of the number of calories recommended for women 15 to 18 years old to the number recommended for women 51 to 74 years old.

Figure 25 Recommended number of calories per day for women and men
Source: Numbers, by Andrea Sutcliffe (HarperCollins)

APPLYING THE CONCEPTS

 The graph in Figure 26 shows the amount of money, in billions of dollars, spent by the United States government for drug prevention in the 1990s.

18. Create a table that shows the total amount spent for foreign and domestic aid for each year from 1991 to 1999.

19. Create a table that shows the difference between the amount spent for foreign and for domestic aid in each year from 1991 to 1999.

Figure 26 Aid provided by the U.S. government for drug prevention
Source: Reprinted by permission of the *San Diego Union-Tribune.*

7.3 Histograms and Frequency Polygons

Objective A **To read a histogram**

A research group measured the fuel usage of 92 cars. The results are recorded in the histogram in Figure 27. A **histogram** is a special type of bar graph. The width of each bar corresponds to a range of numbers called a **class interval.** The height of each bar corresponds to the number of occurrences of data in each class interval and is called the **class frequency.**

Class Intervals (miles per gallon)	Class Frequencies (number of cars)
18–20	12
20–22	19
22–24	24
24–26	17
26–28	15
28–30	5

Figure 27

Twenty-four cars get between 22 and 24 miles per gallon.

A precision tool company has 85 employees. Their hourly wages are recorded in the histogram in Figure 28.

The ratio of the number of employees whose hourly wage is between $14 and $16 to the total number of employees is $\frac{17 \text{ employees}}{85 \text{ employees}} = \frac{1}{5}$.

Figure 28

Example 1

Use Figure 28 to find the number of employees whose hourly wage is between $16 and $20.

Strategy

To find the number of employees:

• Read the histogram to find the number of employees whose hourly wage is between $16 and $18 and the number whose hourly wage is between $18 and $20.
• Add the two numbers.

Solution

Number whose wage is between $16 and $18: 20; between $18 and $20: 14

20 + 14 = 34

34 employees have an hourly wage between $16 and $20.

You Try It 1

Use Figure 28 to find the number of employees whose hourly wage is between $10 and $14.

Your strategy

Your solution

Solution on p. S19

Objective B   **To read a frequency polygon**

The speeds of 70 cars on a highway were measured by radar. The results are recorded in the frequency polygon in Figure 29. A **frequency polygon** is a graph that displays information in a manner similar to a histogram. A dot is placed above the center of each class interval at a height corresponding to that class's frequency. The dots are then connected to form a broken-line graph. The center of a class interval is called the **class midpoint.**

Class Inteval (miles per hour)	Class Midpoint	Class Frequency
30–40	35	7
40–50	45	13
50–60	55	25
60–70	65	21
70–80	75	4

Figure 29

Twenty-five cars were traveling between 50 and 60 miles per hour.

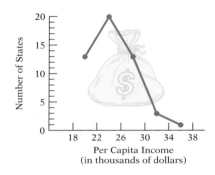

The per capita incomes in a recent year for the 50 states are recorded in the frequency polygon in Figure 30.

The number of states with a per capita income between $22,000 and $26,000 is 20.

Figure 30

Source: Department of Commerce

Example 2

According to Figure 30, what percent of the states have a per capita income between $22,000 and $26,000?

Strategy

To find the percent, solve the basic percent equation for percent. The base is 50. The amount is 20.

Solution

Percent × base = amount

$n \times 50 = 20$

$n = 20 \div 50$

$n = 0.4$

40% of the states have a per capita income between $22,000 and $26,000.

You Try It 2

Use Figure 30 to find the ratio of the number of states with a per capita income between $18,000 and $22,000 to the number with a per capita income between $26,000 and $30,000.

Your strategy

Your solution

Solution on p. S20

7.3 Exercises

Objective A

● The annual tuition for undergraduate college students attending four-year institutions varies depending on the college. The histogram in Figure 31 shows the tuition amounts for a representative sample of 120 students from various parts of the United States.

1. How many students have a tuition that is between $3000 and $6000 per year?

2. What is the ratio of the number of students whose tuition is between $9000 per year and $12,000 per year to the total number of students represented?

3. How many students pay more than $12,000 annually for tuition?

4. What percent of the total number of students spend less than $6000 annually?

● The histogram in Figure 32 is based on data from the American Automobile Manufacturers Association. It shows the ages of a sample of 1000 cars in a typical city in the United States.

5. How many cars are between 6 and 12 years old?

6. Find the ratio of the number of cars between 12 and 15 years old to the total number of cars.

7. Find the number of cars more than 12 years old.

8. Find the percent of cars that are less than 9 years old.

● According to a Maritz AmeriPoll, the average U.S. adult goes to a shopping mall about two times a month. The histogram shows the average time 100 adults spend in the mall per trip.

9. Find the number of adults who spend between 1 and 2 hours at the mall.

10. Find the number of adults who spend between 3 and 4 hours at the mall.

11. What percent of the adults spend less than 1 hour at the mall?

12. What percent of the adults spend 5 or more hours at the mall?

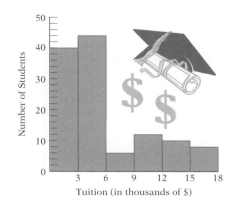

Figure 31
Source: Educational Testing Service

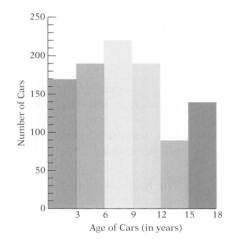

Figure 32
Source: American Automobile Manufacturers Association

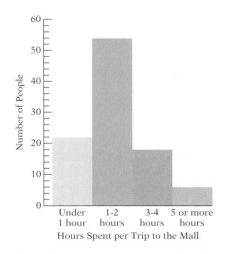

Figure 33
Source: Maritz AmeriPoll

Objective B

The frequency polygon in Figure 34 shows the finishing times of the entrants in the University & College Women's Distance Medley Relay Finals at the Drake Relays in April 2000.

13. How many teams competed in the relay finals?

14. How many teams completed the relay in more than 12 minutes?

15. What percent of the teams completed the relay in less than 12 minutes? Round to the nearest tenth of a percent.

The frequency polygon in Figure 35 is based on data from a Gallup poll survey of 74 people who purchased lottery tickets.

16. How many people purchased between 1 and 10 tickets?

17. What percent of the people purchased between 20 and 30 tickets each month? Round to the nearest tenth of a percent.

18. What percent of the people purchased more than 10 tickets each month? Round to the nearest tenth of a percent.

19. Is it possible to determine from the graph how many people purchased 15 lottery tickets? Explain.

The frequency polygon in Figure 36 shows the distribution of scores of the approximately 1,080,000 students who took an SAT exam.

20. How many students scored between 1200 and 1400 on the exam?

21. What percent of the number of students who took the exam scored between 800 and 1000? Round to the nearest tenth of a percent.

22. How many students scored below 1000?

23. How many students scored above 800?

APPLYING THE CONCEPTS

24. Write a paragraph explaining the difference between a histogram and a bar graph.

25. In your own words, describe a frequency table.

Figure 34

Figure 35

Figure 36
Source: Educational Testing Service

7.4

Statistical Measures

Objective A   **To find the mean, median, and mode of a distribution**

The average score on the math portion of the SAT was 432. The EPA estimates that a 2001 Ford Escort averages 35 miles per gallon on the highway. The average rainfall for portions of Kauai is 350 inches per year. Each of these statements uses one number to describe an entire collection of numbers. Such a number is called an *average.*

In statistics there are various ways to calculate an average. Three of the most common—*mean, median,* and *mode*—are discussed here.

An automotive engineer tests the miles-per-gallon ratings of 15 cars and records the results as follows:

Miles-per-Gallon Ratings of 15 Cars														
25	22	21	27	25	35	29	31	25	26	21	39	34	32	28

The **mean** of the data is the sum of the measurements divided by the number of measurements. The symbol for the mean is $\bar{x}$.

Formula for the Mean

$$\bar{x} = \frac{\text{sum of the data values}}{\text{number of data values}}$$

To find the mean for the data above, add the numbers and then divide by 15.

$$\bar{x} = \frac{25 + 22 + 21 + 27 + 25 + 35 + 29 + 31 + 25 + 26 + 21 + 39 + 34 + 32 + 28}{15}$$

$$= \frac{420}{15} = 28$$

The mean number of miles per gallon for the 15 cars tested was 28 miles per gallon.

The mean is one of the most frequently computed averages. It is the one that is commonly used to calculate a student's performance in a class.

➡ The test scores for a student taking American history were 78, 82, 91, 87, and 93. What was the mean score for this student?

Strategy
To find the mean, divide the sum of the test scores by 5, the number of scores.

Solution
$$\bar{x} = \frac{78 + 82 + 91 + 87 + 93}{5} = \frac{431}{5} = 86.2$$

The mean score for the history student was 86.2.

The **median** of data is the number that separates the data into two equal parts when the numbers are arranged from smallest to largest (or from largest to smallest). There is an equal number of values above the median and below the median.

To find the median of a set of numbers, first arrange the numbers from smallest to largest. The median is the number in the middle.

The result of arranging the miles-per-gallon ratings given on the previous page from smallest to largest is shown below.

21 21 22 25 25 25 26 27 28 29 31 32 34 35 39

7 values below the median Middle number **Median** 7 values above the median

The median is 27 miles per gallon.

If data contain an *even* number of values, the median is the mean of the two middle numbers.

➡ The selling prices of the last six homes sold by a real estate agent were $175,000, $150,000, $250,000, $130,000, $245,000, and $190,000. Find the median selling price of these homes.

Strategy
To find the median, arrange the numbers from smallest to largest. Because there is an even number of values, the median is the mean of the two middle numbers.

Solution
130,000 150,000 175,000 190,000 245,000 250,000

Middle 2 numbers

$$\text{Median} = \frac{175,000 + 190,000}{2} = 182,500$$

The median selling price was $182,500.

The **mode** of a set of numbers is the value that occurs most frequently. If a set of numbers has no number occurring more than once, then the data have no mode.

Here again are the data for the gasoline mileage ratings of 15 cars.

Miles-per-Gallon Ratings of 15 Cars														
25	22	21	27	25	35	29	31	25	26	21	39	34	32	28

25 is the number that occurs most frequently.

The mode is 25 miles per gallon.

Note from the miles-per-gallon example that the mean, median, and mode may be different.

Example 1

Twenty students were asked the number of units in which they were enrolled. The responses were as follows:

15	12	13	15	17	18	13	20	9	16
14	10	15	12	17	16	6	14	15	12

Find the mean number of units taken by these students.

Strategy

To find the mean number of units:

- Find the sum of the 20 numbers.
- Divide the sum by 20.

Solution

15 + 12 + 13 + 15 + 17 + 18 + 13 + 20 + 9 +
  16 + 14 + 10 + 15 + 12 + 17 + 16 + 6 +
  14 + 15 + 12 = 279

$$\bar{x} = \frac{279}{20} = 13.95$$

The mean is 13.95 units.

You Try It 1

The amounts spent by 12 customers at a McDonald's restaurant were as follows:

6.26	8.23	5.09	8.11	7.50	6.69
5.66	4.89	5.25	9.36	6.75	7.05

Find the mean amount spent by these customers. Round to the nearest cent.

Your strategy

Your solution

Example 2

The starting hourly wages for an apprentice electrician for six different work locations are $10.90, $11.25, $10.10, $11.08, $11.56, and $10.55. Find the median starting hourly wage.

Strategy

To find the median starting hourly wage:

- Arrange the numbers from smallest to largest.
- Because there is an even number of values, the median is the mean of the two middle numbers.

Solution

10.10, 10.55, 10.90, 11.08, 11.25, 11.56

$$\text{Median} = \frac{10.90 + 11.08}{2} = 10.99$$

The median starting hourly wage is $10.99.

You Try It 2

The amounts of weight lost, in pounds, by 10 participants in a 6-month weight-reduction program were 22, 16, 31, 14, 27, 16, 29, 31, 40, and 10. Find the median weight loss for these participants.

Your strategy

Your solution

Solutions on p. S20

Objective B To draw a box-and-whiskers plot

Recall from the last objective that an average is one number that helps to describe all the numbers in a set of data. For example, we know from the statement

> The average annual snowfall in Erie, Pennsylvania, is 85 inches.

that Erie gets a lot of snow each winter.

Now look at these two statements.

> The average annual temperature in San Francisco, California, is 57°.
> The average annual temperature in St. Louis, Missouri, is 57°.

The average annual temperature in both cities is the same. However, we do not expect the climate in St. Louis to be like San Francisco's climate. Although both cities have the same average annual temperature, their temperature ranges differ. In fact, the difference between the average monthly high temperatures in July and January in San Francisco is 14°, whereas the difference between the average monthly high temperatures in July and January in St. Louis is 50°.

Note that for this example, a single number (the average annual temperature) does not provide us with a very complete picture of the climate of either of these two cities.

One method used to picture an entire set of data is a box-and-whiskers plot. To prepare a box-and-whiskers plot, we begin by separating a set of data into four parts, called quartiles. We will illustrate this by using the average monthly high temperatures for St. Louis, in degrees Fahrenheit. These are listed below from January through December.

39	47	58	72	81	88	89	89	85	76	49	47

Source: The Weather Channel

First list the numbers in order from smallest to largest and determine the median.

Median = 74

Now find the median of the data values below the median. The median of the data values below the median is called the **first quartile,** symbolized by Q_1. Also find the median of the data values above the median. The median of the data values above the median is called the **third quartile,** symbolized by Q_3.

The first quartile, Q_1, is the number that one-quarter of the data lie below. This means that 25% of the data lie below the first quartile. The third quartile, Q_3, is the number that one-quarter of the data lie above. This means that 25% of the data lie above the third quartile.

The **range** of a set of numbers is the difference between the largest number and the smallest number in the set. The range describes the spread of the data. For the data above,

$$\text{Range} = \text{largest value} - \text{smallest value} = 89 - 39 = 50$$

The **interquartile range** is the difference between the third quartile, Q_3, and the first quartile, Q_1. For the data above,

$$\text{Interquartile range} = Q_3 - Q_1 = 86.5 - 48 = 38.5$$

The interquartile range is the distance that spans the "middle" 50% of the data values. Because it excludes the bottom fourth of the data values and the top fourth of the data values, it excludes any extremes in the numbers in the set.

A **box-and-whiskers plot,** or **boxplot,** is a graph that shows five numbers: the smallest value, the first quartile, the median, the third quartile, and the greatest value. Here are these five values for the data on St. Louis temperatures.

The smallest number	39
The first quartile, Q_1	48
The median	74
The third quartile, Q_3	86.5
The largest number	89

Think of a number line that includes the five values listed above. With this in mind, mark off the five values. Draw a box that spans the distance from Q_1 to Q_3. Draw a vertical line the height of the box at the median.

Listed below are the average monthly high temperatures for San Francisco.

57	60	61	64	68	71	71	73	74	73	60	59

Source: The Weather Channel

We can perform the same calculations on these data to determine the five values needed for the box-and-whiskers plot.

The smallest number	57
The first quartile, Q_1	60
The median	66
The third quartile, Q_3	72
The largest number	74

The box-and-whiskers plot is shown at the right with the same scale used for the data on the St. Louis temperatures.

Note that by comparing the two boxplots, we can see that the range of temperatures in St. Louis is greater than the range of temperatures in San Francisco. For the St. Louis temperatures, there is a greater spread of the data below the median than above the median, whereas the spreads of the data above and below the median of the San Francisco boxplot are nearly equal.

The numbers of avalanche deaths in the United States during each of nine consecutive winters were 8, 24, 29, 13, 28, 30, 22, 26, and 32. (Source: Colorado Avalanche Information Center) Draw a box-and-whiskers plot of the data and determine the interquartile range.

Strategy

To draw the box-and-whiskers plot, arrange the data from smallest to largest. Then find the median, Q_1, and Q_3. Use the smallest value, Q_1, the median, Q_3, and the largest value to draw the box-and-whiskers plot.

To find the interquartile range, find the difference between Q_3 and Q_1.

Solution

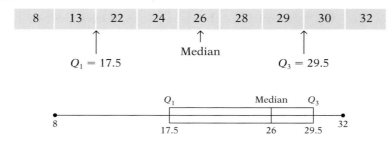

Interquartile range $= Q_3 - Q_1 = 29.5 - 17.5 = 12$

The interquartile range is 12 deaths.

....

Example 3

The average monthly snowfall, in inches, in Buffalo, New York, from October through April is 1, 12, 24, 25, 18, 12, and 3. (Source: The Weather Channel) Draw a box-and-whiskers plot of the data.

Strategy

To draw the box-and-whiskers plot:

- Arrange the data from smallest to largest.
- Find the median, Q_1, and Q_3.
- Use the smallest value, Q_1, the median, Q_3, and the largest value to draw the box-and-whiskers plot.

Solution

You Try It 3

The average monthly snowfall, in inches, in Denver, Colorado, from October through April is 4, 7, 7, 8, 8, 9, and 13. (Source: The Weather Channel) Draw a box-and-whiskers plot of the data. How does the spread of the data within the interquartile range compare with that in Example 3?

Your strategy

Your solution

Solution on p. S20

7.4 Exercises

Objective A

1. State whether the mean, median, or mode is being used.
   a. Half of the houses in the new development are priced under $125,000.
   b. The average bill for lunch at the college union is $7.95.
   c. The college bookstore sells more green college sweatshirts than any other color.
   d. In a recent year, there were as many people age 26 and younger in the world as there were people age 26 and older.
   e. The majority of full-time students carry a load of 12 credit hours per semester.
   f. The average annual return on this investment is 6.5%.

2. The number of big-screen televisions sold each month for one year was recorded by an electronics store. The results were 15, 12, 20, 20, 19, 17, 22, 24, 17, 20, 15, and 27. Calculate the mean, the median, and the mode of the number of televisions sold per month.

3. The number of seats occupied on a jet for 16 transatlantic flights was recorded. The numbers were 309, 422, 389, 412, 401, 352, 367, 319, 410, 391, 330, 408, 399, 387, 411, and 398. Calculate the mean, the median, and the mode of the number of seats occupied per flight.

4. The times, in seconds, for a 100-meter dash at a college track meet were 10.45, 10.23, 10.57, 11.01, 10.26, 10.90, 10.74, 10.64, 10.52, and 10.78. Calculate the mean and the median times for the 100-meter dash.

5. A consumer research group purchased identical items in eight grocery stores. The costs for the purchased items were $45.89, $52.12, $41.43, $40.67, $48.73, $42.45, $47.81, and $45.82. Calculate the mean and the median costs of the purchased items.

6. One measure of a computer's hard-drive speed is called access time; this is measured in milliseconds (thousandths of a second). Find the mean and median for 11 hard drives whose access times were 9, 8.5, 8, 8.5, 9, 9.5, 10, 9.5, 7, 8.5, and 8.5. Round to the nearest tenth.

7. Eight health maintenance organizations (HMOs) presented group health insurance plans to a company. The monthly rates per employee were $423, $390, $405, $396, $426, $355, $404, and $430. Calculate the mean and the median monthly rates for these eight companies.

8. The lengths of the terms, in years, of all the former Supreme Court Chief Justices are given in the table below. Find the mean and median length of term for the Chief Justices.

5	0	4	34	28	8	14	21
10	8	11	4	7	15	17	

9. The life expectancies, in years, in 10 selected Central and South American countries are given at the right. Find the mean and median life expectancies in these countries.

Brazil	62
Chile	75
Costa Rica	78
Ecuador	70
Guatemala	64
Panama	75
Peru	66
Trinidad and Tobago	71
Uruguay	74
Venezuela	73

10. Your scores on six history tests were 78, 92, 95, 77, 94, and 88. If an "average score" of 90 receives an A for the course, which average, the mean or the median, would you prefer that the instructor use?

11. One student received scores of 85, 92, 86, and 89. A second student received scores of 90, 97, 91, and 94 (exactly 5 points more on each exam). Are the means of the two students the same? If not, what is the relationship between the means of the two students?

12. The table below shows the defense expenditures, in billions of dollars, by the federal government for 1965 through 1973, years during which the United States was actively involved in the Vietnam War.
   a. Calculate the mean annual defense expenditure. Round to the nearest tenth of a billion.
   b. Find the median annual defense expenditure.
   c. If the year 1965 were eliminated from the data, how would that affect the mean? The median?

Year	1965	1966	1967	1968	1969	1970	1971	1972	1973
Expenditures	$49.6	$56.8	$70.1	$80.5	$81.2	$80.3	$77.7	$78.3	$76.0

Source: Statistical Abstract of the United States

Objective B

13. **a.** What percent of the data in a set of numbers lie above Q_3?
    b. What percent of the data in a set of numbers lie above Q_1?
    c. What percent of the data in a set of numbers lie below Q_3?
    d. What percent of the data in a set of numbers lie below Q_1?

14. The box-and-whiskers plot below shows the distribution of the ages of presidents of the United States at the time of their inauguration. What is the youngest age in the set of data? The oldest age? The first quartile? The third quartile? The median? Find the range and the interquartile range.

15. The box-and-whiskers plot below shows the distribution of median incomes for 50 states and the District of Columbia. What is the lowest value in the set of data? The highest value? The first quartile? The third quartile? The median? Find the range and the interquartile range.

24,880   29,778   33,858 36,453   47,954

16. An aptitude test was taken by 200 students at the Fairfield Middle School. The box-and-whiskers plot at the right shows the distribution of their scores.

   43   54   72   88   98

   a. How many students scored over 88?
   b. How many students scored below 72?
   c. How many scores are represented in each quartile?
   d. What percent of the students had scores of at least 54?

17. The cholesterol levels for 80 adults were recorded and then displayed in the box-and-whiskers plot shown at the right.

   172   198 217   254   345

   a. How many adults had a cholesterol level above 217?
   b. How many adults had a cholesterol level below 254?
   c. How many cholesterol levels are represented in each quartile?
   d. What percent of the adults had a cholesterol level of not more than 198?

18. The gasoline consumption of 19 cars was tested, and the results were recorded in the table below.
   a. Find the range, the first quartile, the third quartile, and the interquartile range.
   b. Draw a box-and-whiskers plot of the data.
   c. Is the data value 21 in the interquartile range?

Miles per Gallon for 19 Cars

33	21	30	32	20	31	25	20	16	24
22	31	30	28	26	19	21	17	26	

19. Carbon dioxide is among the gases that contribute to global warming. The world's biggest emitters of carbon dioxide are listed below. The figures are emissions in millions of metric tons per year.
   a. Find the range, the first quartile, the third quartile, and the interquartile range.
   b. Draw a box-and-whiskers plot of the data.
   c. What data value is responsible for the long "whisker" at the right?

Carbon Dioxide Emissions (in millions of metric tons per year)			
Canada	0.41	Japan	1.06
China	2.60	Russian Federation	2.10
Germany	0.87	Ukraine	0.61
India	0.76	United Kingdom	0.56
Italy	0.41	United States	4.80

Source: U.S. State Department

20. The average monthly amounts of rainfall, in inches, from January through December for Seattle, Washington, and Houston, Texas, are listed below.
   a. Is the difference between the means greater than 1 inch?
   b. What is the difference between the medians?
   c. Draw a box-and-whiskers plot of each set of data. Use the same scale.
   d. Describe the difference between the distributions of the data for Seattle and Houston.

| Seattle | 6.0 | 4.2 | 3.6 | 2.4 | 1.6 | 1.4 | 0.7 | 1.3 | 2.0 | 3.4 | 5.6 | 6.3 |
| Houston | 3.2 | 3.3 | 2.7 | 4.2 | 4.7 | 4.1 | 3.3 | 3.7 | 4.9 | 3.7 | 3.4 | 3.7 |

Source: The Weather Channel

21. The average monthly amounts of rainfall, in inches, from January through December for Orlando, Florida, and Portland, Oregon, are listed below.
   a. Is the difference between the means greater than 1 inch?
   b. What is the difference between the medians?
   c. Draw a box-and-whiskers plot of each set of data. Use the same scale.
   d. Describe the difference between the distributions of the data for Orlando and Portland.

| Orlando | 2.1 | 2.8 | 3.2 | 2.2 | 4.0 | 7.4 | 7.8 | 6.3 | 5.6 | 2.8 | 1.8 | 1.8 |
| Portland | 6.2 | 3.9 | 3.6 | 2.3 | 2.1 | 1.5 | 0.5 | 1.1 | 1.6 | 3.1 | 5.2 | 6.4 |

Source: The Weather Channel

APPLYING THE CONCEPTS

22. Explain how to determine (a) the mean, (b) the median, and (c) the mode of a set of data.

23. Write a set of data with five data values for which the mean, median, and mode are all 55.

24. A set of data has a mean of 16, a median of 15, and a mode of 14. Which of these numbers must be a value in the data? Explain your answer.

25. Explain each notation.
   a. Q_1
   b. Q_3
   c. $\bar{x}$

26. What values are shown on a box-and-whiskers plot? Explain how each is displayed.

27. The box in a box-and-whiskers plot represents 50%, or one-half, of the data in a set. Why is the box in Example 3 in this section not one-half of the entire length of the box-and-whiskers plot?

28. Create a set of data containing 25 numbers that would correspond to the box-and-whiskers plot shown at the right.

7.5 Introduction to Probability

Objective A   **To calculate the probability of simple events**

A weather forecaster estimates that there is a 75% chance of rain. A state lottery director claims that there is a $\frac{1}{9}$ chance of winning a prize in a new game offered by the lottery. Each of these statements involves uncertainty to some extent. The degree of uncertainty is called **probability.** For the statements above, the probability of rain is 75%, and the probability of winning a prize in the new lottery game is $\frac{1}{9}$.

A probability is determined from an **experiment,** which is any activity that has an observable outcome. Examples of experiments include

> Tossing a coin and observing whether it lands heads or tails
> Interviewing voters to determine their preference for a political candidate
> Drawing a card from a standard deck of 52 cards

All the possible outcomes of an experiment are called the **sample space** of the experiment. The outcomes are listed between braces. For example,

The number cube shown at the left is rolled once. Any of the numbers from 1 to 6 could show on the top of the cube. The sample space is

$$\{1, 2, 3, 4, 5, 6\}$$

A fair coin is tossed once. (A fair coin is one for which heads and tails have an equal chance of landing face up.) If H represents "heads up" and T represents "tails up," then the sample space is

$$\{H, T\}$$

An **event** is one or more outcomes of an experiment. For the experiment of rolling the six-sided cube described above, some possible events are

> The number is even: $\{2, 4, 6\}$
> The number is a multiple of 3: $\{3, 6\}$
> The number is less than 10: $\{1, 2, 3, 4, 5, 6\}$
> Note that in this case, the event is the entire sample space.

⇒ The spinner at the left is spun once. Assume that the spinner does not come to rest on a line.

a. What is the sample space?

b. List the outcomes in the event that the spinner points to an odd number.

a. The arrow could come to rest on any one of the four sectors. The sample space is $\{1, 2, 3, 4\}$.

b. $\{1, 3\}$

In discussing experiments and events, it is convenient to refer to the **favorable outcomes** of an experiment. These are the outcomes of an experiment that satisfy the requirements of a particular event.

For instance, consider the experiment of rolling a fair die once. The sample space is

$$\{1, 2, 3, 4, 5, 6\}$$

and one possible event would be rolling a number that is divisible by 3. The outcomes of the experiment that are favorable to the event are 3 and 6:

$$\{3, 6\}$$

The outcomes of the experiment of tossing a fair coin are *equally likely*. Either one of the outcomes is just as likely as the other. If a fair coin is tossed once, the probability of a head is $\frac{1}{2}$, and the probability of a tail is $\frac{1}{2}$. Both events are equally likely. The theoretical probability formula, given below, applies to experiments for which the outcomes are equally likely.

> **Theoretical Probability Formula**
>
> The theoretical probability of an event is a fraction with the number of favorable outcomes of the experiment in the numerator and the total number of possible outcomes in the denominator.
>
> $$\text{Probability of an event} = \frac{\text{number of favorable outcomes}}{\text{number of possible outcomes}}$$

A probability of an event is a number from 0 to 1 that tells us how likely it is that this outcome will happen.

A probability of 0 means that the event is impossible.
The probability of getting a heads when rolling the die shown at the left is 0.

A probability of 1 means that the event must happen.
The probability of getting either a heads or tails when tossing a coin is 1.

A probability of $\frac{1}{4}$ means that it is expected that the outcome will happen 1 in every 4 times the experiment is performed.

TAKE NOTE

The phrase *at random* means that each card has an equal chance of being drawn.

➡ Each of the letters of the word *TENNESSEE* is written on a card, and the cards are placed in a hat. If one card is drawn at random from the hat, what is the probability that the card has the letter *E* on it?

Count the possible outcomes of the experiment.

There are 9 letters in *TENNESSEE*.
There are 9 possible outcomes of the experiment.

Count the number of outcomes of the experiment that are favorable to the event that a card with the letter *E* on it is drawn.

There are 4 cards with an *E* on them.

Use the probability formula.

$$\text{Probability of the event} = \frac{\text{number of favorable outcomes}}{\text{number of possible outcomes}} = \frac{4}{9}$$

The probability of drawing an *E* is $\frac{4}{9}$.

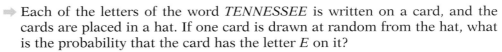

As shown above, calculating the probability of an event requires counting the number of possible outcomes of an experiment and the number of outcomes that are favorable to the event. One way to do this is to list the outcomes of the experiment in a systematic way. Using a table is often helpful.

When two dice are rolled, the sample space for the experiment can be recorded systematically as in the following table.

Point of Interest

Romans called a die that was marked on four faces a *talus*, which meant "anklebone." The anklebone was considered an ideal die because it is roughly a rectangular solid and it has no marrow, so loose anklebones from sheep were more likely than other bones to be lying around after the wolves had left their prey.

Possible Outcomes from Rolling Two Dice

(1, 1)	(2, 1)	(3, 1)	(4, 1)	(5, 1)	(6, 1)
(1, 2)	(2, 2)	(3, 2)	(4, 2)	(5, 2)	(6, 2)
(1, 3)	(2, 3)	(3, 3)	(4, 3)	(5, 3)	(6, 3)
(1, 4)	(2, 4)	(3, 4)	(4, 4)	(5, 4)	(6, 4)
(1, 5)	(2, 5)	(3, 5)	(4, 5)	(5, 5)	(6, 5)
(1, 6)	(2, 6)	(3, 6)	(4, 6)	(5, 6)	(6, 6)

⇒ Two dice are rolled once. Calculate the probability that the sum of the numbers on the two dice is 7.

Use the table above to count the number of possible outcomes of the experiment.

There are 36 possible outcomes.

Count the number of outcomes of the experiment that are favorable to the event that a sum of 7 is rolled.

There are 6 favorable outcomes: (1, 6), (2, 5), (3, 4), (4, 3), (5, 2), and (6, 1)

Use the probability formula.

$$\text{Probability of the event} = \frac{\text{number of favorable outcomes}}{\text{number of possible outcomes}} = \frac{6}{36} = \frac{1}{6}$$

The probability of a sum of 7 is $\frac{1}{6}$.

The probabilities calculated above are **theoretical probabilities** because the calculations are based on theory—for example, that either side of a coin is equally likely to land face up or that each of the six sides of a fair die are equally likely to land face up. Not all probabilities arise from such assumptions.

Empirical probabilities are based on observations of certain events. For instance, a weather forecast of a 75% chance of rain is an empirical probability. From historical records kept by the weather bureau, when a similar weather pattern existed, rain occurred 75% of the time. It is theoretically impossible to predict the weather, and only observations of past weather patterns can be used to predict future weather conditions.

> **Empirical Probability Formula**
>
> The empirical probability of an event is a ratio of the number of observations of the event to the total number of observations.
>
> $$\text{Probability of an event} = \frac{\text{number of observations of the event}}{\text{total number of observations}}$$

For example, suppose the records of an insurance company show that of 2549 claims for theft filed by policy holders, 927 were claims for more than \$5000. The empirical probability that the next claim for theft that this company receives will be a claim for more than \$5000 is the ratio of the number of claims for over \$5000 to the total number of claims. The probability is approximately 0.36.

$$\frac{927}{2549} \approx 0.36$$

Example 1

There are three choices, *a*, *b*, or *c*, for each of the two questions on a multiple-choice quiz. If the instructor randomly chooses which questions will have an answer of *a*, *b*, or *c*, what is the probability that the two correct answers on this quiz will be the same letter?

Strategy

To find the probability:

- List the outcomes of the experiment in a systematic way.
- Count the number of possible outcomes of the experiment.
- Count the number of outcomes of the experiment that are favorable to the event that the two correct answers on the quiz will be the same letter.
- Use the probability formula.

Solution

Possible outcomes: (a, a) (b, a) (c, a)
 (a, b) (b, b) (c, b)
 (a, c) (b, c) (c, c)

There are 9 possible outcomes.

There are 3 favorable outcomes:
(a, a), (b, b), (c, c)

$$\text{Probability} = \frac{\text{number of favorable outcomes}}{\text{number of possible outcomes}}$$

$$= \frac{3}{9} = \frac{1}{3}$$

The probability that the two correct answers will be the same letter is $\frac{1}{3}$.

You Try It 1

A professor writes three true/false questions for a quiz. If the professor randomly chooses which questions will have a true answer and which will have a false answer, what is the probability that the test will have 2 true questions and 1 false question?

Your strategy

Your solution

Solution on pp. S20–S21

7.5 Exercises

Objective A

1. A coin is tossed four times. List all the possible outcomes of the experiment as a sample space.

2. Three cards—one red, one green, and one blue—are to be arranged in a stack. Using R for red, G for green, and B for blue, list all the different stacks that can be formed. (Some computer monitors are called RGB monitors for the colors red, green, and blue.)

3. A tetrahedral die is one with four triangular sides. The sides show the numbers from 1 to 4. Say two tetrahedral dice are rolled. List all the possible outcomes of the experiment as a sample space.

4. A coin is tossed and then a die is rolled. List all the possible outcomes of the experiment as a sample space. [To get you started, (H, 1) is one of the possible outcomes.]

5. The spinner at the right is spun once. Assume that the spinner does not come to rest on a line.
    a. What is the sample space?
    b. List the outcomes in the event that the number is less than 4.

6. A coin is tossed four times.
    a. What is the probability that the outcomes of the tosses are exactly in the order HHTT? (See Exercise 1.)
    b. What is the probability that the outcomes of the tosses consist of two heads and two tails?
    c. What is the probability that the outcomes of the tosses consist of one head and three tails?

7. Two dice are rolled.
    a. What is the probability that the sum of the dots on the upward faces is 5?
    b. What is the probability that the sum of the dots on the upward faces is 15?
    c. What is the probability that the sum of the dots on the upward faces is less than 15?
    d. What is the probability that the sum of the dots on the upward faces is 2?

8. A dodecahedral die has 12 sides numbered from 1 to 12. The die is rolled once.
    a. What is the probability that the upward face shows an 11?
    b. What is the probability that the upward face shows a 5?

9. Two tetrahedral dice are rolled (see Exercise 3).
    a. What is the probability that the sum on the upward faces is 4?
    b. What is the probability that the sum on the upward faces is 6?

10. A dodecahedral die has 12 sides numbered from 1 to 12. The die is rolled once.
    a. What is the probability that the upward face shows a number divisible by 4?
    b. What is the probability that the upward face shows a number that is a multiple of 3?

11. Two dice are rolled. Which has the greater probability, throwing a sum of 10 or throwing a sum of 5?

12. Two dice are rolled once. Calculate the probability that the two numbers on the dice are equal.

13. Each of the letters of the word *MISSISSIPPI* is written on a card, and the cards are placed in a hat. One card is drawn at random from the hat.
   a. What is the probability that the card has the letter *I* on it?
   b. Which is greater, the probability of choosing an *S* or that of choosing a *P*?

14. There are five choices—choices *a* through *e*—for each question on a multiple-choice test. What is the probability of choosing the correct answer for a certain question by just guessing?

15. Three blue marbles, four green marbles, and five red marbles are placed in a bag. One marble is chosen at random.
   a. What is the probability that the marble chosen is green?
   b. Which is greater, the probability of choosing a blue marble or that of choosing a red marble?

16. Which has the greater probability, drawing a jack, queen, or king from a deck of cards or drawing a spade?

17. In a history class, a set of exams earned the following grades: 4 A's, 8 B's, 22 C's, 10 D's, and 3 F's. If a single student's exam is chosen from this class, what is the probability that it received a B?

18. A survey of 95 people showed that 37 preferred (to using a credit card) a cash discount of 2% if an item was purchased using cash or a check. Judging on the basis of this survey, what is the empirical probability that a person prefers a cash discount? Write the answer as a decimal rounded to the nearest hundredth.

19. A survey of 725 people showed that 587 had a group health insurance plan where they worked. On the basis of this survey, what is the empirical probability that an employee has a group health insurance plan? Write the answer as a decimal rounded to the nearest hundredth.

20. A television cable company surveyed some of its customers and asked them to rate the cable service as excellent, satisfactory, average, unsatisfactory, or poor. The results are recorded in the table at the right. What is the probability that a customer who was surveyed rated the service as satisfactory or excellent?

Quality of Service	Number Who Voted
Excellent	98
Satisfactory	87
Average	129
Unsatisfactory	42
Poor	21

APPLYING THE CONCEPTS

21. If the spinner at the right is spun once, is each of the numbers 1 through 5 equally likely? Why or why not?

22. The probability of tossing a coin and having it land heads is $\frac{1}{2}$. Does this mean that if that coin is tossed 100 times, it will land heads 50 times? Explain your answer.

23. Why can the probability of an event not be $\frac{5}{3}$?

Focus on Problem Solving

Inductive Reasoning Suppose that, beginning in January, you save $25 each month. The total amount you have saved at the end of each month can be described by a list of numbers.

25	50	75	100	125	150	175	
Jan	Feb	Mar	Apr	May	June	July	...

The list of numbers that indicates your total savings is an *ordered* list of numbers called a **sequence.** Each of the numbers in a sequence is called a **term** of the sequence. The list is ordered because the position of a number in the list indicates the month in which that total amount has been saved. For example, the 7th term of the sequence (indicating July) is 175. This number means that a total of $175 has been saved by the end of the 7th month.

Now consider a person who has a different savings plan. The total amount saved by this person for the first seven months is given by the sequence

20, 35, 50, 65, 80, 95, 110, ...

The process you use to discover the next number in the above sequence is *inductive reasoning.* **Inductive reasoning** involves making generalizations from specific examples; in other words, we reach a conclusion by making observations about particular facts or cases. In the case of the above sequence, the person saved $15 per month after the first month.

Here is another example of inductive reasoning. Find the next two letters of the sequence A, B, E, F, I, J, ...

By trying different patterns, we can determine that a pattern for this sequence is

<u>A, B,</u> C, D, <u>E, F,</u> G, H, <u>I, J,</u> ...

That is, write two letters, skip two letters, write two letters, skip two letters, and so on. The next two letters are M, N.

Use inductive reasoning to solve the following problems.

1. What is the next term of the sequence, ban, ben, bin, bon, ... ?

2. Using a calculator, determine the decimal representation of several proper fractions that have a denominator of 99. For instance, you may use $\frac{8}{99}$, $\frac{23}{99}$, and $\frac{75}{99}$. Now use inductive reasoning to explain the pattern, and use your reasoning to find the decimal representation of $\frac{53}{99}$ without a calculator.

3. Find the next number in the sequence 1, 1, 2, 3, 5, 8, 13, 21, ...

4. The decimal representation of a number begins 0.10100100010000100000 ... . What are the next 10 digits in this number?

5. The first seven rows of a triangle of numbers called Pascal's triangle are given below. Find the next row.

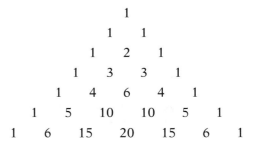

Projects and Group Activities

Deceptive Graphs   A graphical representation of data can sometimes be misleading. Consider the graphs shown below. A financial advisor with an investment firm claims that an investment with the firm grew as shown in the graph on the left, whereas an investment with a competitor grew as shown in the graph on the right. Apparently, one would have accumulated more money by choosing the investment on the left.

Growth of Investment
with Broker A

Growth of Investment
with Broker B

However, these graphs have a serious flaw. There are no labels on the horizontal and vertical axes. Therefore, it is impossible to tell which investment increased more or over what time interval. When labels are not placed on the axes of a graph, the data represented are meaningless. This is one way in which advertisers use a visual impact to distort the true meaning of data.

The graphs below are the same as those drawn above except that scales have been drawn along each axis. Now it is possible to tell how each investment performed. Note that both turned in exactly the same performance!

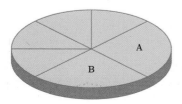

Drawing a circle graph as an oval is another way of distorting data. This is especially true if a three-dimensional representation is given. From the appearance of the circle graph at the left, one would think that region B is larger than region A. However, that isn't true. Measure the angle of each sector to see this for yourself. As you read newspapers and magazines, find examples of graphs that may distort the actual data. Discuss how these graphs should be drawn to be more accurate.

Collecting, Organizing, Displaying, and Analyzing Data

Before standardized units of measurement, measurements were made in terms of the human body. For example, the cubit was the distance from the end of the elbow to the tips of the fingers. The yard was the distance from the tip of the nose to the tip of the fingers on an outstretched arm.

For each student in the class, find the measure from the tip of the nose to the tip of the fingers on an outstretched arm. Round each measure to the nearest centimeter. Record all the measurements on the board.

1. From the data collected, determine each of the following.

Mean _____   First quartile, Q_1 _____
Median _____   Third quartile, Q_3 _____
Mode _____   Interquartile range _____
Range _____

2. Prepare a box-and-whiskers plot of the data.

3. Write a description of the spread of the data.

4. Explain why we need standardized units of measurement.

Chapter Summary

Key Words

Statistics is the branch of mathematics concerned with *data,* or numerical information. [p. 289]

A *graph* is a pictorial representation of data. [p. 289]

A *pictograph* uses symbols to represent information. [p. 289]

A *circle graph* represents data by the size of the sectors. [p. 291]

A *bar graph* represents data by the height of the bars. [p. 297]

A *broken-line graph* represents data by the position of the lines and shows trends and comparisons. [p. 298]

A *histogram* is a special kind of bar graph. In a histogram, the width of each bar corresponds to a range of numbers called a *class interval.* The height of each bar corresponds to the number of occurrences of data in each class interval and is called the *class frequency.* [p. 301]

A *frequency polygon* is a graph that displays information in a manner similar to a histogram. A dot is placed above the center of each class interval at a height corresponding to that class's frequency. The dots are connected to form a broken-line graph. The center of a class interval is called the *class midpoint.* [p. 302]

The *mean* of a set of data is the sum of the numbers in the set divided by the number of items in the set. The symbol for mean is $\bar{x}$. [p. 305]

The *median* of a set of data is the number that separates the data into two equal parts when the numbers are arranged from smallest to largest. There is an equal number of values above the median and below the median. [p. 306]

The *mode* of a set of numbers is the value that occurs most frequently. [p. 306]

The *range* of a set of numbers is the difference between the largest number and the smallest number in the set. The range describes the spread of the data. [p. 309]

The *first quartile*, symbolized by Q_1, is the number that one-quarter of the data lie below. The *third quartile*, symbolized by Q_3, is the number that one-quarter of the data lie above. [p. 308]

The *interquartile range* is the difference between the third quartile, Q_3, and the first quartile, Q_1. [p. 309]

A *box-and-whiskers plot*, or *boxplot*, is a graph that shows five numbers of a data set: the lowest number, the first quartile, the median, the third quartile, and the highest number. [p. 309]

An *experiment* is any activity that has an observable outcome. All the possible outcomes of an experiment are called the *sample space* of the experiment. [p. 315]

An *event* is one or more outcomes of an experiment. *Favorable outcomes* of an experiment are the outcomes that satisfy the requirements of a particular event. [pp. 315–316]

The *probability* of an event is a number from 0 to 1 that tells us how likely it is that a certain outcome will happen. [pp. 315–316]

Essential Rules

To Find the Mean
Divide the sum of the numbers by the number of addends.

$$\bar{x} = \frac{\text{sum of all the data values}}{\text{number of data values}} \quad [\text{p. 305}]$$

To Find the Median
1. Arrange the numbers from smallest to largest.
2. If there is an *odd* number of values in the set of data, the median is the middle number. If there is an *even* number of values in the set of data, the median is the mean of the two middle numbers. [p. 306]

To Find the Mode
Find the number in the set of data that occurs more often than any other number in the set. [p. 306]

To Find the Range of a Set of Data
Subtract the smallest number in the set from the largest number in the set. Range = largest value − smallest value [p. 309]

To Find Q_1
Arrange the numbers from smallest to largest and locate the median. Q_1 is the median of the lower half of the data. [p. 308]

To Find Q_3
Arrange the numbers from smallest to largest and locate the median. Q_3 is the median of the upper half of the data. [p. 308]

To Find the Interquartile Range
Interquartile range = $Q_3 - Q_1$ [p. 309]

Theoretical Probability Formula

$$\text{Probability of an event} = \frac{\text{number of favorable outcomes}}{\text{number of possible outcomes}} \quad [\text{p. 316}]$$

Empirical Probability Formula

$$\text{Probability of an event} = \frac{\text{number of observations of the event}}{\text{total number of observations}} \quad [\text{p. 318}]$$

Chapter Review

 The circle graph in Figure 37 shows the approximate amount of money that government agencies spent on maintaining Internet web sites for a 3-year period.

1. Find the total amount of money that these agencies spent on maintaining web sites.

2. What is the ratio of the amount spent by the Department of Commerce to the amount spent by the EPA?

3. What percent of the total money spent did NASA spend? Round to the nearest tenth of a percent.

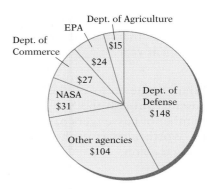

Figure 37 Millions of dollars that federal agencies spent on web sites
Source: General Accounting Office

 The double-line graph in Figure 38 shows the populations of California and Texas.

4. In 1900, which state had the larger population?

5. In 2000, approximately how much greater was the population of California than the population of Texas?

6. During which 25-year period did the population of Texas increase the least?

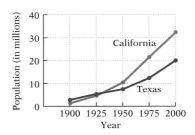

Figure 38 Populations of California and Texas

 The frequency polygon in Figure 39 shows the range of scores for the first 80 games of a season for the New York Knicks basketball team.

7. Find the number of games in which fewer than 100 points were scored by the Knicks.

8. What is the ratio of the number of games in which 90 to 100 points were scored to the number of games in which 110 to 120 points were scored?

9. In what percent of the games were 110 points or more scored? Round to the nearest tenth of a percent.

Figure 39
Source: Sports Illustrated web site at
http://CNNSI.com/

The pictograph in Figure 40 shows the numbers of passengers annually passing through the five busiest airports in the United States.

10. How many more passengers pass through O'Hare each year than pass through the Los Angeles airport each year?

11. What is the ratio of the number of passengers passing through the San Francisco airport to the number of passengers passing through the Dallas/ Ft. Worth airport each year? Write your answer using a colon.

O'Hare
Hartsfield
Dallas/Ft. Worth
Los Angeles
San Francisco

= 10 million passengers per year

Figure 40 The busiest U.S. airports
Source: Airports Association Council International–North America; Air Transport Association of America

The double-bar graph in Figure 41 shows the total days open and the days of full operation of ski resorts in different regions of the country.

12. Find the difference between the total days open and the days of full operation for Midwest ski areas.

13. What percent of the total days open were the days of full operation for the Rocky Mountain ski areas?

14. Which region had the lowest number of days of full operation? How many days of full operation did this region have?

15. A coin is tossed four times. What is the probability that the outcomes of the tosses consist of one tail and three heads?

Based on a Gallup poll, the numbers of hours that the 46 people surveyed slept during a typical weekday night are shown in the histogram in Figure 42.

16. How many people slept 8 hours or more?

17. What percent of the people surveyed slept 7 hours? Round to the nearest tenth of a percent.

18. The heart rates of 24 women tennis players were measured after each of them had run one-quarter of a mile. The results are listed in the table below. Find the mean, median, and mode for the data. Round to the nearest tenth.

| 80 | 82 | 99 | 91 | 93 | 87 | 103 | 94 | 73 | 96 | 86 | 80 |
| 97 | 94 | 108 | 81 | 100 | 109 | 91 | 84 | 78 | 96 | 96 | 100 |

19. For the data in Exercise 18, find the range and the interquartile range.

Figure 41
Source: Economic Analysis of United States Ski Areas

Hours of Sleep per Night

Figure 42

Chapter Test

Forty college students were surveyed to see how much money they spent each week on dining out in restaurants. The results are recorded in the frequency polygon shown in Figure 43.

Figure 43

1. How many students spent between $15 and $25 per week?

2. Find the ratio of the number of students who spent between $10 and $15 to the number who spent between $15 and $20.

3. What percent of the students surveyed spent less than $15 per week?

The pictograph in Figure 44 is based on the results of a Gallup poll survey of married couples. Each individual was asked to give a letter grade to the marriage.

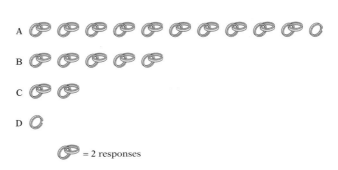

Figure 44 Survey of married couples rating their marriage

4. Find the total number of people who were surveyed.

5. Find the ratio of the number of people who gave their marriage a B to the number who gave it a C.

6. What percent of the total number of people surveyed gave their marriage an A?

The bar graph in Figure 45 shows the number of fatalities that occurred in accidents on amusement rides in the 1990s in the United States.

Figure 45 Number of fatalities in accidents on amusement rides

Source: USA Today, April 7, 2000

7. During which two consecutive years were the numbers of fatalities the same?

8. Find the total number of fatalities on amusement rides during 1991 through 1999.

9. How many more such fatalities occurred during the years 1995 through 1998 than occurred during the years 1991 through 1994?

The circle graph in Figure 46 categorizes the 655 films released during a recent year by their ratings.

10. How many more R-rated films were released than PG films?

11. How many times more PG-13 films were released than NC-17?

12. What percent of the films released were rated G? Round to the nearest tenth of a percent.

Figure 46 Ratings of films released
Source: MPA Worldwide Market Research

The histogram in Figure 47 gives information about the per capita incomes, by state, in the United States.

13. How many states have per capita incomes between $16,000 and $22,000?

14. What percent of the states have a per capita income that is between $19,000 and $22,000?

15. What percent of the states have a per capita income that is $25,000 or more?

16. A box contains 50 balls, of which 15 are red. If 1 ball is randomly selected from the box, what is the probability of the ball's being red?

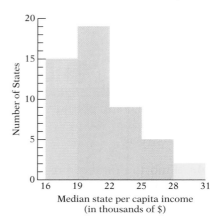

Figure 47
Source: Bureau of Labor Statistics

The line graph in Figure 48 shows the number of students enrolled in colleges.

17. During which decade did the student population increase the least?

18. Approximate the increase in the college enrollment from 1960 to 2000.

19. The length of time (in days) that various batteries operated a portable CD player continuously are given in the table below. Find the mean and median for the data.

2.9	2.4	3.1	2.5	2.6	2.0	3.0	2.3	2.4	2.7
2.0	2.4	2.6	2.7	2.1	2.9	2.8	2.4	2.0	2.8

20. For the data in Exercise 19, draw a box-and-whiskers plot.

Figure 48 Student enrollment in public and private colleges
Source: National Center for Educational Statistics

Cumulative Review

1. Simplify: $2^2 \cdot 3^3 \cdot 5$

2. Simplify: $3^2 \cdot (5 - 2) \div 3 + 5$

3. Find the LCM of 24 and 40.

4. Write $\frac{60}{144}$ in simplest form.

5. Find the total of $4\frac{1}{2}$, $2\frac{3}{8}$, and $5\frac{1}{5}$.

6. Subtract: $12\frac{5}{8} - 7\frac{11}{12}$

7. Multiply: $\frac{5}{8} \times 3\frac{1}{5}$

8. Find the quotient of $3\frac{1}{5}$ and $4\frac{1}{4}$.

9. Simplify: $\frac{5}{8} \div \left(\frac{3}{4} - \frac{2}{3}\right) + \frac{3}{4}$

10. Write two hundred nine and three hundred five thousandths in standard form.

11. Find the product of 4.092 and 0.69.

12. Convert $16\frac{2}{3}$ to a decimal. Round to the nearest hundredth.

13. Write "330 miles on 12.5 gallons of gas" as a unit rate.

14. Solve the proportion: $\frac{n}{5} = \frac{16}{25}$

15. Write $\frac{4}{5}$ as a percent.

16. 8 is 10% of what?

17. What is 38% of 43?

18. What percent of 75 is 30?

19. Tanim Kamal, a salesperson at a department store, receives $100 per week plus 2% commission on sales. Find the income for a week in which Tanim had $27,500 in sales.

20. A life insurance policy costs $4.15 for every $1000 of insurance. At this rate, what is the cost for $50,000 of life insurance?

21. A contractor borrowed $125,000 for 6 months at an annual simple interest rate of 11%. Find the interest due on the loan.

22. A compact disc player with a cost of $180 is sold for $279. Find the mark-up rate.

23. The circle graph in Figure 49 shows how a family's monthly income of $3000 is budgeted. How much is budgeted for food?

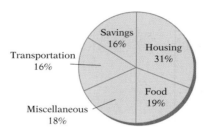

Figure 49 Budget for a monthly income of $3000

24. The double-broken-line graph in Figure 50 shows two students' scores on 5 math tests of 30 problems each. Find the difference between the numbers of problems that the two students answered correctly on Test 1.

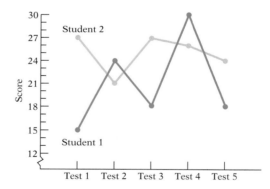

Figure 50

25. The average daily high temperatures, in degrees Fahrenheit, for a week in Newtown were 56°, 72°, 80°, 75°, 68°, 62°, and 74°. Find the mean high temperature for the week. Round to the nearest tenth of a degree.

26. Two dice are rolled. What is the probability that the sum of the dots on the upward faces is 8?

Chapter

8

U.S. Customary Units of Measurement

Police officers investigating the scene of a traffic accident have a number of tools at their disposal. One such tool is a nomograph, such as the one in the **project on page 352.** A nomograph is used to determine the speed of a car given the length of the skid marks left by the car and the road conditions at the time. For example, a car that skids 200 ft on wet concrete was traveling 60 mph.

Need help? For on-line student resources, such as section quizzes, visit this textbook's web site at **math.college.hmco.com/students.**

1. 485
   + 217

2.   145
    − 87

3. $36 \times \dfrac{1}{9}$

4. $\dfrac{5}{3} \times 6$

5. $400 \times \dfrac{1}{8} \times \dfrac{1}{2}$

6. $5\dfrac{3}{4} \times 8$

7. $3\overline{)714}$

8. $12\overline{)18}$

Go Figure

Mandy walks to work at a constant rate. One-third of the way to work, she passes a bank. Three-fourths of the way to work, she passes a book store. At the bank her watch reads 7:52 A.M. and at the book store it reads 8:02 A.M. At what time is Mandy one-half of the way to work?

8.1 Length

Objective A   **To convert measurements of length in the U.S. Customary System**

A **measurement** includes a number and a unit.

$$
\begin{array}{cc}
3 & \text{feet} \\
7 & \text{miles} \\
\underline{12} & \underline{\text{yards}} \\
\text{Number} & \text{Unit}
\end{array}
$$

Standard units of measurement have been established to simplify trade and commerce.

The unit of length, or distance, that is called the **yard** was originally defined as the length of a specified bronze bar located in London.

The standard U.S. Customary System units of length are **inch, foot, yard,** and **mile.** Equivalences between units of length in the U.S. Customary System are

$$
\begin{aligned}
12 \text{ inches (in.)} &= 1 \text{ foot (ft)} \\
3 \text{ ft} &= 1 \text{ yard (yd)} \\
36 \text{ in.} &= 1 \text{ yard (yd)} \\
5280 \text{ ft} &= 1 \text{ mile (mi)}
\end{aligned}
$$

These equivalences can be used to form conversion rates to change one unit of measurement to another. For example, because 3 ft = 1 yd, the conversion rates $\frac{3 \text{ ft}}{1 \text{ yd}}$ and $\frac{1 \text{ yd}}{3 \text{ ft}}$ are both equivalent to 1.

➡ Convert 27 ft to yards.

$$27 \text{ ft} = 27 \text{ ft} \times \boxed{\dfrac{1 \text{ yd}}{3 \text{ ft}}}$$

$$= 27 \, \cancel{\text{ft}} \times \dfrac{1 \text{ yd}}{3 \, \cancel{\text{ft}}}$$

$$= \dfrac{27 \text{ yd}}{3}$$

$$= 9 \text{ yd}$$

➡ Convert 5 yd to feet.

$$5 \text{ yd} = 5 \text{ yd} \times \boxed{\dfrac{3 \text{ ft}}{1 \text{ yd}}}$$

$$= 5 \, \cancel{\text{yd}} \times \dfrac{3 \text{ ft}}{1 \, \cancel{\text{yd}}}$$

$$= \dfrac{15 \text{ ft}}{1}$$

$$= 15 \text{ ft}$$

Note that in the conversion rate chosen, the unit in the numerator is the same as the unit desired in the answer. The unit in the denominator is the same as the unit in the given measurement.

Example 1   Convert 40 in. to feet.

Solution $40 \text{ in.} = 40 \text{ in.} \times \dfrac{1 \text{ ft}}{12 \text{ in.}}$

$= \dfrac{40 \text{ ft}}{12} = 3\dfrac{1}{3} \text{ ft}$

You Try It 1   Convert 14 ft to yards.

Your solution

Example 2

Convert $3\dfrac{1}{4}$ yd to feet.

Solution $3\dfrac{1}{4} \text{ yd} = \dfrac{13}{4} \text{ yd} = \dfrac{13}{4} \text{ yd} \times \dfrac{3 \text{ ft}}{1 \text{ yd}}$

$= \dfrac{39 \text{ ft}}{4} = 9\dfrac{3}{4} \text{ ft}$

You Try It 2   Convert 9240 ft to miles.

Your solution

Solutions on p. S21

Objective B   **To perform arithmetic operations with measurements of length**

When performing arithmetic operations with measurements of length, write the answer in simplest form. For example, 1 ft 14 in. should be written as 2 ft 2 in.

➡ Convert: 50 in. = _____ ft _____ in.

$$\begin{array}{r} 4 \text{ ft } 2 \text{ in.} \\ 12\overline{)\ 50} \\ -48 \\ \hline 2 \end{array}$$

• Because 12 in. = 1 ft, divide 50 in. by 12. The whole-number part of the quotient is the number of feet. The remainder is the number of inches.

50 in. = 4 ft 2 in.

Example 3   Convert: 17 in. = _____ ft _____ in.

Solution $$\begin{array}{r} 1 \text{ ft } 5 \text{ in.} \\ 12\overline{)\ 17} \\ -12 \\ \hline 5 \end{array}$$

• 12 in. = 1 ft

17 in. = 1 ft 5 in.

You Try It 3   Convert:
42 in. = _____ ft _____ in.

Your solution

Example 4   Convert: 31 ft = _____ yd _____ ft

Solution $$\begin{array}{r} 10 \text{ yd } 1 \text{ ft} \\ 3\overline{)\ 31} \\ -30 \\ \hline 1 \end{array}$$

• 3 ft = 1 yd

31 ft = 10 yd 1 ft

You Try It 4   Convert: 14 ft = _____ yd _____ ft

Your solution

Solutions on p. S21

Example 5 Find the sum of 4 ft 4 in. and 1 ft 11 in.

Solution
$$
\begin{array}{r}
4 \text{ ft } \ 4 \text{ in.} \\
+ \ 1 \text{ ft } 11 \text{ in.} \\
\hline
5 \text{ ft } 15 \text{ in.}
\end{array}
$$
• 15 in. = 1 ft 3 in.

5 ft 15 in. = 6 ft 3 in.

You Try It 5 Find the sum of 3 ft 5 in. and 4 ft 9 in.

Your solution

Example 6 Subtract: 9 ft 6 in. − 3 ft 8 in.

Solution

^{8 ft} ^{18 in.}

$$
\begin{array}{r}
\cancel{9} \text{ ft } \ \cancel{6} \text{ in.} \\
- \ 3 \text{ ft } \ 8 \text{ in.} \\
\hline
5 \text{ ft } 10 \text{ in.}
\end{array}
$$

• Borrow 1 ft (12 in.) from 9 ft and add to 6 in.

You Try It 6 Subtract: 4 ft 2 in. − 1 ft 8 in.

Your solution

Example 7 Multiply: 3 yd 2 ft × 4

Solution
$$
\begin{array}{r}
3 \text{ yd } 2 \text{ ft} \\
\times \qquad 4 \\
\hline
12 \text{ yd } 8 \text{ ft}
\end{array}
$$
• 8 ft = 2 yd 2 ft

12 yd 8 ft = 14 yd 2 ft

You Try It 7 Multiply: 4 yd 1 ft × 8

Your solution

Example 8 Find the quotient of 4 ft 3 in. and 3.

Solution
$$
\begin{array}{r}
1 \text{ ft } \quad 5 \text{ in.} \\
3 \overline{)\ 4 \text{ ft } \quad 3 \text{ in.}} \\
- \ 3 \text{ ft} \\
\hline
1 \text{ ft} = \underline{12 \text{ in.}} \\
15 \text{ in.} \\
\underline{-15 \text{ in.}} \\
0
\end{array}
$$

You Try It 8 Find the quotient of 7 yd 1 ft and 2.

Your solution

Example 9 Multiply: $2\frac{3}{4}$ ft × 3

Solution
$$
\begin{aligned}
2\frac{3}{4} \text{ ft} \times 3 &= \frac{11}{4} \text{ ft} \times 3 \\
&= \frac{33}{4} \text{ ft} \\
&= 8\frac{1}{4} \text{ ft}
\end{aligned}
$$

You Try It 9 Subtract: $6\frac{1}{4}$ ft − $3\frac{2}{3}$ ft

Your solution

Solutions on p. S21

Objective C   **To solve application problems**

Example 10

A concrete block is 9 in. high. How many rows of blocks are required for a retaining wall that is 6 ft high?

Strategy

To find the number of rows of blocks, convert 9 in. to feet. Then divide the height of the wall (6 ft) by the height of each block.

Solution

$$9 \text{ in.} = \frac{9 \text{ in.}}{1} \cdot \frac{1 \text{ ft}}{12 \text{ in.}} = \frac{9 \text{ ft}}{12} = 0.75 \text{ ft}$$

$$\frac{6 \text{ ft}}{0.75 \text{ ft}} = 8$$

The wall will have 8 rows of blocks.

You Try It 10

The floor of a storage room is being tiled. Eight tiles, each a 9-inch square, fit across the width of the floor. Find the width, in feet, of the storage room.

Your strategy

Your solution

Example 11

A plumber used 3 ft 9 in., 2 ft 6 in., and 11 in. of copper tubing to install a sink. Find the total length of copper tubing used.

Strategy

To find the total length of copper tubing used, add the three lengths of copper tubing (3 ft 9 in., 2 ft 6 in., and 11 in.).

Solution

```
    3 ft   9 in.
    2 ft   6 in.
+         11 in.
───────────────
    5 ft  26 in.      •  26 in. = 2 ft 2 in.
```

5 ft 26 in. = 7 ft 2 in.

The plumber used 7 ft 2 in. of copper tubing.

You Try It 11

A board 9 ft 8 in. is cut into four pieces of equal length. How long is each piece?

Your strategy

Your solution

Solutions on p. S21

8.1 Exercises

Objective A

Convert.

1. 6 ft = _____ in.

2. 9 ft = _____ in.

3. 30 in. = _____ ft

4. 64 in. = _____ ft

5. 13 yd = _____ ft

6. $4\frac{1}{2}$ yd = _____ ft

7. 16 ft = _____ yd

8. $4\frac{1}{2}$ ft = _____ yd

9. $2\frac{1}{3}$ yd = _____ in.

10. 5 yd = _____ in.

11. 120 in. = _____ yd

12. 66 in. = _____ yd

13. 2 mi = _____ ft

14. $1\frac{1}{2}$ mi = _____ ft

15. $7\frac{1}{2}$ in. = _____ ft

Objective B

Perform the arithmetic operation.

16. 100 in. =
    ___ ft ___ in.

17. 6400 ft =
    ___ mi ___ ft

18. 15 in. =
    ___ ft ___ in.

19.   6 ft 7 in.
    + 3 ft 4 in.

20.   9 ft 11 in.
    + 3 ft 6 in.

21.   5 ft 3 in.
    − 2 ft 6 in.

22.   9 yd 1 ft
    − 3 yd 2 ft

23.   2 ft 5 in.
    ×   6

24. $3\frac{2}{3}$ ft × 4

25. $2\overline{)5\ \text{ft}\ 4\ \text{in.}}$

26. $12\frac{1}{2}$ in. ÷ 3

27. $4\frac{2}{3}$ ft + $6\frac{1}{2}$ ft

28.   3 yd 2 ft
    + 6 yd 2 ft

29.   1 mi 4200 ft
    + 2 mi 3600 ft

30.   5 yd 1 ft
    − 2 yd 2 ft

31. A kitchen counter is to be covered with tile that is 4 in. square. How many tiles can be placed along one row of a counter top that is 4 ft 8 in. long?

32. Thirty-two yards of material were used for making pleated draperies. How many feet of material were used?

33. Find the missing dimension.

34. Find the total length of the shaft.

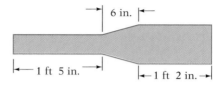

35. What length of material is needed to drill two holes 3 in. in diameter and leave $\frac{1}{2}$ in. between the holes and on either side as shown in the diagram?

36. Find the missing dimension in the figure.

37. A board $6\frac{2}{3}$ ft long is cut into four equal pieces. How long is each piece?

38. How long must a board be if four pieces, each 3 ft 4 in. long, are to be cut from it?

39. A picture is 1 ft 9 in. high and 1 ft 6 in. wide. Find the length of framing needed to frame the picture.

40. You bought 32 ft of baseboard to install in the kitchen of your house. How many inches of baseboard did you purchase?

41. Forty-five bricks, each 9 in. long, are laid end to end to make the base for a wall. Find the length of the wall in feet.

42. A roof is constructed with nine rafters, each 8 ft 4 in. long. Find the total number of feet of material needed to build the rafters.

APPLYING THE CONCEPTS

43. There are approximately 200,000,000 adults living in the United States. Assume that the average adult is 19 in. wide from shoulder to shoulder. If all the adults in the United States are standing shoulder to shoulder, could they reach around Earth at the equator, a distance of approximately 25,000 mi?

44. How good are you at estimating lengths or distances? Estimate the length of a pencil, the width of your room, the length of a block, and the distance to the grocery store. Then measure these lengths and compare the results with your estimates.

8.2 Weight

Objective A   **To convert measurements of weight in the U.S. Customary System**

Weight is a measure of how strongly Earth is pulling on an object. The unit of weight called the **pound** is defined as the weight of a standard solid kept at the Bureau of Standards in Washington, D.C. The U.S. Customary System units of weight are **ounce, pound,** and **ton.**

Equivalences between units of weight in the U.S. Customary System are

$$16 \text{ ounces (oz)} = 1 \text{ pound (lb)}$$
$$2000 \text{ lb} = 1 \text{ ton}$$

These equivalences can be used to form conversion rates to change one unit of measurement to another. For example, because 16 oz = 1 lb, the conversion rates $\frac{16 \text{ oz}}{1 \text{ lb}}$ and $\frac{1 \text{ lb}}{16 \text{ oz}}$ are both equivalent to 1.

➡ Convert 62 oz to pounds.

$$62 \text{ oz} = 62 \text{ oz} \times \boxed{\frac{1 \text{ lb}}{16 \text{ oz}}}$$

$$= \frac{62 \text{ o\!z}}{1} \times \frac{1 \text{ lb}}{16 \text{ o\!z}}$$

$$= \frac{62 \text{ lb}}{16}$$

$$= 3\frac{7}{8} \text{ lb}$$

• The conversion rate must contain lb (the unit desired in the answer) in the numerator and must contain oz (the original unit) in the denominator.

Example 1

Convert $3\frac{1}{2}$ tons to pounds.

Solution

$$3\frac{1}{2} \text{ tons} = \frac{7}{2} \text{ ton\!s} \times \frac{2000 \text{ lb}}{1 \text{ to\!n}}$$

$$= \frac{14{,}000 \text{ lb}}{2} = 7000 \text{ lb}$$

You Try It 1   Convert 3 lb to ounces.

Your solution

Example 2   Convert 42 oz to pounds.

Solution

$$42 \text{ oz} = 42 \text{ o\!z} \times \frac{1 \text{ lb}}{16 \text{ o\!z}}$$

$$= \frac{42 \text{ lb}}{16} = 2\frac{5}{8} \text{ lb}$$

You Try It 2   Convert 4200 lb to tons.

Your solution

Solutions on p. S22

Objective B **To perform arithmetic operations with measurements of weight**

When performing arithmetic operations with measurements of weight, write the answer in simplest form. For example, 1 lb 22 oz should be written 2 lb 6 oz.

Example 3 Find the difference between 14 lb 5 oz and 8 lb 14 oz.

Solution

$$
\begin{array}{r}
\overset{13\,lb}{\cancel{14\,lb}} \quad \overset{21\,oz}{\cancel{5\,oz}} \\
-\quad 8\;lb\;14\;oz \\
\hline
5\;lb\quad 7\;oz
\end{array}
$$

• **Borrow 1 lb (16 oz) from 14 lb and add it to 5 oz.**

You Try It 3 Find the difference between 7 lb 1 oz and 3 lb 4 oz.

Your solution

Example 4 Divide: 7 lb 14 oz ÷ 3

Solution

$$
\begin{array}{r}
\;2\;lb\quad 10\;oz \\
3\overline{)\;7\;lb\quad 14\;oz} \\
-6\;lb \\
\hline
1\;lb = 16\;oz \\
\hline
30\;oz \\
-30\;oz \\
\hline
0
\end{array}
$$

You Try It 4 Multiply: 3 lb 6 oz × 4

Your solution

Solutions on p. S22

Objective C **To solve application problems**

Example 5

Sirina Jasper purchased 4 lb 8 oz of oat bran and 2 lb 11 oz of wheat bran. She plans to blend the two brans and then repackage the mixture in 3-ounce packages for a diet supplement. How many 3-ounce packages can she make?

Strategy

To find the number of 3-ounce packages:

• Add the amount of oat bran (4 lb 8 oz) to the amount of wheat bran (2 lb 11 oz).
• Convert the sum to ounces.
• Divide the total ounces by the weight of each package (3 oz).

Solution

$$
\begin{array}{r}
4\;lb\quad 8\;oz \\
+\;2\;lb\;11\;oz \\
\hline
6\;lb\;19\;oz = 7\;lb\;3\;oz = 115\;oz
\end{array}
$$

$$
\frac{115\;oz}{3\;oz} \approx 38.3
$$

She can make 38 packages.

You Try It 5

Find the weight in pounds of 12 bars of soap. Each bar weighs 7 oz.

Your strategy

Your solution

Solution on p. S22

8.2 Exercises

Objective A

Convert.

1. 64 oz = _____ lb

2. 36 oz = _____ lb

3. 8 lb = _____ oz

4. 7 lb = _____ oz

5. 3200 lb = _____ tons

6. 7000 lb = _____ tons

7. 6 tons = _____ lb

8. $1\frac{1}{4}$ tons = _____ lb

9. 66 oz = _____ lb

10. 90 oz = _____ lb

11. $1\frac{1}{2}$ lb = _____ oz

12. $2\frac{5}{8}$ lb = _____ oz

13. $1\frac{3}{10}$ tons = _____ lb

14. $\frac{4}{5}$ ton = _____ lb

15. 500 lb = _____ ton

16. 5000 lb = _____ tons

17. 180 oz = _____ lb

18. 12 oz = _____ lb

Objective B

Perform the arithmetic operation.

19. 9000 lb =
_____ tons _____ lb

20. 85 oz =
_____ lb _____ oz

21. 40 oz =
_____ lb _____ oz

22.   4 lb 7 oz
+ 3 lb 12 oz

23.   1 ton   800 lb
+ 3 tons 1600 lb

24.   7 lb 5 oz
− 3 lb 8 oz

25.   3 tons 500 lb
− 1 ton   800 lb

26.   3 lb 6 oz
×   4

27. $5\frac{1}{2}$ lb × 6

28. 2)3 lb 8 oz

29. $4\frac{2}{3}$ lb × 3

30.   7 lb 7 oz
+ 6 lb 9 oz

31. $6\frac{1}{2}$ oz
+ $2\frac{1}{2}$ oz

32. $6\frac{3}{8}$ lb
− $2\frac{5}{6}$ lb

33. 5 lb 12 oz ÷ 4

Objective C   *Application Problems*

34. A machinist has 25 iron rods to mill. Each rod weighs 20 oz. Find the total weight of the rods in pounds.

35. A fireplace brick weighs $2\frac{1}{2}$ lb. What is the weight of a load of 800 bricks?

36. A college bookstore received 1200 textbooks, each weighing 9 oz. Find the total weight of the 1200 textbooks in pounds.

37. A 4 × 4-inch tile weighs 7 oz. Find the weight, in pounds, of a package of 144 tiles.

38. A farmer ordered 20 tons of feed for 100 cattle. After 15 days, the farmer has 5 tons of feed left. How many pounds of food has each cow eaten per day?

39. A case of soft drink contains 24 cans, each weighing 6 oz. Find the weight, in pounds, of the case of soft drink.

40. A baby weighed 7 lb 8 oz at birth. At 6 months of age, the baby weighed 15 lb 13 oz. Find the baby's increase in weight during the 6 months.

41. Shampoo weighing 5 lb 4 oz is divided equally and poured into four containers. How much shampoo is in each container?

42. A steel rod weighing 16 lb 11 oz is cut into three pieces. Find the weight of each piece of steel rod.

43. Find the cost of a ham roast that weighs 5 lb 10 oz if the price per pound is $3.20.

44. A candy store buys candy weighing 12 lb for $14.40. The candy is repackaged and sold in 6-ounce packages for $1.15 each. Find the markup on the 12 lb of candy.

45. A manuscript weighing 2 lb 3 oz is mailed at the parcel post rate of $.25 per ounce. Find the cost of mailing the manuscript.

APPLYING THE CONCEPTS

46. Write a paragraph describing the growing need for precision in our measurements as civilization progressed. Include a discussion of the need for precision in the space industry.

47. Estimate the weight of a nickel, a textbook, a friend, and a car. Then find the actual weights and compare them with your estimates.

8.3 Capacity

Objective A

To convert measurements of capacity in the U.S. Customary System

Liquid substances are measured in units of **capacity.** The standard U.S. Customary units of capacity are the **fluid ounce, cup, pint, quart,** and **gallon.** Equivalences between units of capacity in the U.S. Customary System are

$$8 \text{ fluid ounces (fl oz)} = 1 \text{ cup (c)}$$
$$2 \text{ c} = 1 \text{ pint (pt)}$$
$$2 \text{ pt} = 1 \text{ quart (qt)}$$
$$4 \text{ qt} = 1 \text{ gallon (gal)}$$

These equivalences can be used to form conversion rates to change one unit of measurement to another. For example, because 8 fl oz = 1 c, the conversion rates $\frac{8 \text{ fl oz}}{1 \text{ c}}$ and $\frac{1 \text{ c}}{8 \text{ fl oz}}$ are both equivalent to 1.

➡ Convert 36 fl oz to cups.

$$36 \text{ fl oz} = 36 \text{ fl oz} \times \boxed{\frac{1 \text{ c}}{8 \text{ fl oz}}}$$

$$= \frac{36 \ \cancel{\text{fl oz}}}{1} \times \frac{1 \text{ c}}{8 \ \cancel{\text{fl oz}}}$$

$$= \frac{36 \text{ c}}{8}$$

$$= 4\frac{1}{2} \text{ c}$$

- The conversion rate must contain c in the numerator and fl oz in the denominator.

➡ Convert 3 qt to cups.

$$3 \text{ qt} = 3 \text{ qt} \times \boxed{\frac{2 \text{ pt}}{1 \text{ qt}}} \times \boxed{\frac{2 \text{ c}}{1 \text{ pt}}}$$

$$= \frac{3 \ \cancel{\text{qt}}}{1} \times \frac{2 \ \cancel{\text{pt}}}{1 \ \cancel{\text{qt}}} \times \frac{2 \text{ c}}{1 \ \cancel{\text{pt}}}$$

$$= \frac{12 \text{ c}}{1}$$

$$= 12 \text{ c}$$

- The direct equivalence is not given above. Use two conversion rates. First convert quarts to pints, and then convert pints to cups. The unit in the denominator of the second conversion rate and the unit in the numerator of the first conversion rate must be the same in order to cancel.

Example 1   Convert 42 c to quarts.

Solution $42 \text{ c} = 42 \ \cancel{\text{c}} \times \dfrac{1 \ \cancel{\text{pt}}}{2 \ \cancel{\text{c}}} \times \dfrac{1 \text{ qt}}{2 \ \cancel{\text{pt}}}$

$\qquad = \dfrac{42 \text{ qt}}{4} = 10\dfrac{1}{2} \text{ qt}$

You Try It 1   Convert 18 pt to gallons.

Your solution

Solution on p. S22

Objective B

To perform arithmetic operations with measurements of capacity

When performing arithmetic operations with measurements of capacity, write the answer in simplest form. For example, 1 c 12 fl oz should be written as 2 c 4 fl oz.

Example 2 What is 4 gal 1 qt decreased by 2 gal 3 qt?

Solution

$$\begin{array}{r} \overset{3 \text{ gal}}{\cancel{4 \text{ gal}}}\ \overset{5 \text{ qt}}{\cancel{1 \text{ qt}}} \\ -\ 2 \text{ gal } 3 \text{ qt} \\ \hline 1 \text{ gal } 2 \text{ qt} \end{array}$$

• Borrow 1 gal (4 qt) from 4 gal and add to 1 qt.

You Try It 2 Find the quotient of 4 gal 2 qt and 3.

Your solution

Solution on p. S22

Objective C

To solve application problems

Example 3
A can of apple juice contains 25 fl oz. Find the number of quarts of apple juice in a case of 24 cans.

Strategy
To find the number of quarts of apple juice in one case:

• Multiply the number of cans (24) by the number of fluid ounces per can (25) to find the total number of fluid ounces in the case.
• Convert the number of fluid ounces in the case to quarts.

Solution
24 × 25 fl oz = 600 fl oz

$$600 \text{ fl oz} = \frac{600 \ \cancel{\text{fl oz}}}{1} \cdot \frac{1 \ \cancel{c}}{8 \ \cancel{\text{fl oz}}} \cdot \frac{1 \ \cancel{\text{pt}}}{2 \ \cancel{c}} \cdot \frac{1 \text{ qt}}{2 \ \cancel{\text{pt}}}$$

$$= \frac{600 \text{ qt}}{32} = 18\frac{3}{4} \text{ qt}$$

One case of apple juice contains $18\frac{3}{4}$ qt.

You Try It 3
Five students are going backpacking in the desert. Each student requires 3 qt of water per day. How many gallons of water should they take for a 3-day trip?

Your strategy

Your solution

Solution on p. S22

8.3 Exercises

Objective A

Convert.

1. 60 fl oz = _____ c

2. 48 fl oz = _____ c

3. 3 c = _____ fl oz

4. $2\frac{1}{2}$ c = _____ fl oz

5. 8 c = _____ pt

6. 5 c = _____ pt

7. $3\frac{1}{2}$ pt = _____ c

8. 12 pt = _____ qt

9. 22 qt = _____ gal

10. 10 qt = _____ gal

11. $2\frac{1}{4}$ gal = _____ qt

12. 7 gal = _____ qt

13. $7\frac{1}{2}$ pt = _____ qt

14. $3\frac{1}{2}$ qt = _____ pt

15. 20 fl oz = _____ pt

16. $1\frac{1}{2}$ pt = _____ fl oz

17. 17 c = _____ qt

18. $1\frac{1}{2}$ qt = _____ c

Objective B

Perform the arithmetic operation.

19. 14 qt =
_____ gal _____ qt

20. 9 pt =
_____ qt _____ pt

21. 5 pt =
_____ qt _____ pt

22.   3 gal 2 qt
     + 4 gal 3 qt

23.   4 qt 1 pt
     + 2 qt 1 pt

24.   3 gal 1 qt
     − 1 gal 2 qt

25.   3 c 3 fl oz
     − 2 c 5 fl oz

26.   2 qt 1 pt
     ×   5

27. $3\frac{1}{2}$ pt × 5

28. 5⟌6 gal 1 qt

29. $3\frac{1}{2}$ gal ÷ 4

30.   5 c 3 fl oz
     + 3 c 6 fl oz

31.   3 gal 3 qt
     + 1 gal 2 qt

32.   4 c 6 fl oz
     − 2 c 7 fl oz

33.   3 gal
     − 1 gal 2 qt

34. $1\frac{1}{2}$ pt $+ 2\frac{2}{3}$ pt

35. $4\frac{1}{2}$ gal $- 1\frac{3}{4}$ gal

36. $2\overline{)3 \text{ gal } 2 \text{ qt}}$

Objective C *Application Problems*

37. Sixty adults will attend a church social. Each adult will drink 2 c of coffee. How many gallons of coffee should be prepared?

38. The Bayside Playhouse serves punch during intermission. Assume that 200 people will each drink 1 c of punch. How many gallons of punch should be ordered?

39. A solution needed for a chemistry class required 72 fl oz of water, 16 fl oz of one solution, and 48 fl oz of another solution. Find the number of quarts of the final solution.

40. A cafeteria sold 124 cartons of milk in 1 day. Each carton contained 1 c of milk. How many quarts of milk were sold that day?

41. A farmer changed the oil in a tractor seven times during the year. Each oil change required 5 qt of oil. How many gallons of oil did the farmer use in the seven oil changes?

42. There are 24 cans in a case of tomato juice. Each can contains 10 fl oz of tomato juice. Find the number of 1-cup servings in the case of tomato juice.

43. One brand of orange juice costs $1.59 for 1 qt. Another brand costs $1.25 for 24 fl oz. Which is the more economical purchase?

44. Mandy carried 6 qt of water for 3 days of desert camping. Water weighs $8\frac{1}{3}$ lb per gallon. Find the weight of water that she carried.

45. A department store bought hand lotion in 5-quart containers and then repackaged the lotion in 8-fluid-ounce bottles. The lotion and bottles cost $41.50, and each 8-fluid-ounce bottle was sold for $4.25. How much profit was made on each 5-quart package of lotion?

46. Orlando bought oil in 50-gallon containers for changing the oil in his customers' cars. He paid $240 for the 50 gal of oil and charged customers $2.10 per quart. Find the profit Orlando made on one 50-gallon container of oil.

APPLYING THE CONCEPTS

47. Define the following units: grain, dram, furlong, and rod. Give an example where each would be used.

48. Assume that you wanted to invent a new measuring system. Discuss some of the features that would have to be incorporated into the system.

8.4 Energy and Power

Objective A   **To use units of energy in the U.S. Customary System**

Energy can be defined as the ability to do work. Energy is stored in coal, in gasoline, in water behind a dam, and in one's own body.

One **foot-pound** (ft · lb) of energy is the amount of energy necessary to lift 1 pound a distance of 1 foot.

To lift 50 lb a distance of 5 ft requires
50 × 5 = 250 ft · lb of energy.

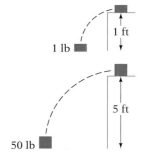

Consumer items that use energy, such as furnaces, stoves, and air conditioners, are rated in **British thermal units** (Btu). For example, a furnace might have a rating of 35,000 Btu per hour, which means that it releases 35,000 Btu of energy in one hour (1 h).

Because 1 Btu = 778 ft · lb, the following conversion rate, equivalent to 1, can be written:

$$\frac{778 \text{ ft} \cdot \text{lb}}{1 \text{ Btu}} = 1$$

Example 1
Convert 250 Btu to foot-pounds.

Solution

250 Btu = 250 B̶t̶u̶ × $\frac{778 \text{ ft} \cdot \text{lb}}{1 \text{ B̶t̶u̶}}$

= 194,500 ft · lb

You Try It 1
Convert 4.5 Btu to foot-pounds.

Your solution

Example 2
Find the energy required for a 125-pound person to climb a mile-high mountain.

Solution
In climbing the mountain, the person is lifting 125 lb a distance of 5280 ft.

Energy = 125 lb × 5280 ft
= 660,000 ft · lb

You Try It 2
Find the energy required for a motor to lift 800 lb a distance of 16 ft.

Your solution

Solutions on p. S22

Example 3

A furnace is rated at 80,000 Btu per hour. How many foot-pounds of energy are released in 1 h?

Solution

$$80,000 \text{ Btu} = 80,000 \text{ Btu} \times \frac{778 \text{ ft} \cdot \text{lb}}{1 \text{ Btu}}$$

$$= 62,240,000 \text{ ft} \cdot \text{lb}$$

You Try It 3

A furnace is rated at 56,000 Btu per hour. How many foot-pounds of energy are released in 1 h?

Your solution

Solution on p. S22

Objective B   **To use units of power in the U.S. Customary System**

Power is the rate at which work is done, or the rate at which energy is released.

Power is measured in **foot-pounds per second** $\left(\frac{\text{ft} \cdot \text{lb}}{\text{s}} \right)$. In each of the following examples, the amount of energy released is the same, but the time taken to release the energy is different; thus the power is different.

100 lb is lifted 10 ft in 10 s.

$$\text{Power} = \frac{10 \text{ ft} \times 100 \text{ lb}}{10 \text{ s}} = 100 \frac{\text{ft} \cdot \text{lb}}{\text{s}}$$

100 lb is lifted 10 ft in 5 s.

$$\text{Power} = \frac{10 \text{ ft} \times 100 \text{ lb}}{5 \text{ s}} = 200 \frac{\text{ft} \cdot \text{lb}}{\text{s}}$$

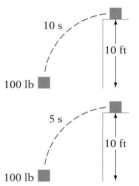

The U.S. Customary unit of power is the **horsepower.** A horse doing average work can pull 550 lb a distance of 1 ft in 1 s and can continue this work all day.

$$1 \text{ horsepower (hp)} = 550 \frac{\text{ft} \cdot \text{lb}}{\text{s}}$$

Example 4

Find the power needed to raise 300 lb a distance of 30 ft in 15 s.

Solution $\text{Power} = \dfrac{30 \text{ ft} \times 300 \text{ lb}}{15 \text{ s}}$

$$= 600 \frac{\text{ft} \cdot \text{lb}}{\text{s}}$$

You Try It 4

Find the power needed to raise 1200 lb a distance of 90 ft in 24 s.

Your solution

Example 5

A motor has a power of 2750 $\frac{\text{ft} \cdot \text{lb}}{\text{s}}$. Find the horsepower of the motor.

Solution $\dfrac{2750}{550} = 5 \text{ hp}$

You Try It 5

A motor has a power of 3300 $\frac{\text{ft} \cdot \text{lb}}{\text{s}}$. Find the horsepower of the motor.

Your solution

Solutions on p. S22

8.4 Exercises

Objective A

1. Convert 25 Btu to foot-pounds.

2. Convert 6000 Btu to foot-pounds.

3. Convert 25,000 Btu to foot-pounds.

4. Convert 40,000 Btu to foot-pounds.

5. Find the energy required to lift 150 lb a distance of 10 ft.

6. Find the energy required to lift 300 lb a distance of 16 ft.

7. Find the energy required to lift a 3300-pound car a distance of 9 ft.

8. Find the energy required to lift a 3680-pound elevator a distance of 325 ft.

9. Three tons are lifted 5 ft. Find the energy required in foot-pounds.

10. Seven tons are lifted 12 ft. Find the energy required in foot-pounds.

11. A construction worker carries 3-pound blocks up a 10-foot flight of stairs. How many foot-pounds of energy are required to carry 850 blocks up the stairs?

12. A crane lifts an 1800-pound steel beam to the roof of a building 36 ft high. Find the amount of energy the crane requires in lifting the beam.

13. A furnace is rated at 45,000 Btu per hour. How many foot-pounds of energy are released by the furnace in 1 h?

14. A furnace is rated at 22,500 Btu per hour. How many foot-pounds of energy does the furnace release in 1 h?

15. Find the amount of energy in foot-pounds given off when 1 lb of coal is burned. One pound of coal gives off 12,000 Btu of energy when burned.

16. Find the amount of energy in foot-pounds given off when 1 lb of gasoline is burned. One pound of gasoline gives off 21,000 Btu of energy when burned.

17. Convert 1100 $\frac{\text{ft} \cdot \text{lb}}{\text{s}}$ to horsepower.

18. Convert 6050 $\frac{\text{ft} \cdot \text{lb}}{\text{s}}$ to horsepower.

19. Convert 4400 $\frac{\text{ft} \cdot \text{lb}}{\text{s}}$ to horsepower.

20. Convert 1650 $\frac{\text{ft} \cdot \text{lb}}{\text{s}}$ to horsepower.

21. Convert 5 hp to foot-pounds per second.

22. Convert 4 hp to foot-pounds per second.

23. Convert 7 hp to foot-pounds per second.

24. Convert 2 hp to foot-pounds per second.

25. Find the power in foot-pounds per second needed to raise 125 lb a distance of 12 ft in 3 s.

26. Find the power in foot-pounds per second needed to raise 500 lb a distance of 60 ft in 8 s.

27. Find the power in foot-pounds per second needed to raise 3000 lb a distance of 40 ft in 25 s.

28. Find the power in foot-pounds per second of an engine that can raise 12,000 lb to a height of 40 ft in 60 s.

29. Find the power in foot-pounds per second of an engine that can raise 180 lb to a height of 40 ft in 5 s.

30. Find the power in foot-pounds per second of an engine that can raise 1200 lb to a height of 18 ft in 30 s.

31. A motor has a power of 4950 $\frac{\text{ft} \cdot \text{lb}}{\text{s}}$. Find the horsepower of the motor.

32. A motor has a power of 16,500 $\frac{\text{ft} \cdot \text{lb}}{\text{s}}$. Find the horsepower of the motor.

33. A motor has a power of 6600 $\frac{\text{ft} \cdot \text{lb}}{\text{s}}$. Find the horsepower of the motor.

APPLYING THE CONCEPTS

34. Pick a source of energy and write an article about it. Include the source, possible pollution problems, and future prospects associated with this form of energy.

Focus on Problem Solving

Applying Solutions to Other Problems

Problem solving in the previous chapters concentrated on solving specific problems. After a problem is solved, however, there is an important question to be asked: "Does the solution to this problem apply to other types of problems?"

To illustrate this extension of problem solving, we will consider *triangular numbers,* which were studied by ancient Greek mathematicians. The numbers 1, 3, 6, 10, 15, and 21 are the first six triangular numbers. What is the next triangular number?

To answer this question, note in the diagram below that a triangle can be formed using the number of dots that correspond to a triangular number.

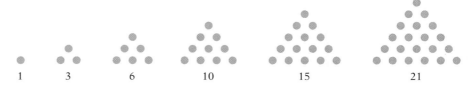

Observe that the number of dots in each row is one more than the number of dots in the row above. The total number of dots can be found by addition.

The pattern suggests that the next triangular number (the seventh one) is the sum of the first 7 natural numbers. The seventh triangular number is 28. The diagram at the right shows the seventh triangular number.

Using the pattern for triangular numbers, it is easy to determine that the tenth triangular number is

$$1 + 2 + 3 + 4 + 5 + 6 + 7 + 8 + 9 + 10 = 55$$

Now consider a situation that may seem to be totally unrelated to triangular numbers. Suppose you are in charge of scheduling softball games for a league. There are seven teams in the league, and each team must play every other team once. How many games must be scheduled?

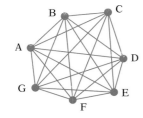

We label the teams A, B, C, D, E, F, and G. (See the figure at the left.) A line between two teams indicates that the two teams play each other. Beginning with A, there are 6 lines for the 6 teams that A must play. There are 6 teams that B must play, but the line between A and B has already been drawn, so there are only 5 remaining games to schedule for B. Now move on to C. The lines between C and A and between C and B have already been drawn, so there are only 4 additional lines to be drawn to represent the teams C will play. Moving on to D, we see that the lines between D and A, D and B, and D and C have already been drawn, so there are 3 more lines to be drawn to represent the teams D will play.

Note that each time we move from team to team, one fewer line needs to be drawn. When we reach F, there is only one line to be drawn, the one between F and G. The total number of lines drawn is 6 + 5 + 4 + 3 + 2 + 1 = 21, the sixth triangular number. For a league with 7 teams, the number of games that must be scheduled so that each team plays every other team once is the sixth triangular number. If there were 10 teams in the league, the number of games that must be scheduled would be the ninth triangular number, which is 45.

A college chess team wants to schedule a match so that each of its 15 members plays each other member of the team once. How many matches must be scheduled?

Projects and Group Activities

Astronomical Distances

Although one mile may seem like a long distance, for astronomers this measurement is inadequate for the vastness of space. Other measurements of distance are more useful. An **astronomical unit** is the average distance between Earth and the sun. A **light year** is the distance a ray of light travels in one year. To measure even greater distances within the universe, a **parsec** is used.

Use a reference to find the distance of one astronomical unit, one light year, and one parsec. Then write equivalences between these three units.

Nomographs

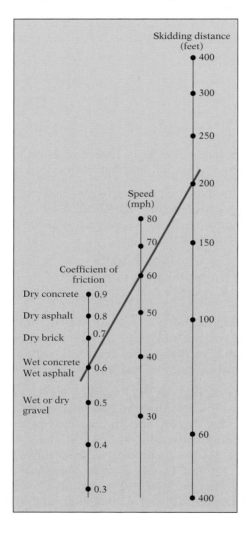

A chart is another tool that is used in problem solving. The chart at the left is a nomograph. A **nomograph** is a chart that represents numerical relationships among variables.

One of the details a traffic accident investigator checks when looking into a car accident is the length of the skid marks of the car. This length can help the investigator determine the speed of the car when the brakes were applied.

The nomograph at the left can be used to determine the speed of a car under given conditions. It shows the relationship among the speed of the car, the skidding distance, and the *coefficient of friction*.

The coefficient of friction is an experimentally obtained value that reflects how easy or hard it is to drag one object over another. For instance, it is easier to drag a box across ice than to drag it across a carpet. The coefficient of friction is smaller for the box and ice than it is for the box and carpet.

To use the nomograph at the left, an investigator would draw a line from the coefficient of friction to the skidding distance. The point at which the line crosses the speed line shows how fast the car was going when the brakes were applied. The line from 0.6 to 200 intersects the speed line at 60. This indicates that a car that skidded 200 feet on wet asphalt or wet concrete was traveling at 60 mph.

1. Use the nomograph to determine the speed of a car when the brakes were applied for a car traveling on gravel and for skid marks of 100 ft.

2. Use the nomograph to determine the speed of a car when the brakes were applied for a car traveling on dry concrete and for skid marks of 150 ft.

3. Suppose a car is traveling 80 mi/h when the brakes are applied. Find the difference in skidding distance if the car is traveling on wet concrete rather than dry concrete.

Averages   If two towns are 150 mi apart and you drive between the two towns in 3 h, then your

$$\text{Average speed} = \frac{\text{total distance}}{\text{total time}} = \frac{150}{3} = 50 \text{ mi/h}$$

It is highly unlikely that your speed was *exactly* 50 mi/h the entire time of the trip. Sometimes you will have traveled faster than 50 mi/h, and other times you will have traveled slower than 50 mi/h. Dividing the total distance you traveled by the total time it took to go that distance is an example of calculating an average.

There are many other averages that may be calculated. For instance, the Environmental Protection Agency calculates an estimated miles per gallon (mpg) for new cars. Miles per gallon is an average calculated from the formula

$$\frac{\text{Miles traveled}}{\text{Gallons of gasoline}}$$

For instance, the miles per gallon for a car that travels 308 mi on 11 gal of gas is $\frac{308}{11} = 28$ mpg.

A pilot would not use miles per gallon as a measure of fuel efficiency. Rather, pilots use gallons per hour. A plane that travels 5 h and uses 400 gal of fuel has an average that is calculated as

$$\frac{\text{Gallons of fuel}}{\text{Hours flown}} = \frac{400}{5} = 80 \text{ gal/h}$$

Using the examples above, calculate the following averages.

1. Determine the average speed of a car that travels 355 mi in 6 h. Round to the nearest tenth.

2. Determine the miles per gallon of a car that can travel 405 mi on 12 gal of gasoline. Round to the nearest tenth.

3. If a plane flew 2000 mi in 5 h and used 1000 gal of fuel, determine the average number of gallons per hour that the plane used.

Another type of average is grade-point average (GPA). It is calculated by multiplying the units for each class by the grade point for that class, adding the results, and dividing by the total number of units taken. Here is an example using the grading scale $A = 4$, $B = 3$, $C = 2$, $D = 1$, and $F = 0$.

Class	Units	Grade
Math	4	B (= 3)
English	3	A (= 4)
French	5	C (= 2)
Biology	3	B (= 3)

$$\text{GPA} = \frac{4 \cdot 3 + 3 \cdot 4 + 5 \cdot 2 + 3 \cdot 3}{4 + 3 + 5 + 3} = \frac{43}{15} \approx 2.87$$

4. A grading scale that provides for plus or minus grades uses A = 4, A− = 3.7, B+ = 3.3, B = 3, B− = 2.7, C+ = 2.3, C = 2, C− = 1.7, D+ = 1.3, D = 1, D− = 0.7, and F = 0. Calculate the GPA of the student whose grades are given below.

Class	Units	Grade
Math	5	B +
English	3	C +
Spanish	5	A −
Physical Science	3	B −

Chapter Summary

Key Words

A *measurement* includes a number and a unit. [p. 333]

The U.S. Customary units of length are *inch, foot, yard,* and *mile.* [p. 333]

Weight is a measure of how strongly Earth is pulling on an object. The U.S. Customary units of weight are *ounce, pound,* and *ton.* [p. 339]

Liquid substances are measured in units of *capacity.* The U.S. Customary units of capacity are *fluid ounce, cup, pint, quart,* and *gallon.* [p. 343]

Energy is the ability to do work. One *foot-pound* of energy is the amount of energy required to lift 1 pound a distance of 1 foot. [p. 347]

Furnaces, stoves, and air conditioners are rated in energy units called *British thermal units* (Btu). [p. 347]

Power is the rate at which work is done. The U.S. Customary unit of power is the *horsepower* (hp). [p. 348]

Essential Rules

Equivalences Between Units of Length

12 in. = 1 ft
3 ft = 1 yd
36 in. = 1 yd
5280 ft = 1 mi [p. 333]

Equivalences Between Units of Weight

16 oz = 1 lb
2000 lb = 1 ton [p. 339]

Equivalences Between Units of Capacity

8 fl oz = 1 c
2 c = 1 pt
2 pt = 1 qt
4 qt = 1 gal [p. 343]

Equivalences Between Units of Energy

1 Btu = 778 ft · lb [p. 347]

Equivalences Between Units of Power

$1 \text{ hp} = 550 \dfrac{\text{ft} \cdot \text{lb}}{\text{s}}$ [p. 348]

A conversion rate is used to convert one unit of measurement to another. For example, the conversion rate $\dfrac{12 \text{ in.}}{1 \text{ ft}}$ is used to convert feet to inches. [p. 333]

Chapter Review

1. Convert 4 ft to inches.

2. What is 7 ft 6 in. divided by 3?

3. Find the energy needed to lift 200 lb a distance of 8 ft.

4. Convert $2\frac{1}{2}$ pt to fluid ounces.

5. Convert 14 ft to yards.

6. Convert 2400 lb to tons.

7. Find the quotient of 7 lb 5 oz and 3.

8. Convert $3\frac{3}{8}$ lb to ounces.

9. Add: 3 ft 9 in.
          + 5 ft 6 in.

10. Subtract: 3 tons 500 lb
              − 1 ton 1500 lb

11. Add: 5 lb 11 oz
          + 3 lb 8 oz

12. Subtract: 5 yd 1 ft
              − 3 yd 2 ft

13. Convert 12 c to quarts.

14. Find the product of 2 ft 8 in. and 5.

15. Convert 2.5 hp to foot-pounds per second. $\left(1\ hp = 550\ \dfrac{ft \cdot lb}{s}.\right)$

16. Multiply: 5 lb 8 oz
$\underline{\times\qquad 8}$

17. Convert 50 Btu to foot-pounds. (1 Btu = 778 ft · lb.)

18. Convert $3850\ \dfrac{ft \cdot lb}{s}$ to horsepower. $\left(1\ hp = 550\ \dfrac{ft \cdot lb}{s}.\right)$

19. A board 6 ft 11 in. long is cut from a board 10 ft 5 in. long. Find the length of the remaining piece of board.

20. A book weighing 2 lb 3 oz is mailed at the parcel post rate of $.18 per ounce. Find the cost of mailing the book.

21. A can of pineapple juice contains 18 fl oz. Find the number of quarts in a case of 24 cans.

22. A cafeteria sold 256 cartons of milk in one school day. Each carton contains 1 c of milk. How many gallons of milk were sold that day?

23. A furnace is rated at 35,000 Btu per hour. How many foot-pounds of energy does the furnace release in 1 h? (1 Btu = 778 ft · lb.)

24. Find the power in foot-pounds per second of an engine that can raise 800 lb to a height of 15 ft in 25 s.

Chapter Test

1. Convert $2\frac{1}{2}$ ft to inches.

2. Subtract: 4 ft 2 in. − 1 ft 9 in.

3. A board $6\frac{2}{3}$ ft long is cut into five equal pieces. How long is each piece?

4. Seventy-two bricks, each 8 in. long, are laid end to end to make the base for a wall. Find the length of the wall in feet.

5. Convert $2\frac{7}{8}$ lb to ounces.

6. Convert: 40 oz = ___ lb ___ oz

7. Find the sum of 9 lb 6 oz and 7 lb 11 oz.

8. Divide: 6 lb 12 oz ÷ 4

9. A college bookstore received 1000 workbooks, each weighing 12 oz. Find the total weight of the 1000 workbooks in pounds.

10. An elementary school class gathered 800 aluminum cans for recycling. Four aluminum cans weigh 3 oz. Find the amount the class received if the rate of pay was $.75 per pound for the aluminum cans. Round to the nearest cent.

11. Convert 13 qt to gallons.

12. Convert $3\frac{1}{2}$ gal to pints.

13. What is $1\frac{3}{4}$ gal times 7?   **14.** Add: 5 gal 2 qt + 2 gal 3 qt

15. A can of grapefruit juice contains 20 fl oz. Find the number of cups of grapefruit juice in a case of 24 cans.

16. Nick, a mechanic, bought oil in 40-gallon containers for changing the oil in customers' cars. He paid $200 for the 40 gallons of oil and charged customers $2.15 per quart. Find the profit Nick made on one 40-gallon container of oil.

17. Find the energy required to lift 250 lb a distance of 15 ft.

18. A furnace is rated at 40,000 Btu per hour. How many foot-pounds of energy are released by the furnace in 1 h? (1 Btu $= 778$ ft $\cdot$ lb.)

19. Find the power needed to lift 200 lb a distance of 20 ft in 25 s.

20. A motor has a power of 2200 $\frac{\text{ft} \cdot \text{lb}}{\text{s}}$. Find the motor's horsepower. $\left(1 \text{ hp} = 550 \frac{\text{ft} \cdot \text{lb}}{\text{s}}.\right)$

Cumulative Review

1. Find the LCM of 9, 12, and 15.

2. Write $\frac{43}{8}$ as a mixed number.

3. Subtract: $5\frac{7}{8} - 2\frac{7}{12}$

4. What is $5\frac{1}{3}$ divided by $2\frac{2}{3}$?

5. Simplify: $\frac{5}{8} \div \left(\frac{3}{8} - \frac{1}{4}\right) - \frac{5}{8}$

6. Round 2.0972 to the nearest hundredth.

7. Multiply: $\begin{array}{r} 0.0792 \\ \times \quad 0.49 \\ \hline \end{array}$

8. Solve the proportion: $\frac{n}{12} = \frac{44}{60}$

9. Find $2\frac{1}{2}\%$ of 50.

10. 18 is 42% of what? Round to the nearest hundredth.

11. A 7.2-pound roast costs $29.88. Find the unit cost.

12. Add: $3\frac{2}{5}$ in. $+ 5\frac{1}{3}$ in.

13. Convert: 24 oz = ___ lb ___ oz

14. Multiply: 3 lb 8 oz × 9

15. Subtract: $4\frac{1}{3}$ qt $- 1\frac{5}{6}$ qt

16. Find 2 lb 10 oz less than 4 lb 6 oz.

17. An investor receives a dividend of $56 from 40 shares of stock. At the same rate, find the dividend that 200 shares of stock would yield.

18. Anna had a balance of $578.56 in her checkbook. She wrote checks of $216.98 and $34.12 and made a deposit of $315.33. What is her new checking account balance?

19. An account executive receives a salary of $800 per month plus a commission of 2% on all sales over $25,000. Find the total monthly income of an account executive who has monthly sales of $140,000.

20. A health inspector found that 3% of a 2500-pound shipment of carrots were spoiled and could not be sold. Find the amount of carrots that could be sold.

21. The scores on the final exam of a trigonometry class are recorded in the histogram in the accompanying figure. What percent of the class received a score between 80% and 90%? Round to the nearest percent.

22. Hayes Department Store uses a markup rate of 40% on all merchandise. What is the selling price of a compact disk player that cost the store $220?

23. A construction firm received a loan of $200,000 for 8 months at a simple interest rate of 11%. Find the interest paid on the loan.

24. Six college students spent several weeks panning for gold during their summer vacation. The students obtained 1 lb 3 oz of gold, which they sold for $400 per ounce. How much money did each student receive if they shared the money equally? Round to the nearest dollar.

25. Four books were mailed at the postal rate of $.28 per ounce. The books weighed 1 lb 3 oz, 13 oz, 1 lb 8 oz, and 1 lb. Find the cost of mailing the books.

26. One brand of yogurt costs $.79 for 8 oz, and 36 oz of another brand can be bought for $2.98. Which purchase is the better buy?

27. Two dice are rolled. What is the probability that the sum of the dots on the upward faces is 9?

28. Find the energy required to lift 400 lb a distance of 8 ft.

29. Find the power, in foot-pounds per second, needed to raise 600 lb a distance of 8 ft in 12 s.

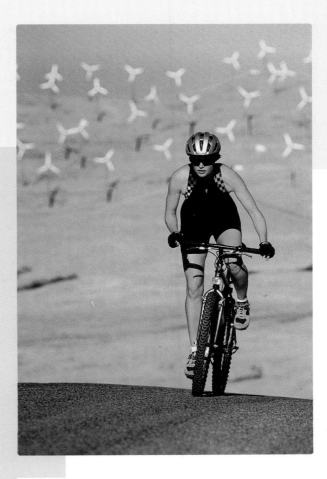

Objectives

Section 9.1
A To convert units of length in the metric system of measurement
B To solve application problems

Section 9.2
A To convert units of mass in the metric system of measurement
B To solve application problems

Section 9.3
A To convert units of capacity in the metric system of measurement
B To solve application problems

Section 9.4
A To use units of energy in the metric system of measurement

Section 9.5
A To convert U.S. Customary units to metric units
B To convert metric units to U.S. Customary units

We have all heard the rules for good health. Eating right, watching our weight, and exercising regularly are vital components of physical well being. Understanding the relationship between the number of Calories we eat and the number of Calories we burn during exercise can help us maintain our ideal weight. **Exercises 1 to 10 on page 377** focus on these topics.

Need help? For on-line student resources, such as section quizzes, visit this textbook's web site at **math.college.hmco.com/students.**

1. $3.732 \times 10{,}000$

2. 65.9×10^4

3. $41.07 \div 1000$

4. $28{,}496 \div 10^3$

5. $6 - 0.875$

6. $5 + 0.96$

7. 3.25×0.04

8. $35 \times \dfrac{1.61}{1}$

9. $1.67 \times \dfrac{1}{3.34}$

10. $4\dfrac{1}{2} \times 150$

Go Figure

Suppose you threw six darts and all six hit the target shown. Which of the following could be your score?

4, 15, 58, 28, 29, 31

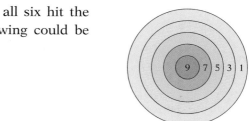

9.1 Length

Objective A

To convert units of length in the metric system of measurement

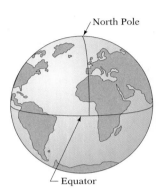

North Pole

Equator

Point of Interest

Originally the meter (spelled *metre* in some countries) was defined as $\dfrac{1}{10,000,000}$ of the distance from the equator to the north pole. Modern scientists have redefined the meter as 1,650,763.73 wavelengths of the orange-red light given off by the element krypton.

In 1789, an attempt was made to standardize units of measurement internationally in order to simplify trade and commerce between nations. A commission in France developed a system of measurement known as the **metric system.** The basic unit of length in the metric system is the **meter.** One meter is approximately the distance from a doorknob to the floor.

All units of length in the metric system are derived from the meter. Prefixes to the basic unit denote the length of each unit. For example, the prefix "centi-" means one-hundredth, so 1 centimeter is 1 one-hundredth of a meter.

kilo- = 1000	1 kilometer (km) = 1000 meters (m)
hecto- = 100	1 hectometer (hm) = 100 m
deca- = 10	1 decameter (dam) = 10 m
	1 meter (m) = 1 m
deci- = 0.1	1 decimeter (dm) = 0.1 m
centi- = 0.01	1 centimeter (cm) = 0.01 m
milli- = 0.001	1 millimeter (mm) = 0.001 m

Conversion between units of length in the metric system involves moving the decimal point to the right or to the left. Listing the units in order from largest to smallest will indicate how many places to move the decimal point and in which direction.

To convert 4200 cm to meters, write the units in order from largest to smallest.

km hm dam m dm cm mm

2 positions

4200 cm = 42.00 m

2 places

Converting cm to m requires moving 2 positions to the left.

Move the decimal point the same number of places and in the same direction.

A metric measurement that involves two units is customarily written in terms of one unit. Convert the smaller unit to the larger unit and then add.

To convert 8 km 32 m to kilometers, first convert 32 m to kilometers.

km hm dam m dm cm mm

• Converting m to km requires moving 3 positions to the left.

32 m = 0.032 km

• Move the decimal point the same number of places and in the same direction.

8 km 32 m = 8 km + 0.032 km
= 8.032 km

• Add the result to 8 km.

Example 1   Convert 0.38 m to millimeters.

Solution   0.38 m = 380 mm

You Try It 1   Convert 3.07 m to centimeters.

Your solution

Example 2   Convert 4 m 62 cm to meters.

Solution   62 cm = 0.62 m

4 m 62 cm = 4 m + 0.62 m
          = 4.62 m

You Try It 2   Convert 3 km 750 m to kilometers.

Your solution

Solutions on p. S23

Objective B   **To solve application problems**

In the application problems in this section, we perform arithmetic operations with the measurements of length in the metric system. It is important to remember that before measurements can be added or subtracted, they must be expressed in terms of the same unit. In this textbook, unless otherwise stated, the units should be changed to the larger unit before the arithmetic operation is performed.

To subtract 85 cm from 2 m, convert 85 cm to meters.

$$2 \text{ m} - 85 \text{ cm} = 2 \text{ m} - 0.85 \text{ m}$$
$$= 1.15 \text{ m}$$

Example 3

A piece measuring 142 cm is cut from a board 4.20 m long. Find the length of the remaining piece.

Strategy
To find the length of the remaining piece:

• Convert the length of the piece cut (142 cm) to meters.
• Subtract the length of the piece cut from the original length.

Solution
142 cm = 1.42 m

4.20 m − 142 cm = 4.20 m − 1.42 m
               = 2.78 m

The length of the remaining piece is 2.78 m.

You Try It 3

A bookcase 175 cm long has four shelves. Find the cost of the shelves when the price of the lumber is $15.75 per meter.

Your strategy

Your solution

Solution on p. S23

9.1 Exercises

Objective A

Convert.

1. 42 cm = _____ mm

2. 62 cm = _____ mm

3. 81 mm = _____ cm

4. 68.2 mm = _____ cm

5. 6804 m = _____ km

6. 3750 m = _____ km

7. 2.109 km = _____ m

8. 32.5 km = _____ m

9. 432 cm = _____ m

10. 61.7 cm = _____ m

11. 0.88 m = _____ cm

12. 3.21 m = _____ cm

13. 7038 m = _____ km

14. 2589 m = _____ km

15. 3.5 km = _____ m

16. 9.75 km = _____ m

17. 260 cm = _____ m

18. 705 cm = _____ m

19. 1.685 m = _____ cm

20. 0.975 m = _____ cm

21. 14.8 cm = _____ mm

22. 6 m 42 cm = _____ m

23. 62 m 7 cm = _____ m

24. 42 cm 6 mm = _____ cm

25. 31 cm 9 mm = _____ cm

26. 62 km 482 m = _____ km

27. 8 km 75 m = _____ km

Objective B   *Application Problems*

28. How many shelves, each 140 cm long, can be cut from a board that is 4.20 m in length? Find the length of the board remaining after the shelves are cut.

29. Find the missing dimension, in centimeters, in the diagram at the right.

30. A walk-a-thon had two checkpoints. One checkpoint was 1400 m from the starting point. The second checkpoint was 1200 m from the first checkpoint. The second checkpoint was 1800 m from the finish line. How long was the walk? Express the answer in kilometers.

31. Twenty rivets are used to fasten two steel plates together. The plates are 3.4 m long, and the rivets are equally spaced, with a rivet at each end. Find the distance between the rivets. Round to the nearest tenth of a centimeter.

32. Find the total length, in centimeters, of the shaft in the diagram at the right.

33. You purchase a 50-meter roll of fencing, at a cost of $14.95 per meter, in order to build a dog run that is 340 cm wide and 1380 cm long. After you cut the four pieces of fencing from the roll, how much of the fencing is left on the roll?

34. Carmine is a member of a group that has adopted 10 km of highway. During this week, Carmine cleaned up litter along the highway for 2500 m, 1500 m, 1200 m, 1300 m, and 1400 m. Find the average number of kilometers Carmine cleaned up on each of the 5 days this week.

35. The distance between Earth and the sun is 150,000,000 km. Light travels 300,000,000 m in 1 s. How long does it take for light to reach Earth from the sun?

36. The circumference of Earth is 40,000 km. How long would it take to travel the circumference of Earth at a speed of 85 km per hour? Round to the nearest tenth.

37. Light travels 300,000 km in 1 s. How far does light travel in 1 day?

APPLYING THE CONCEPTS

38. Other prefixes in the metric system are becoming more commonly used as a result of technological advances. Find the meaning of the following prefixes: tera-, giga-, mega-, micro-, nano-, and pico.

39. Write a short history of the metric system.

9.2 Mass

Objective A   **To convert units of mass in the metric system of measurement**

Mass and weight are closely related. Weight is a measure of how strongly Earth is pulling on an object. Therefore, an object's weight is less in space than on Earth's surface. However, the amount of material in the object, its **mass,** remains the same. On the surface of Earth, mass and weight can be used interchangeably.

The basic unit of mass in the metric system is the **gram.** If a box that is 1 cm long on each side is filled with water, then the mass of that water is 1 gram.

1 gram = the mass of water in the box

Point of Interest

An average snowflake weighs about $\frac{1}{300}$ g and contains approximately 100,000,000,000,000,000,000 water molecules. You may have heard the expression "No two snowflakes are the same." It is this large number of water molecules, and the number of their possible arrangements, that led to that statement.

The gram is a very small unit of mass. A paper clip weighs about 1 gram. The kilogram (1000 grams) is a more useful unit of mass in consumer applications. This textbook weighs about 1 kilogram.

The units of mass in the metric system have the same prefixes as the units of length:

$$1 \text{ kilogram (kg)} = 1000 \text{ grams (g)}$$
$$1 \text{ hectogram (hg)} = 100 \text{ g}$$
$$1 \text{ decagram (dag)} = 10 \text{ g}$$
$$1 \text{ gram (g)} = 1 \text{ g}$$
$$1 \text{ decigram (dg)} = 0.1 \text{ g}$$
$$1 \text{ centigram (cg)} = 0.01 \text{ g}$$
$$1 \text{ milligram (mg)} = 0.001 \text{ g}$$

Conversion between units of mass in the metric system involves moving the decimal point to the right or to the left. Listing the units in order from largest to smallest will indicate how many places to move the decimal point and in which direction.

To convert 324 g to kilograms, first write the units in order from largest to smallest.

kg hg dag g dg cg mg   Converting g to kg requires moving
         3 positions   3 positions to the left.

324 g = 0.324 kg   Move the decimal point the same number of
       3 places   places and in the same direction.

Example 1   Convert 4.23 g to milligrams.

Solution   4.23 g = 4230 mg

You Try It 1   Convert 42.3 mg to grams.

Your solution

Solution on p. S23

Example 2   Convert 2 kg 564 g to kilograms.

Solution   564 g = 0.564 kg

2 kg 564 g = 2 kg + 0.564 kg
             = 2.564 kg

You Try It 2   Convert 3 g 54 mg to grams.

Your solution

Solution on p. S23

Objective B   **To solve application problems**

TAKE NOTE

Although in this text we will always change units to the larger unit, it is possible to perform the calculation by changing to the smaller unit.

3 kg − 750 g
   = 3000 g − 750 g
   = 2250 g

Note that

2250 g = 2.250 kg.

In the application problems in this section, we perform arithmetic operations with the measurements of mass in the metric system. Remember that before measurements can be added or subtracted, they must be expressed in terms of the same unit. In this textbook, unless otherwise stated, the units should be changed to the larger unit before the arithmetic operation is performed.

To subtract 750 g from 3 kg, convert 750 g to kilograms.

$$3 \text{ kg} - 750 \text{ g} = 3 \text{ kg} - 0.750 \text{ kg}$$
$$= 2.250 \text{ kg}$$

Example 3

Find the cost of three packages of ground meat weighing 540 g, 670 g, and 890 g if the price per kilogram is $9.89. Round to the nearest cent.

Strategy

To find the cost of the meat:

• Find the total weight of the three packages.
• Convert the total weight to kilograms.
• Multiply the weight by the cost per kilogram ($9.89).

Solution

540 g + 670 g + 890 g = 2100 g

2100 g = 2.1 kg

2.1 × 9.89 = 20.769

The cost of the meat is $20.77.

You Try It 3

How many kilograms of fertilizer are required to fertilize 400 trees in an apple orchard if 300 g of fertilizer are used for each tree?

Your strategy

Your solution

Solution on p. S23

9.2 Exercises

Objective A

Convert.

1. 420 g = _____ kg

2. 7421 g = _____ kg

3. 127 mg = _____ g

4. 43 mg = _____ g

5. 4.2 kg = _____ g

6. 0.027 kg = _____ g

7. 0.45 g = _____ mg

8. 325 g = _____ mg

9. 1856 g = _____ kg

10. 8900 g = _____ kg

11. 4057 mg = _____ g

12. 1970 mg = _____ g

13. 1.37 kg = _____ g

14. 5.1 kg = _____ g

15. 0.0456 g = _____ mg

16. 0.2 g = _____ mg

17. 18,000 g = _____ kg

18. 0.87 kg = _____ g

19. 3 kg 922 g = _____ kg

20. 1 kg 47 g = _____ kg

21. 7 g 891 mg = _____ g

22. 209 g 42 mg = _____ g

23. 4 kg 63 g = _____ kg

24. 18 g 5 mg = _____ g

Objective B   *Application Problems*

25. A 1.19-kilogram container of Quaker Oats contains 30 servings. Find the number of grams in one serving of the oatmeal. Round to the nearest gram.

26. A patient is advised to supplement her diet with 2 g of calcium per day. The calcium tablets she purchases contain 500 mg of calcium per tablet. How many tablets per day should the patient take?

27. **a.** One egg contains 274 mg of cholesterol. How many grams of cholesterol are in a dozen eggs?
b. One glass of milk contains 33 mg of cholesterol. How many grams of cholesterol are in four glasses of milk?

28. A carat is a unit of weight equal to 200 mg. Find the weight in grams of a 10-carat precious stone.

29. The nutrition label for a corn bread mix is shown at the right.
a. How many kilograms of mix are in the package?
b. How many grams of sodium are contained in two servings of the corn bread?

30. Find the cost of three packages of ground meat weighing 470 g, 680 g, and 590 g if the price per kilogram is $8.40.

31. Eighty grams of grass seed are used for every 100 m² of lawn. How many kilograms of grass seed are needed to cover 2000 m²?

32. A commuter flight charges $9.95 for each kilogram or part of a kilogram over 15 kg of luggage weight. How much extra must be paid for three pieces of luggage weighing 6450 g, 5850 g, and 7500 g?

33. A health food store buys nuts in 10-kilogram containers and repackages the nuts for resale. The store packages the nuts in 200-gram bags, costing $.04 each, and sells them for $2.89 per bag. Find the profit on a 10-kilogram container of nuts costing $75.

34. A train car is loaded with 15 automobiles weighing 1405 kg each. Find the total weight of the automobiles.

35. During one year the United States exported 37,141 million kg of wheat, 2680 million kg of rice, and 40,365 million kg of corn. What percent of the total of these grain exports was corn? Round to the nearest tenth of a percent.

APPLYING THE CONCEPTS

36. Define a metric ton. Convert the weights in Exercise 35 to metric tons.

37. Discuss the advantages and disadvantages of the U.S. Customary System and the metric system of measurement.

Nutrition Facts		
Serving Size ⅙ pkg. (31g mix)		
Servings Per Container 6		
Amount Per Serving	**Mix**	**Prepared**
Calories	110	160
Calories from Fat	10	50
		% Daily Value*
Total Fat 1g	1%	9%
Saturated Fat 0g	0%	7%
Cholesterol 0mg	0%	12%
Sodium 210mg	9%	11%
Total Carbohydrate 24g	8%	8%
Sugars 6g		
Protein 2g		

9.3 Capacity

Objective A **To convert units of capacity in the metric system of measurement**

The basic unit of capacity in the metric system is the liter. One **liter** is defined as the capacity of a box that is 10 cm long on each side.

The units of capacity in the metric system have the same prefixes as the units of length:

1 kiloliter (kl) = 1000 L
1 hectoliter (hl) = 100 L
1 decaliter (dal) = 10 L
1 liter (L) = 1 L
1 deciliter (dl) = 0.1 L
1 centiliter (cl) = 0.01 L
1 milliliter (ml) = 0.001 L

The milliliter is equal to 1 **cubic centimeter** (cm^3).

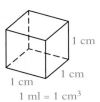

1 ml = 1 cm^3

Conversion between units of capacity in the metric system involves moving the decimal point to the right or to the left. Listing the units in order from largest to smallest will indicate how many places to move the decimal point and in which direction.

To convert 824 ml to liters, first write the units in order from largest to smallest.

kl hl dal L dl cl ml
        3 positions

Converting ml to L requires moving 3 positions to the left.

824 ml = 0.824 L
        3 places

Move the decimal point the same number of places and in the same direction.

Example 1 Convert 4 L 32 ml to liters.

Solution 32 ml = 0.032 L

4 L 32 ml = 4 L + 0.032 L
          = 4.032 L

You Try It 1 Convert 2 kl 167 L to liters.

Your solution

Solution on p. S23

Example 2 Convert 1.23 L to cubic centimeters.

Solution 1.23 L = 1230 ml = 1230 cm³

Solution on p. S23

You Try It 2 Convert 325 cm³ to liters.

Your solution

Objective B   **To solve application problems**

TAKE NOTE
Although in this text we will always change units to the larger unit, it is possible to perform the calculation by changing to the smaller unit.

2.5 kl + 875 L
= 2500 L + 875 L
= 3375 L
Note that
3375 L = 3.375 kl.

In the application problems in this section, we perform arithmetic operations with the measurements of capacity in the metric system. Remember that before measurements can be added or subtracted, they must be expressed in terms of the same unit. In this textbook, unless otherwise stated, the units should be changed to the larger unit before the arithmetic operation is performed.

To add 2.5 kl and 875 L, convert 875 L to kiloliters.

$$2.5 \text{ kl} + 875 \text{ L} = 2.5 \text{ kl} + 0.875 \text{ kl}$$
$$= 3.375 \text{ kl}$$

Example 3
A laboratory assistant is in charge of ordering acid for three chemistry classes of 30 students each. Each student requires 80 ml of acid. How many liters of acid should be ordered? (The assistant must order by the whole liter.)

Strategy
To find the number of liters to be ordered:
• Find the number of milliliters of acid needed by multiplying the number of classes (3) by the number of students per class (30) by the number of milliliters of acid required by each student (80).
• Convert milliliters to liters.
• Round up to the nearest whole number.

Solution
3(30)(80) = 7200 ml

7200 ml = 7.2 L

7.2 rounded up to the nearest whole number is 8.

The assistant should order 8 L of acid.

You Try It 3
For $299.50, a cosmetician buys 5 L of moisturizer and repackages it in 125-milliliter jars. Each jar costs the cosmetician $.85. Each jar of moisturizer is sold for $19.95. Find the profit on the 5 L of moisturizer.

Your strategy

Your solution

Solution on p. S23

9.3 Exercises

Objective A

Convert.

1. 4200 ml = _____ L

2. 7.5 ml = _____ L

3. 3.42 L = _____ ml

4. 0.037 L = _____ ml

5. 423 ml = _____ cm^3

6. 0.32 ml = _____ cm^3

7. 642 cm^3 = _____ ml

8. 0.083 cm^3 = _____ ml

9. 42 cm^3 = _____ L

10. 3075 cm^3 = _____ L

11. 0.435 L = _____ cm^3

12. 2.57 L = _____ cm^3

13. 4.62 kl = _____ L

14. 0.035 kl = _____ L

15. 1423 L = _____ kl

16. 897 L = _____ kl

17. 1.267 L = _____ cm^3

18. 4.105 L = _____ cm^3

19. 3 L 42 ml = _____ L

20. 1 L 127 ml = _____ L

21. 3 kl 4 L = _____ kl

22. 6 kl 32 L = _____ kl

23. 8 L 200 ml = _____ L

24. 9 kl 505 L = _____ kl

Objective B   *Application Problems*

25. The air in Earth's atmosphere is 78% nitrogen and 21% oxygen.
    a. Without calculating, determine whether the amount of oxygen in 50 L of air is more or less than 25 L.
    b. Find the amount of oxygen in 50 L of air.

26. A can of tomato juice contains 1.36 L. How many 170-milliliter servings are in one can of tomato juice?

27. An athletic club uses 800 ml of chlorine each day for its swimming pool. How many liters of chlorine are used in a month of 30 days?

28. The printed label from a container of milk is shown at the right. How many 230-milliliter servings are in the container? Round to the nearest whole number.

1 GAL. (3.78 L)

29. A flu vaccine is being given for the coming winter season. A medical corporation buys 12 L of flu vaccine. How many patients can be immunized if each person receives 3 cm³ of the vaccine?

30. A chemistry experiment requires 12 ml of an acid solution. How many liters of acid should be ordered when 4 classes of 90 students each are going to perform the experiment? (The acid must be ordered by the whole liter.)

31. A case of 12 one-liter bottles of apple juice costs $19.80. A case of 24 cans, each can containing 340 ml of apple juice, costs $14.50. Which case of apple juice is the better buy?

32. For $195, a pharmacist purchases 5 L of cough syrup and repackages it in 250-milliliter bottles. Each bottle costs the pharmacist $.55. Each bottle of cough syrup is sold for $17.89. Find the profit on the 5 L of cough syrup.

33. A service station operator bought 85 kl of gasoline for $23,750. The gasoline was sold for $.379 per liter. Find the profit on the 85 kl of gasoline.

34. A wholesale distributor purchased 32 kl of cooking oil for $44,480. The wholesaler repackaged the cooking oil in 1.25-liter bottles. The bottles cost $.21 each. Each bottle of cooking oil was sold for $2.97. Find the distributor's profit on the 32 kl of cooking oil.

APPLYING THE CONCEPTS

35. After a 280-milliliter serving is taken from a 3-liter bottle of water, how much water remains in the container? Write the answer in three different ways.

36. Write an essay describing problems in trade and manufacturing that arise between the United States and Europe because they use different systems of measurement.

9.4 Energy

Objective A   **To use units of energy in the metric system of measurement**

Two commonly used units of energy in the metric system are the calorie and the watt-hour.

Heat is generally measured in units called calories or in larger units called Calories (with a capital C). A **Calorie** is 1000 calories and should be called a kilocalorie, but it is common practice in nutritional references and food labeling to simply call it a Calorie. A Calorie is the amount of heat required to raise the temperature of 1 kg of water 1 degree Celsius. One Calorie is also the energy required to lift 1 kg a distance of 427 m.

➡ Swimming uses 480 Calories per hour. How many Calories are used by swimming $\frac{1}{2}$ h each day for 30 days?

Strategy To find the number of Calories used:

- Find the number of hours spent swimming.
- Multiply the number of hours spent swimming by the Calories used per hour.

Solution $\frac{1}{2} \times 30 = 15$

$15(480) = 7200$

7200 Calories are used by swimming $\frac{1}{2}$ h each day for 30 days.

The **watt-hour** is used for measuring electrical energy. One watt-hour is the amount of energy required to lift 1 kg a distance of 370 m. A light bulb rated at 100 watts (W) will emit 100 watt-hours (Wh) of energy each hour.

1000 watt-hours (Wh) = 1 kilowatt-hour (kWh)

➡ A 150-watt bulb is on for 8 h. At 8¢ per kilowatt-hour, find the cost of the energy used.

Strategy To find the cost:

- Find the number of watt-hours used.
- Convert to kilowatt-hours.
- Multiply the number of kilowatt-hours used by the cost per kilowatt-hour.

Solution $150 \times 8 = 1200$

$1200 \text{ Wh} = 1.2 \text{ kWh}$

$1.2 \times 0.08 = 0.096$

The cost of the energy used is $.096.

Example 1

Walking uses 180 Calories per hour. How many Calories will you burn off by walking $5\frac{1}{4}$ h during one week?

Strategy

To find the number of Calories, multiply the number of hours spent walking by the Calories used per hour.

Solution

$$5\frac{1}{4} \times 180 = \frac{21}{4} \times 180 = 945$$

You will burn off 945 Calories.

You Try It 1

Housework requires 240 Calories per hour. How many Calories are burned off by doing $4\frac{1}{2}$ h of housework?

Your strategy

Your solution

Example 2

A clothes iron is rated at 1200 W. If the iron is used for 1.5 h, how much energy, in kilowatt-hours, is used?

Strategy

To find the energy used:

• Find the number of watt-hours used.
• Convert watt-hours to kilowatt-hours.

Solution

$1200 \times 1.5 = 1800$

$1800 \text{ Wh} = 1.8 \text{ kWh}$

1.8 kWh of energy are used.

You Try It 2

Find the number of kilowatt-hours of energy used when a 150-watt light bulb burns for 200 h.

Your strategy

Your solution

Example 3

A TV set rated at 1800 W is on for an average of 3.5 h per day. At 7.2¢ per kilowatt-hour, find the cost of operating the set for 1 week.

Strategy

To find the cost:

• Multiply to find the total number of hours the set is used.
• Multiply the product by the number of watts to find the watt-hours.
• Convert watt-hours to kilowatt-hours.
• Multiply the number of kilowatt-hours by the cost per kilowatt-hour.

Solution

$3.5 \times 7 = 24.5$

$24.5 \times 1800 = 44{,}100$

$44{,}100 \text{ Wh} = 44.1 \text{ kWh}$

$44.1 \times 0.072 = 3.1752$

The cost is $3.1752.

You Try It 3

A microwave oven rated at 500 W is used an average of 20 min per day. At 8.7¢ per kilowatt-hour, find the cost of operating the oven for 30 days.

Your strategy

Your solution

Solutions on p. S24

9.4 Exercises

Objective A   *Application Problems*

1. How many Calories can you eliminate from your diet by omitting 1 slice of bread per day for 30 days? One slice of bread contains 110 Calories.

2. How many Calories can you eliminate from your diet in 2 weeks by omitting 400 Calories per day?

3. A nutrition label from a package of crisp bread is shown at the right.
   a. How many Calories are in $1\frac{1}{2}$ servings?
   b. How many Calories from fat are in 6 slices of the bread?

4. Moderately active people need 20 Calories per pound of body weight to maintain their weight. How many Calories should a 150-pound, moderately active person consume per day to maintain that weight?

5. People whose daily activity level would be described as light need 15 Calories per pound of body weight to maintain their weight. How many Calories should a 135-pound, lightly active person consume per day to maintain that weight?

6. For a healthy diet, it is recommended that 55% of the daily intake of Calories come from carbohydrates. Find the daily intake of Calories from carbohydrates that is appropriate if you want to limit your Calorie intake to 1600 Calories.

7. Playing singles tennis requires 450 Calories per hour. How many Calories do you burn in 30 days playing 45 min per day?

8. After playing golf for 3 h, Ruben had a banana split containing 550 Calories. Playing golf uses 320 Calories per hour.
   a. Without doing the calculations, did the banana split contain more or fewer Calories than Ruben burned off playing golf?
   b. Find the number of Calories Ruben gained or lost from these two activities.

9. Hiking requires approximately 315 Calories per hour. How many hours would you have to hike to burn off the Calories in a 375-Calorie sandwich, a 150-Calorie soda, and a 280-Calorie ice cream cone? Round to the nearest tenth.

10. Riding a bicycle requires 265 Calories per hour. How many hours would Shawna have to ride a bicycle to burn off the Calories in a 320-Calorie milkshake, a 310-Calorie cheeseburger, and a 150-Calorie apple? Round to the nearest tenth.

Nutrition Facts
Serving Size 2 Slices (18g)
Servings Per Container about 15

Amount Per Serving

Calories 60   Calories from Fat 10

	% Daily Value*
Total Fat 1g	2%
Saturated Fat 0g	0%
Polyunsaturated Fat 0.5g	
Monounsaturated Fat 0.5g	
Cholesterol 0mg	0%
Sodium 60mg	3%
Total Carbohydrate 10g	3%
Dietary Fiber 3g	10%
Sugars 1g	
Protein 2g	

Vitamin A 0%	•	Vitamin C 0%
Calcium 0%	•	Iron 4%

* Percent Daily Values are based on a 2,000 calorie diet. Your daily values may be higher or lower depending on your calorie needs.

		Calories:	2,000	2,500
Total Fat	Less than		65g	80g
Saturated Fat	Less than		20g	25g
Cholesterol	Less than		300mg	300mg
Sodium	Less than		2,400mg	2,400mg
Total Carbohydrate			300g	375g
Dietary Fiber			25g	30g

Calories per gram:
Fat 9 • Carbohydrate 4 • Protein 4

11. An oven uses 500 W of energy. How many watt-hours of energy are used to cook a 5-kilogram roast for $2\frac{1}{2}$ h?

12. A 21-inch color TV set is rated at 90 W. The TV is used an average of $3\frac{1}{2}$ h each day for a week. How many kilowatt-hours of energy are used during the week?

13. A fax machine is rated at 9 W when the machine is in standby mode and at 36 W when in operation. How many kilowatt-hours of energy are used during a week in which the fax machine is in standby mode for 39 h and in operation for 6 h?

14. A 120-watt CD player is on an average of 2 h a day. Find the cost of listening to the CD player for 2 weeks at a cost of 9.4¢ per kilowatt-hour. Round to the nearest cent.

15. How much does it cost to run a 2200-watt air conditioner for 8 h at 9¢ per kilowatt-hour? Round to the nearest cent.

16. A space heater is used for 3 h. The heater uses 1400 W per hour. Electricity costs 11.1¢ per kilowatt-hour. Find the cost of using the electric heater. Round to the nearest cent.

17. A 60-watt Sylvania Long Life Soft White Bulb has a light output of 835 lumens and an average life of 1250 h. A 34-watt Sylvania Energy Saver Bulb has a light output of 400 lumens and an average life of 1500 h.
   a. Is the light output of the Energy Saver Bulb more or less than half that of the Long Life Soft White Bulb?
   b. If electricity costs 10.8¢ per kilowatt-hour, what is the difference in cost between using the Long Life Soft White Bulb for 150 h and using the Energy Saver Bulb for 150 h? Round to the nearest cent.

18. A house is insulated to save energy. The house used 265 kWh of electrical energy per month before insulation and saves 45 kWh of energy per month after insulation. What percent decrease does this amount represent? Round to the nearest tenth.

19. A welder uses 6.5 kW of energy each hour. Find the cost of using the welder for 6 h a day for 30 days. The cost is 9.4¢ per kilowatt-hour.

APPLYING THE CONCEPTS

20. Write an essay on how to improve the energy efficiency of a home.

21. A maintenance intake of Calories allows a person to neither gain nor lose weight. Consult a book on nutrition in order to make a table of weights and the corresponding maintenance intake of Calories. Then for each weight, add another column indicating the appropriate Calorie intake when an individual at that weight wants to lose 1 lb per week.

Conversion Between the U.S. Customary and the Metric Systems of Measurement

Objective A   **To convert U.S. Customary units to metric units**

More than 90% of the world's population use the metric system of measurement. Therefore, converting U.S. Customary units to metric units is essential in trade and commerce—for example, in importing foreign goods and exporting domestic goods. Approximate equivalences between the two systems follow.

Units of Length	Units of Weight	Units of Capacity
1 in. = 2.54 cm	1 oz ≈ 28.35 g	1 L ≈ 1.06 qt
1 m ≈ 3.28 ft	1 lb ≈ 454 g	1 gal ≈ 3.79 L
1 m ≈ 1.09 yd	1 kg ≈ 2.2 lb	
1 mi ≈ 1.61 km		

Point of Interest

Back in September of 1999, CNN reported that "NASA lost a $125 million Mars orbiter because a Lockheed Martin engineering team used English units of measurement while the agency's team used the . . . metric system for a key spacecraft operation." One team used miles, the other kilometers. This resulted in the orbiter breaking up when it hit the Martian atmosphere because the approach was too near the surface.

These equivalences can be used to form conversion rates to change one unit of measurement to another. For example, because 1 mi ≈ 1.61 km, the conversion rates $\frac{1 \text{ mi}}{1.61 \text{ km}}$ and $\frac{1.61 \text{ km}}{1 \text{ mi}}$ are both approximately equal to 1.

➡ Convert 55 mi to kilometers.

$$55 \text{ mi} \approx 55 \text{ mi} \times \boxed{\frac{1.61 \text{ km}}{1 \text{ mi}}}$$

$$= \frac{55 \text{ mi}}{1} \times \frac{1.61 \text{ km}}{1 \text{ mi}}$$

$$= \frac{88.55 \text{ km}}{1}$$

$55 \text{ mi} \approx 88.55 \text{ km}$

- The conversion rate must contain km in the numerator and mi in the denominator.

- -

Example 1

Convert 45 mi/h to kilometers per hour.

Solution

$$\frac{45 \text{ mi}}{\text{h}} \approx \frac{45 \text{ mi}}{\text{h}} \times \frac{1.61 \text{ km}}{1 \text{ mi}}$$

$$\approx \frac{72.45 \text{ km}}{1 \text{ h}}$$

$\approx 72.45 \text{ km/h}$

You Try It 1

Convert 60 ft/s to meters per second. Round to the nearest hundredth.

Your solution

Solution on p. S24

Example 2

The price of gasoline is \$1.59/gal. Find the cost per liter. Round to the nearest tenth of a cent.

Solution

$$\frac{\$1.59}{\text{gal}} \approx \frac{\$1.59}{\text{gal}} \times \frac{1 \text{ gal}}{3.79 \text{ L}} = \frac{\$1.59}{3.79 \text{ L}}$$

$$\approx \$.420/\text{L}$$

You Try It 2

The price of milk is \$2.69/gal. Find the cost per liter. Round to the nearest cent.

Your solution

Solution on p. S24

Objective B **To convert metric units to U.S. Customary units**

Metric units are used in the United States. Cereal is sold by the gram, 35-mm film is available, and soda is sold by the liter. The same conversion rates used in Objective A are used for converting metric units to U.S. Customary units.

Example 3

Convert 200 m to feet.

Solution

$$200 \text{ m} \approx 200 \text{ m} \times \frac{3.28 \text{ ft}}{1 \text{ m}}$$

$$\approx 656 \text{ ft}$$

You Try It 3

Convert 45 cm to inches. Round to the nearest hundredth.

Your solution

Example 4

Convert 90 km/h to miles per hour. Round to the nearest hundredth.

Solution

$$\frac{90 \text{ km}}{\text{h}} \approx \frac{90 \text{ km}}{\text{h}} \times \frac{1 \text{ mi}}{1.61 \text{ km}} = \frac{90 \text{ mi}}{1.61 \text{ h}}$$

$$\approx 55.90 \text{ mi/h}$$

You Try It 4

Express 75 km/h in miles per hour. Round to the nearest hundredth.

Your solution

Example 5

The price of gasoline is \$.392/L. Find the cost per gallon. Round to the nearest cent.

Solution

$$\frac{\$.392}{1 \text{ L}} \approx \frac{\$.392}{1 \text{ L}} \times \frac{3.79 \text{ L}}{1 \text{ gal}} \approx \frac{\$1.49}{1 \text{ gal}}$$

$$\approx \$1.49/\text{gal}$$

You Try It 5

The price of ice cream is \$1.75/L. Find the cost per gallon. Round to the nearest cent.

Your solution

Solutions on p. S24

9.5 Exercises

Objective A

Convert. Round to the nearest hundredth.

1. Convert the 100-yard dash to meters.

2. Find the weight in kilograms of a 145-pound person.

3. Find the height in meters of a person 5 ft 8 in. tall.

4. Find the number of liters in 2 c of soda.

5. How many kilograms does a 15-pound turkey weigh?

6. Find the number of liters in 14.3 gal of gasoline.

7. Find the number of milliliters in 1 c.

8. The winning long jump at a track meet was 29 ft 2 in. Convert this distance to meters.

9. Express 65 mi/h in kilometers per hour.

10. Express 30 mi/h in kilometers per hour.

11. Fat-free hot dogs cost $3.49/lb. Find the cost per kilogram.

12. Seedless watermelon costs $.59/lb. Find the cost per kilogram.

13. The cost of gasoline is $1.47/gal. Find the cost per liter.

14. Deck stain costs $24.99/gal. Find the cost per liter.

15. The distance around Earth is 24,887 mi. Convert this distance to kilometers.

16. The distance from Earth to the sun is 93,000,000 mi. Convert this distance to kilometers.

Objective B

Convert. Round to the nearest hundredth.

17. Convert the 100-meter dash to feet.

18. Find the weight in pounds of an 86-kilogram person.

19. Find the number of gallons in 6 L of antifreeze.

20. Find the height in inches of a person 1.85 m tall.

21. Find the distance of the 1500-meter race in feet.

22. Find the weight in ounces of 327 g of cereal.

23. How many gallons of water does a 24-liter aquarium hold?

24. Find the width in inches of 35-mm film.

25. Express 80 km/h in miles per hour.

26. Express 30 m/s in feet per second.

27. Gasoline costs $.385/L. Find the cost per gallon.

28. A 5-kilogram ham costs $10/kg. Find the cost per pound.

29. A backpack tent weighs 2.1 kg. Find its weight in pounds.

30. A 2.5-kilogram bag of grass seed costs $7.89. Find the cost per pound.

31. Gary is planning a 5-day backpacking trip and decides to hike an average of 5 h each day. Hiking requires an extra 320 Calories per hour. How many pounds will Gary lose during the trip if he consumes an extra 900 Calories each day? (3500 Calories is equivalent to 1 lb.)

32. Swimming requires 550 Calories per hour. How many pounds could be lost by swimming $1\frac{1}{2}$ h each day for 5 days if no extra calories were consumed? (3500 Calories is equivalent to 1 lb.)

33. The speed of light is 300,000 km/s. Convert this speed to miles per second.

APPLYING THE CONCEPTS

34. For the following U.S. Customary units, make an estimate of the metric equivalent. Then perform the conversion and see how close you came to the actual measurement.

60 mi/h ≈	120 lb ≈	6 ft ≈
1 mi ≈	1 gal ≈	1 quarter-mile ≈

35. Determine whether the statement is true or false.
   a. A liter is more than a gallon.
   b. A meter is less than a yard.
   c. 30 mi/h is less than 60 km/h.
   d. A kilogram is greater than a pound.
   e. An ounce is less than a gram.

36. Should the United States keep the U.S. Customary System or convert to the metric system? Justify your position.

Focus on Problem Solving

Working Backward

Sometimes the solution to a problem can be found by *working backward*. This problem-solving technique can be used to find a winning strategy for a game called Nim.

There are many variations of this game. For our game, there are two players, Player A and Player B, who alternately place 1, 2, or 3 matchsticks in a pile. The object of the game is to place the 32nd matchstick in the pile. Is there a strategy that Player A can use to guarantee winning the game?

Working backward, if there are 29, 30, or 31 matchsticks in the pile when it is A's turn to play, A can win by placing 3 matchsticks (29 + 3 = 32), 2 matchsticks, (30 + 2 = 32), or 1 matchstick (31 + 1 = 32) on the pile. If there are to be 29, 30, or 31 matchsticks in the pile when it is A's turn, there must be 28 matchsticks in the pile when it is B's turn.

Working backward from 28, if there are to be 28 matches in the pile at B's turn, there must be 25, 26, or 27 at A's turn. Player A can then add 3 matchsticks, 2 matchsticks, or 1 matchstick to the pile to bring the number to 28. For there to be 25, 26, or 27 matchsticks in the pile at A's turn, there must be 24 matchsticks at B's turn.

Now working backward from 24, if there are to be 24 matches in the pile at B's turn, there must be 21, 22, or 23 at A's turn. Player A can then add 3 matchsticks, 2 matchsticks, or 1 matchstick to the pile to bring the number to 24. For there to be 21, 22, or 23 matchsticks in the pile at A's turn, there must be 20 matchsticks at B's turn.

So far, we have found that for Player A to win, there must be 28, 24, or 20 matchsticks in the pile when it is B's turn to play. Note that each time, the number is decreasing by 4. Continuing this pattern, Player A will win if there are 16, 12, 8, or 4 matchsticks in the pile when it is B's turn.

Player A can guarantee winning by making sure that the number of matchsticks in the pile is a multiple of 4. To ensure this, Player A allows Player B to go first and then adds exactly enough matchsticks to the pile to bring the total to a multiple of 4.

For example, suppose B places 3 matchsticks in the pile; then A places 1 matchstick (3 + 1 = 4). Now B places 2 matchsticks in the pile. The total is now 6 matchsticks. Player A then places 2 matchsticks in the pile to bring the total to 8, a multiple of 4. If play continues in this way, Player A will win.

Here are some variations of Nim. See whether you can develop a winning strategy for Player A.

1. Suppose the goal is to place the last matchstick in a pile of 30 matches.

2. Suppose the players make two piles of matchsticks, with the final number of matchsticks in each pile to be 20.

3. In this variation of Nim, there are 40 matchsticks in a pile. Each player alternately removes 1, 2, or 3 matches from the pile. The player who removes the last match wins.

Projects and Group Activities

Name That Metric Unit

What unit in the metric system would be used to measure each of the following? If you are working in a group, be sure that each member agrees on the unit to be used and understands why that unit is used before going on to the next item.

1. The distance from Los Angeles to New York

2. The weight of a truck

3. A person's waist

4. The amount of coffee in a mug

5. The weight of a thumbtack

6. The amount of water in a swimming pool

7. The distance a baseball player hits a baseball

8. A person's hat size

9. The amount of protein needed daily

10. A person's weight

11. The amount of maple syrup served with pancakes

12. The amount of water in a water cooler

13. The amount of medication in an aspirin

14. The distance to the grocery store

15. The width of a hair

16. A person's height

17. The weight of a lawn mower

18. The amount of water a family uses monthly

19. The contents of a bottle of salad dressing

20. The newspapers collected at a recycling center

Deductive Reasoning

Suppose that during the last week of your math class, your instructor tells you that if you receive an A on the final exam, you will earn an A in the course. When the final exam grades are posted, you learn that you received an A on the final exam. You can then assume that you will earn an A in the course.

The process used to determine your grade in the math course is deductive reasoning. **Deductive reasoning** involves drawing a conclusion that is based on given facts. The problems below require deductive reasoning.

1. Given that $\triangle\triangle\triangle = \diamond\diamond\diamond\diamond$ and $\diamond\diamond\diamond\diamond = \acute{O}\acute{O}$, then $\triangle\triangle\triangle\triangle\triangle\triangle =$ how many $\acute{O}$s?

2. Given that ‡‡ = ●●● and ●●● = Λ, then ‡‡‡‡ = how many Λs?

3. Given that ÓÓÓ = ΩΩ and ¤ = ΩΩ, then ¤¤ = how many Ós?

4. Given that ∫∫∫∫∫∫ = ∂∂ and ∂∂∂∂ = ¥¥¥, then ¥¥¥¥¥¥ = how many ∫s?

5. Given that ÔÔÔÔÔ = □□□ and □□□□□□ = §§§§, then §§§§§§ = how many Ôs?

6. Chris, Dana, Leslie, and Pat are neighbors. Each drives a different type of vehicle: a compact car, a sedan, a sports car, or a station wagon. From the following statements, determine which type of vehicle each of the neighbors drives. It may be helpful to use the chart provided below.
   a. Although the vehicle owned by Chris has more mileage on it than does either the sedan or the sports car, it does not have the highest mileage of all four cars.
   b. Pat and the owner of the sports car live on one side of the street, and Leslie and the owner of the compact car live on the other side of the street.
   c. Leslie owns the vehicle with the most mileage on it.

TAKE NOTE

To use the chart to solve this problem, write an X in a box to indicate that a possibility has been eliminated. Write a √ to show that a match has been found. When a row or column has 3 X's, a √ is written in the remaining open box in that row or column of the chart.

	Compact	*Sedan*	*Sports Car*	*Wagon*
Chris				
Dana				
Leslie				
Pat				

7. Four neighbors, Anna, Kay, Michelle, and Nicole, each plant a different vegetable (beans, cucumbers, squash, or tomatoes) in their garden. From the following statements, determine which vegetable each neighbor plants.
   a. Nicole's garden is bigger than the one that has tomatoes but smaller than the one that has cucumbers.
   b. Anna, who planted the largest garden, didn't plant the beans.
   c. The person who planted the beans has a garden the same size as Nicole's.
   d. Kay and the person who planted the tomatoes also have flower gardens.

8. The Ontkeans, Kedrovas, McIvers, and Levinsons are neighbors. Each of the four families specializes in a different national cuisine (Chinese, French, Italian, or Mexican). From the following statements, determine which cuisine each family specializes in.
   a. The Ontkeans invited the family that specializes in Chinese cuisine and the family that specializes in Mexican cuisine for dinner last night.
   b. The McIvers live between the family that specializes in Italian cuisine and the Ontkeans. The Levinsons live between the Kedrovas and the family that specializes in Chinese cuisine.
   c. The Kedrovas and the family that specializes in Italian cuisine both subscribe to the same culinary magazine.

Search the World Wide Web

Use the Internet to find information regarding the metric system. You might enter *metric* as the word to search for.

Find professional organizations whose members are in careers related to your major. For example, we found information regarding the metric system and architecture, traffic safety services, cooking, surveying, and transportation.

The process of changing our system of measurement to the metric system is called **metrication** or **metric transition.** Find at least two associations involved in metrication.

Describe one way in which a country that has changed to the metric system helped its citizens in the transition to the metric system. For example, one web site shows metric-conversion postal stamps printed in Australia.

Chapter Summary

Key Words

The *metric system of measurement* is a system of measurement based on the decimal system. [p. 363]

The basic unit of measurement of length in the metric system is the *meter.* [p. 363]

The basic unit of mass in the metric system is the *gram.* [p. 367]

The basic unit of capacity in the metric system is the *liter.* [p. 371]

Heat is commonly measured in units called *Calories.* [p. 375]

The *watt-hour* is used in the metric system for measuring electrical energy. [p. 375]

Essential Rules

Prefixes to the basic unit denote the magnitude of each unit in the metric system.

kilo-	1000
hecto-	100
deca-	10
deci-	0.1
centi-	0.01
milli-	0.001 [p. 363]

Conversion between units in the metric system involves moving the decimal point to the right or to the left. Listing the units in order from largest to smallest will indicate how many places to move the decimal point.

1. When converting from a larger unit to a smaller unit, move the decimal point to the **right.**

2. When converting from a smaller unit to a larger unit, move the decimal point to the **left.** [p. 363]

Approximate equivalences between units in the metric system and the U.S. Customary System are used to form conversion rates to change one unit of measurement to another. [p. 379]

Chapter Review

1. Convert 1.25 km to meters.

2. Convert 0.450 g to milligrams.

3. Convert 0.0056 L to milliliters.

4. Convert the 1000-meter run to yards. (1 m ≈ 1.09 yd)

5. Convert 79 mm to centimeters.

6. Convert 5 m 34 cm to meters.

7. Convert 990 g to kilograms.

8. Convert 2550 ml to liters.

9. Convert 4870 m to kilometers.

10. Convert 0.37 cm to millimeters.

11. Convert 6 g 829 mg to grams.

12. Convert 1.2 L to cubic centimeters.

13. Convert 4.050 kg to grams.

14. Convert 8.7 m to centimeters.

15. Convert 192 ml to cubic centimeters.

16. Convert 356 mg to grams.

17. Convert 372 cm to meters.

18. Convert 8.3 kl to liters.

19. Convert 2 L 89 ml to liters.

20. Convert 5410 cm^3 to liters.

21. Convert 3792 L to kiloliters.

22. Convert 468 cm^3 to milliliters.

23. Three pieces of wire are cut from a 50-meter roll. The three pieces measure 240 cm, 560 cm, and 480 cm. How much wire is left on the roll after the three pieces are cut?

24. Find the total cost of three packages of chicken weighing 790 g, 830 g, and 655 g if the cost is $5.59 per kilogram.

25. Cheese costs $3.40 per pound. Find the cost per kilogram. (1 kg ≈ 2.2 lb)

26. One hundred twenty-five guests are expected to attend a reception. Assuming that each person drinks 400 ml of coffee, how many liters of coffee should be prepared?

27. A large egg contains approximately 90 Calories. How many Calories can be eliminated from your diet in a 30-day month by eliminating one large egg per day from your usual breakfast?

28. A TV uses 240 W of energy. The set is on an average of 5 h a day in a 30-day month. At a cost of 9.5¢ per kilowatt-hour, how much does it cost to run the set for 30 days?

29. A backpack weighs 1.90 kg. Find the weight in pounds. Round to the nearest hundredth. (1 kg ≈ 2.2 lb)

30. Cycling burns up approximately 400 Calories per hour. How many hours of cycling are necessary to lose 1 lb? (3500 Calories is equivalent to 1 lb.)

31. Six liters of liquid soap were bought for $11.40 per liter. The soap was repackaged in 150-milliliter plastic containers. The cost of each container was $.26. Each container of soap sold for $3.29 per bottle. Find the profit on the 6 L of liquid soap.

32. A color TV is rated at 80 W. The TV is used an average of 2 h each day for a week. How many kilowatt-hours of energy are used during the week?

33. How many kilograms of fertilizer are necessary to fertilize 500 trees in an orchard if 250 g of fertilizer is used for each tree?

Chapter Test

1. Convert 2.96 km to meters.

2. Convert 0.378 g to milligrams.

3. Convert 0.046 L to milliliters.

4. Convert 919 cm³ to milliliters.

5. Convert 42.6 mm to centimeters.

6. Convert 7 m 96 cm to meters.

7. Convert 847 g to kilograms.

8. Convert 3920 ml to liters.

9. Convert 5885 m to kilometers.

10. Convert 1.5 cm to millimeters.

11. Convert 3 g 89 mg to grams.

12. Convert 1.6 L to cubic centimeters.

13. Convert 3.29 kg to grams.

14. Convert 4.2 m to centimeters.

15. Convert 96 ml to cubic centimeters.

16. Convert 1375 mg to grams.

17. Convert 402 cm to meters.

18. Convert 8.92 kl to liters.

19. Convert 5 km 38 m to meters.

20. Convert 6020 L to kiloliters.

21. A carpenter needs 30 rafters, each 380 cm long. Find the total length of the rafters in meters.

22. A tile measuring 20 × 20 cm weighs 250 g. Find the weight, in kilograms, of a box of 144 tiles.

23. The community health clinic is giving flu shots for the coming flu season. Each flu shot contains 2 cm³ of vaccine. How many liters of vaccine are needed to inoculate 2600 people?

24. Convert 35 mi/h to kilometers per hour. Round to the nearest tenth. (1 mi ≈ 1.61 km)

25. Twenty-five rivets are used to fasten two steel plates together. The plates are 4.20 m long, and the rivets are equally spaced, with a rivet at each end. Find the distance, in centimeters, between the rivets.

26. Two hundred grams of fertilizer are used for each tree in an orchard containing 1200 trees. At $2.75 per kilogram of fertilizer, how much does it cost to fertilize the orchard?

27. An air conditioner rated at 1600 W is operated an average of 4 h per day. Electrical energy costs 8.5¢ per kilowatt-hour. How much does it cost to operate the air conditioner for 30 days?

28. A laboratory assistant is in charge of ordering acid for three chemistry classes of 40 students each. Each student requires 90 ml of acid. How many liters of acid should be ordered? (The assistant must order by the whole liter.)

29. The longest ski jump in Olympic history in the men's competition was 137 m. Convert this distance to feet. Round to the nearest hundredth. (1 m ≈ 3.28 ft)

30. At present, the rules in ski jumping allow skis that are up to 31.5 in. longer than the jumper's height, with a maximum length of 110 in. Convert the maximum length of the skis to centimeters. (1 in. = 2.54 cm)

Cumulative Review

1. Simplify: $12 - 8 \div (6 - 4)^2 \cdot 3$

2. Find the total of $5\frac{3}{4}$, $1\frac{5}{6}$, and $4\frac{7}{9}$.

3. Subtract: $4\frac{2}{9} - 3\frac{5}{12}$

4. Divide: $5\frac{3}{8} \div 1\frac{3}{4}$

5. Simplify: $\left(\frac{2}{3}\right)^4 \cdot \left(\frac{9}{4}\right)^2$

6. Subtract: $12.0072 - 9.937$

7. Solve the proportion $\frac{5}{8} = \frac{n}{50}$. Round to the nearest tenth.

8. Write $1\frac{3}{4}$ as a percent.

9. 6.09 is 4.2% of what number?

10. Convert 18 pt to gallons.

11. Convert 875 cm to meters.

12. Convert 3420 m to kilometers.

13. Convert 5.05 kg to grams.

14. Convert 3 g 672 mg to grams.

15. Convert 6 L to milliliters.

16. Convert 2.4 kl to liters.

17. The Guerrero family has a monthly income of $3244 per month. The family spends one-fourth of its monthly income on rent. How much money is left after the rent is paid?

18. The state income tax on a business is $620 plus 0.08 times the profit the business makes. The business made a profit of $82,340.00 last year. Find the amount of state income tax the business paid.

19. The property tax on a $45,000 home is $900. At the same rate, what is the property tax on a home worth $75,000?

20. A car dealer offers new-car buyers a 12% rebate on some models. What rebate would a new-car buyer receive on a car that cost $23,500?

21. Rob Akullian received a dividend of $533 on an investment of $8200. What percent of the investment is the dividend?

22. You received grades of 78, 92, 45, 80, and 85 on five English exams. Find your mean grade.

23. Karla Perella, a ski instructor, receives a salary of $22,500. Her salary will increase by 12% next year. What will her salary be next year?

24. A sporting goods store has regularly priced $80 fishing rods on sale for $62.40. What is the discount rate?

25. Forty-eight blocks, each 9 in. long, are laid end to end to make the base for a wall. Find the length of the wall in feet.

26. A bottle of apple juice contains 24 oz. Find the number of quarts of apple juice in a case of 16 bottles.

27. A garage mechanic bought oil in a 40-gallon container. The mechanic bought the oil for $4.88 per gallon and sold the oil for $1.99 per quart. Find the profit on the 40-gallon container of oil.

28. A school swimming pool uses 1200 ml of chlorine each school day. How many liters of chlorine are used for 20 days during the month?

29. A 1200-watt hair dryer is used an average of 30 min a day. At a cost of 10.5¢ per kilowatt-hour, how much does it cost to operate the hair dryer for 30 days?

30. Convert 60 mi/h to kilometers per hour. Round to the nearest tenth. (1.61 km ≈ 1 mi)

Chapter 10 Rational Numbers

The mountaineer in this photograph is in the Himalayas in Asia. Mt. Everest, the highest peak in the world, is in the background. **Exercises 92 to 94 on page 406** reference the highest and lowest points on four continents on Earth. These places are measured relative to sea level. Those below sea level are written as negative numbers, one of the topics in this chapter.

Need help? For on-line student resources, such as section quizzes, visit this textbook's web site at **math.college.hmco.com/students.**

Prep Test

1. Place the correct symbol, $<$ or $>$, between the two numbers.

    54   45

2. What is the distance from 4 to 8 on the number line?

For Exercises 3 to 14, add, subtract, multiply, or divide.

3. $7654 + 8193$

4. $6097 - 2318$

5. 472×56

6. $\dfrac{144}{24}$

7. $\dfrac{2}{3} + \dfrac{3}{5}$

8. $\dfrac{3}{4} - \dfrac{5}{16}$

9. $0.75 + 3.9 + 6.408$

10. $5.4 - 1.619$

11. $\dfrac{3}{4} \times \dfrac{8}{15}$

12. $\dfrac{5}{12} \div \dfrac{3}{4}$

13. 23.5×0.4

14. $0.96 \div 2.4$

15. Simplify: $(8 - 6)^2 + 12 \div 4 \cdot 3^2$

Go Figure

Super Yeast causes bread to double in volume each minute. If it takes one loaf of bread made with Super Yeast 30 minutes to fill the oven, how long does it take two loaves of bread made with Super Yeast to fill one-half the oven?

10.1 Introduction to Integers

Objective A   **To identify the order relation between two integers**

Thus far in the text, we have encountered only zero and the numbers greater than zero. The numbers greater than zero are called **positive numbers.** However, the phrases "12 degrees below zero," "$25 in debt," and "15 feet below sea level" refer to numbers less than zero. These numbers are called **negative numbers.**

The **integers** are ..., -4, -3, -2, -1, 0, 1, 2, 3, 4, ...

Each integer can be shown on a number line. The integers to the left of zero on the number line are called **negative integers** and are represented by a negative sign $(-)$ placed in front of the number. The integers to the right of zero are called **positive integers.** The positive integers are also called natural numbers. Zero is neither a positive nor a negative integer.

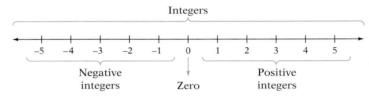

A number line can be used to visualize the order relation between two integers. A number that appears to the left of a given number is less than $(<)$ the given number. A number that appears to the right of a given number is greater than $(>)$ the given number.

2 is greater than negative 4.

$2 > -4$

Negative 5 is less than negative 3.

$-5 < -3$

Example 1

The temperature at the North Pole was recorded as 87 degrees below zero. Represent this temperature as an integer.

Solution -87 degrees

You Try It 1

The surface of the Salton Sea is 232 ft below sea level. Represent this depth as an integer.

Your solution

Example 2

Graph -2 on the number line.

Solution   ⟨—+—+—●—+—+—+—+—+—⟩
                 -4 -3 -2 -1 0 1 2 3 4

You Try It 2

Graph -4 on the number line.

Your solution   ⟨—+—+—+—+—+—+—+—+—⟩
                     -4 -3 -2 -1 0 1 2 3 4

Example 3

Place the correct symbol, $<$ or $>$, between the two numbers.

        a. -5 -7 **b.** 1 -2

Solution   **a.** $-5 > -7$ **b.** $1 > -2$

You Try It 3

Place the correct symbol, $<$ or $>$, between the two numbers.

        a. -12 -8 **b.** -5 0

Your solution

Solutions on p. S24

Objective B | **To evaluate expressions that contain the absolute value symbol**

Two numbers that are the same distance from zero on the number line but on opposite sides of zero are called **opposites.**

−4 is the opposite of 4

and

4 is the opposite of −4.

Note that a negative sign can be read as "the opposite of."

−(4) = −4   The opposite of positive 4 is negative 4.

−(−4) = 4   The opposite of negative 4 is positive 4.

The **absolute value** of a number is the distance between zero and the number on the number line. Therefore, the absolute value of a number is a positive number or zero. The symbol for absolute value is │ │.

The distance from 0 to 4 is 4. Thus |4| = 4 (the absolute value of 4 is 4).

The distance from 0 to −4 is 4. Thus |−4| = 4 (the absolute value of −4 is 4).

The absolute value of a positive number is the number itself. The absolute value of a negative number is the opposite of the negative number. The absolute value of zero is zero.

Example 4   Find the absolute value of 2 and −3.

Solution |2| = 2

|−3| = 3

You Try It 4   Find the absolute value of −7 and 21.

Your solution

Example 5   Evaluate |−34| and |0|.

Solution |−34| = 34

|0| = 0

You Try It 5   Evaluate |2| and |−9|.

Your solution

Example 6   Evaluate −|−4|.

Solution   −|−4| = −4

The minus sign *in front of* the absolute value sign is not affected by the absolute value sign.

You Try It 6   Evaluate −|−12|.

Your solution

Solutions on p. S25

10.1 Exercises

Objective A

Represent the quantity as an integer.

1. A lake 120 ft below sea level

2. A temperature that is 15° below zero

3. A share of stock up 2 dollars

4. A loss of 324 dollars

Graph the numbers on the number line.

5. 3 and −3

6. −2 and 0

7. −4 and 1

8. 4 and −1

Place the correct symbol between the two numbers.

9. −2   −5

10. −6   −1

11. −16   1

12. −2   13

13. 3   −7

14. 5   −6

15. −11   −8

16. −4   −10

17. 35   28

18. 42   19

19. −42   27

20. −36   49

21. 21   −34

22. 53   −46

23. −27   −39

24. −51   −20

25. −87   63

26. −75   92

27. 86   −79

28. 95   −71

29. −62   −84

30. −91   −70

31. −131   101

32. 127   −150

Objective B

Find the opposite number.

33. 4   **34.** 16   **35.** -2 **36.** -3 **37.** 22

38. 45   **39.** -31 **40.** -59 **41.** 70   **42.** -88

Evaluate.

43. $|2|$ **44.** $|-2|$ **45.** $|-6|$ **46.** $|6|$ **47.** $|8|$

48. $|5|$ **49.** $|-9|$ **50.** $|-1|$ **51.** $-|-1|$ **52.** $-|-5|$

53. $-|0|$ **54.** $|16|$ **55.** $|19|$ **56.** $|-12|$ **57.** $|-22|$

58. $-|29|$ **59.** $-|20|$ **60.** $-|-14|$ **61.** $-|-18|$ **62.** $|-15|$

63. $|-23|$ **64.** $-|33|$ **65.** $-|27|$ **66.** $|32|$ **67.** $|25|$

68. $-|-42|$ **69.** $|-74|$ **70.** $|-61|$ **71.** $-|88|$ **72.** $-|52|$

APPLYING THE CONCEPTS

73. The graph at the right shows the lowest recorded temperatures, in degrees Fahrenheit, for selected states in the United States. Which state has the lowest recorded temperature?

74. a. Name two numbers that are 5 units from 3 on the number line.
    b. Name two numbers that are 3 units from -1 on the number line.

75. a. Find a number that is halfway between -7 and -5.
    b. Find a number that is halfway between -10 and -6.
    c. Find a number that is one-third of the way between -12 and -3.

76. In your own words, describe **a.** the opposite of a number, and **b.** the absolute value of a number.

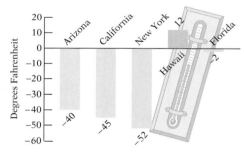

Lowest Recorded Temperatures

Sources: National Climatic Data Center, NESDIS, NOAA, U.S. Dept. of Commerce

10.2 Addition and Subtraction of Integers

Objective A   To add integers

An integer can be graphed as a dot on a number line, as shown in the last section. An integer also can be represented anywhere along a number line by an arrow. A positive number is represented by an arrow pointing to the right. A negative number is represented by an arrow pointing to the left. The absolute value of the number is represented by the length of the arrow. The integers 5 and -4 are shown on the number line in the figure below.

The sum of two integers can be shown on a number line. To add two integers, use arrows to represent the addends, with the first arrow starting at zero. The sum is the number directly below the tip of the arrow that represents the second addend.

$4 + 2 = 6$

$-4 + (-2) = -6$

$-4 + 2 = -2$

$4 + (-2) = 2$

The sums of the integers shown above can be categorized by the signs of the addends.

The addends have the same sign.

$4 + 2$ *positive* 4 plus *positive* 2
$-4 + (-2)$ *negative* 4 plus *negative* 2

The addends have different signs.

$-4 + 2$ *negative* 4 plus *positive* 2
$4 + (-2)$ *positive* 4 plus *negative* 2

The rule for adding two integers depends on whether the signs of the addends are the same or different.

> **Rule for Adding Two Numbers**
>
> To add numbers with the same sign, add the absolute values of the numbers. Then attach the sign of the addends.
>
> To add numbers with different signs, find the difference between the absolute values of the numbers. Then attach the sign of the addend with the larger absolute value.

Point of Interest

Although mathematical symbols are fairly standard in every country, that has not always been true. Italian mathematicians in the 15th century used a "p" to indicate plus. The "p" was from the Italian word *piu*.

➡ Add: $(-4) + (-9)$

$$|-4| = 4, |-9| = 9$$
$$4 + 9 = 13$$

$$(-4) + (-9) = -13$$

- Because the signs of the addends are the same, add the absolute values of the numbers.
- Then attach the sign of the addends.

➡ Add: $6 + (-13)$

$$|6| = 6, |-13| = 13$$
$$13 - 6 = 7$$

$$6 + (-13) = -7$$

- Because the signs of the addends are different, subtract the smaller absolute value from the larger absolute value.
- Then attach the sign of the number with the larger absolute value. Because $|-13| > |6|$, attach the negative sign.

➡ Add: $162 + (-247)$

$$162 + (-247) = -85$$

- Because the signs are different, find the difference between the absolute values of the numbers and attach the sign of the number with the greater absolute value.

CALCULATOR NOTE

To add $-14 + (-47)$ on your calculator, enter the following:

14 $\boxed{+/-}$ $\boxed{+}$ 47 $\boxed{+/-}$ $\boxed{=}$

➡ Find the sum of -14 and -47.

$$-14 + (-47) = -61$$

- Because the signs are the same, add the absolute values of the numbers and attach the sign of the addends.

When adding more than two integers, start from the left and add the first two numbers. Then add the sum to the third number. Continue this process until all the numbers have been added.

➡ Add: $(-4) + (-6) + (-8) + 9$

$$(-4) + (-6) + (-8) + 9 = (-10) + (-8) + 9$$
$$= (-18) + 9$$
$$= -9$$

- Add the first two numbers.
- Add the sum to the next number.
- Continue adding until all numbers have been added.

Example 1 What is -162 added to 98?

Solution $-162 + 98 = -64$

You Try It 1   Add: $-154 + (-37)$

Your solution

Solution on p. S25

Example 2

Add: $-2 + (-7) + 4 + (-6)$

Solution

$$-2 + (-7) + 4 + (-6) = -9 + 4 + (-6)$$
$$= -5 + (-6)$$
$$= -11$$

You Try It 2

Add: $-5 + (-2) + 9 + (-3)$

Your solution

Solution on p. S25

Objective B   **To subtract integers**

Before the rules for subtracting two integers are explained, look at the translation into words of an expression that is the difference of two integers:

$9 - 3$	positive 9 minus positive 3
$(-9) - 3$	negative 9 minus positive 3
$9 - (-3)$	positive 9 minus negative 3
$(-9) - (-3)$	negative 9 minus negative 3

Note that the sign $-$ is used in two different ways. One way is as a negative sign, as in (-9), *negative* 9. The second way is to indicate the operation of subtraction, as in $9 - 3$, 9 *minus* 3.

Look at the next four subtraction expressions and decide whether the second number in each expression is a positive number or a negative number.

1. $(-10) - 8$ **2.** $(-10) - (-8)$ **3.** $10 - (-8)$ **4.** $10 - 8$

In expressions 1 and 4, the second number is a positive 8. In expressions 2 and 3, the second number is a negative 8.

Rule for Subtracting Two Numbers

To subtract two numbers, add the opposite of the second number to the first number.

This rule states that to subtract two integers, we rewrite the subtraction expression as the sum of the first number and the opposite of the second number.

Here are some examples:

First number	$-$	second number	$=$	first number	$+$	the opposite of the second number	
8	$-$	15	$=$	8	$+$	(-15)	$= -7$
8	$-$	(-15)	$=$	8	$+$	15	$= 23$
(-8)	$-$	15	$=$	(-8)	$+$	(-15)	$= -23$
(-8)	$-$	(-15)	$=$	(-8)	$+$	15	$= 7$

⇒ Subtract: $(-15) - 75$

$$(-15) - 75 = (-15) + (-75)$$
$$= -90$$

• To subtract, add the opposite of the second number to the first number.

CALCULATOR NOTE

The $\boxed{+/-}$ key on your calculator is used to find the opposite of a number. The $\boxed{-}$ is used to perform the operation of subtraction.

➡ Subtract: $27 - (-32)$

$$27 - (-32) = 27 + 32$$
$$= 59$$

● To subtract, add the opposite of the second number to the first number.

When subtraction occurs several times in an expression, rewrite each subtraction as addition of the opposite and then add.

➡ Subtract: $-13 - 5 - (-8)$

$$-13 - 5 - (-8) = -13 + (-5) + 8$$
$$= -18 + 8$$
$$= -10$$

● Rewrite each subtraction as the addition of the opposite and then add.

Example 3
Find 8 less than -12.

Solution
$-12 - 8 = -12 + (-8)$
$ = -20$

You Try It 3
Find -8 less 14.

Your solution

Example 4
Subtract: $6 - (-20)$

Solution
$6 - (-20) = 6 + 20$
$ = 26$

You Try It 4
Subtract: $3 - (-15)$

Your solution

Example 5
Subtract: $-8 - 30 - (-12) - 7$

Solution
$-8 - 30 - (-12) - 7$
$\quad = -8 + (-30) + 12 + (-7)$
$\quad = -38 + 12 + (-7)$
$\quad = -26 + (-7)$
$\quad = -33$

You Try It 5
Subtract: $4 - (-3) - 12 - (-7) - 20$

Your solution

Solutions on p. S25

Objective C   **To solve application problems**

Example 6
Find the temperature after an increase of 9°C from -6°C.

Strategy
To find the temperature, add the increase (9) to the previous temperature (-6).

Solution
$-6 + 9 = 3$

The temperature is 3°C.

You Try It 6
Find the temperature after an increase of 12°C from -10°C.

Your strategy

Your solution

Solution on p. S25

10.2 Exercises

Objective A

Name the negative integers in the list of numbers.

1. $-14, 28, 0, -\dfrac{5}{7}, -364, -9.5$

2. $-37, 90, -\dfrac{7}{10}, -88.8, 42, -561$

Add.

3. $3 + (-5)$ **4.** $-4 + 2$ **5.** $8 + 12$ **6.** $16 + 23$

7. $-3 + (-8)$ **8.** $-12 + (-1)$ **9.** $-4 + (-5)$ **10.** $-12 + (-12)$

11. $6 + (-9)$ **12.** $4 + (-9)$ **13.** $-6 + 7$ **14.** $-12 + 6$

15. $2 + (-3) + (-4)$ **16.** $7 + (-2) + (-8)$ **17.** $-3 + (-12) + (-15)$

18. $9 + (-6) + (-16)$ **19.** $-17 + (-3) + 29$ **20.** $13 + 62 + (-38)$

21. $-3 + (-8) + 12$ **22.** $-27 + (-42) + (-18)$

23. $13 + (-22) + 4 + (-5)$ **24.** $-14 + (-3) + 7 + (-6)$

25. $-22 + 10 + 2 + (-18)$ **26.** $-6 + (-8) + 13 + (-4)$

27. $-16 + (-17) + (-18) + 10$ **28.** $-25 + (-31) + 24 + 19$

29. $-126 + (-247) + (-358) + 339$ **30.** $-651 + (-239) + 524 + 487$

31. What is -8 more than -12?   **32.** What is -5 more than 3?

33. What is -7 added to -16?   **34.** What is 7 added to -25?

35. What is −4 plus 2?

36. What is −22 plus −17?

37. Find the sum of −2, 8, and −12.

38. Find the sum of 4, −4, and −6.

39. What is the total of 2, −3, 8, and −13?

40. What is the total of −6, −8, 13, and −2?

Objective B

Translate the expression into words. Represent each number as a positive or negative.

41. −6 − 4   **42.** −6 − (−4)   **43.** 6 − (−4)   **44.** 6 − 4

Rewrite the subtraction as the sum of the first number and the opposite of the second number.

45. 9 − (−5)   **46.** −3 − 7   **47.** 1 − 8   **48.** −2 − (−10)

Subtract.

49. 16 − 8   **50.** 12 − 3   **51.** 7 − 14   **52.** 6 − 9

53. −7 − 2   **54.** −9 − 4   **55.** 7 − (−29)   **56.** 3 − (−4)

57. −6 − (−3)   **58.** −4 − (−2)   **59.** 6 − (−12)   **60.** −12 − 16

61. −4 − 3 − 2   **62.** 4 − 5 − 12   **63.** 12 − (−7) − 8

64. −12 − (−3) − (−15)   **65.** 4 − 12 − (−8)   **66.** 13 − 7 − 15

67. −6 − (−8) − (−9)   **68.** 7 − 8 − (−1)

69. −30 − (−65) − 29 − 4   **70.** 42 − (−82) − 65 − 7

71. −16 − 47 − 63 − 12   **72.** 42 − (−30) − 65 − (−11)

73. 47 − (−67) − 13 − 15   **74.** −18 − 49 − (−84) − 27

75. $167 - 432 - (-287) - 359$

76. $-521 - (-350) - 164 - (-299)$

77. Subtract -8 from -4.

78. Subtract -12 from 3.

79. What is the difference between -8 and 4?

80. What is the difference between 8 and -3?

81. What is -4 decreased by 8?

82. What is -13 decreased by 9?

83. Find -2 less than 1.

84. Find -3 less than -5.

Objective C   *Application Problems*

85. Find the temperature after a rise of 7°C from -8°C.

86. Find the temperature after a rise of 5°C from -19°C.

87. During a card game of Hearts, Nick had a score of 11 points before his opponent "shot the moon," subtracting a score of 26 from Nick's total. What was Nick's score after his opponent shot the moon?

88. In a card game of Hearts, Monique had a score of -19 before she "shot the moon," entitling her to add 26 points to her score. What was Monique's score after she shot the moon?

89. The price of Byplex Corporation's stock fell each trading day of the first week of June. Use the figure at the right to find the change in the price of Byplex stock over the week's time.

Change in Price, in Dollars, of Byplex Corporation

90. The daytime temperature on the moon can reach 266°F, and the nighttime temperature can go as low as -292°F. Find the difference between these extremes.

91. The average temperature throughout Earth's stratosphere is -70°F. The average temperature on Earth's surface is 45°F. Find the difference between these average temperatures.

The elevation, or height, of places on Earth is measured in relation to sea level, or the average level of the ocean's surface. The table below shows height above sea level as a positive number and depth below sea level as a negative number. Use the table for Exercises 92 to 94.

Continent	Highest Elevation (in feet)		Lowest Elevation (in feet)	
Africa	Mt. Kilimanjaro	19,340	Lake Assal	−512
Asia	Mt. Everest	29,028	Dead Sea	−1312
North America	Mt. McKinley	20,320	Death Valley	−282
South America	Mt. Aconcagua	22,834	Valdes Peninsula	−131

92. What is the difference in elevation between Mt. Kilimanjaro and Lake Assal?

93. What is the difference in elevation between Mt. Aconcagua and the Valdes Peninsula?

94. For which continent shown is the difference between the highest and lowest elevations **a.** greatest? **b.** smallest?

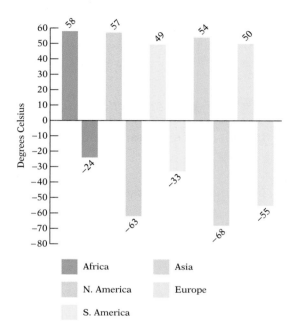

The figure at the right shows the highest and lowest temperatures ever recorded for selected regions of the world. Use this graph for Exercises 95 to 97.

95. What is the difference between the highest and lowest temperatures recorded in Africa?

96. What is the difference between the highest and lowest temperatures recorded in South America?

97. What is the difference between the lowest temperature recorded in Europe and the lowest temperature recorded in Asia?

Highest and Lowest Temperatures Recorded (in degrees Celsius)

Source: National Climatic Data Center

APPLYING THE CONCEPTS

98. Consider the numbers 4, −7, −5, 13, and −9. What is the largest difference that can be obtained by subtracting one number in the list from another number in the list? Find the smallest positive difference.

99. Fill in the blank squares at the right with integers so that the sum of the integers along any row, column, or diagonal is zero.

100. The sum of two negative integers is −8. Find the integers.

101. Explain the difference between the words *negative* and *minus*.

10.3 Multiplication and Division of Integers

Objective A To multiply integers

Multiplication is the repeated addition of the same number.

Several different symbols are used to indicate multiplication:

$$3 \times 2 = 6 \qquad\qquad 3 \cdot 2 = 6 \qquad\qquad (3)(2) = 6$$

When 5 is multiplied by a sequence of decreasing integers, the products decrease by 5.

$$5 \times 3 = 15$$
$$5 \times 2 = 10$$
$$5 \times 1 = 5$$
$$5 \times 0 = 0$$

The pattern developed can be continued so that 5 is multiplied by a sequence of negative numbers. The resulting products must be negative in order to maintain the pattern of decreasing by 5.

$$5 \times (-1) = -5$$
$$5 \times (-2) = -10$$
$$5 \times (-3) = -15$$
$$5 \times (-4) = -20$$

This example illustrates that the product of a positive number and a negative number is negative.

When −5 is multiplied by a sequence of decreasing integers, the products increase by 5.

$$-5 \times 3 = -15$$
$$-5 \times 2 = -10$$
$$-5 \times 1 = -5$$
$$-5 \times 0 = 0$$

The pattern developed can be continued so that −5 is multiplied by a sequence of negative numbers. The resulting products must be positive in order to maintain the pattern of increasing by 5.

$$-5 \times (-1) = 5$$
$$-5 \times (-2) = 10$$
$$-5 \times (-3) = 15$$
$$-5 \times (-4) = 20$$

This example illustrates that the product of two negative numbers is positive.

The pattern for multiplication shown above is summarized in the following rules for multiplying integers.

CALCULATOR NOTE

To multiply $(-4)(-8)$ on your calculator, enter the following:

4 $\boxed{+/-}$ $\boxed{\times}$ 8 $\boxed{+/-}$ $\boxed{=}$

> **Rule for Multiplying Two Numbers**
>
> To multiply numbers with the same sign, multiply the absolute values of the factors. The product is positive.
>
> To multiply numbers with different signs, multiply the absolute values of the factors. The product is negative.

$$4 \cdot 8 = 32$$
$$(-4)(-8) = 32$$
$$-4 \cdot 8 = -32$$
$$(4)(-8) = -32$$

⇒ Multiply: $2(-3)(-5)(-7)$

$2(-3)(-5)(-7) = -6(-5)(-7)$

$= 30(-7)$

$= -210$

- To multiply more than two numbers, multiply the first two numbers.
- Then multiply the product by the third number.
- Continue until all the numbers have been multiplied.

Example 1
Multiply: $(-2)(6)$

Solution
$(-2)(6) = -12$
- The signs are different. The product is negative.

You Try It 1
Multiply: $(-3)(-5)$

Your solution

Example 2
Find the product of -42 and 62.

Solution
$-42 \cdot 62 = -2604$

You Try It 2
Find -38 multiplied by 51.

Your solution

Example 3
Multiply: $-5(-4)(6)(-3)$

Solution
$-5(-4)(6)(-3) = 20(6)(-3)$
$= 120(-3)$
$= -360$

You Try It 3
Multiply: $-7(-8)(9)(-2)$

Your solution

Solutions on p. S25

Objective B   **To divide integers**

For every division problem, there is a related multiplication problem.

Division: $\dfrac{8}{2} = 4$ Related multiplication: $4 \cdot 2 = 8$

This fact can be used to illustrate the rules for dividing signed numbers.

Rule for Dividing Two Numbers

To divide numbers with the same sign, divide the absolute values of the numbers. The quotient is positive.

To divide numbers with different signs, divide the absolute values of the numbers. The quotient is negative.

$\dfrac{8}{2} = 4$ because $4 \cdot 2 = 8$

$\dfrac{-8}{-2} = 4$ because $4(-2) = -8$

$\dfrac{8}{-2} = -4$ because $-4(-2) = 8$

$\dfrac{-8}{2} = -4$ because $-4(2) = -8$

Note that $\frac{8}{-2}$, $\frac{-8}{2}$, and $-\frac{8}{2}$ are all equal to -4.

If a and b are two integers, then $\frac{a}{-b} = \frac{-a}{b} = -\frac{a}{b}$.

> **Properties of Zero and One in Division**
>
> Zero divided by any number other than zero is zero.
>
> Any number other than zero divided by itself is 1.
>
> Any number divided by 1 is the number.
>
> Division by zero is not defined.

$\frac{0}{a} = 0$ because $0 \cdot a = 0$

$\frac{a}{a} = 1$ because $1 \cdot a = a$

$\frac{a}{1} = a$ because $a \cdot 1 = a$

$\frac{4}{0} = ?$ $? \cdot 0 = 4$

There is no number whose product with zero is 4.

The examples below illustrate these properties of division.

$$\frac{0}{-8} = 0 \qquad \frac{-7}{-7} = 1 \qquad \frac{-9}{1} = -9 \qquad \frac{-3}{0} \text{ is undefined.}$$

Example 4
Divide: $(-120) \div (-8)$

Solution
$(-120) \div (-8) = 15$ • The signs are the same. The quotient is positive.

You Try It 4
Divide: $(-135) \div (-9)$

Your solution

Example 5
Divide: $95 \div (-5)$

Solution
$95 \div (-5) = -19$ • The signs are different. The quotient is negative.

You Try It 5
Divide: $84 \div (-6)$

Your solution

Example 6
Find the quotient of -81 and 3.

Solution
$-81 \div 3 = -27$

You Try It 6
What is -72 divided by 4?

Your solution

Example 7
Divide: $0 \div (-24)$

Solution
$0 \div (-24) = 0$ • Zero divided by a nonzero number is zero.

You Try It 7
Divide: $-39 \div 0$

Your solution

Solutions on p. S25

> *Objective C* **To solve application problems** 〈 10 〉

. . . .

Example 8

The combined scores of the top five golfers in a tournament equaled −10 (10 under par). What was the average score of the five golfers?

Strategy

To find the average score, divide the combined scores (−10) by the number of golfers (5).

Solution

−10 ÷ 5 = −2

The average score was −2.

You Try It 8

The melting point of mercury is −38°C. The melting point of argon is five times the melting point of mercury. Find the melting point of argon.

Your strategy

Your solution

. . . .

Example 9

The daily high temperatures during one week were recorded as follows: −9°, 3°, 0°, −8°, 2°, 1°, 4°. Find the average daily high temperature for the week.

Strategy

To find the average daily high temperature:

• Add the seven temperature readings.
• Divide by 7.

Solution

−9 + 3 + 0 + (−8) + 2 + 1 + 4 = −7

−7 ÷ 7 = −1

The average daily high temperature was −1°.

You Try It 9

The daily low temperatures during one week were recorded as follows: −6°, −7°, 1°, 0°, −5°, −10°, −1°. Find the average daily low temperature for the week.

Your strategy

Your solution

Solutions on p. S25

10.3 Exercises

Objective A

State whether the operation in the expression is addition, subtraction, or multiplication.

1. $5 - (-6)$ **2.** $4(-9)$ **3.** $-8(-5)$ **4.** $-3 + (-7)$

Multiply.

5. 14×3

6. 62×9

7. $-4 \cdot 6$

8. $-7 \cdot 3$

9. $-2 \cdot (-3)$

10. $-5 \cdot (-1)$

11. $(9)(2)$

12. $(3)(8)$

13. $5(-4)$

14. $4(-7)$

15. $-8(2)$

16. $-9(3)$

17. $(-5)(-5)$

18. $(-3)(-6)$

19. $(-7)(0)$

20. -32×4

21. -24×3

22. $19 \cdot (-7)$

23. $6(-17)$

24. $-8(-26)$

25. $-4(-35)$

26. $-5 \cdot (23)$

27. $-6 \cdot (38)$

28. $9(-27)$

29. $8(-40)$

30. $-7(-34)$

31. $-4(39)$

32. $4 \cdot (-8) \cdot 3$

33. $5 \times 7 \times (-2)$

34. $8 \cdot (-6) \cdot (-1)$

35. $(-9)(-9)(2)$

36. $-8(-7)(-4)$

37. $-5(8)(-3)$

38. $(-6)(5)(7)$

39. $-1(4)(-9)$

40. $6(-3)(-2)$

41. $4(-4) \cdot 6(-2)$ **42.** $-5 \cdot 9(-7) \cdot 3$ **43.** $-9(4) \cdot 3(1)$

44. $8(8)(-5)(-4)$ **45.** $(-6) \cdot 7 \cdot (-10)(-5)$ **46.** $-9(-6)(11)(-2)$

47. What is -5 multiplied by -4?   **48.** What is 6 multiplied by -5?

49. What is -8 times 6?   **50.** What is -8 times -7?

51. Find the product of -4, 7, and -5.   **52.** Find the product of -2, -4, and -7.

> **Objective B**

Write the related multiplication problem.

53. $\dfrac{-36}{-12} = 3$ **54.** $\dfrac{28}{-7} = -4$ **55.** $\dfrac{-55}{11} = -5$ **56.** $\dfrac{-20}{-10} = 2$

Divide.

57. $12 \div (-6)$ **58.** $18 \div (-3)$ **59.** $(-72) \div (-9)$

60. $(-64) \div (-8)$ **61.** $0 \div (-6)$ **62.** $-49 \div 7$

63. $45 \div (-5)$ **64.** $-24 \div 4$ **65.** $-36 \div 4$

66. $-56 \div 7$ **67.** $-81 \div (-9)$ **68.** $-40 \div (-5)$

69. $72 \div (-3)$ **70.** $44 \div (-4)$ **71.** $-60 \div 5$

72. $-66 \div 6$ **73.** $-93 \div (-3)$ **74.** $-98 \div (-7)$

75. $(-85) \div (-5)$ **76.** $(-60) \div (-4)$ **77.** $120 \div 8$

78. $144 \div 9$ **79.** $78 \div (-6)$ **80.** $84 \div (-7)$

81. $-72 \div 4$ **82.** $-80 \div 5$ **83.** $-114 \div (-6)$

84. $-91 \div (-7)$ **85.** $-104 \div (-8)$ **86.** $-126 \div (-9)$

87. $57 \div (-3)$ **88.** $162 \div (-9)$ **89.** $-136 \div (-8)$

90. $-128 \div 4$ **91.** $-130 \div (-5)$ **92.** $(-280) \div 8$

93. $(-92) \div (-4)$ **94.** $-196 \div (-7)$ **95.** $-150 \div (-6)$

96. $(-261) \div 9$ **97.** $204 \div (-6)$ **98.** $165 \div (-5)$

99. $-132 \div (-12)$ **100.** $-156 \div (-13)$ **101.** $-182 \div 14$

102. $-144 \div 12$ **103.** $143 \div 11$ **104.** $168 \div 14$

105. $-180 \div (-15)$ **106.** $-169 \div (-13)$ **107.** $154 \div (-11)$

108. Find the quotient of -132 and -11.   **109.** Find the quotient of 182 and -13.

110. What is -60 divided by -15?   **111.** What is 144 divided by -24?

112. Find the quotient of -135 and 15.   **113.** Find the quotient of -88 and 22.

114. The daily low temperatures during one week were recorded as follows: 4°, −5°, 8°, −1°, −12°, −14°, −8°. Find the average daily low temperature for the week.

115. The daily high temperatures during one week were recorded as follows: −6°, −11°, 1°, 5°, −3°, −9°, −5°. Find the average daily high temperature for the week.

116. The graph at the right shows the boiling points of three chemical elements. The boiling point of neon is seven times the highest boiling point shown in the table.

    a. Without actually calculating the boiling point, determine whether the boiling point of neon is above 0°C or below 0°C.

    b. What is the boiling point of neon?

117. The combined scores of the top ten golfers in a tournament equaled −20 (20 under par). What was the average score of the ten golfers?

118. The combined scores of the top four golfers in a tournament equaled −12 (12 under par). What was the average score of the four golfers?

119. The wind chill factor when the temperature is −15°F and the wind is blowing at 20 mph is five times the wind chill factor when the temperature is 25°F and the wind is blowing at 35 mph. If the wind chill factor at 25°F with a 35-mph wind is −12°F, what is the wind chill factor at −15°F with a 20-mph wind?

120. To discourage guessing on a multiple-choice exam, an instructor graded the test by giving 5 points for a correct answer, −2 points for an answer left blank, and −5 points for an incorrect answer. How many points did a student score who answered 20 questions correctly, answered 5 questions incorrectly, and left 2 questions blank?

APPLYING THE CONCEPTS

121.   **a.** Find the largest possible product of two negative integers whose sum is −10.

    b. Find the smallest possible sum of two negative integers whose product is 16.

122. Use repeated addition to show that the product of two integers with different signs is a negative number.

123. Determine whether the statement is true or false.

    a. The product of a nonzero number and its opposite is negative.

    b. The square of a negative number is a positive number.

124. In your own words, describe the rules for multiplying and dividing integers.

Operations with Rational Numbers

To add or subtract rational numbers

In this section, operations with rational numbers are discussed. A **rational number** is the quotient of two integers.

> **Rational Numbers**
>
> A rational number is a number that can be written in the form $\frac{a}{b}$, where a and b are integers and $b \neq 0$.

Each of the three numbers shown at the right is a rational number.

$$\frac{3}{4} \qquad \frac{-2}{9} \qquad \frac{13}{-5}$$

An integer can be written as the quotient of the integer and 1. Therefore, **every integer is a rational number.**

$$6 = \frac{6}{1} \qquad -8 = \frac{-8}{1}$$

A mixed number can be written as the quotient of two integers. Therefore, **every mixed number is a rational number.**

$$1\frac{4}{7} = \frac{11}{7} \qquad 3\frac{2}{5} = \frac{17}{5}$$

Recall that every fraction can be written as a decimal by dividing the numerator of the fraction by the denominator. The result is either a terminating decimal or a repeating decimal.

We can write the fraction $\frac{3}{4}$ as the terminating decimal 0.75.

<table>
<tr><td>

TAKE NOTE

The fraction bar can be read "divided by."

$\frac{3}{4} = 3 \div 4$

</td></tr>
</table>

$$
\begin{array}{r}
0.75 \\
4\overline{)\ 3.00} \\
-2\ 8 \\
\hline
20 \\
-20 \\
\hline
0
\end{array}
$$

This is a **terminating decimal.**

The remainder is zero.

We can write the fraction $\frac{2}{3}$ as the repeating decimal $0.\overline{6}$.

$$
\begin{array}{r}
0.666 = 0.\overline{6} \\
3\overline{)\ 2.000} \\
-1\ 8 \\
\hline
20 \\
-18 \\
\hline
20 \\
-18 \\
\hline
2
\end{array}
$$

This is a **repeating decimal.**
The bar over the digit 6 in $0.\overline{6}$ is used to show that this digit repeats.

The remainder is never zero.

All terminating and repeating decimals are rational numbers.

To add or subtract rational numbers in fractional form, first find the least common multiple (LCM) of the denominators.

⟹ Add: $-\dfrac{7}{8} + \dfrac{5}{6}$

$$8 = 2 \cdot 2 \cdot 2$$
$$6 = 2 \cdot 3$$
$$\text{LCM} = 2 \cdot 2 \cdot 2 \cdot 3 = 24$$

- Find the LCM of the denominators.

$$-\dfrac{7}{8} + \dfrac{5}{6} = -\dfrac{21}{24} + \dfrac{20}{24}$$

- Rewrite each fraction using the LCM of the denominators as the common denominator.

$$= \dfrac{-21 + 20}{24}$$

- Add the numerators.

$$= \dfrac{-1}{24} = -\dfrac{1}{24}$$

⟹ Subtract: $-\dfrac{7}{9} - \dfrac{5}{12}$

$$9 = 3 \cdot 3$$
$$12 = 2 \cdot 2 \cdot 3$$
$$\text{LCM} = 2 \cdot 2 \cdot 3 \cdot 3 = 36$$

- Find the LCM of the denominators.

$$-\dfrac{7}{9} - \dfrac{5}{12} = -\dfrac{28}{36} - \dfrac{15}{36}$$

- Rewrite each fraction using the LCM of the denominators as the common denominator.

$$= \dfrac{-28}{36} + \dfrac{-15}{36}$$

- Rewrite subtraction as addition of the opposite. Rewrite negative fractions with the negative sign in the numerator.

$$= \dfrac{-28 + (-15)}{36}$$

- Add the numerators.

$$= \dfrac{-43}{36} = -\dfrac{43}{36} = -1\dfrac{7}{36}$$

To add or subtract rational numbers in decimal form, use the sign rules for adding integers.

⟹ Add: $47.034 + (-56.91)$

$$\begin{array}{r} 56.910 \\ -\ 47.034 \\ \hline 9.876 \end{array}$$

- The signs are different. Find the difference between the absolute values of the numbers.

$$47.034 + (-56.91)$$
$$= -9.876$$

- Attach the sign of the number with the greater absolute value.

➡ Subtract: $-39.09 - 102.98$

$-39.09 - 102.98$
$= -39.09 + (-102.98)$

- Rewrite subtraction as addition of the opposite number.

$\begin{array}{r} 39.09 \\ + 102.98 \\ \hline 142.07 \end{array}$

- The signs of the addends are the same. Find the sum of the absolute values of the numbers.

$-39.09 - 102.98 = -142.07$

- Attach the sign of the addends.

Example 1

Subtract: $\dfrac{5}{16} - \dfrac{7}{40}$

Solution
The LCM of 16 and 40 is 80.

$$\dfrac{5}{16} - \dfrac{7}{40} = \dfrac{25}{80} - \dfrac{14}{80}$$
$$= \dfrac{25}{80} + \dfrac{-14}{80}$$
$$= \dfrac{25 + (-14)}{80} = \dfrac{11}{80}$$

You Try It 1

Subtract: $\dfrac{5}{9} - \dfrac{11}{12}$

Your solution

Example 2

Simplify: $-\dfrac{3}{4} + \dfrac{1}{6} - \dfrac{5}{8}$

Solution
The LCM of 4, 6, and 8 is 24.

$$-\dfrac{3}{4} + \dfrac{1}{6} - \dfrac{5}{8} = -\dfrac{18}{24} + \dfrac{4}{24} - \dfrac{15}{24}$$
$$= \dfrac{-18}{24} + \dfrac{4}{24} + \dfrac{-15}{24}$$
$$= \dfrac{-18 + 4 + (-15)}{24}$$
$$= \dfrac{-29}{24} = -\dfrac{29}{24} = -1\dfrac{5}{24}$$

You Try It 2

Simplify: $-\dfrac{7}{8} - \dfrac{5}{6} + \dfrac{2}{3}$

Your solution

Example 3

Subtract: $42.987 - 98.61$

Solution
$42.987 - 98.61 = 42.987 + (-98.61)$

$\begin{array}{r} 98.610 \\ - 42.987 \\ \hline 55.623 \end{array}$

$42.987 - 98.61 = -55.623$

You Try It 3

Subtract: $16.127 - 67.91$

Your solution

Solutions on pp. S25–S26

Example 4

Simplify: $1.02 + (-3.6) + 9.24$

Solution

$$1.02 + (-3.6) + 9.24 = -2.58 + 9.24$$
$$= 6.66$$

You Try It 4

Simplify: $2.7 + (-9.44) + 6.2$

Your solution

Solution on p. S26

Objective B   **To multiply or divide rational numbers**

The product of two rational numbers written as fractions is the product of the numerators over the product of the denominators. Use the sign rules for multiplying integers.

➡ Simplify: $-\frac{3}{8} \times \frac{12}{17}$

$$-\frac{3}{8} \times \frac{12}{17} = -\left(\frac{3 \cdot 12}{8 \cdot 17}\right) = -\frac{9}{34}$$

- The signs are different. The product is negative.

To divide rational numbers written as fractions, invert the divisor and then multiply. Use the sign rules for dividing integers.

➡ Simplify: $-\frac{3}{10} \div \left(-\frac{18}{25}\right)$

$$-\frac{3}{10} \div \left(-\frac{18}{25}\right) = \frac{3}{10} \times \frac{25}{18} = \frac{3 \cdot 25}{10 \cdot 18} = \frac{5}{12}$$

- The signs are the same. The quotient is positive.

To multiply or divide rational numbers written in decimal form, use the sign rules for integers.

➡ Simplify: $(-6.89) \times (-0.00035)$

$$
\begin{array}{rl}
6.89 & \text{2 decimal places} \\
\times \quad 0.00035 & \text{5 decimal places} \\
\hline
3445 & \\
2067 \quad\quad & \\
\hline
0.0024115 & \text{7 decimal places}
\end{array}
$$

- The signs are the same. Multiply the absolute values.

$$(-6.89) \times (-0.00035) = 0.0024115$$

- The product is positive.

➡ Divide 1.32 ÷ (−0.27). Round to the nearest tenth.

$$\begin{array}{r} 4.88 \approx 4.9 \\ 0.27.\overline{)1.32.00} \\ -1\ 08 \\ \hline 24\ 0 \\ -21\ 6 \\ \hline 2\ 40 \\ -2\ 16 \\ \hline 24 \end{array}$$

• Divide the absolute values. Move the decimal point two places in the divisor and then in the dividend. Place the decimal point in the quotient.

$$1.32 \div (-0.27) \approx -4.9$$

• The signs are different. The quotient is negative.

Example 5

Multiply: $-\dfrac{7}{12} \times \dfrac{9}{14}$

Solution The product is negative.

$$-\frac{7}{12} \times \frac{9}{14} = -\left(\frac{7 \cdot 9}{12 \cdot 14}\right)$$
$$= -\frac{3}{8}$$

You Try It 5

Multiply: $\left(-\dfrac{2}{3}\right)\left(-\dfrac{9}{10}\right)$

Your solution

Example 6

Divide: $-\dfrac{3}{8} \div \left(-\dfrac{5}{12}\right)$

Solution The quotient is positive.

$$-\frac{3}{8} \div \left(-\frac{5}{12}\right) = \frac{3}{8} \times \frac{12}{5}$$
$$= \frac{3 \cdot 12}{8 \cdot 5}$$
$$= \frac{9}{10}$$

You Try It 6

Divide: $-\dfrac{5}{8} \div \dfrac{5}{40}$

Your solution

Example 7 Multiply: -4.29×8.2

Solution The product is negative.

$$\begin{array}{r} 4.29 \\ \times\ 8.2 \\ \hline 858 \\ 3432 \\ \hline 35.178 \end{array}$$

$$-4.29 \times 8.2 = -35.178$$

You Try It 7 Multiply: -5.44×3.8

Your solution

Solutions on p. S26

Example 8 Multiply: $-3.2 \times (-0.4) \times 6.9$

Solution $-3.2 \times (-0.4) \times 6.9$
$= 1.28 \times 6.9$
$= 8.832$

You Try It 8 Multiply: $3.44 \times (-1.7) \times 0.6$

Your solution

Example 9 Divide: $-0.0792 \div (-0.42)$
Round to the nearest hundredth.

Solution

$$
\begin{array}{r}
0.188 \approx 0.19 \\
0.42.\overline{)0.07.920} \\
-4\,2 \\
\hline
3\,72 \\
-3\,36 \\
\hline
360 \\
-336 \\
\hline
24
\end{array}
$$

$-0.0792 \div (-0.42) \approx 0.19$

You Try It 9 Divide: $-0.394 \div 1.7$
Round to the nearest hundredth.

Your solution

Solutions on p. S26

Objective C **To solve application problems**

Example 10

In Fairbanks, Alaska, the average temperature during the month of July is 61.5°F. During the month of January, the average temperature in Fairbanks is −12.7°F. What is the difference between the average temperature in Fairbanks during July and the average temperature during January?

Strategy
To find the difference, subtract the average temperature in January (−12.7) from the average temperature in July (61.5).

Solution
$61.5 - (-12.7) = 61.5 + 12.7 = 74.2$

The difference between the average temperature during July and the average temperature during January in Fairbanks is 74.2°F.

You Try It 10

On January 10, 1911, in Rapid City, South Dakota, the temperature fell from 12.78°C at 7:00 A.M. to −13.33°C at 7:15 A.M. How many degrees did the temperature fall during the 15-minute period?

Your strategy

Your solution

Solution on p. S26

10.4 Exercises

Objective A

Simplify.

1. $\dfrac{5}{8} - \dfrac{5}{6}$

2. $\dfrac{1}{9} - \dfrac{5}{27}$

3. $-\dfrac{5}{12} - \dfrac{3}{8}$

4. $-\dfrac{5}{6} - \dfrac{5}{9}$

5. $-\dfrac{6}{13} + \dfrac{17}{26}$

6. $-\dfrac{7}{12} + \dfrac{5}{8}$

7. $-\dfrac{5}{8} - \left(-\dfrac{11}{12}\right)$

8. $-\dfrac{7}{12} - \left(-\dfrac{7}{8}\right)$

9. $\dfrac{5}{12} - \dfrac{11}{15}$

10. $\dfrac{2}{5} - \dfrac{14}{15}$

11. $-\dfrac{3}{4} - \dfrac{5}{8}$

12. $-\dfrac{2}{3} - \dfrac{5}{8}$

13. $-\dfrac{5}{2} - \left(-\dfrac{13}{4}\right)$

14. $-\dfrac{7}{3} - \left(-\dfrac{3}{2}\right)$

15. $-\dfrac{3}{8} - \dfrac{5}{12} - \dfrac{3}{16}$

16. $-\dfrac{5}{16} + \dfrac{3}{4} - \dfrac{7}{8}$

17. $\dfrac{1}{2} - \dfrac{3}{8} - \left(-\dfrac{1}{4}\right)$

18. $\dfrac{3}{4} - \left(-\dfrac{7}{12}\right) - \dfrac{7}{8}$

19. $\dfrac{1}{3} - \dfrac{1}{4} - \dfrac{1}{5}$

20. $\dfrac{5}{16} + \dfrac{1}{8} - \dfrac{1}{2}$

21. $\dfrac{1}{2} + \left(-\dfrac{3}{8}\right) + \dfrac{5}{12}$

22. $-\dfrac{3}{8} + \dfrac{3}{4} - \left(-\dfrac{3}{16}\right)$

23. $3.4 + (-6.8)$

24. $-4.9 + 3.27$

25. $-8.32 + (-0.57)$

26. $-3.5 + 7$

27. $-4.8 + (-3.2)$

28. $6.2 + (-4.29)$

29. $-4.6 + 3.92$

30. $7.2 + (-8.42)$

31. $-45.71 + (-135.8)$

32. $-35.274 + 12.47$

33. $4.2 + (-6.8) + 5.3$

34. $6.7 + 3.2 + (-10.5)$

35. $-4.5 + 3.2 + (-19.4)$

36. $2.09 - 6.72 - 5.4$

37. $-18.39 + 4.9 - 23.7$

38. $19 - (-3.72) - 82.75$

39. $-3.09 - 4.6 - 27.3$

40. $-3.89 + (-2.9) + 4.723 + 0.2$

41. $-4.02 + 6.809 - (-3.57) - (-0.419)$

42. $0.0153 + (-1.0294) + (-1.0726)$

43. $0.27 + (-3.5) - (-0.27) + (-5.44)$

Simplify.

44. $\dfrac{1}{2} \times \left(-\dfrac{3}{4}\right)$

45. $-\dfrac{2}{9} \times \left(-\dfrac{3}{14}\right)$

46. $\left(-\dfrac{3}{8}\right)\left(-\dfrac{4}{15}\right)$

47. $\left(-\dfrac{3}{4}\right)\left(-\dfrac{8}{27}\right)$

48. $-\dfrac{1}{2} \times \dfrac{8}{9}$

49. $\dfrac{5}{12} \times \left(-\dfrac{8}{15}\right)$

50. $\left(-\dfrac{5}{12}\right)\left(\dfrac{42}{65}\right)$

51. $\left(\dfrac{3}{8}\right)\left(-\dfrac{15}{41}\right)$

52. $\left(-\dfrac{15}{8}\right)\left(-\dfrac{16}{3}\right)$

53. $\left(-\dfrac{5}{7}\right)\left(-\dfrac{14}{15}\right)$

54. $\dfrac{5}{8} \times \left(-\dfrac{7}{12}\right) \times \dfrac{16}{25}$

55. $\left(\dfrac{1}{2}\right)\left(-\dfrac{3}{4}\right)\left(-\dfrac{5}{8}\right)$

56. $\dfrac{1}{3} \div \left(-\dfrac{1}{2}\right)$

57. $-\dfrac{3}{8} \div \dfrac{7}{8}$

58. $\left(-\dfrac{3}{4}\right) \div \left(-\dfrac{7}{40}\right)$

59. $\dfrac{5}{6} \div \left(-\dfrac{3}{4}\right)$

60. $-\dfrac{5}{12} \div \dfrac{15}{32}$

61. $-\dfrac{5}{16} \div \left(-\dfrac{3}{8}\right)$

62. $\left(-\dfrac{3}{8}\right) \div \left(-\dfrac{5}{12}\right)$

63. $\left(-\dfrac{8}{19}\right) \div \dfrac{7}{38}$

64. $\left(-\dfrac{2}{3}\right) \div 4$

65. $-6 \div \dfrac{4}{9}$

66. $-6.7 \times (-4.2)$

67. $-8.9 \times (-3.5)$

68. -1.6×4.9

69. -14.3×7.9

70. $(-0.78)(-0.15)$

71. $(-1.21)(-0.03)$

72. $(-8.919) \div (-0.9)$

73. $-77.6 \div (-0.8)$

74. $59.01 \div (-0.7)$

75. $(-7.04) \div (-3.2)$

76. $(-84.66) \div 1.7$

77. $-3.312 \div (0.8)$

78. $1.003 \div (-0.59)$

79. $26.22 \div (-6.9)$

Divide.

80. $(-19.08) \div 0.45$

81. $21.792 \div (-0.96)$

82. $(-38.665) \div (-9.5)$

83. $(-3.171) \div (-45.3)$

84. $27.738 \div (-60.3)$

85. $(-13.97) \div (-25.4)$

Objective C *Application Problems*

86. On January 23, 1916, the temperature in Browing, Montana, was 6.67°C. On January 24, 1916, the temperature in Browing was −48.9°C. Find the difference between the temperatures in Browing on these two days.

87. On January 22, 1943, in Spearfish, South Dakota, the temperature fell from 12.22°C at 9 A.M. to −20°C at 9:27 A.M. How many degrees did the temperature fall during the 27-minute period?

88. The boiling point of nitrogen is −195.8°C and the melting point is −209.86°C. Find the difference between the boiling point and the melting point of nitrogen.

89. The boiling point of oxygen is −182.962°C. Oxygen's melting point is −218.4°C. What is the difference between the boiling point and the melting point of oxygen?

The chart at the right shows the closing price of a share of stock on October 6, 2000, for each of five companies. Also shown is the change in the closing price from the previous day. To find the closing price on the previous day, subtract the change in price from the closing price on October 6. Use this chart for Exercises 90 and 91.

Company	Closing Price	Change in Price
Campbell Soup	$25\frac{7}{8}$	$-\frac{5}{8}$
Coca Cola Co.	$58\frac{5}{16}$	$-\frac{13}{16}$
Kellogg Co.	$24\frac{13}{16}$	$-\frac{1}{2}$
Unilever PLC	$27\frac{1}{4}$	$+\frac{5}{8}$
Wm. Wrigley Jr.	$75\frac{3}{8}$	$-1\frac{5}{16}$

90. Find the closing price on the previous day for
a. Campbell Soup **b.** Wm. Wrigley Jr.

91. Find the closing price on the previous day for
a. Kellogg Co. **b.** Unilever PLC

APPLYING THE CONCEPTS

92. Determine whether the statement is true or false.
a. Every integer is a rational number.
b. Every whole number is an integer.
c. Every integer is a positive number.
d. Every rational number is an integer.

93. Find a rational number between $-\frac{3}{4}$ and $-\frac{2}{3}$.

94. Find a rational number between **a.** 0.1 and 0.2, **b.** 1 and 1.1, and **c.** 0 and 0.005.

95. Given any two different rational numbers, is it always possible to find a rational number between them? If so, explain how. If not, give an example of two different rational numbers for which there is no rational number between them.

Scientific Notation and the Order of Operations Agreement

To write a number in scientific notation

Scientific notation uses negative exponents. Therefore, we will discuss that topic before presenting scientific notation.

Look at the powers of 10 shown at the right. Note the pattern: The exponents are decreasing by 1, and each successive number on the right is one-tenth of the number above it. $(100,000 \div 10 = 10,000; \quad 10,000 \div 10 = 1000;$ etc.)

$$10^5 = 100,000$$
$$10^4 = 10,000$$
$$10^3 = 1000$$
$$10^2 = 100$$
$$10^1 = 10$$

If we continue this pattern, the next exponent on 10 is $1 - 1 = 0$, and the number on the right side is $10 \div 10 = 1$.

$$10^0 = 1$$

The next exponent on 10 is $0 - 1 = -1$, and 10^{-1} is equal to $1 \div 10 = 0.1$.

$$10^{-1} = 0.1$$

The pattern is continued on the right. Note that a negative exponent does not indicate a negative number. Rather, each power of 10 with a negative exponent is equal to a number between 0 and 1. Also note that as the exponent on 10 decreases, so does the number it is equal to.

$$10^{-2} = 0.01$$
$$10^{-3} = 0.001$$
$$10^{-4} = 0.0001$$
$$10^{-5} = 0.00001$$
$$10^{-6} = 0.000001$$

Very large and very small numbers are encountered in the natural sciences. For example, the mass of an electron is 0.0000000000000000000000000000911 kg. Numbers such as this are difficult to read, so a more convenient system called **scientific notation** is used. In scientific notation, a number is expressed as the product of two factors, one a number between 1 and 10, and the other a power of 10.

To express a number in scientific notation, write it in the form $a \times 10^n$, where a is a number between 1 and 10 and n is an integer.

For numbers greater than 10, move the decimal point to the right of the first digit. The exponent n is positive and equal to the number of places the decimal point has been moved.

$$240,000 = 2.4 \times 10^5$$
$$93,000,000 = 9.3 \times 10^7$$

TAKE NOTE

There are two steps in writing a number in scientific notation: (1) determine the number between 1 and 10, and (2) determine the exponent on 10.

For numbers less than 1, move the decimal point to the right of the first nonzero digit. The exponent n is negative. The absolute value of the exponent is equal to the number of places the decimal point has been moved.

$$0.0003 = 3 \times 10^{-4}$$

$$0.0000832 = 8.32 \times 10^{-5}$$

Changing a number written in scientific notation to decimal notation also requires moving the decimal point.

When the exponent on 10 is positive, move the decimal point to the right the same number of places as the exponent.

$$3.45 \times 10^9 = 3,450,000,000$$

$$2.3 \times 10^8 = 230,000,000$$

When the exponent on 10 is negative, move the decimal point to the left the same number of places as the absolute value of the exponent.

$$8.1 \times 10^{-3} = 0.0081$$

$$6.34 \times 10^{-6} = 0.00000634$$

Example 1   Write 824,300,000,000 in scientific notation.

Solution   The number is greater than 10. Move the decimal point 11 places to the left. The exponent on 10 is 11.

$$824,300,000,000 = 8.243 \times 10^{11}$$

You Try It 1   Write 0.000000961 in scientific notation.

Your solution

Example 2   Write 6.8×10^{-10} in decimal notation.

Solution   The exponent on 10 is negative. Move the decimal point 10 places to the left.

$$6.8 \times 10^{-10} = 0.00000000068$$

You Try It 2   Write 7.329×10^6 in decimal notation.

Your solution

Solutions on p. S26

Objective B   **To use the Order of Operations Agreement to simplify expressions**

The Order of Operations Agreement has been used throughout this book. In simplifying expressions with rational numbers, the same Order of Operations Agreement is used. This agreement is restated here.

> **The Order of Operations Agreement**
>
> **Step 1** Do all operations inside parentheses.
> **Step 2** Simplify any number expressions containing exponents.
> **Step 3** Do multiplication and division as they occur from left to right.
> **Step 4** Rewrite subtraction as addition of the opposite. Then do additions as they occur from left to right.

TAKE NOTE

In $(-3)^2$, we are squaring -3; multiply -3 times -3. In -3^2, we are finding the opposite of 3^2. The expression -3^2 is the same as $-(3^2)$.

Exponents may be confusing in expressions with signed numbers.

$$(-3)^2 = (-3) \times (-3) = 9$$
$$-3^2 = -(3)^2 = -(3 \times 3) = -9$$

Note that -3 is squared only when the negative sign is *inside* the parentheses.

CALCULATOR NOTE

As shown above, the value of -3^2 is different from the value of $(-3)^2$. The keystrokes to evaluate each of these on your calculator are different. To evaluate -3^2, enter

3 $\boxed{x^2}$ $\boxed{+/-}$

To evaluate $(-3)^2$, enter

3 $\boxed{+/-}$ $\boxed{x^2}$

➡ Simplify: $(-3)^2 - 2 \times (8 - 3) + (-5)$

$$(-3)^2 - 2 \times \underbrace{(8 - 3)} + (-5)$$
$$\underbrace{(-3)^2} - 2 \times 5 + (-5)$$
$$9 - \underbrace{2 \times 5} + (-5)$$
$$9 - 10 + (-5)$$

$$\underbrace{9 + (-10)} + (-5)$$
$$\underbrace{(-1) + (-5)}$$
$$-6$$

1. Perform operations inside parentheses.

2. Simplify expressions with exponents.

3. Do multiplications and divisions as they occur from left to right.

4. Rewrite subtraction as the addition of the opposite. Then add from left to right.

➡ Simplify: $\left(\dfrac{1}{4} - \dfrac{1}{2}\right)^2 \div \dfrac{3}{8}$

$$\underbrace{\left(\dfrac{1}{4} - \dfrac{1}{2}\right)^2} \div \dfrac{3}{8}$$
$$\underbrace{\left(-\dfrac{1}{4}\right)^2} \div \dfrac{3}{8}$$
$$\dfrac{1}{16} \div \dfrac{3}{8}$$
$$\dfrac{1}{16} \times \dfrac{8}{3}$$
$$\dfrac{1}{6}$$

1. Perform operations inside parentheses.

2. Simplify expressions with exponents.

3. Do multiplication and division as they occur from left to right.

Example 3   Simplify: $8 - 4 \div (-2)$

Solution $8 - 4 \div (-2) = 8 - (-2)$
$$= 8 + 2$$
$$= 10$$

You Try It 3   Simplify: $9 - 9 \div (-3)$

Your solution

Solution on p. S26

Example 4   Simplify: $12 \div (-2)^2 + 5$

Solution
$$12 \div (-2)^2 + 5$$
$$= 12 \div 4 + 5$$
$$= 3 + 5$$
$$= 8$$

You Try It 4   Simplify: $8 \div 4 \cdot 4 - (-2)^2$

Your solution

Example 5   Simplify: $12 - (-10) \div (8 - 3)$

Solution
$$12 - (-10) \div (8 - 3)$$
$$= 12 - (-10) \div 5$$
$$= 12 - (-2)$$
$$= 12 + 2$$
$$= 14$$

You Try It 5   Simplify: $8 - (-15) \div (2 - 7)$

Your solution

Example 6   Simplify:
$(-3)^2 \times (5 - 7)^2 - (-9) \div 3$

Solution
$$(-3)^2 \times (5 - 7)^2 - (-9) \div 3$$
$$= (-3)^2 \times (-2)^2 - (-9) \div 3$$
$$= 9 \times 4 - (-9) \div 3$$
$$= 36 - (-9) \div 3$$
$$= 36 - (-3)$$
$$= 36 + 3$$
$$= 39$$

You Try It 6   Simplify:
$(-2)^2 \times (3 - 7)^2 - (-16) \div (-4)$

Your solution

Example 7   Simplify: $3 \div \left(\frac{1}{2} - \frac{1}{4}\right) - 3$

Solution
$$3 \div \left(\frac{1}{2} - \frac{1}{4}\right) - 3$$
$$= 3 \div \frac{1}{4} - 3$$
$$= 3 \times \frac{4}{1} - 3$$
$$= 12 - 3$$
$$= 12 + (-3)$$
$$= 9$$

You Try It 7   Simplify: $7 \div \left(\frac{1}{7} - \frac{3}{14}\right) - 9$

Your solution

Solutions on p. S26

10.5 Exercises

Objective A

Write the number in scientific notation.

1. 2,370,000   **2.** 75,000   **3.** 0.00045   **4.** 0.000076

5. 309,000   **6.** 819,000,000   **7.** 0.000000601   **8.** 0.00000000096

9. 57,000,000,000   **10.** 934,800,000,000   **11.** 0.000000017   **12.** 0.0000009217

Write the number in decimal notation.

13. 7.1×10^5 **14.** 2.3×10^7 **15.** 4.3×10^{-5} **16.** 9.21×10^{-7}

17. 6.71×10^8 **18.** 5.75×10^9 **19.** 7.13×10^{-6} **20.** 3.54×10^{-8}

21. 5×10^{12} **22.** 1.0987×10^{11} **23.** 8.01×10^{-3} **24.** 4.0162×10^{-9}

Solve.

25. Light travels approximately 16,000,000,000 mi in one day. Write this number in scientific notation.

26. Write the mass of the planet Earth, which is approximately 5,980,000,000,000,000,000,000,000 kg, in scientific notation.

27. The graph at the right shows the monetary cost of four wars. Write the monetary cost of World War II in scientific notation.

Monetary Cost of War

Source: Congressional Research Service, using numbers from the *Statistical Abstract of the United States*

28. The electric charge on an electron is 0.00000000000000000016 coulomb. Write this number in scientific notation.

29. The length of an infrared light wave is approximately 0.0000037 m. Write this number in scientific notation.

30. One unit used to measure the speed of a computer is the nano-second. One nanosecond is 0.000000001 of a second. Write one nanosecond in scientific notation.

Objective B

Simplify.

31. $8 \div 4 + 2$

32. $3 - 12 \div 2$

33. $4 + (-7) + 3$

34. $-16 \div 2 + 8$

35. $4^2 - 4$

36. $6 - 2^2$

37. $2 \times (3 - 5) - 2$

38. $2 - (8 - 10) \div 2$

39. $4 - (-3)^2$

40. $(-2)^2 - 6$

41. $4 - (-3) - 5$

42. $6 + (-8) - (-3)$

43. $4 - (-2)^2 + (-3)$

44. $-3 + (-6)^2 - 1$

45. $3^2 - 4 \times 2$

46. $9 \div 3 - (-3)^2$

47. $3 \times (6 - 2) \div 6$

48. $4 \times (2 - 7) \div 5$

49. $2^2 - (-3)^2 + 2$

50. $3 \times (8 - 5) + 4$

51. $6 - 2 \times (1 - 5)$

52. $4 \times 2 \times (3 - 6)$

53. $(-2)^2 - (-3)^2 + 1$

54. $4^2 - 3^2 - 4$

55. $6 - (-3) \times (-3)^2$

56. $4 - (-5) \times (-2)^2$

57. $4 \times 2 - 3 \times 7$

58. $16 \div 2 - 9 \div 3$

59. $(-2)^2 - 5 \times 3 - 1$

60. $4 - 2 \times 7 - 3^2$

61. $7 \times 6 - 5 \times 6 + 3 \times 2 - 2 + 1$

62. $3 \times 2^2 + 5 \times (3 + 2) - 17$

63. $-4 \times 3 \times (-2) + 12 \times (3 - 4) + (-12)$

64. $3 \times 4^2 - 16 - 4 + 3 - (1 - 2)^2$

65. $-12 \times (6 - 8) + 1^2 \times 3^2 \times 2 - 6 \times 2$

66. $-3 \times (-2)^2 \times 4 \div 8 - (-12)$

67. $10 \times 9 - (8 + 7) \div 5 + 6 - 7 + 8$

68. $-27 - (-3)^2 - 2 - 7 + 6 \times 3$

69. $3^2 \times (4 - 7) \div 9 + 6 - 3 - 4 \times 2$

70. $16 - 4 \times 8 + 4^2 - (-18) \div (-9)$

71. $(-3)^2 \times (5 - 7)^2 - (-9) \div 3$

72. $-2 \times 4^2 - 3 \times (2 - 8) - 3$

73. $4 - 6(2 - 5)^3 \div (17 - 8)$

74. $5 + 7(3 - 8)^2 \div (-14 + 9)$

75. $(1.2)^2 - 4.1 \times 0.3$

76. $2.4 \times (-3) - 2.5$

77. $1.6 - (-1.6)^2$

78. $4.1 \times 8 \div (-4.1)$

79. $(4.1 - 3.9) - 0.7^2$

80. $1.8 \times (-2.3) - 2$

81. $(-0.4)^2 \times 1.5 - 2$

82. $(6.2 - 1.3) \times (-3)$

83. $4.2 - (-3.9) - 6$

84. $-\dfrac{1}{2} + \dfrac{3}{8} \div \left(-\dfrac{3}{4}\right)$

85. $\left(\dfrac{3}{4}\right)^2 - \dfrac{3}{8}$

86. $\left(\dfrac{1}{2}\right)^2 - \left(-\dfrac{1}{2}\right)^2$

87. $\dfrac{5}{16} - \dfrac{3}{8} + \dfrac{1}{2}$

88. $\dfrac{2}{7} \div \dfrac{5}{7} - \dfrac{3}{14}$

89. $\dfrac{1}{2} \times \dfrac{1}{4} \times \dfrac{1}{2} - \dfrac{3}{8}$

90. $\dfrac{2}{3} \times \dfrac{5}{8} \div \dfrac{2}{7}$

91. $\dfrac{1}{2} - \left(\dfrac{3}{4} - \dfrac{3}{8}\right) \div \dfrac{1}{3}$

92. $\dfrac{3}{8} \div \left(-\dfrac{1}{2}\right)^2 + 2$

APPLYING THE CONCEPTS

93. Place the correct symbol, $<$ or $>$, between the two numbers.
    a. $3.45 \times 10^{-14} \ \square \ 3.45 \times 10^{-15}$
    b. $5.23 \times 10^{18} \ \square \ 5.23 \times 10^{17}$
    c. $3.12 \times 10^{12} \ \square \ 3.12 \times 10^{11}$

94. Light travels 3×10^8 m in 1 s. How far does light travel in 1 year? (Astronomers refer to this distance as 1 light year.)

95. a. Evaluate $1^3 + 2^3 + 3^3 + 4^3$.
    b. Evaluate $(-1)^3 + (-2)^3 + (-3)^3 + (-4)^3$.
    c. Evaluate $1^3 + 2^3 + 3^3 + 4^3 + 5^3$.
    d. On the basis of your answers to parts a, b, and c, evaluate $(-1)^3 + (-2)^3 + (-3)^3 + (-4)^3 + (-5)^3$.

96. Evaluate $2^{(3^2)}$ and $(2^3)^2$. Are the answers the same? If not, which is larger?

97. Abdul, Becky, Carl, and Diana were being questioned by their teacher. One of the students had left an apple on the teacher's desk, but the teacher did not know which one. Abdul said it was either Becky or Diana. Diana said it was neither Becky nor Carl. If both those statements are false, who left the apple on the teacher's desk? Explain how you arrived at your solution.

98. In your own words, explain how you know that a number is written in scientific notation.

99. Express the mass of the sun and the mass of a neutron in kilograms using scientific notation.

Focus on Problem Solving

Drawing Diagrams   How do you best remember something? Do you remember best what you hear? The word *aural* means "pertaining to the ear"; people with a strong aural memory remember best those things that they hear. The word *visual* means "pertaining to the sense of sight"; people with a strong visual memory remember best that which they see written down. Some people claim that their memory is in their writing hand—they remember something only if they write it down! The method by which you best remember something is probably also the method by which you can best learn something new.

In problem-solving situations, try to capitalize on your strengths. If you tend to understand the material better when you hear it spoken, read application problems aloud or have someone else read them to you. If writing helps you to organize ideas, rewrite application problems in your own words.

No matter what your main strength, visualizing a problem can be a valuable aid in problem solving. A drawing, sketch, diagram, or chart can be a useful tool in problem solving, just as calculators and computers are tools. A diagram can be helpful in gaining an understanding of the relationships inherent in a problem-solving situation. A sketch will help you to organize the given information and can lead to your being able to focus on the method by which the solution can be determined.

➡ A tour bus drives 5 mi south, then 4 mi west, then 3 mi north, then 4 mi east. How far is the tour bus from the starting point?

Draw a diagram of the given information.

From the diagram, we can see that the solution can be determined by subtracting 3 from 5: $5 - 3 = 2$.

The bus is 2 mi from the starting point.

➡ If you roll two ordinary six-sided dice and multiply the two numbers that appear on top, how many different possible products are there?

Make a chart of the possible products. In the chart below, repeated products are marked with an asterisk.

$1 \cdot 1 = 1$	$2 \cdot 1 = 2$ (*)	$3 \cdot 1 = 3$ (*)	$4 \cdot 1 = 4$ (*)	$5 \cdot 1 = 5$ (*)	$6 \cdot 1 = 6$ (*)
$1 \cdot 2 = 2$	$2 \cdot 2 = 4$ (*)	$3 \cdot 2 = 6$ (*)	$4 \cdot 2 = 8$ (*)	$5 \cdot 2 = 10$ (*)	$6 \cdot 2 = 12$ (*)
$1 \cdot 3 = 3$	$2 \cdot 3 = 6$ (*)	$3 \cdot 3 = 9$	$4 \cdot 3 = 12$ (*)	$5 \cdot 3 = 15$ (*)	$6 \cdot 3 = 18$ (*)
$1 \cdot 4 = 4$	$2 \cdot 4 = 8$	$3 \cdot 4 = 12$ (*)	$4 \cdot 4 = 16$	$5 \cdot 4 = 20$ (*)	$6 \cdot 4 = 24$ (*)
$1 \cdot 5 = 5$	$2 \cdot 5 = 10$	$3 \cdot 5 = 15$	$4 \cdot 5 = 20$	$5 \cdot 5 = 25$	$6 \cdot 5 = 30$ (*)
$1 \cdot 6 = 6$	$2 \cdot 6 = 12$	$3 \cdot 6 = 18$	$4 \cdot 6 = 24$	$5 \cdot 6 = 30$	$6 \cdot 6 = 36$

By counting the products that are not repeats, we can see that there are 18 different possible products.

Look at Sections 1 and 2 in this chapter. You will notice that number lines are used to help you visualize the integers, as an aid in ordering integers, to help you understand the concepts of opposite and absolute value, and to illustrate addition of integers. As you begin your work with integers, you may find that sketching a number line proves helpful in coming to understand a problem or in working through a calculation that involves integers.

Projects and Group Activities

Time Zones

In 1884, a system of standard time was adopted by the International Meridian Conference. The prime meridian is a semicircle passing through Greenwich, England, and labeled 0°, as shown in the diagram at the right.

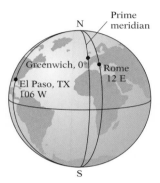

The other meridians are 15° apart, and each 15° width determines a time zone. The time zones to the east of the prime meridian are negative, and the zones to the west are positive, as shown below.

The table below gives the time zones for selected cities.

City	Time Zone	City	Time Zone
Athens	−2	London	0
Beijing	−8	Moscow	−3
Houston	6	Oslo	−1
Honolulu	11	Rio de Janeiro	3

To find the time for a city that is not in your time zone, subtract the time zone of the other city from the time zone of your city. Add that difference to the current time in your time zone. For example, say it is 2:00 P.M. in Athens. What time is it in Houston?

From the table above, the time zone in Athens is −2 and the time zone of Houston is 6.

Subtract 6 from −2. $-2 - 6 = -8$

Add that number to the current time $2:00 \text{ P.M.} + (-8) = 6:00 \text{ A.M.}$
in your time zone.

It is 6:00 A.M. in Houston.

1. Find the time in Beijing when the time in Oslo is 11:00 P.M.

2. Find the time in Moscow when the time in Honolulu is 8:00 A.M.

3. Find the time in Rio de Janeiro when the time in Beijing is 3:00 P.M.

4. An office in Athens is open from 9:00 A.M. to 5:00 P.M. What are the times in Houston that a call to Athens can be made during Athens office hours?

5. An office in Honolulu is open from 8:00 A.M. to 4:00 P.M. What are the times in Moscow that a call to Honolulu can be made during Honolulu office hours?

Closure   The whole numbers are said to be *closed* with respect to addition because when two whole numbers are added, the result is a whole number. The whole numbers are not closed with respect to subtraction because, for example, 4 and 7 are whole numbers but $4 - 7 = -3$, and -3 is not a whole number. Complete the table below by entering a Y if the operation is closed for those numbers and an N if it is not closed. When we discuss whether multiplication and division are closed, zero is not included because division by zero is not defined.

	Add	*Subtract*	*Multiply*	*Divide*
Whole numbers	Y	N		
Integers				
Rational numbers				

www.fedstats.gov

Information regarding the history of the federal budget can be found on the World Wide Web. Go to the web site **www.fedstats.gov**. Find the Frequently Requested Tables. You will see a table entitled "Federal Budget—Summary." When the federal budget table appears on the screen, look for the column that lists each year's surplus or deficit. You will see that a negative sign (−) is used to show a deficit. Note that the figures are in millions of dollars.

1. From 1930 through 2000, during which years was there a surplus?

2. After 1932, what was the first year during which the deficit was less than −$1 billion?

3. In the 1980s, during which years was the deficit reduced from the previous year?

4. Find the lowest number in the column. During which year did this deficit occur?

5. Find the difference between the surplus or deficit this year and the deficit noted in Exercise 4.

6. Use the figures in the table to write three word problems. Ask a classmate to solve them.

Chapter Summary

Key Words

Positive numbers are numbers greater than zero. [p. 395]

Negative numbers are numbers less than zero. [p. 395]

The integers are ..., $-4, -3, -2, -1, 0, 1, 2, 3, 4, \ldots$. [p. 395]

Negative integers are integers to the left of zero on the number line. [p. 395]

Positive integers are integers to the right of zero on the number line. [p. 395]

Two numbers that are the same distance from zero on the number line but on opposite sides of zero are called *opposites*. [p. 396]

The *absolute value* of a number is its distance from zero on a number line. [p. 396]

A *rational number* is a number that can be written as the quotient of two integers. [p. 415]

In *scientific notation*, a number is expressed in the form $a \times 10^n$, where a is a number between 1 and 10 and n is an integer. [p. 425]

Essential Rules

Addition of Numbers with the Same Sign
To add numbers with the same sign, add the absolute values of the numbers. Then attach the sign of the addends. [p. 400]

Addition of Numbers with Different Signs
To add numbers with different signs, find the difference between the absolute values of the numbers. Then attach the sign of the number with the greater absolute value. [p. 400]

Subtraction of Numbers
To subtract one number from another, add the opposite of the second number to the first number. [p. 401]

Multiplication of Numbers with the Same Sign
To multiply numbers with the same sign, multiply the absolute values of the numbers. The product is positive. [p. 407]

Multiplication of Numbers with Different Signs
To multiply numbers with different signs, multiply the absolute values of the numbers. The product is negative. [p. 407]

Division of Numbers with the Same Sign
The quotient of two numbers with the same sign is positive. [p. 408]

Division of Numbers with Different Signs
The quotient of two numbers with different signs is negative. [p. 408]

Order of Operations Agreement

Step 1   Perform operations inside parentheses.

Step 2   Simplify exponential expressions.

Step 3   Do multiplication and division as they occur from left to right.

Step 4   Rewrite subtraction as the addition of the opposite. Then do additions as they occur from left to right. [p. 427]

Chapter Review

1. Find the opposite of 22.

2. Subtract: $-8 - (-2) - (-10) - 3$

3. Subtract: $\dfrac{5}{8} - \dfrac{5}{6}$

4. Simplify: $-0.33 + 1.98 - 1.44$

5. Multiply: $\left(-\dfrac{2}{3}\right)\left(\dfrac{6}{11}\right)\left(-\dfrac{22}{25}\right)$

6. Multiply: -0.08×16

7. Simplify: $12 - 6 \div 3$

8. Simplify: $\left(\dfrac{2}{3}\right)^2 - \dfrac{5}{6}$

9. Find the opposite of -4.

10. Place the correct symbol, $<$ or $>$, between the two numbers.
$0 \quad -3$

11. Evaluate $-|-6|$.

12. Divide: $-18 \div (-3)$

13. Add: $-\dfrac{3}{8} + \dfrac{5}{12} + \dfrac{2}{3}$

14. Multiply: $\dfrac{1}{3} \times \left(-\dfrac{3}{4}\right)$

15. Divide: $-\dfrac{7}{12} \div \left(-\dfrac{14}{39}\right)$

16. Simplify: $16 \div 4(8 - 2)$

17. Add: $-22 + 14 + (-18)$

18. Simplify: $3^2 - 9 + 2$

19. Write 0.0000397 in scientific notation.

20. Divide: $-1.464 \div 18.3$

21. Simplify: $-\dfrac{5}{12} + \dfrac{7}{9} - \dfrac{1}{3}$

22. Multiply: $\dfrac{6}{34} \times \dfrac{17}{40}$

23. Multiply: $1.2 \times (-0.035)$

24. Simplify: $-\dfrac{1}{2} + \dfrac{3}{8} \div \dfrac{9}{20}$

25. Evaluate $|-5|$.

26. Place the correct symbol, $<$ or $>$, between the two numbers.
$$2 \quad -40$$

27. Find 2 times -13.

28. Simplify: $-0.4 \times 5 - (-3.33)$

29. Add: $\dfrac{5}{12} + \left(-\dfrac{2}{3}\right)$

30. Simplify: $-33.4 + 9.8 - (-16.2)$

31. Divide: $\left(-\dfrac{3}{8}\right) \div \left(-\dfrac{4}{5}\right)$

32. Write 2.4×10^5 in decimal notation.

33. Find the temperature after a rise of $18°$ from $-22°$.

34. To discourage guessing on a multiple-choice exam, an instructor graded the test by giving 3 points for a correct answer, -1 point for an answer left blank, and -2 points for an incorrect answer. How many points did a student score who answered 38 questions correctly, answered 4 questions incorrectly, and left 8 questions blank?

35. The boiling point of mercury is 356.58°C. The melting point of mercury is -38.87°C. Find the difference between the boiling point and the melting point of mercury.

Chapter Test

1. Subtract: $-5 - (-8)$

2. Evaluate $-|-2|$.

3. Add: $-\frac{2}{5} + \frac{7}{15}$

4. Find the product of 0.032 and -1.9.

5. Place the correct symbol, $<$ or $>$, between the two numbers.
$-8 \quad -10$

6. Add: $1.22 + (-3.1)$

7. Simplify: $4 \times (4 - 7) \div (-2) - 4 \times 8$

8. Multiply: $-5 \times (-6) \times 3$

9. What is -1.004 decreased by 3.01?

10. Divide: $-72 \div 8$

11. Find the sum of -2, 3, and -8.

12. Add: $-\frac{3}{8} + \frac{2}{3}$

13. Write 87,600,000,000 in scientific notation.

14. Find the product of -4 and 12.

15. Divide: $\frac{0}{-17}$

16. Subtract: $16 - 4 - (-5) - 7$

17. Find the quotient of $-\frac{2}{3}$ and $\frac{5}{6}$.

18. Place the correct symbol, $<$ or $>$, between the two numbers.

        0   -4

19. Add: $16 + (-10) + (-20)$

20. Simplify: $(-2)^2 - (-3)^2 \div (1 - 4)^2 \times 2 - 6$

21. Subtract: $-\frac{2}{5} - \left(-\frac{7}{10}\right)$

22. Write 9.601×10^{-8} in decimal notation.

23. Divide: $-15.64 \div (-4.6)$

24. Find the sum of $-\frac{1}{2}, \frac{1}{3}$, and $\frac{1}{4}$.

25. Multiply: $\frac{3}{8} \times \left(-\frac{5}{6}\right) \times \left(-\frac{4}{15}\right)$

26. Subtract: $2.113 - (-1.1)$

27. Find the temperature after a rise of 11°C from -4°C.

28. The melting point of radon is -71°C. The melting point of oxygen is three times the melting point of radon. Find the melting point of oxygen.

29. On December 24, 1924, in Fairfield, Montana, the temperature fell from 17.22°C at noon to -29.4°C at midnight. How many degrees did the temperature fall in the 12-hour period?

30. The daily low temperature readings for a 3-day period were as follows: $-7°$, $9°$, $-8°$. Find the average low temperature for the 3-day period.

Cumulative Review

1. Simplify: $16 - 4 \cdot (3 - 2)^2 \cdot 4$

2. Find the difference between $8\frac{1}{2}$ and $3\frac{4}{7}$.

3. Divide: $3\frac{7}{8} \div 1\frac{1}{2}$

4. Simplify: $\frac{3}{8} \div \left(\frac{3}{8} - \frac{1}{4}\right) \div \frac{7}{3}$

5. Subtract: $2.907 - 1.09761$

6. Solve the proportion $\frac{7}{12} = \frac{n}{32}$.
    Round to the nearest hundredth.

7. 22 is 160% of what number?

8. Convert: 7 qt = _____ gal _____ qt

9. Convert: 6692 ml = _____ L

10. Convert 4.2 ft to meters. Round to the nearest hundredth. (1 m = 3.28 ft.)

11. Find 32% of 180.

12. Convert $3\frac{2}{5}$ to a percent.

13. Add: $-8 + 5$

14. Add: $3\frac{1}{4} + \left(-6\frac{5}{8}\right)$

15. Subtract: $-6\frac{1}{8} - 4\frac{5}{12}$

16. Simplify: $-12 - (-7) - 3(-8)$

17. What is -3.2 times -1.09?

18. Multiply: $-6 \times 7 \times \left(-\frac{3}{4}\right)$

19. Find the quotient of 42 and −6.

20. Divide: $-2\frac{1}{7} \div \left(-3\frac{3}{5}\right)$

21. Simplify: $3 \times (3 - 7) \div 6 - 2$

22. Simplify: $4 - (-2)^2 \div (1 - 2)^2 \times 3 + 4$

23. A board $5\frac{2}{3}$ ft long is cut from a board 8 ft long. What is the length of the board remaining?

24. Nimisha had a balance of $763.56 in her checkbook before writing checks for $135.88 and $47.81 and making a deposit of $223.44. Find her new checkbook balance.

25. A suit that regularly sells for $165 is on sale for $120. Find the percent decrease in price. Round to the nearest tenth of a percent.

26. A reception is planned for 80 guests. How many gallons of coffee should be prepared to provide 2 c of coffee for each guest?

27. A stock selling for $82\frac{5}{8}$ per share paid a dividend of $1.50 per share before the dividend was increased by 12%. Find the dividend per share after the increase.

28. The hourly wages for five job classifications at a company are $10.40, $8.32, $14.25, $9.73, and $12.10. Find the median hourly pay.

29. A pre-election survey showed that 5 out of every 8 registered voters would cast ballots in a city election. At this rate, how many people would vote in a city of 960,000 registered voters?

30. The daily high temperature readings for a 4-day period were recorded as follows: −19°, −7°, 1°, and 9°. Find the average high temperature for the 4-day period.

Chapter 11

Introduction to Algebra

Objectives

Section 11.1

A To evaluate variable expressions
B To simplify variable expressions containing no parentheses
C To simplify variable expressions containing parentheses

Section 11.2

A To determine whether a given value is a solution of an equation
B To solve an equation of the form $x + a = b$
C To solve an equation of the form $ax = b$
D To solve application problems

Section 11.3

A To solve an equation of the form $ax + b = c$
B To solve application problems

Section 11.4

A To solve an equation of the form $ax + b = cx + d$
B To solve an equation containing parentheses

Section 11.5

A To translate a verbal expression into a mathematical expression given the variable
B To translate a verbal expression into a mathematical expression by assigning the variable

Section 11.6

A To translate a sentence into an equation and solve
B To solve application problems

This chapter introduces the concept of variables. You are already familiar with variables if you have ever used a formula. In the **Example on page 460,** a formula is used to determine the sale price given the regular price and the discount rate. Formulas are used in **Exercises 102 to 106 on page 464** to calculate selling price given either the cost and the markup or the cost and the markup rate.

Need help? For on-line student resources, such as section quizzes, visit this textbook's web site at **math.college.hmco.com/students.**

Prep Test

1. $2 - 9$

2. $-5(4)$

3. $-16 + 16$

4. $\dfrac{-7}{-7}$

5. $-\dfrac{3}{8}\left(-\dfrac{8}{3}\right)$

6. $\left(\dfrac{3}{5}\right)^3 \cdot \left(\dfrac{5}{9}\right)^2$

7. $\dfrac{2}{3} + \left(\dfrac{3}{4}\right)^2 \cdot \dfrac{2}{9}$

8. $-8 \div (-2)^2 + 6$

9. $4 + 5(2 - 7)^2 \div (-8 + 3)$

Go Figure

In the addition at the right, each letter stands for a different digit. If N = 1, I = 5, U = 7, F = 2, and T = 3, what is the value of S?

$$\begin{array}{r} FUN \\ IN \\ + THE \\ \hline SUN \end{array}$$

11.1 Variable Expressions

Objective A

To evaluate variable expressions

Often we discuss a quantity without knowing its exact value—for example, next year's inflation rate, the price of gasoline next summer, or the interest rate on a new-car loan next fall. In mathematics, a letter of the alphabet is used to stand for a quantity that is unknown or that can change, or *vary*. The letter is called a **variable.** An expression that contains one or more variables is called a **variable expression.**

A company's business manager has determined that the company will make a $2 profit on each radio it sells. The manager wants to describe the company's total profit from the sale of radios. Because the number of radios that the company will sell is unknown, the manager lets the variable n stand for that number. Then the variable expression $2 \cdot n$, or simply $2n$, describes the company's profit from selling n radios.

The company's profit from selling n radios is $\$2 \cdot n = \$2n$.

If the company sells 12 radios, its profit is $\$2 \cdot 12 = \$24$.

If the company sells 75 radios, its profit is $\$2 \cdot 75 = \$150$.

Replacing the variable or variables in a variable expression and then simplifying the resulting numerical expression is called **evaluating the variable expression.**

⇒ Evaluate $3x^2 + xy - z$ when $x = -2$, $y = 3$, and $z = -4$.

$3x^2 + xy - z$

$3(-2)^2 + (-2)(3) - (-4)$ • **Replace each variable in the expression with the number it stands for.**

$= 3 \cdot 4 + (-2)(3) - (-4)$ • **Use the Order of Operations Agreement to simplify the resulting numerical expression.**

$= 12 + (-6) - (-4)$

$= 12 + (-6) + 4$

$= 6 + 4$

$= 10$

The value of the variable expression $3x^2 + xy - z$ when $x = -2$, $y = 3$, and $z = -4$ is 10.

Example 1

Evaluate $3x - 4y$ when $x = -2$ and $y = 3$.

Solution

$3x - 4y$

$3(-2) - 4(3) = -6 - 12$
$\qquad\qquad\quad = -6 + (-12) = -18$

You Try It 1

Evaluate $6a - 5b$ when $a = -3$ and $b = 4$.

Your solution

Example 2

Evaluate $-x^2 - 6 \div y$ when $x = -3$ and $y = 2$.

Solution

$-x^2 - 6 \div y$

$-(-3)^2 - 6 \div 2 = -9 - 6 \div 2$
$\qquad\qquad\qquad\ = -9 - 3$
$\qquad\qquad\qquad\ = -9 + (-3) = -12$

You Try It 2

Evaluate $-3s^2 - 12 \div t$ when $s = -2$ and $t = 4$.

Your solution

Example 3

Evaluate $-\dfrac{1}{2}y^2 - \dfrac{3}{4}z$ when $y = 2$ and $z = -4$.

Solution

$-\dfrac{1}{2}y^2 - \dfrac{3}{4}z$

$-\dfrac{1}{2}(2)^2 - \dfrac{3}{4}(-4) = -\dfrac{1}{2} \cdot 4 - \dfrac{3}{4}(-4)$
$\qquad\qquad\qquad\qquad = -2 - (-3)$
$\qquad\qquad\qquad\qquad = -2 + (3) = 1$

You Try It 3

Evaluate $-\dfrac{2}{3}m + \dfrac{3}{4}n^3$ when $m = 6$ and $n = 2$.

Your solution

Example 4

Evaluate $-2ab + b^2 + a^2$ when $a = -\dfrac{3}{5}$ and $b = \dfrac{2}{5}$.

Solution

$-2ab + b^2 + a^2$

$-2\left(-\dfrac{3}{5}\right)\left(\dfrac{2}{5}\right) + \left(\dfrac{2}{5}\right)^2 + \left(-\dfrac{3}{5}\right)^2$

$= -2\left(-\dfrac{3}{5}\right)\left(\dfrac{2}{5}\right) + \left(\dfrac{4}{25}\right) + \left(\dfrac{9}{25}\right)$

$= \dfrac{12}{25} + \dfrac{4}{25} + \dfrac{9}{25} = \dfrac{25}{25} = 1$

You Try It 4

Evaluate $-3yz - z^2 + y^2$ when $y = -\dfrac{2}{3}$ and $z = \dfrac{1}{3}$.

Your solution

Solutions on p. S27

Objective B   **To simplify variable expressions containing no parentheses**

The **terms** of a variable expression are the addends of the expression. The variable expression at the right has four terms.

$$\overbrace{7x^2 \ + \ \underbrace{(-6xy) \ + \ x}_{\text{Variable terms}} \ + \ \underbrace{(-8)}_{\substack{\text{Constant} \\ \text{term}}}}^{\text{4 terms}}$$

Three of the terms are **variable terms:** $7x^2$, $-6xy$, and x.

One of the terms is a **constant term:** -8. A constant term has no variables.

Each variable term is composed of a **numerical coefficient** and a **variable part** (the variable or variables and their exponents). When the numerical coefficient is 1, the 1 is usually not written. ($1x = x$)

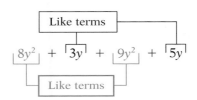

Like terms of a variable expression are the terms with the same variable part. (Because $y^2 = y \cdot y$, y^2 and y are not like terms.)

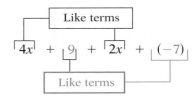

In variable expressions that contain constant terms, the constant terms are like terms.

The Commutative and Associative Properties of Addition are used to simplify variable expressions. These properties can be stated in general form using variables.

TAKE NOTE

With the Commutative Property, the order in which the numbers appear changes. With the Associative Property, the order in which the numbers appear remains the same.

Commutative Property of Addition

If a and b are two numbers, then $a + b = b + a$.

Associative Property of Addition

If a, b, and c are three numbers, then $a + (b + c) = (a + b) + c$.

To **simplify a variable expression,** *combine* like terms by adding their numerical coefficients. For example, to simplify $2y + 3y$, think

$$2y + 3y = (y + y) + (y + y + y) = 5y$$

➡ Simplify: $8z - 5 + 2z$

$8z - 5 + 2z = 8z + 2z - 5$
$\qquad\qquad\quad = 10z - 5$

- **Use the Commutative and Associative Properties of Addition to group like terms. Combine the like terms $8z + 2z$.**

➡ Simplify: $12a - 4b - 8a + 2b$

$12a - 4b - 8a + 2b = 12a + (-4)b + (-8)a + 2b$ • **Change subtraction to addition of the opposite.**

$= 12a + (-8)a + (-4)b + 2b$ • **Use the Commutative and Associative Properties of Addition to group like terms. Combine like terms.**
$= 4a + (-2)b$

$= 4a - 2b$ • **Recall that $a + (-b) = a - b$.**

➡ Simplify: $6z^2 + 3 - z^2 - 7$

$6z^2 + 3 - z^2 - 7 = 6z^2 + 3 + (-1)z^2 + (-7)$ • **Change subtraction to addition of the opposite.**

$= 6z^2 + (-1)z^2 + 3 + (-7)$ • **Use the Commutative and Associative Properties of Addition to group like terms. Combine like terms.**
$= 5z^2 + (-4)$
$= 5z^2 - 4$

Example 5

Simplify: $6xy - 8x + 5x - 9xy$

Solution

$6xy - 8x + 5x - 9xy$
$= 6xy + (-8)x + 5x + (-9)xy$
$= 6xy + (-9)xy + (-8)x + 5x$
$= -3xy + (-3)x$
$= -3xy - 3x$

You Try It 5

Simplify: $5a^2 - 6b^2 + 7a^2 - 9b^2$

Your solution

Example 6

Simplify: $-4z^2 + 8 + 5z^2 - 3$

Solution

$-4z^2 + 8 + 5z^2 - 3$
$= -4z^2 + 8 + 5z^2 + (-3)$
$= -4z^2 + 5z^2 + 8 + (-3)$
$= z^2 + 5$

You Try It 6

Simplify: $-6x + 7 + 9x - 10$

Your solution

Example 7

Simplify: $\frac{1}{4}m^2 - \frac{1}{2}n^2 + \frac{1}{2}m^2$

Solution

$\frac{1}{4}m^2 - \frac{1}{2}n^2 + \frac{1}{2}m^2 = \frac{1}{4}m^2 + \left(-\frac{1}{2}\right)n^2 + \frac{1}{2}m^2$

$= \frac{1}{4}m^2 + \frac{1}{2}m^2 + \left(-\frac{1}{2}\right)n^2$

$= \frac{1}{4}m^2 + \frac{2}{4}m^2 + \left(-\frac{1}{2}\right)n^2$

$= \frac{3}{4}m^2 + \left(-\frac{1}{2}\right)n^2$

$= \frac{3}{4}m^2 - \frac{1}{2}n^2$

You Try It 7

Simplify: $\frac{3}{8}w + \frac{1}{2} - \frac{1}{4}w - \frac{2}{3}$

Your solution

Solutions on p. S27

Objective C   **To simplify variable expressions containing parentheses**   〈 11 〉

The Commutative and Associative Properties of Multiplication and the Distributive Property are used to simplify variable expressions that contain parentheses. These properties can be stated in general form using variables.

> **Commutative Property of Multiplication**
>
> If a and b are two numbers, then $a \cdot b = b \cdot a$.

> **Associative Property of Multiplication**
>
> If a, b, and c are three numbers, then $a \cdot (b \cdot c) = (a \cdot b) \cdot c$.

The Associative and Commutative Properties of Multiplication are used to simplify variable expressions such as the following.

➡ Simplify: $-5(4x)$

$$-5(4x) = (-5 \cdot 4)x$$ • Use the Associative Property of Multiplication.
$$= -20x$$

➡ Simplify: $(6y) \cdot 5$

$$(6y) \cdot 5 = 5 \cdot (6y)$$ • Use the Commutative Property of Multiplication.
$$= (5 \cdot 6)y = 30y$$ • Use the Associative Property of Multiplication.

The Distributive Property is used to remove parentheses from variable expressions that contain both multiplication and addition.

> **Distributive Property**
>
> If a, b, and c are three numbers, then $a(b + c) = ab + ac$.

➡ Simplify: $4(z + 5)$

$$4(z + 5) = 4z + 4(5)$$ • The Distributive Property is used to rewrite the
$$= 4z + 20$$ variable expression without parentheses.

➡ Simplify: $-3(2x + 7)$

$$-3(2x + 7) = -3(2x) + (-3)(7)$$ • Use the Distributive Property.
$$= -6x + (-21)$$
$$= -6x - 21$$ • Recall that $a + (-b) = a - b$.

The Distributive Property can also be stated in terms of subtraction.

$$a(b - c) = ab - ac$$

➡ Simplify: $8(2r - 3s)$

$$8(2r - 3s) = 8(2r) - 8(3s)$$
$$= 16r - 24s$$

• Use the Distributive Property.

➡ Simplify: $-5(2x - 4y)$

$$-5(2x - 4y) = (-5)(2x) - (-5)(4y)$$
$$= -10x - (-20y)$$
$$= -10x + 20y$$

• Use the Distributive Property.

• Recall that $a - (-b) = a + b$.

➡ Simplify: $12 - 5(m + 2) + 2m$

$$12 - 5(m + 2) + 2m = 12 - 5m + (-5)(2) + 2m$$
$$= 12 - 5m + (-10) + 2m$$

• Use the Distributive Property to simplify the expression $-5(m + 2)$.

$$= -5m + 2m + 12 + (-10)$$

• Use the Commutative and Associative Properties to group like terms.

$$= -3m + 2$$

• Combine like terms by adding their numerical coefficients. Add constant terms.

The answer $-3m + 2$ can also be written as $2 - 3m$. In this text, we will write answers with variable terms first, followed by the constant term.

Example 8
Simplify: $4(x - 3)$

Solution
$$4(x - 3) = 4x - 4(3)$$
$$= 4x - 12$$

You Try It 8
Simplify: $5(a - 2)$

Your solution

Example 9
Simplify: $5n - 3(2n - 4)$

Solution
$$5n - 3(2n - 4) = 5n - 3(2n) - (-3)(4)$$
$$= 5n - 6n - (-12)$$
$$= 5n - 6n + 12$$
$$= -n + 12$$

You Try It 9
Simplify: $8s - 2(3s - 5)$

Your solution

Example 10
Simplify: $3(c - 2) + 2(c + 6)$

Solution
$$3(c - 2) + 2(c + 6) = 3c - 3(2) + 2c + 2(6)$$
$$= 3c - 6 + 2c + 12$$
$$= 3c + 2c - 6 + 12$$
$$= 5c + 6$$

You Try It 10
Simplify: $4(x - 3) - 2(x + 1)$

Your solution

Solutions on p. S27

11.1 Exercises

Objective A

Evaluate the variable expression when $a = -3$, $b = 6$, and $c = -2$.

1. $5a - 3b$ **2.** $4c - 2b$ **3.** $2a + 3c$ **4.** $2c + 4a$

5. $-c^2$ **6.** $-a^2$ **7.** $b - a^2$ **8.** $b - c^2$

9. $ab - c^2$ **10.** $bc - a^2$ **11.** $2ab - c^2$ **12.** $3bc - a^2$

13. $a - (b \div a)$ **14.** $c - (b \div c)$ **15.** $2ac - (b \div a)$ **16.** $4ac \div (b \div a)$

17. $b^2 - c^2$ **18.** $b^2 - a^2$ **19.** $b^2 \div (ac)$ **20.** $3c^2 \div (ab)$

21. $c^2 - (b \div c)$ **22.** $a^2 - (b \div a)$ **23.** $a^2 + b^2 + c^2$ **24.** $a^2 - b^2 - c^2$

25. $ac + bc + ab$ **26.** $ac - bc - ab$ **27.** $a^2 + b^2 - ab$ **28.** $b^2 + c^2 - bc$

29. $2b - (3c + a^2)$ **30.** $\dfrac{2}{3}b + \left(\dfrac{1}{2}c - a \right)$ **31.** $\dfrac{1}{3}a + \left(\dfrac{1}{2}b - \dfrac{2}{3}a \right)$

32. $-\dfrac{2}{3}b - \left(\dfrac{1}{2}c + a \right)$ **33.** $\dfrac{1}{6}b + \dfrac{1}{3}(c + a)$ **34.** $\dfrac{1}{2}c + \left(\dfrac{1}{3}b - a \right)$

Evaluate the variable expression when $a = -\dfrac{1}{2}$, $b = \dfrac{3}{4}$, and $c = \dfrac{1}{4}$.

35. $4a + (3b - c)$ **36.** $2b + (c - 3a)$ **37.** $2a - b^2 \div c$ **38.** $b \div (-c) + 2a$

 Evaluate the variable expression when $a = 3.72$, $b = -2.31$, and $c = -1.74$.

39. $a^2 - b^2$ **40.** $a^2 - b \cdot c$ **41.** $3ac - (c \div a)$ **42.** $2c + (b^2 - c)$

Name the terms of the variable expression. Then underline the constant term.

43. $2x^2 + 3x - 4$ **44.** $-4y^2 + 5$ **45.** $3a^2 - 4a + 8$ **46.** $7 - b$

Name the variable terms of the expression. Then underline the coefficients of the variable terms.

47. $3x^2 - 4x + 9$ **48.** $-5a^2 + a - 4$ **49.** $y^2 + 6a - 1$ **50.** $8 - c$

Simplify.

51. $7z + 9z$ **52.** $6x + 5x$ **53.** $12m - 3m$

54. $5y - 12y$ **55.** $5at + 7at$ **56.** $12mn + 11mn$

57. $-4yt + 7yt$ **58.** $-12yt + 5yt$ **59.** $-3x - 12y$

60. $-12y - 7y$ **61.** $3t^2 - 5t^2$ **62.** $7t^2 + 8t^2$

63. $6c - 5 + 7c$ **64.** $7x - 5 + 3x$ **65.** $2t + 3t - 7t$

66. $9x^2 - 5 - 3x^2$ **67.** $7y^2 - 2 - 4y^2$ **68.** $3w - 7u + 4w$

69. $6w - 8u + 8w$ **70.** $4 - 6xy - 7xy$ **71.** $10 - 11xy - 12xy$

72. $7t^2 - 5t^2 - 4t^2$ **73.** $3v^2 - 6v^2 - 8v^2$ **74.** $5ab - 7a - 10ab$

75. $-10ab - 3a + 2ab$ **76.** $-4x^2 - x + 2x^2$ **77.** $-3y^2 - y + 7y^2$

78. $4x^2 - 8y - x^2 + y$ **79.** $2a - 3b^2 - 5a + b^2$ **80.** $8y - 4z - y + 2z$

81. $3x^2 - 7x + 4x^2 - x$ **82.** $5y^2 - y + 6y^2 - 5y$ **83.** $6s - t - 9s + 7t$

84. $5w - 2v - 9w + 5v$ **85.** $4m + 8n - 7m + 2n$ **86.** $z + 9y - 4z + 3y$

87. $-5ab + 7ac + 10ab - 3ac$

88. $-2x^2 - 3x - 11x^2 + 14x$

89. $\dfrac{4}{9}a^2 - \dfrac{1}{5}b^2 + \dfrac{2}{9}a^2 + \dfrac{4}{5}b^2$

90. $\dfrac{6}{7}x^2 + \dfrac{2}{5}x - \dfrac{3}{7}x^2 - \dfrac{4}{5}x$

91. $4.235x - 0.297x + 3.056x$

92. $8.092y - 3.0793y + 0.063y$

93. $7.81m + 3.42n - 6.25m - 7.19n$

94. $8.34y^2 - 4.21y - 6.07y^2 - 5.39y$

Objective C

Simplify.

95. $5(x + 4)$

96. $3(m + 6)$

97. $(y - 3)4$

98. $(z - 3)7$

99. $-2(a + 4)$

100. $-5(b + 3)$

101. $3(5x + 10)$

102. $2(4m - 7)$

103. $5(3c - 5)$

104. $-4(w - 3)$

105. $-3(y - 6)$

106. $3m + 4(m + z)$

107. $5x + 2(x + 7)$

108. $6z - 3(z + 4)$

109. $8y - 4(y + 2)$

110. $7w - 2(w - 3)$

111. $9x - 4(x - 6)$

112. $-5m + 3(m + 4)$

113. $-2y + 3(y - 2)$

114. $5m + 3(m + 4) - 6$

115. $4n + 2(n + 1) - 5$

116. $8z - 2(z - 3) + 8$

117. $9y - 3(y - 4) + 8$

118. $6 - 4(a + 4) + 6a$

119. $3x + 2(x + 2) + 5x$

120. $7x + 4(x + 1) + 3x$

121. $-7t + 2(t - 3) - t$

122. $-3y + 2(y - 4) - y$ **123.** $z - 2(1 - z) - 2z$ **124.** $2y - 3(2 - y) + 4y$

125. $3(y - 2) - 2(y - 6)$ **126.** $7(x + 2) + 3(x - 4)$ **127.** $2(t - 3) + 7(t + 3)$

128. $3(y - 4) - 2(y - 3)$ **129.** $3t - 6(t - 4) + 8t$ **130.** $5x + 3(x - 7) - 9x$

APPLYING THE CONCEPTS

131. The square and the rectangle at the right can be used to illustrate algebraic expressions. The illustration below represents the expression $2x + 1$.

a. Using similar squares and rectangles, draw figures that represent the expressions $3 + 2x$, $4x + 6$, $3x + 2$, and $2x + 4$.

b. The illustration below represents the expression $3(x + 1)$. Rearrange these rectangles so that the x's are together and the 1's are together. Write a mathematical expression for the rearranged figure.

| x | 1 | x | 1 | x | 1 |

132. Using squares and rectangles similar to those in Exercise 131, draw figures that represent the expressions $2 + 3x$, $2(2x + 3)$, $4x + 3$, and $4x + 6$.
a. Does the figure $2(2x + 3)$ equal the figure $4x + 6$? Explain how this is related to the Distributive Property.
b. Does the figure $2 + 3x$ equal the figure $5x$? How is this related to combining like terms?

133. **a.** Simplifying variable expressions requires combining like terms. Give some examples of how this applies to everyday experience.
b. It was stated in this section that the variable terms y^2 and y are not like terms. Use measurements of area and distance to show that these terms would not be combined as measurements.

134. Explain why the simplification of the expression $2 + 3(2x + 4)$ shown at the right is incorrect. What is the correct simplification?

Why is this incorrect?
$$2 + 3(2x + 4) = 5(2x + 4)$$
$$= 10x + 20$$

11.2 Introduction to Equations

Objective A   **To determine whether a given value is a solution of an equation**

An **equation** expresses the equality of two mathematical expressions. These expressions can be either numerical or variable expressions.

$$5 + 4 = 9$$
$$3x + 13 = x - 8$$
$$y^2 + 4 = 6y + 1$$
$$x = -3$$

Equations

In the equation at the right, if the variable is replaced by 4, the equation is true.

$$x + 3 = 7$$
$$4 + 3 = 7 \quad \text{A true equation}$$

If the variable is replaced by 6, the equation is false.

$$6 + 3 = 7 \quad \text{A false equation}$$

A **solution of an equation** is a value of the variable that results in a true equation. 4 is a solution of the equation $x + 3 = 7$. 6 is not a solution of the equation $x + 3 = 7$.

➡ Is -2 a solution of the equation $x^2 + 1 = 2x + 9$?

$$\begin{array}{c|c} \multicolumn{2}{c}{x^2 + 1 = 2x + 9} \\ \hline (-2)^2 + 1 & 2(-2) + 9 \\ 4 + 1 & -4 + 9 \\ \multicolumn{2}{c}{5 = 5} \end{array}$$

- Replace the variable by the given value.
- Evaluate the numerical expressions.
- Compare the results. If the results are equal, the given value is a solution. If the results are not equal, the given value is not a solution.

Yes, -2 is a solution of the equation $x^2 + 1 = 2x + 9$.

Example 1

Is $\frac{1}{2}$ a solution of

$2x(x + 2) = 3x + 1$?

Solution

$$\begin{array}{c|c} \multicolumn{2}{c}{2x(x + 2) = 3x + 1} \\ \hline 2\left(\dfrac{1}{2}\right)\left(\dfrac{1}{2} + 2\right) & 3\left(\dfrac{1}{2}\right) + 1 \\ 2\left(\dfrac{1}{2}\right)\left(\dfrac{5}{2}\right) & 3\left(\dfrac{1}{2}\right) + 1 \\ \multicolumn{2}{c}{\dfrac{5}{2} = \dfrac{5}{2}} \end{array}$$

Yes, $\frac{1}{2}$ is a solution.

You Try It 1 Is -2 a solution of $x(x + 3) = 4x + 6$?

Your solution

Solution on p. S27

Example 2 Is 5 a solution of
$(x - 2)^2 = x^2 - 4x + 2?$

Solution

$$\begin{array}{c|l}
\multicolumn{2}{c}{(x - 2)^2 = x^2 - 4x + 2} \\
\hline
(5 - 2)^2 & 5^2 - 4(5) + 2 \\
3^2 & 25 - 4(5) + 2 \\
9 & 25 - 20 + 2 \\
& 25 + (-20) + 2 \\
9 \neq 7 & (\neq \text{ means "is not} \\
& \text{equal to")}
\end{array}$$

No, 5 is not a solution.

You Try It 2 Is -3 a solution of
$x^2 - x = 3x + 7?$

Your solution

Solution on p. S27

Objective B **To solve an equation of the form $x + a = b$**

A solution of an equation is a value of the variable that, when substituted in the equation, results in a true equation. To **solve an equation** means to find a solution of the equation.

The simplest equation to solve is an equation of the form *variable = constant*. The constant is the solution of the equation.

If $x = 7$, then 7 is the solution of the equation because $7 = 7$ is a true equation.

In solving an equation of the form $x + a = b$, the goal is to simplify the given equation to one of the form *variable = constant*. The Addition Properties that follow are used to simplify equations to this form.

Addition Property of Zero

The sum of a term and zero is the term.
$$a + 0 = a \qquad 0 + a = a$$

Addition Property of Equations

If a, b, and c are algebraic expressions, then the equations $a = b$ and $a + c = b + c$ have the same solutions.

The Addition Property of Equations states that the same quantity can be added to each side of an equation without changing the solution of the equation.

The Addition Property of Equations is used to rewrite an equation in the form *variable = constant*. Remove a term from one side of the equation by adding the opposite of that term to each side of the equation.

➡ Solve: $x - 7 = -2$

$$x - 7 = -2$$

$$x - 7 + 7 = -2 + 7$$
$$x + 0 = 5$$
$$x = 5$$

- The goal is to simplify the equation to one of the form *variable = constant*.

- Add the opposite of the constant term -7 to each side of the equation. After we simplify and use the Addition Property of Zero, the equation will be in the form *variable = constant*.

The solution is 5.

Because subtraction is defined in terms of addition, the Addition Property of Equations allows the same number to be subtracted from each side of an equation.

➡ Solve: $x + 8 = 5$

$$x + 8 = 5$$

$$x + 8 - 8 = 5 - 8$$
$$x + 0 = -3$$
$$x = -3$$

- The goal is to simplify the equation to one of the form *variable = constant*.

- Add the opposite of the constant term 8 to each side of the equation. This procedure is equivalent to subtracting 8 from each side of the equation.

The solution is -3. You should check this solution.

Example 3   Solve: $4 + m = -2$

Solution
$$4 + m = -2$$
$$4 - 4 + m = -2 - 4$$
$$0 + m = -6$$
$$m = -6$$
The solution is -6.

You Try It 3   Solve: $-2 + y = -5$

Your solution

Example 4   Solve: $3 = y - 2$

Solution
$$3 = y - 2$$
$$3 + 2 = y - 2 + 2$$
$$5 = y + 0$$
$$5 = y$$
The solution is 5.

You Try It 4   Solve: $7 = y + 8$

Your solution

Example 5
Solve: $\dfrac{2}{7} = \dfrac{5}{7} + t$

Solution
$$\frac{2}{7} = \frac{5}{7} + t$$
$$\frac{2}{7} - \frac{5}{7} = \frac{5}{7} - \frac{5}{7} + t$$
$$-\frac{3}{7} = 0 + t$$
$$-\frac{3}{7} = t$$

The solution is $-\dfrac{3}{7}$.

You Try It 5
Solve: $\dfrac{1}{5} = z + \dfrac{4}{5}$

Your solution

Solutions on p. S27

Objective C   **To solve an equation of the form $ax = b$**

In solving an equation of the form $ax = b$, the goal is to simplify the given equation to one of the form *variable = constant*. The Multiplication Properties that follow are used to simplify equations to this form.

Multiplication Property of Reciprocals

The product of a nonzero term and its reciprocal equals 1.

Because $a = \frac{a}{1}$, the reciprocal of a is $\frac{1}{a}$. $a\left(\frac{1}{a}\right) = 1$ $\frac{1}{a}(a) = 1$

The reciprocal of $\frac{a}{b}$ is $\frac{b}{a}$. $\left(\frac{a}{b}\right)\left(\frac{b}{a}\right) = 1$ $\left(\frac{b}{a}\right)\left(\frac{a}{b}\right) = 1$

Multiplication Property of One

The product of a term and 1 is the term.
$$a \cdot 1 = a \qquad 1 \cdot a = a$$

Multiplication Property of Equations

If a, b, and c are algebraic expressions and $c \neq 0$, then the equation $a = b$ has the same solutions as the equation $ac = bc$.

The Multiplication Property of Equations states that each side of an equation can be multiplied by the same nonzero number without changing the solutions of the equation.

Recall that the goal of solving an equation is to rewrite the equation in the form *variable = constant*. The Multiplication Property of Equations is used to rewrite an equation in this form by multiplying each side of the equation by the reciprocal of the coefficient.

➡ Solve: $\frac{2}{3}x = 8$

$$\frac{2}{3}x = 8$$
$$\left(\frac{3}{2}\right)\left(\frac{2}{3}\right)x = \left(\frac{3}{2}\right)8$$ • Multiply each side of the equation by $\frac{3}{2}$, the
$$1 \cdot x = 12$$ reciprocal of $\frac{2}{3}$. After simplifying, the equation
$$x = 12$$ will be in the form *variable = constant*.

Check: $\frac{2}{3}x = 8$

$$\left(\frac{2}{3}\right)12 \ \Big|\ 8$$
$$8 = 8$$

The solution is 12.

Because division is defined in terms of multiplication, the Multiplication Property of Equations allows each side of an equation to be divided by the same nonzero quantity.

⇒ Solve: $-4x = 24$

$$-4x = 24$$

- The goal is to rewrite the equation in the form *variable = constant*.

$$\frac{-4x}{-4} = \frac{24}{-4}$$
$$1x = -6$$
$$x = -6$$

- Multiply each side of the equation by the reciprocal of -4. This is equivalent to dividing each side of the equation by -4. Then simplify.

The solution is -6. You should check this solution.

In using the Multiplication Property of Equations, it is usually easier to multiply each side of the equation by the reciprocal of the coefficient when the coefficient is a fraction. Divide each side of the equation by the coefficient when the coefficient is an integer or a decimal.

Example 6 Solve: $-2x = 6$

Solution
$$-2x = 6$$
$$\frac{-2x}{-2} = \frac{6}{-2}$$
$$1x = -3$$
$$x = -3$$

The solution is -3.

You Try It 6 Solve: $4z = -20$

Your solution

Example 7
Solve: $-9 = \frac{3}{4}y$

Solution
$$-9 = \frac{3}{4}y$$
$$\left(\frac{4}{3}\right)(-9) = \left(\frac{4}{3}\right)\left(\frac{3}{4}y\right)$$
$$-12 = 1y$$
$$-12 = y$$

The solution is -12.

You Try It 7
Solve: $8 = \frac{2}{5}n$

Your solution

Example 8 Solve: $6z - 8z = -5$

Solution
$$6z - 8z = -5 \quad \text{Combine like terms.}$$
$$-2z = -5$$
$$\frac{-2z}{-2} = \frac{-5}{-2}$$
$$1z = \frac{5}{2}$$
$$z = \frac{5}{2} = 2\frac{1}{2}$$

The solution is $2\frac{1}{2}$.

You Try It 8
Solve: $\frac{2}{3}t - \frac{1}{3}t = -2$

Your solution

Solutions on p. S28

Objective D   **To solve application problems** ‹ 11 ›

Example 9

An accountant for an auto parts store found that the weekly profit for the store was $1700 and that the total amount spent during the week was $2400. Use the formula $P = R - C$, where P is the profit, R is the revenue, and C is the amount spent, to find the revenue for the week.

Strategy

To find the revenue for the week, replace the variables P and C in the formula by the given values, and solve for R.

Solution

$$P = R - C$$
$$1700 = R - 2400$$
$$1700 + 2400 = R - 2400 + 2400$$
$$4100 = R + 0$$
$$4100 = R$$

The revenue for the week was $4100.

You Try It 9

A clothing store's sale price for a pair of slacks is $44. This is a discount of $16 off the regular price. Use the formula $S = R - D$, where S is the sale price, R is the regular price, and D is the discount, to find the regular price.

Your strategy

Your solution

Example 10

A store manager uses the formula $S = R - d \cdot R$, where S is the sale price, R is the regular price, and d is the discount rate. During a clearance sale, all items are discounted 20%. Find the regular price of a jacket that is on sale for $120.

Strategy

To find the regular price of the jacket, replace the variables S and d in the formula by the given values, and solve for R.

Solution

$$S = R - d \cdot R$$
$$120 = R - 0.20R$$
$$120 = 0.80R \qquad \bullet \; R - 0.20R = 1R - 0.20R$$
$$\frac{120}{0.80} = \frac{0.80R}{0.80}$$
$$150 = R$$

The regular price of the jacket is $150.

You Try It 10

Find the monthly payment when the total amount paid on a loan is $6840 and the loan is paid off in 24 months. Use the formula $A = MN$, where A is the total amount paid on a loan, M is the monthly payment, and N is the number of monthly payments.

Your strategy

Your solution

Solutions on p. S28

11.2 Exercises

Objective A

1. Is -3 a solution of
$2x + 9 = 3$?

2. Is -2 a solution of
$5x + 7 = 12$?

3. Is 2 a solution of
$4 - 2x = 8$?

4. Is 4 a solution of
$5 - 2x = 4x$?

5. Is 3 a solution of
$3x - 2 = x + 4$?

6. Is 2 a solution of
$4x + 8 = 4 - 2x$?

7. Is 3 a solution of
$x^2 - 5x + 1 = 10 - 5x$?

8. Is -5 a solution of
$x^2 - 3x - 1 = 9 - 6x$?

9. Is -1 a solution of
$2x(x - 1) = 3 - x$?

10. Is 2 a solution of
$3x(x - 3) = x - 8$?

11. Is 2 a solution of
$x(x - 2) = x^2 - 4$?

12. Is -4 a solution of
$x(x + 4) = x^2 + 16$?

13. Is $-\frac{2}{3}$ a solution of
$3x + 6 = 4$?

14. Is $\frac{1}{2}$ a solution of
$2x - 7 = -3$?

15. Is $\frac{1}{4}$ a solution of
$2x - 3 = 1 - 14x$?

16. Is $-\frac{1}{3}$ a solution of
$5x - 2 = 1 - 2x$?

17. Is $\frac{3}{4}$ a solution of
$3x(x - 2) = x - 4$?

18. Is $\frac{2}{5}$ a solution of
$5x(x + 1) = x + 3$?

19. Is 1.32 a solution of
$x^2 - 3x = -0.8776 - x$?

20. Is -1.9 a solution of
$x^2 - 3x = x + 3.8$?

21. Is 1.05 a solution of
$x^2 + 3x = x(x + 3)$?

Objective B

Solve.

22. $x + 3 = 9$

23. $x + 7 = 5$

24. $y - 6 = 16$

25. $z - 4 = 10$

26. $3 + n = 4$

27. $6 + x = 8$

28. $z + 7 = 2$

29. $w + 9 = 5$

30. $x - 3 = -7$ **31.** $m - 4 = -9$ **32.** $y + 6 = 6$ **33.** $t - 3 = -3$

34. $v - 7 = -4$ **35.** $x - 3 = -1$ **36.** $1 + x = 0$ **37.** $3 + y = 0$

38. $x - 10 = 5$ **39.** $y - 7 = 3$ **40.** $x + 4 = -7$ **41.** $t - 3 = -8$

42. $w + 5 = -5$ **43.** $z + 6 = -6$ **44.** $x + 7 = -8$ **45.** $x + 2 = -5$

46. $x + \dfrac{1}{2} = -\dfrac{1}{2}$ **47.** $x - \dfrac{5}{6} = -\dfrac{1}{6}$ **48.** $y + \dfrac{7}{11} = -\dfrac{3}{11}$ **49.** $\dfrac{2}{5} + x = -\dfrac{3}{5}$

50. $\dfrac{7}{8} + y = -\dfrac{1}{8}$ **51.** $\dfrac{1}{3} + x = \dfrac{2}{3}$ **52.** $x + \dfrac{1}{2} = -\dfrac{1}{3}$ **53.** $y + \dfrac{3}{8} = \dfrac{1}{4}$

54. $y + \dfrac{2}{3} = -\dfrac{3}{8}$ **55.** $t + \dfrac{1}{4} = -\dfrac{1}{2}$ **56.** $x + \dfrac{1}{3} = \dfrac{5}{12}$ **57.** $y + \dfrac{2}{3} = -\dfrac{5}{12}$

Objective C

Solve.

58. $3y = 12$ **59.** $5x = 30$ **60.** $5z = -20$ **61.** $3z = -27$

62. $-2x = 6$ **63.** $-4t = 20$ **64.** $-5x = -40$ **65.** $-2y = -28$

66. $40 = 8x$ **67.** $24 = 3y$ **68.** $-24 = 4x$ **69.** $-21 = 7y$

70. $\dfrac{x}{3} = 5$ **71.** $\dfrac{y}{2} = 10$ **72.** $\dfrac{n}{4} = -2$ **73.** $\dfrac{y}{7} = -3$

74. $-\dfrac{x}{4} = 1$ **75.** $-\dfrac{y}{3} = 5$ **76.** $\dfrac{2}{3}w = 4$ **77.** $\dfrac{5}{8}x = 10$

78. $\dfrac{3}{4}v = -3$ **79.** $\dfrac{2}{7}x = -12$ **80.** $-\dfrac{1}{3}x = -2$ **81.** $-\dfrac{1}{5}y = -3$

82. $\dfrac{3}{8}x = -24$ **83.** $\dfrac{5}{12}y = -16$ **84.** $-4 = -\dfrac{2}{3}z$ **85.** $-8 = -\dfrac{5}{6}x$

86. $-12 = -\dfrac{3}{8}y$ **87.** $-9 = \dfrac{5}{6}t$ **88.** $\dfrac{2}{3}x = -\dfrac{2}{7}$ **89.** $\dfrac{3}{7}y = \dfrac{5}{6}$

90. $4x - 2x = 7$ **91.** $3a - 6a = 8$ **92.** $\dfrac{4}{5}m - \dfrac{1}{5}m = 9$ **93.** $\dfrac{1}{3}b - \dfrac{2}{3}b = -1$

Objective D   *Application Problems*

In Exercises 94 to 97, use the formula $A = P + I$, where A is the value of the investment after 1 year, P is the original investment, and I is the increase in value of the investment.

94. The value of an investment in a high-tech company after 1 year was $17,700. The increase in value during the year was $2700. Find the amount of the original investment.

95. The value of an investment in a software company after 1 year was $26,440. The increase in value during the year was $2830. Find the amount of the original investment.

96. The original investment in a mutual fund was $8000. The value of the mutual fund after 1 year was $11,420. Find the increase in value of the investment.

97. The original investment in a money market fund was $7500. The value of the mutual fund after 1 year was $8690. Find the increase in value of the investment.

In Exercises 98 to 101, use the formula $D = M \cdot G$, where D is the distance, M is the miles per gallon, and G is the number of gallons. Round to the nearest tenth.

98. Julio, a sales executive, averages 28 mi/gal on a 621-mile trip. Find the number of gallons of gasoline used on the trip.

99. Over a three-day weekend, you take a 592-mile trip. If you average 32 mi/gal on the trip, how many gallons of gasoline did you use?

100. The manufacturer of a subcompact car estimates that the car can travel 560 mi on a 15-gallon tank of gas. Find the miles per gallon.

101. You estimate that your car can travel 410 mi on 12 gal of gasoline. Find the miles per gallon.

In Exercises 102 and 103, use the formula $S = C + M$, where S is the selling price, C is the cost, and M is the markup.

102. A computer store sells a computer for $2240. The computer has a markup of $420. Find the cost of the computer.

103. A department store buys shirts for $23.50 and sells the shirts for $39.80. Find the markup on each shirt.

In Exercises 104 and 105, use the formula $S = C + R \cdot C$, where S is the selling price, C is cost, and R is the markup rate.

104. A store manager uses a markup rate of 24% on all appliances. Find the cost of a blender that sells for $77.50.

105. A music store uses a markup rate of 30%. Find the cost of a compact disc that sells for $18.85.

APPLYING THE CONCEPTS

106. Write out the steps for solving the equation $x - 3 = -5$. Identify each property of real numbers and each property of equations as you use it.

107. Write out the steps for solving the equation $\frac{3}{4}x = 6$. Identify each property of real numbers and each property of equations as you use it.

108. Is 2 a solution of $x = x + 4$? Try -2, 0, 3, 6, and 10. Do you think there is a solution of this equation? Why or why not?

109. Write an equation of the form $x + a = b$ that has -4 as its solution.

110. Write an equation of the form $a - x = b$ that has -2 as its solution.

111. In your own words, state **a.** the Addition Property of Equations and **b.** the Multiplication Property of Equations.

11.3 General Equations: Part I

Objective A   To solve an equation of the form $ax + b = c$

Point of Interest

Evariste Galois, despite being killed in a duel at the age of 21, made significant contributions to solving equations. In fact, there is a branch of mathematics called Galois theory, showing what kinds of equations can and cannot be solved.

To solve an equation of the form $ax + b = c$, it is necessary to use both the Addition and the Multiplication Properties to simplify the equation to one of the form $variable = constant$.

➡ Solve: $\dfrac{x}{4} - 1 = 3$

$$\dfrac{x}{4} - 1 = 3$$

- The goal is to simplify the equation to one of the form *variable = constant*.

$$\dfrac{x}{4} - 1 + 1 = 3 + 1$$
$$\dfrac{x}{4} + 0 = 4$$
$$\dfrac{x}{4} = 4$$

- Add the opposite of the constant term −1 to each side of the equation. Then simplify (Addition Properties).

$$4 \cdot \dfrac{x}{4} = 4 \cdot 4$$
$$1x = 16$$
$$x = 16$$

- Multiply each side of the equation by the reciprocal of the numerical coefficient of the variable term. Then simplify (Multiplication Properties).

The solution is 16.

- Write the solution.

TAKE NOTE

$\dfrac{x}{4} = \dfrac{1}{4}x$

The reciprocal of $\dfrac{1}{4}$ is 4.

Example 1   Solve: $3x + 7 = 2$

Solution
$$3x + 7 = 2$$
$$3x + 7 - 7 = 2 - 7$$
$$3x = -5$$
$$\dfrac{3x}{3} = \dfrac{-5}{3}$$
$$x = -\dfrac{5}{3} = -1\dfrac{2}{3}$$

The solution is $-1\dfrac{2}{3}$.

You Try It 1   Solve: $5x + 8 = 6$

Your solution

Example 2   Solve: $5 - x = 6$

Solution
$$5 - x = 6$$
$$5 - 5 - x = 6 - 5$$
$$-x = 1$$
$$(-1)(-x) = (-1) \cdot 1$$
$$x = -1$$

The solution is -1.

You Try It 2   Solve: $7 - x = 3$

Your solution

Solutions on p. S28

> **Objective B**   **To solve application problems**   ‹ 11 ›

Example 3

Find the Celsius temperature when the Fahrenheit temperature is 212°. Use the formula $F = \frac{9}{5}C + 32$, where F is the Fahrenheit temperature and C is the Celsius temperature.

Strategy

To find the Celsius temperature, replace the variable F in the formula by the given value and solve for C.

Solution

$$F = \frac{9}{5}C + 32$$

$$212 = \frac{9}{5}C + 32$$

$$212 - 32 = \frac{9}{5}C + 32 - 32$$

$$180 = \frac{9}{5}C$$

$$\frac{5}{9} \cdot 180 = \frac{5}{9} \cdot \frac{9}{5}C$$

$$100 = C$$

The Celsius temperature is 100°.

You Try It 3

Find the Celsius temperature when the Fahrenheit temperature is −22°. Use the formula $F = \frac{9}{5}C + 32$, where F is the Fahrenheit temperature and C is the Celsius temperature.

Your strategy

Your solution

Example 4

To find the total cost of production, an economist uses the formula $T = U \cdot N + F$, where T is the total cost, U is the cost per unit, N is the number of units made, and F is the fixed cost. Find the number of units made during a week when the total cost was $8000, the cost per unit was $16, and the fixed costs were $2000.

Strategy

To find the number of units made, replace the variables T, U, and F in the formula by the given values and solve for N.

Solution

$$T = U \cdot N + F$$
$$8000 = 16 \cdot N + 2000$$
$$8000 - 2000 = 16 \cdot N + 2000 - 2000$$
$$6000 = 16 \cdot N$$
$$\frac{6000}{16} = \frac{16 \cdot N}{16}$$
$$375 = N$$

The number of units made was 375.

You Try It 4

Find the cost per unit during a week when the total cost was $4500, the number of units produced was 250, and the fixed costs were $1500. Use the formula $T = U \cdot N + F$, where T is the total cost, U is the cost per unit, N is the number of units made, and F is the fixed cost.

Your strategy

Your solution

Solutions on p. S28

11.3 Exercises

Objective A

Solve.

1. $3x + 5 = 14$

2. $5z + 6 = 31$

3. $2n - 3 = 7$

4. $4y - 4 = 20$

5. $5w + 8 = 3$

6. $3x + 10 = 1$

7. $3z - 4 = -16$

8. $6x - 1 = -13$

9. $5 + 2x = 7$

10. $12 + 7x = 33$

11. $6 - x = 3$

12. $4 - x = -2$

13. $3 - 4x = 11$

14. $2 - 3x = 11$

15. $5 - 4x = 17$

16. $8 - 6x = 14$

17. $3x + 6 = 0$

18. $5x - 20 = 0$

19. $-3x - 4 = -1$

20. $-7x - 22 = -1$

21. $12x - 30 = 6$

22. $9x - 7 = 2$

23. $3x + 7 = 4$

24. $8x + 13 = 5$

25. $-2x + 11 = -3$

26. $-4x + 15 = -1$

27. $14 - 5x = 4$

28. $7 - 3x = 4$

29. $-8x + 7 = -9$

30. $-7x + 13 = -8$

31. $9x + 13 = 13$

32. $-2x + 7 = 7$

33. $7x - 14 = 0$

34. $5x + 10 = 0$

35. $4x - 4 = -4$

36. $-13x - 1 = -1$

37. $3x + 5 = 7$

38. $4x + 6 = 9$

39. $6x - 1 = 16$

40. $12x - 3 = 7$

41. $2x - 3 = -8$ **42.** $5x - 3 = -12$ **43.** $-6x + 2 = -7$ **44.** $-3x + 9 = -1$

45. $-2x - 3 = -7$ **46.** $-5x - 7 = -4$ **47.** $3x + 8 = 2$ **48.** $2x - 9 = 8$

49. $3x - 7 = 0$ **50.** $7x - 2 = 0$ **51.** $-2x + 9 = 12$ **52.** $-7x + 3 = 1$

53. $\frac{1}{2}x - 2 = 3$ **54.** $\frac{1}{3}x + 1 = 4$ **55.** $\frac{3}{5}w - 1 = 2$ **56.** $\frac{2}{5}w + 5 = 6$

57. $\frac{2}{9}t - 3 = 5$ **58.** $\frac{5}{9}t - 3 = 2$ **59.** $\frac{y}{3} - 6 = -8$ **60.** $\frac{y}{2} - 2 = 3$

61. $\frac{x}{3} - 2 = -5$ **62.** $\frac{x}{4} - 3 = 5$ **63.** $\frac{5}{8}v + 6 = 3$ **64.** $\frac{2}{3}v - 4 = 3$

65. $\frac{4}{7}z + 10 = 5$ **66.** $\frac{3}{8}v - 3 = 4$ **67.** $\frac{2}{9}x - 3 = 5$ **68.** $\frac{1}{2}x + 3 = -8$

69. $\frac{3}{4}x - 5 = -4$ **70.** $\frac{2}{3}x - 5 = -8$ **71.** $1.5x - 0.5 = 2.5$

72. $2.5w - 1.3 = 3.7$ **73.** $0.8t + 1.1 = 4.3$ **74.** $0.3v + 2.4 = 1.5$

75. $0.4x - 2.3 = 1.3$ **76.** $1.2t + 6.5 = 2.9$ **77.** $3.5y - 3.5 = 10.5$

78. $1.9x - 1.9 = -1.9$ **79.** $0.32x + 4.2 = 3.2$ **80.** $5x - 3x + 2 = 8$

81. $6m + 2m - 3 = 5$

82. $4a - 7a - 8 = 4$

83. $3y - 8y - 9 = 6$

84. $x - 4x + 5 = 11$

85. $-2y + y - 3 = 6$

86. $-4y - y - 8 = 12$

87. $0.032x - 0.0194 = 0.139$

88. $-3.256x + 42.38 = -16.9$

89. $6.09x + 17.33 = 16.805$

90. $1.925x + 32.87 = -16.994$

 Objective B   *Application Problems*

In Exercises 91 and 92, use the relationship between Fahrenheit temperature and Celsius temperature, which is given by the formula $F = \frac{9}{5}C + 32$, where F is the Fahrenheit temperature and C is the Celsius temperature.

91. Find the Celsius temperature when the Fahrenheit temperature is $-40°$.

92. Find the Celsius temperature when the Fahrenheit temperature is $72°$. Round to the nearest tenth of a degree.

In Exercises 93 and 94, use the formula $V = V_0 + 32t$, where V is the final velocity of a falling object, V_0 is the starting velocity of a falling object, and t is the time for the object to fall.

93. Find the time required for an object to increase in velocity from 8 ft/s to 472 ft/s.

94. Find the time required for an object to increase in velocity from 16 ft/s to 128 ft/s.

In Exercises 95 and 96, use the formula $T = U \cdot N + F$, where T is the total cost, U is the cost per unit, N is the number of units made, and F is the fixed cost.

95. Find the number of units made during a week when the total cost was $25,000, the cost per unit was $8, and the fixed costs were $5000.

96.   Find the cost per unit during a week when the total cost was $80,000, the total number of units produced was 500, and the fixed costs were $15,000.

In Exercises 97 and 98, use the formula $T = I \cdot R + B$, where T is the monthly tax, I is the monthly income, R is the income tax rate, and B is the base monthly tax.

97.   The monthly tax that a mechanic pays is $476. The mechanic's monthly tax rate is 22%, and the base monthly tax is $80. Find the mechanic's monthly income.

98.   The monthly tax that Marcy, a teacher, pays is $770. Her monthly income is $3100, and the base monthly tax is $150. Find Marcy's income tax rate.

In Exercises 99 to 102, use the formula $M = S \cdot R + B$, where M is the monthly earnings, S is total sales, R is the commission rate, and B is the base monthly salary.

99.   A book representative earns a base monthly salary of $600 plus a 9% commission on total sales. Find the total sales during a month in which the representative earned $3480.

100.   A sales executive earns a base monthly salary of $1000 plus a 5% commission on total sales. Find the total sales during a month in which the executive earned $2800.

101.   Miguel earns a base monthly salary of $750. Find his commission rate during a month in which total sales were $42,000 and he earned $2640.

102.   Tina earns a base monthly salary of $500. Find her commission rate during a month when total sales were $42,500 and her earnings were $3560.

APPLYING THE CONCEPTS

103.   Explain in your own words the steps you would take to solve the equation $\frac{2}{3}x - 4 = 10$. State the property of real numbers or the property of equations that is used at each step.

104.   Make up an equation of the form $ax + b = c$ that has -3 as its solution.

105.   Does the sentence "Solve $3x + 4(x - 3)$" make sense? Why or why not?

11.4 General Equations: Part II

Objective A   To solve an equation of the form $ax + b = cx + d$

When a variable occurs on each side of an equation, the Addition Properties are used to rewrite the equation so that variable terms are on one side of the equation and constant terms are on the other side of the equation. Then the Multiplication Properties are used to simplify the equation to one of the form *variable = constant*.

➡ Solve: $4x - 6 = 8 - 3x$

$$4x - 6 = 8 - 3x$$

- The goal is to write the equation in the form *variable = constant*.

$$4x + 3x - 6 = 8 - 3x + 3x$$
$$7x - 6 = 8 + 0$$
$$7x - 6 = 8$$

- Add $3x$ to each side of the equation. Then simplify (Addition Properties). Now only one variable term occurs in the equation.

$$7x - 6 + 6 = 8 + 6$$
$$7x + 0 = 14$$
$$7x = 14$$

- Add 6 to each side of the equation. Then simplify (Addition Properties). Now only one constant term occurs in the equation.

$$\frac{7x}{7} = \frac{14}{7}$$
$$1x = 2$$
$$x = 2$$

- Divide each side of the equation by the numerical coefficient of the variable term. Then simplify (Multiplication Properties).

The solution is 2.

- Write the solution.

- -

Example 1

Solve: $\frac{2}{9}x - 3 = \frac{7}{9}x + 2$

Solution

$$\frac{2}{9}x - 3 = \frac{7}{9}x + 2$$
$$\frac{2}{9}x - \frac{7}{9}x - 3 = \frac{7}{9}x - \frac{7}{9}x + 2$$
$$-\frac{5}{9}x - 3 = 2$$
$$-\frac{5}{9}x - 3 + 3 = 2 + 3$$
$$-\frac{5}{9}x = 5$$
$$\left(-\frac{9}{5}\right)\left(-\frac{5}{9}\right)x = \left(-\frac{9}{5}\right)5$$
$$x = -9$$

The solution is -9.

You Try It 1

Solve: $\frac{1}{5}x - 2 = \frac{2}{5}x + 4$

Your solution

Solution on pp. S28–S29

Objective B   **To solve an equation containing parentheses**   ⑪

When an equation contains parentheses, one of the steps in solving the equation requires use of the Distributive Property.

$$a(b + c) = ab + ac$$

The Distributive Property is used to rewrite a variable expression without parentheses.

➡ Solve: $4(3 + x) - 2 = 2(x - 4)$

$4(3 + x) - 2 = 2(x - 4)$ • The goal is to write the equation in the form *variable = constant*.

$12 + 4x - 2 = 2x - 8$
$10 + 4x = 2x - 8$ • Use the Distributive Property to rewrite the equation without parentheses. Combine like terms.

$10 + 4x - 2x = 2x - 2x - 8$ • Subtract 2x from each side of the equation.
$10 + 2x = -8$

$10 - 10 + 2x = -8 - 10$ • Subtract 10 from each side of the equation.
$2x = -18$

$\dfrac{2x}{2} = \dfrac{-18}{2}$ • Divide each side of the equation by the numerical coefficient of the variable term.
$x = -9$

The solution is -9.   • Write the solution.

....

Example 2   Solve: $3(x + 2) - x = 11$

Solution
$$3(x + 2) - x = 11$$
$$3x + 6 - x = 11$$
$$2x + 6 = 11$$
$$2x + 6 - 6 = 11 - 6$$
$$2x = 5$$
$$\frac{2x}{2} = \frac{5}{2}$$
$$x = \frac{5}{2}$$
$$x = 2\frac{1}{2}$$

The solution is $2\frac{1}{2}$.

You Try It 2   Solve: $4(x - 1) - x = 5$

Your solution

Solution on p. S29

11.4 Exercises

Objective A

Solve.

1. $6x + 3 = 2x + 5$

2. $7x + 1 = x + 19$

3. $3x + 3 = 2x + 2$

4. $6x + 3 = 3x + 6$

5. $5x + 4 = x - 12$

6. $3x - 12 = x - 8$

7. $7x - 2 = 3x - 6$

8. $2x - 9 = x - 8$

9. $9x - 4 = 5x - 20$

10. $8x - 7 = 5x + 8$

11. $2x + 1 = 16 - 3x$

12. $3x + 2 = -23 - 2x$

13. $5x - 2 = -10 - 3x$

14. $4x - 3 = 7 - x$

15. $2x + 7 = 4x + 3$

16. $7x - 6 = 10x - 15$

17. $x + 4 = 6x - 11$

18. $x - 6 = 4x - 21$

19. $3x - 7 = x - 7$

20. $2x + 6 = 7x + 6$

21. $3 - 4x = 5 - 3x$

22. $6 - 2x = 9 - x$

23. $7 + 3x = 9 + 5x$

24. $12 + 5x = 9 - 3x$

25. $5 + 2x = 7 + 5x$

26. $9 + x = 2 + 3x$

27. $8 - 5x = 4 - 6x$

28. $9 - 4x = 11 - 5x$

29. $6x + 1 = 3x + 2$

30. $7x + 5 = 4x + 7$

31. $5x + 8 = x + 5$

32. $9x + 1 = 3x - 4$

33. $2x - 3 = 6x - 4$

34. $4 - 3x = 4 - 5x$

35. $6 - 3x = 6 - 5x$

36. $2x + 7 = 4x - 3$

37. $6x - 2 = 2x - 9$

38. $4x - 7 = -3x + 2$

39. $6x - 3 = -5x + 8$

40. $7x - 5 = 3x + 9$

41. $-6x - 2 = -8x - 4$

42. $-7x + 2 = 3x - 8$

43. $-3 - 4x = 7 - 2x$

44. $-8 + 5x = 8 + 6x$

45. $3 - 7x = -2 + 5x$

46. $3x - 2 = 7 - 5x$

47. $5x + 8 = 4 - 2x$

48. $4 - 3x = 6x - 8$

49. $12x - 9 = 3x + 12$

50. $4x + 13 = -6x + 9$

51. $\dfrac{5}{7}x - 3 = \dfrac{2}{7}x + 6$

52. $\dfrac{4}{5}x - 1 = \dfrac{1}{5}x + 5$

53. $\dfrac{3}{7}x + 5 = \dfrac{5}{7}x - 1$

54. $\dfrac{3}{4}x + 2 = \dfrac{1}{4}x - 9$

Objective B

Solve.

55. $6x + 2(x - 1) = 14$

56. $3x + 2(x + 4) = 13$

57. $-3 + 4(x + 3) = 5$

58. $8x - 3(x - 5) = 30$

59. $6 - 2(x + 4) = 6$

60. $5 - 3(x + 2) = 8$

61. $5 + 7(x + 3) = 20$

62. $6 - 3(x - 4) = 12$

63. $2x + 3(x - 5) = 10$

64. $3x - 4(x + 3) = 9$

65. $3(x - 4) + 2x = 3$

66. $4 + 3(x - 9) = -12$

67. $2x - 3(x - 4) = 12$

68. $4x - 2(x - 5) = 10$

69. $2x + 3(x + 4) = 7$

70. $3(x + 2) + 7 = 12$

71. $3(x - 2) + 5 = 5$

72. $4(x - 5) + 7 = 7$

73. $3x + 7(x - 2) = 5$

74. $-3x - 3(x - 3) = 3$

75. $4x - 2(x + 9) = 8$

76. $3x - 6(x - 3) = 9$

77. $3x + 5(x - 2) = 10$

78. $3x - 5(x - 1) = -5$

79. $3x + 4(x + 2) = 2(x + 9)$

80. $5x + 3(x + 4) = 4(x + 2)$

81. $2x - 3(x - 4) = 2(x + 6)$

82. $3x - 4(x - 1) = 3(x - 2)$

83. $7 - 2(x - 3) = 3(x - 1)$

84. $4 - 3(x + 2) = 2(x - 4)$

85. $6x - 2(x - 3) = 11(x - 2)$

86. $9x - 5(x - 3) = 5(x + 4)$

87. $6x - 3(x + 1) = 5(x + 2)$

88. $2x - 7(x - 2) = 3(x - 4)$

89. $7 - (x + 1) = 3(x + 3)$

90. $12 + 2(x - 9) = 3(x - 12)$

91. $2x - 3(x + 4) = 2(x - 5)$

92. $3x + 2(x - 7) = 7(x - 1)$

93. $x + 5(x - 4) = 3(x - 8) - 5$

94. $2x - 2(x - 1) = 3(x - 2) + 7$

95. $9x - 3(x - 4) = 13 + 2(x - 3)$

96. $3x - 4(x - 2) = 15 - 3(x - 2)$

97. $3(x - 4) + 3x = 7 - 2(x - 1)$

98. $2(x - 6) + 7x = 5 - 3(x - 2)$

99. $3.67x - 5.3(x - 1.932) = 6.99$

100. $4.06x + 4.7(x + 3.22) = 1.774$

101. $8.45(x - 10) = 3(x - 3.854)$

102. $4(x - 1.99) - 3.92 = 3(x - 1.77)$

APPLYING THE CONCEPTS

103. If $2x - 2 = 4x + 6$, what is the value of $3x^2$?

104. If $3 + 2(4a - 3) = 4$ and $4 - 3(2 - 3b) = 11$, which is larger, a or b?

105. Explain what is wrong with the demonstration at the right, which suggests that $4 = 5$.

$$5x + 7 = 4x + 7$$
$$5x + 7 - 7 = 4x + 7 - 7 \quad \bullet \text{ Subtract 7 from each side of the equation.}$$
$$5x = 4x$$
$$\frac{5x}{x} = \frac{4x}{x} \quad \bullet \text{ Divide each side of the equation by } x.$$
$$5 = 4$$

106. The equation $x = x + 1$ has no solution, whereas the solution of the equation $2x + 3 = 3$ is zero. Is there a difference between no solution and a solution of zero? Explain.

Translating Verbal Expressions into Mathematical Expressions

Objective A **To translate a verbal expression into a mathematical expression given the variable**

One of the major skills required in applied mathematics is to translate a verbal expression into a mathematical expression. Doing so requires recognizing the verbal phrases that translate into mathematical operations. Following is a partial list of the verbal phrases used to indicate the different mathematical operations.

Addition	more than	5 more than x	$x + 5$
	the sum of	the sum of w and 3	$w + 3$
	the total of	the total of 6 and z	$6 + z$
	increased by	x increased by 7	$x + 7$
Subtraction	less than	5 less than y	$y - 5$
	the difference between	the difference between w and 3	$w - 3$
	decreased by	8 decreased by a	$8 - a$
Multiplication	times	3 times c	$3c$
	the product of	the product of 4 and t	$4t$
	of	two-thirds of v	$\dfrac{2}{3}v$
	twice	twice d	$2d$
Division	divided by	n divided by 3	$\dfrac{n}{3}$
	the quotient of	the quotient of z and 4	$\dfrac{z}{4}$
	the ratio of	the ratio of s to 6	$\dfrac{s}{6}$

Translating phrases that contain the words *sum*, *difference*, *product*, and *quotient* can sometimes cause a problem. In the examples at the right, note where the operation symbol is placed.

the *sum* of x and y $x + y$

the *difference* between x and y $x - y$

the *product* of x and y $x \cdot y$

the *quotient* of x and y $\dfrac{x}{y}$

Note where we place the fraction bar when translating the word *ratio*.

the *ratio* of x to y $\dfrac{x}{y}$

⇒ Translate "the quotient of n and the sum of n and 6" into a mathematical expression.

the *quotient* of n and the *sum* of n and 6 $\dfrac{n}{n + 6}$

Example 1 Translate "the sum of 5 and the product of 4 and n" into a mathematical expression.

Solution $5 + 4n$

You Try It 1 Translate "the difference between 8 and twice t" into a mathematical expression.

Your solution

Example 2 Translate "the product of 3 and the difference between z and 4" into a mathematical expression.

Solution $3(z - 4)$

You Try It 2 Translate "the quotient of 5 and the product of 7 and x" into a mathematical expression.

Your solution

Solutions on p. S29

Objective B **To translate a verbal expression into a mathematical expression by assigning the variable**

In most applications that involve translating phrases into mathematical expressions, the variable to be used is not given. To translate these phrases, we must assign a variable to the unknown quantity before writing the mathematical expression.

→ Translate "the difference between seven and twice a number" into a mathematical expression.

The difference between seven and twice a number

- Identify the phrases that indicate the mathematical operations.

The unknown number: n

- Assign a variable to one of the unknown quantities.

Twice the number: $2n$

- Use the assigned variable to write an expression for any other unknown quantity.

$7 - 2n$

- Use the identified operations to write the mathematical expression.

Example 3
Translate "the total of a number and the square of the number" into a mathematical expression.

Solution
The *total* of a number and the *square* of the number

The unknown number: x
The square of the number: x^2

$x + x^2$

You Try It 3
Translate "the product of a number and one-half of the number" into a mathematical expression.

Your solution

Solution on p. S29

11.5 Exercises

Objective A

Translate into a mathematical expression.

1. 9 less than y

2. w divided by 7

3. z increased by 3

4. the product of -2 and x

5. the sum of two-thirds of n and n

6. the difference between the square of r and r

7. the quotient of m and the difference between m and 3

8. v increased by twice v

9. the product of 9 and 4 more than x

10. the total of a and the quotient of a and 7

11. the difference between n and the product of -5 and n

12. x decreased by the quotient of x and 2

13. the product of c and one-fourth of c

14. the quotient of 3 less than z and z

15. the total of the square of m and twice the square of m

16. the product of y and the sum of y and 4

17. 2 times the sum of t and 6

18. the quotient of r and the difference between 8 and r

19. x divided by the total of 9 and x

20. the sum of z and the product of 6 and z

21. three times the sum of b and 6

22. the ratio of w to the sum of w and 8

Objective B

Translate into a mathematical expression.

23. the square of a number

24. five less than some number

25. a number divided by twenty

26. the difference between a number and twelve

27. four times some number

28. the quotient of five and a number

29. three-fourths of a number

30. the sum of a number and seven

31. four increased by some number

32. the ratio of a number to nine

33. the difference between five times a number and the number

34. six less than the total of three and a number

35. the product of a number and two more than the number

36. the quotient of six and the sum of nine and a number

37. seven times the total of a number and eight

38. the difference between ten and the quotient of a number and two

39. the square of a number plus the product of three and the number

40. a number decreased by the product of five and the number

41. the sum of three more than a number and one-half of the number

42. eight more than twice the sum of a number and seven

43. the quotient of three times a number and the number

44. the square of a number divided by the sum of the number and twelve

APPLYING THE CONCEPTS

45. Translate the expressions $2x + 3$ and $2(x + 3)$ into phrases.

46. Translate the expressions $\frac{2x}{7}$ and $\frac{2 + x}{7}$ into phrases.

47. In your own words, explain how variables are used.

48. The chemical formula for water is H_2O. This formula means that there are two hydrogen atoms and one oxygen atom in each molecule of water. If x represents the number of atoms of oxygen in a glass of pure water, express the number of hydrogen atoms in the glass of water.

49. The chemical formula for one molecule of glucose (sugar) is $C_6H_{12}O_6$, where C is carbon, H is hydrogen, and O is oxygen. If x represents the number of atoms of hydrogen in a sample of pure sugar, express the number of carbon atoms and the number of oxygen atoms in the sample in terms of x.

$$\begin{array}{c} H \diagdown \quad \diagup O \\ C \\ | \\ H - C - OH \\ | \\ HO - C - H \\ | \\ H - C - OH \\ | \\ H - C - OH \\ | \\ CH_2OH \end{array}$$

11.6 Translating Sentences into Equations and Solving

Objective A   **To translate a sentence into an equation and solve**

Point of Interest

Number puzzle problems similar to the one on this page have appeared in textbooks for hundreds of years. Here is one from a 1st-century Chinese textbook: "When a number is divided by 3, the remainder is 2; when it is divided by 5, the remainder is 3; when it is divided by 7, the remainder is 2. Find the number." There are actually infinitely many solutions to this problem. See whether you can find one of them.

An equation states that two mathematical expressions are equal. Therefore, to translate a sentence into an equation requires recognition of the words or phrases that mean "equals." Some of these phrases are

$$\left.\begin{array}{l} \text{equals} \\ \text{is} \\ \text{is equal to} \\ \text{amounts to} \\ \text{represents} \end{array}\right\} \text{translate to} =$$

Once the sentence is translated into an equation, the equation can be simplified to one of the form *variable* = *constant* and the solution can be found.

➡ Translate "three more than twice a number is seventeen" into an equation and solve.

The unknown number: n

- Assign a variable to the unknown quantity.

Three more than twice a number	is	seventeen

- Find two verbal expressions for the same value.

$$2n + 3 = 17$$
$$2n + 3 - 3 = 17 - 3$$
$$2n = 14$$
$$\frac{2n}{2} = \frac{14}{2}$$
$$n = 7$$

- Write a mathematical expression for each verbal expression. Write the equals sign.
- Solve the resulting equation.

The number is 7.

....

Example 1

Translate "a number decreased by six equals fifteen" into an equation and solve.

Solution

The unknown number: x

A number decreased by six	equals	fifteen

$$x - 6 = 15$$
$$x - 6 + 6 = 15 + 6$$
$$x = 21$$

The number is 21.

You Try It 1

Translate "a number increased by four equals twelve" into an equation and solve.

Your solution

Solution on p. S29

Example 2

The quotient of a number and six is five. Find the number.

Solution

The unknown number: z

The quotient of a number and six	is	five

$$\frac{z}{6} = 5$$
$$6 \cdot \frac{z}{6} = 6 \cdot 5$$
$$z = 30$$

The number is 30.

Example 3

Eight decreased by twice a number is four. Find the number.

Solution

The unknown number: t

Eight decreased by twice a number	is	four

$$8 - 2t = 4$$
$$8 - 8 - 2t = 4 - 8$$
$$-2t = -4$$
$$\frac{-2t}{-2} = \frac{-4}{-2}$$
$$t = 2$$

The number is 2.

Example 4

Three less than the ratio of a number to seven is one. Find the number.

Solution

The unknown number: x

Three less than the ratio of a number to seven	is	one

$$\frac{x}{7} - 3 = 1$$
$$\frac{x}{7} - 3 + 3 = 1 + 3$$
$$\frac{x}{7} = 4$$
$$7 \cdot \frac{x}{7} = 7 \cdot 4$$
$$x = 28$$

The number is 28.

You Try It 2

The product of two and a number is ten. Find the number.

Your solution

You Try It 3

The sum of three times a number and six equals four. Find the number.

Your solution

You Try It 4

Three more than one-half of a number is nine. Find the number.

Your solution

Solutions on p. S29

Objective B   **To solve application problems**

Example 5

The cost of a television with remote control is $649. This amount is $125 more than the cost of the same television without remote control. Find the cost of the television without remote control.

Strategy

To find the cost of the television without remote control, write and solve an equation using C to represent the cost of the television without remote control.

Solution

| $649 | is | $125 more than the television without remote control |

$$649 = C + 125$$
$$649 - 125 = C + 125 - 125$$
$$524 = C$$

The cost of the television without remote control is $524.

Example 6

By purchasing a fleet of cars, a company receives a discount of $1972 on each car purchased. This amount is 8% of the regular price. Find the regular price.

Strategy

To find the regular price, write and solve an equation using P to represent the regular price of the car.

Solution

| $1972 | is | 8% of the regular price |

$$1972 = 0.08 \cdot P$$
$$\frac{1972}{0.08} = \frac{0.08P}{0.08}$$
$$24,650 = P$$

The regular price is $24,650.

You Try It 5

The sale price of a pair of slacks is $38.95. This amount is $11 less than the regular price. Find the regular price.

Your strategy

Your solution

You Try It 6

At a certain speed, the engine rpm (revolutions per minute) of a car in fourth gear is 2500. This is two-thirds of the rpm of the engine in third gear. Find the rpm of the engine when it is in third gear.

Your strategy

Your solution

Solutions on pp. S29–S30

Example 7

Ron Sierra charged $815 for plumbing repairs in an office building. This charge included $90 for parts and $25 per hour for labor. Find the number of hours he worked in the office building.

Strategy

To find the number of hours worked, write and solve an equation using N to represent the number of hours worked.

Solution

$815	included	$90 for parts and $25 per hour for labor

$$815 = 90 + 25N$$
$$815 - 90 = 90 - 90 + 25N$$
$$725 = 25N$$
$$\frac{725}{25} = \frac{25N}{25}$$
$$29 = N$$

Ron worked 29 h.

You Try It 7

Natalie Adams earned $2500 last month for temporary work. This amount was the sum of a base monthly salary of $800 and an 8% commission on total sales. Find Natalie's total sales for the month.

Your strategy

Your solution

Example 8

The state income tax for Tim Fong last month was $256. This amount is $5 more than 8% of his monthly salary. Find Tim's monthly salary.

Strategy

To find Tim's monthly salary, write and solve an equation using S to represent his monthly salary.

Solution

$256	is	$5 more than 8% of the monthly salary

$$256 = 0.08 \cdot S + 5$$
$$256 - 5 = 0.08 \cdot S + 5 - 5$$
$$251 = 0.08 \cdot S$$
$$\frac{251}{0.08} = \frac{0.08 \cdot S}{0.08}$$
$$3137.50 = S$$

Tim's monthly salary is $3137.50.

You Try It 8

The total cost to make a model ZY television is $300. The cost includes $100 for materials plus $12.50 per hour for labor. How many hours of labor are required to make a model ZY television?

Your strategy

Your solution

Solutions on p. S30

11.6 Exercises

Objective A

Write an equation and solve.

1. The sum of a number and seven is twelve. Find the number.

2. A number decreased by seven is five. Find the number.

3. The product of three and a number is eighteen. Find the number.

4. The quotient of a number and three is one. Find the number.

5. Five more than a number is three. Find the number.

6. A number divided by four is six. Find the number.

7. Six times a number is fourteen. Find the number.

8. Seven less than a number is three. Find the number.

9. Five-sixths of a number is fifteen. Find the number.

10. The total of twenty and a number is five. Find the number.

11. The sum of three times a number and four is eight. Find the number.

12. The sum of one-third of a number and seven is twelve. Find the number.

13. Seven less than one-fourth of a number is nine. Find the number.

14. The total of a number divided by four and nine is two. Find the number.

15. The ratio of a number to nine is fourteen. Find the number.

16. Five increased by the product of five and a number is equal to 30. Find the number.

17. Six less than the quotient of a number and four is equal to negative two. Find the number.

18. The product of a number plus three and two is eight. Find the number.

19. The difference between seven and twice a number is thirteen. Find the number.

20. Five more than the product of three and a number is eight. Find the number.

21. Nine decreased by the quotient of a number and two is five. Find the number.

22. The total of ten times a number and seven is twenty-seven. Find the number.

23. The sum of three-fifths of a number and eight is two. Find the number.

24. Five less than two-thirds of a number is three. Find the number.

25. The difference between a number divided by 4.186 and 7.92 is 12.529. Find the number.

26. The total of 5.68 times a number and 132.7 is the number minus 29.265. Find the number.

Objective B   *Application Problems*

27. Sears has a shoe on sale for $72.50. This amount is $4.25 less than the shoe sells for at Target. Find the price at Target.

28. As a restaurant manager, Uechi Kim is paid a salary of $832 a week. This is $58 more a week than the salary paid last year. Find the weekly salary paid to Uechi last year.

29. The value of a sport utility vehicle this year is $16,000, which is four-fifths of what its value was last year. Find the value of the vehicle last year.

30. This year the value of a lakefront summer cottage is $175,000. This amount is twice the value of the cottage 6 years ago. What was its value 6 years ago?

31. In 2000, McDonalds Restaurants' share of the quick-service restaurant market was approximately $19 billion. This represented 7.3% of the total market. (Source: Multex.com) What was the total quick-service restaurant market in 2000? Round to the nearest billion.

32. Sonia Parker works the night shift at Palomar Hospital and receives a salary of $3200 per month. This is four times the salary she was making 10 years ago. Find her salary 10 years ago.

33. Each month the Manzanares family spends $860 for their house payment and utilities, which amounts to one-fourth of the family's monthly income. Find the family's monthly income.

34. The cost of a graphics calculator is now three-fourths of what the calculator cost 5 years ago. The cost of the graphics calculator is now $72. Find the cost of the calculator 5 years ago.

35. Assume that the Dell Computer Corp. has increased its output of computers by 400 computers per month. This amount represents an 8% increase over last year's production. Find the monthly output last year.

36. The population of a midwestern city has decreased by 12,000 in the last 5 years. This amount represents a 3% decrease. Find the city's population 5 years ago.

37. The nutrition label on a bag of Baked Tostitos tortilla chips lists the sodium content of one serving as 200 mg, which is 8% of the recommended daily intake of sodium. What is the recommended daily intake of sodium? Express the answer in grams.

38. The price of a pair of skis at the Solitude Ski Shop is $340. This price includes the store's cost for the skis plus a markup at the rate of 25%. Find Solitude's cost for the skis.

39. Budget Plumbing charged $385 for a water softener and installation. The charge included $310 for the water softener and $25 per hour for labor. How many hours of labor were charged?

40. Sandy's monthly salary as a sales representative was $2580. This amount included her base monthly salary of $600 plus a 3% commission on total sales. Find her total sales for the month.

41. In Central America and Mexico, 1184 plants and animals are at risk of extinction. This represents approximately 10.7% of all the species at risk of extinction on Earth. (Source: World Conservation Union) Approximately how many plants and animals are at risk of extinction in the world?

42. When you use your car for business, you are able to deduct the expense on your income tax return. In 2000 the deduction was 32.5 cents per mile driven. How many business-related miles did a taxpayer who deducted a total of $1560 drive?

43. Americans spend approximately $295 million a year on remedies for cockroaches. The table at the right shows the top U.S. cities for sales of roach insecticides. What percent of the total is spent in New York? Round to the nearest tenth of a percent.

City	Roach Insecticide Sales
Los Angeles	$16.8 million
New York	$9.8 million
Houston	$6.7 million

Source: IRI InfoScan for Combat

44. McPherson Cement sells cement for $75 plus $24 for each yard of cement. How many yards of cement can be purchased for $363?

45. A water flow restrictor has reduced the flow of water to 2 gal/min. This amount is 1 gal/min less than three-fifths the original flow rate. Find the original rate.

46. The Fahrenheit temperature equals the sum of 32 and nine-fifths of the Celsius temperature. Find the Celsius temperature when the Fahrenheit temperature is 104°.

47. Assume that Paine Webber sales executives receive a base monthly salary of $600 plus an 8.25% commission on total sales per month. Find the total sales of a sales executive who receives a total of $4109.55 for the month.

The graph at the right shows the cost of lunches at the executive dining room and the employee lunch room. Use data from this graph for Exercises 48 and 49. Round to the nearest tenth of a percent.

48. Find the amount subsidized in the executive dining room. What percent of the cost of the meal is subsidized?

49. Find the percent of the lunch room meal that is subsidized.

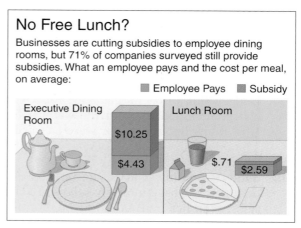

No Free Lunch?

Businesses are cutting subsidies to employee dining rooms, but 71% of companies surveyed still provide subsidies. What an employee pays and the cost per meal, on average:

■ Employee Pays   ■ Subsidy

Executive Dining Room $10.25 $4.43

Lunch Room $.71 $2.59

Source: Copyright © *USA Today*. Reprinted with permission.

APPLYING THE CONCEPTS

50. A man's boyhood lasted $\frac{1}{6}$ of his life, he played football for the next $\frac{1}{8}$ of his life, and he married 5 years after quitting football. A daughter was born after he had been married $\frac{1}{12}$ of his life. The daughter lived $\frac{1}{2}$ as many years as her father. The man died 6 years after his daughter. How old was the man when he died? Use a number line to illustrate the time. Then write an equation and solve it.

51. It is always important to check the answer to an application problem to be sure the answer makes sense. Consider the following problem. "A 4-quart mixture of fruit juices is made from apple juice and cranberry juice. There are 6 more quarts of apple juice than of cranberry juice. Write and solve an equation for the number of quarts of each juice used." Does the answer to this question make sense? Explain.

52. A formula is an equation that relates variables in a known way. Find two examples of formulas that are used in your college major. Explain what each of the variables represents.

Focus on Problem Solving

From Concrete to Abstract

As you progress in your study of algebra, you will find that the problems become less concrete and more abstract. Problems that are concrete provide information pertaining to a specific instance. Abstract problems are theoretical; they are stated without reference to a specific instance. Let's look at an example of an abstract problem.

How many cents are in d dollars?

How can you solve this problem? Are you able to solve the same problem if the information given is concrete?

How many cents are in 5 dollars?

You know that there are 100 cents in 1 dollar. To find the number of cents in 5 dollars, multiply 5 by 100.

$100 \cdot 5 = 500$ There are 500 cents in 5 dollars.

Use the same procedure to find the number of cents in d dollars: multiply d by 100.

$100 \cdot d = 100d$ There are $100d$ cents in d dollars.

This problem might be taken a step further:

If one pen costs c cents, how many pens can be purchased with d dollars?

Consider the same problem using numbers in place of the variables.

If one pen costs 25 cents, how many pens can be purchased with 2 dollars?

To solve this problem, you need to calculate the number of cents in 2 dollars (multiply 2 by 100) and divide the result by the cost per pen (25 cents).

$\dfrac{100 \cdot 2}{25} = \dfrac{200}{25} = 8$ If one pen costs 25 cents, 8 pens can be purchased with 2 dollars.

Use the same procedure to solve the related abstract problem. Calculate the number of cents in d dollars (multiply d by 100) and divide the result by the cost per pen (c cents).

$\dfrac{100 \cdot d}{c} = \dfrac{100d}{c}$ If one pen costs c cents, $\dfrac{100d}{c}$ pens can be purchased with d dollars.

At the heart of the study of algebra is the use of variables. It is the variables in the problems above that make them abstract. But it is variables that allow us to generalize situations and state rules about mathematics.

Try the following problems.

1. How many nickels are in d dollars?

2. How long can you talk on a pay phone if you have only d dollars and the call costs c cents per minute?

3. If you travel m miles on one gallon of gasoline, how far can you travel on g gallons of gasoline?

4. If you walk a mile in x minutes, how far can you walk in h hours?

5. If one photocopy costs n nickels, how many photocopies can you make for q quarters?

Projects and Group Activities

Averages   We often discuss temperature in terms of average high or average low temperature. Temperatures collected over a period of time are analyzed to determine, for example, the average high temperature for a given month in your city or state. The following activity is planned to help you better understand the concept of "average."

1. Choose two cities in the United States. We will refer to them as City X and City Y. Over an eight-day period, record the daily high temperature each day in each city.

2. Determine the average high temperature for City X for the eight-day period. (Add the eight numbers, and then divide the sum by 8.) Do not round your answer.

3. Subtract the average high temperature for City X from each of the eight daily high temperatures for City X. You should have a list of eight numbers; the list should include positive numbers, negative numbers, and possibly zero.

4. Find the sum of the list of eight differences recorded in Step 3.

5. Repeat Steps 2 through 4 for City Y.

6. Compare the two sums found in Steps 4 and 5 for City X and City Y.

7. If you were to conduct this activity again, what would you expect the outcome to be? Use the results to explain what an average high temperature means. In your own words, explain what "average" means.

Addition and Multiplication Properties

The chart below is an addition table. Use it to answer Exercises 1 to 7.

+	Δ	‡	◊
Δ	‡	◊	Δ
‡	◊	Δ	‡
◊	Δ	‡	◊

1. Find the sum of Δ and ‡.

2. What is ◊ plus ◊?

3. In our number system, 0 can be added to any number without changing that number; 0 is called the **additive identity.** What is the additive identity for the system in the chart above? Explain your answer.

4. Does the Commutative Property of Addition apply to this system? Explain your answer.

5. What is −Δ (the opposite of Δ) equal to ? Explain your answer.

6. What is −‡ (the opposite of ‡) equal to? Explain your answer.

7. Simplify −Δ + ‡ − ◊. Explain how you arrived at your answer.

The chart below is a multiplication table. Use it to answer Exercises 8 to 14.

×	£	¿	&
£	£	¿	&
¿	¿	¿	¿
&	&	¿	*

8. Find the product of £ and &.

9. What is ¿ times £?

10. Find the square of &.

11. Does the Commutative Property of Multiplication apply to this system? Explain your answer.

12. In our number system, the product of a number and 0 is 0. Is there an element in this system that corresponds to 0 in our system? Explain your answer.

13. In our number system, 1 can be multiplied by any number without changing that number; 1 is called the **multiplicative identity.** What is the multiplicative identity for the system in the chart above? Explain your answer.

14. Simplify & ÷ £ × ¿. Explain how you arrived at your answer.

Chapter Summary

Key Words

A *variable* is a letter that is used to stand for a quantity that is unknown. [p. 445]

A *variable expression* is an expression that contains one or more variables. [p. 445]

Replacing the variable or variables in a variable expression and then simplifying the resulting numerical expression is called *evaluating the variable expression*. [p. 445]

The *terms* of a variable expression are the addends of the expression. [p. 446]

A *variable term* is composed of a numerical coefficient and a variable part. [p. 447]

Like terms of a variable expression are the terms with the same variable part. [p. 447]

An *equation* expresses the equality of two mathematical expressions. [p. 455]

A *solution of an equation* is a value of the variable that results in a true equation. [p. 455]

To *solve an equation* means to find a solution of the equation. In solving an equation, the goal is to simplify the given equation to one of the form *variable = constant*. [p. 456]

Essential Rules

The Commutative Property of Addition
If a and b are two numbers, then $a + b = b + a$. [p. 447]

The Associative Property of Addition
If a, b, and c are three numbers, then $a + (b + c) = (a + b) + c$. [p. 447]

The Commutative Property of Multiplication
If a and b are two numbers, then $a \cdot b = b \cdot a$. [p. 449]

The Associative Property of Multiplication
If a, b, and c are three numbers, then $a(b \cdot c) = (a \cdot b)c$. [p. 449]

The Distributive Property
If a, b, and c are three numbers, then $a(b + c) = ab + ac$. [p. 449]

Addition Property of Zero
The sum of a term and zero is the term. $a + 0 = a, 0 + a = a$ [p. 456]

Addition Property of Equations
If a, b, and c are algebraic expressions, then the equation $a = b$ has the same solution as the equation $a + c = b + c$. [p. 456]

Multiplication Property of Equations
If a, b, and c are algebraic expressions and $c \neq 0$, then the equation $a = b$ has the same solution as the equation $ac = bc$. [p. 458]

Multiplication Property of One
The product of a term and 1 is the term. $a \cdot 1 = a, 1 \cdot a = a$ [p. 458]

Multiplication Property of Reciprocals
The product of a nonzero term and its reciprocal equals 1. [p. 458]

Chapter Review

1. Simplify: $-2(a - b)$

2. Is -2 a solution of the equation $3x - 2 = -8$?

3. Solve: $x - 3 = -7$

4. Solve: $-2x + 5 = -9$

5. Evaluate $a^2 - 3b$ when $a = 2$ and $b = -3$.

6. Solve: $-3x = 27$

7. Solve: $\frac{2}{3}x + 3 = -9$

8. Simplify: $3x - 2(3x - 2)$

9. Solve: $6x - 9 = -3x + 36$

10. Solve: $x + 3 = -2$

11. Is 5 a solution of the equation $3x - 5 = -10$?

12. Evaluate $a^2 - (b \div c)$ when $a = -2$, $b = 8$, and $c = -4$.

13. Solve: $3(x - 2) + 2 = 11$

14. Solve: $35 - 3x = 5$

15. Simplify: $6bc - 7bc + 2bc - 5bc$

16. Solve: $7 - 3x = 2 - 5x$

17. Solve: $-\frac{3}{8}x = -\frac{15}{32}$

18. Simplify: $\frac{1}{2}x^2 - \frac{1}{3}x^2 + \frac{1}{5}x^2 + 2x^2$

19. Solve: $5x - 3(1 - 2x) = 4(2x - 1)$ **20.** Solve: $\frac{5}{6}x - 4 = 5$

21. A tourist drove a rental car 621 mi on 27 gal of gas. Find the number of miles per gallon of gas. Use the formula $D = M \cdot G$, where D is distance, M is miles per gallon, and G is the number of gallons.

22. Find the Celsius temperature when the Fahrenheit temperature is 100°. Use the formula $F = \frac{9}{5}C + 32$, where F is the Fahrenheit temperature and C is the Celsius temperature. Round to the nearest tenth.

23. Translate "the total of n and the quotient of n and five" into a mathematical expression.

24. Translate "the sum of five more than a number and one-third of the number" into a mathematical expression.

25. The difference between nine and twice a number is five. Find the number.

26. The product of five and a number is fifty. Find the number.

27. A compact disc player is now on sale for $228. This is 60% of the regular price. Find the regular price of the CD player.

28. A farmer harvested 28,336 bushels of corn. This amount represents a 12% increase over last year's crop. How many bushels of corn did the farmer harvest last year?

Chapter Test

1. Solve: $\frac{x}{5} - 12 = 7$

2. Solve: $x - 12 = 14$

3. Simplify: $3y - 2x - 7y - 9x$

4. Solve: $8 - 3x = 2x - 8$

5. Solve: $3x - 12 = -18$

6. Evaluate $c^2 - (2a + b^2)$ when $a = 3$, $b = -6$, and $c = -2$.

7. Is 3 a solution of the equation $x^2 + 3x - 7 = 3x - 2$?

8. Simplify: $9 - 8ab - 6ab$

9. Solve: $-5x = 14$

10. Simplify: $3y + 5(y - 3) + 8$

11. Solve: $3x - 4(x - 2) = 8$

12. Solve: $5 = 3 - 4x$

13. Evaluate $\frac{x^2}{y} - \frac{y^2}{x}$ for $x = 3$ and $y = -2$.

14. Solve: $\frac{5}{8}x = -10$

15. Solve: $y - 4y + 3 = 12$

16. Solve: $2x + 4(x - 3) = 5x - 1$

17. A loan of $6600 is to be paid in 48 equal monthly installments. Find the monthly payment. Use the formula $L = P \cdot N$, where L is the loan amount, P is the monthly payment, and N is the number of months.

18. A clock manufacturer's fixed costs per month are $5000. The unit cost for each clock is $15. Find the number of clocks made during a month in which the total cost was $65,000. Use the formula $T = U \cdot N + F$, where T is the total cost, U is the cost per unit, N is the number of units made, and F is the fixed costs.

19. Find the time required for a falling object to increase in velocity from 24 ft/s to 392 ft/s. Use the formula $V = V_0 + 32t$, where V is the final velocity of a falling object, V_0 is the starting velocity of a falling object, and t is the time for the object to fall.

20. Translate "the sum of x and one-third of x" into a mathematical expression.

21. Translate "five times the sum of a number and three" into a mathematical expression.

22. Translate "three less than two times a number is seven" into an equation and solve.

23. The total of five and three times a number is the number minus two. Find the number.

24. Eduardo Santos earned $3600 last month. This salary is the sum of the base monthly salary of $1200 and a 6% commission on total sales. Find his total sales for the month.

25. Your mechanic charges you $278 for performing a 30,000-mile checkup on your car. This charge includes $152 for parts and $42 per hour for labor. Find the number of hours the mechanic worked on your car.

Cumulative Review

1. Simplify: $6^2 - (18 - 6) \div 4 + 8$

2. Subtract: $3\frac{1}{6} - 1\frac{7}{15}$

3. Simplify: $\left(\frac{3}{8} - \frac{1}{4}\right) \div \frac{3}{4} + \frac{4}{9}$

4. Multiply: 9.67×0.0049

5. Write "$84 earned in 20 hours" as a unit rate.

6. Solve the proportion $\frac{2}{3} = \frac{n}{40}$. Round to the nearest hundredth.

7. Write $5\frac{1}{3}\%$ as a fraction.

8. What percent of 30 is 42?

9. 8 is 125% of what number?

10. Multiply: 3 ft 9 in. $\times$ 5

11. Convert $1\frac{3}{8}$ lb to ounces.

12. Convert 282 mg to grams.

13. Add: $-2 + 5 + (-8) + 4$

14. Find -6 less than 13.

15. Simplify: $(-2)^2 - (-8) \div (3 - 5)^2$

16. Evaluate $3ab - 2ac$ when $a = -2$, $b = 6$, and $c = -3$.

17. Simplify: $3z - 2x + 5z - 8x$

18. Simplify: $6y - 3(y - 5) + 8$

19. Solve: $2x - 5 = -7$

20. Solve: $7x - 3(x - 5) = -10$

21. Solve: $-\frac{2}{3}x = 5$ **22.** Solve: $\frac{x}{3} - 5 = -12$

23. In a mathematics class of 34 students, 6 received an A grade. Find the percent of the students in the mathematics class who received an A grade. Round to the nearest tenth of a percent.

24. The manager of a pottery store used a markup rate of 40%. Find the price of a piece of pottery that cost the store $28.50.

25. A department store has a suit regularly priced at $450 on sale for $369.
    a. What is the discount?
    b. What is the discount rate?

26. A toy store borrowed $80,000 at a simple interest rate of 11% for 4 months. What is the simple interest due on the loan? Round to the nearest cent.

27. Translate "the sum of three times a number and 4" into a mathematical expression.

28. A tetrahedral die is one with four triangular sides. If two tetrahedral dice are rolled, what is the probability that the sum of the upward faces is 7?

29. Sunah Yee, a sales executive, receives a base salary of $800 plus an 8% commission on total sales. Find the total sales during a month in which Sunah earned $3400. Use the formula $M = S \cdot R + B$, where M is the monthly earnings, S is the total sales, R is the commission rate, and B is the base monthly salary.

30. Three less than eight times a number is three more than five times the number. Find the number.

Chapter 12 Geometry

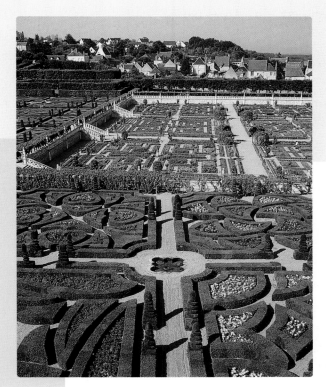

Villandray is a large, formal garden in the Loire Valley in France. The best way to appreciate the different shapes sculptured in the garden is to view it from overhead. Each geometric shape combines with the others to form the entire garden. **Exercise 35 on page 530** illustrates how to use a geometric formula to determine the size of an area and to calculate how much grass seed is needed for an area that size.

Need help? For on-line student resources, such as section quizzes, visit this textbook's web site at **math.college.hmco.com/students.**

1. Solve: $x + 47 = 90$

2. Solve: $32 + 97 + x = 180$

3. Simplify: $2(18) + 2(10)$

4. Evaluate abc when $a = 2$, $b = 3.14$, and $c = 9$.

5. Evaluate xyz^3 when $x = \dfrac{4}{3}$, $y = 3.14$, and $z = 3$.

6. Solve: $\dfrac{5}{12} = \dfrac{6}{x}$

Go Figure

Draw the figure that would come next.

Angles, Lines, and Geometric Figures

Objective A   **To define and describe lines and angles**

The word *geometry* comes from the Greek words for "earth" (*geo*) and "measure." The original purpose of geometry was to measure land. Today geometry is used in many sciences, such as physics, chemistry, and geology, and in applied fields such as mechanical drawing and astronomy. Geometric form is used in art and design.

Two basic geometric concepts are plane and space.

A **plane** is a flat surface, such as a tabletop or a blackboard. Figures that can lie totally in a plane are called **plane figures.**

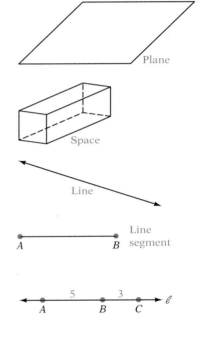

Plane

Space extends in all directions. Objects in space, such as trees, ice cubes, and doors, are called **solids.**

Space

A **line** extends indefinitely in two directions in a plane. A line has no width.

Line

A **line segment** is part of a line and has two endpoints. The line segment *AB* is shown in the figure.

Line segment

The length of a line segment is the distance between the endpoints of the line segment. The length of a line segment may be expressed as the sum of two or more shorter line segments, as shown. For this example, $AB = 5$, $BC = 3$, and $AC = AB + BC = 5 + 3 = 8$.

➡ Given that $AB = 22$ and $AC = 31$, find the length of *BC*.

$$AC = AB + BC$$
$$31 = 22 + BC$$
$$31 - 22 = 22 - 22 + BC$$
$$9 = BC$$

• Substitute 22 for *AB* and 31 for *AC*, and solve for *BC*.

Lines in a plane can be parallel or intersecting. **Parallel lines** never meet; the distance between them is always the same. **Intersecting lines** cross at a point in the plane.

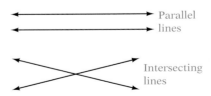

Parallel lines

Intersecting lines

The symbol ∥ means "is parallel to." In the accompanying figure, $AB \parallel CD$ and $p \parallel q$. Note that line p contains line segment AB and that line q contains line segment CD. Parallel lines contain parallel line segments.

A **ray** starts at a point and extends indefinitely in one direction.

An **angle** is formed when two rays start from the same point. Rays r_1 and r_2 start from point B. The common endpoint is called the **vertex** of the angle.

If A and C are points on rays r_1 and r_2 above, respectively, then the angle is called $\angle ABC$, $\angle CBA$, or $\angle B$, where $\angle$ is the symbol for angle. Note that an angle is named by giving three points, with the vertex as the second point listed, or by giving the point at the vertex.

An angle can also be named by writing a variable between the rays close to the vertex. In the figure, $\angle x = \angle QRS = \angle SRQ$ and $\angle y = \angle SRT = \angle TRS$. Note that in this figure, more than two rays meet at the vertex. In this case, the vertex cannot be used to name the angle.

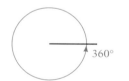

An angle is measured in **degrees.** The symbol for degree is °. One complete revolution is 360° (360 degrees).

Point of Interest

The Babylonians knew that the Earth was in approximately the same position in the sky every 365 days. Historians suggest that the reason one complete revolution of a circle is 360° is that 360 is the closest number to 365 that is divisible by many numbers.

$\frac{1}{4}$ of a revolution is 90°. A 90° angle is called a **right angle.** The symbol ∟ represents a right angle.

Perpendicular lines are intersecting lines that form right angles.

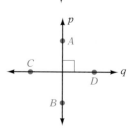

The symbol ⊥ means "is perpendicular to." In the accompanying figure, $AB \perp CD$ and $p \perp q$. Note that line p contains line segment AB and line q contains line segment CD. Perpendicular lines contain perpendicular line segments.

Complementary angles are two angles
whose sum is 90°.

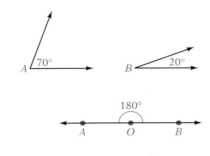

$\angle A + \angle B = 70° + 20° = 90°$
$\angle A$ and $\angle B$ are complementary angles.

$\frac{1}{2}$ of a revolution is 180°. A 180° angle is
called a **straight angle.** $\angle AOB$ in the figure is a straight angle.

Supplementary angles are two angles
whose sum is 180°.

$\angle C + \angle D = 130° + 50° = 180°$
$\angle C$ and $\angle D$ are supplementary angles.

An **acute angle** is an angle whose measure is between 0° and 90°. $\angle D$ in the figure above is an acute angle. An **obtuse angle** is an angle whose measure is
between 90° and 180°. $\angle C$ in the figure above is an obtuse angle.

In the accompanying figure, $\angle DAC = 45°$
and $\angle CAB = 55°$.

$\angle DAB = \angle DAC + \angle CAB$
$\qquad\quad = 45° + 55° = 100°$

Example 1
Given that $MN = 15$, $NO = 18$, and
$MP = 48$, find the length of OP.

Solution
$\quad MP = MN + NO + OP$
$\quad\ 48 = 15 + 18 + OP$
$\quad\ 48 = 33 + OP$
$48 - 33 = 33 - 33 + OP$
$\quad\ 15 = OP$

You Try It 1
Given that $QR = 24$, $ST = 17$, and $QT = 62$,
find the length of RS.

Your solution

Example 2
Find the complement of a 32° angle.

Solution
Let x represent the complement of 32°.

$\quad\quad x + 32° = 90°$ • The sum of
$x + 32° - 32° = 90° - 32°$ complementary
$\quad\quad\quad\quad x = 58°$ angles is 90°.

58° is the complement of 32°.

You Try It 2
Find the supplement of a 32° angle.

Your solution

Solutions on p. S30

Example 3
Find the measure of ∠x.

Solution

$$\angle x + 47° = 90°$$
$$\angle x + 47° - 47° = 90° - 47°$$
$$\angle x = 43°$$

You Try It 3
Find the measure of ∠a.

Your solution

Solution on p. S30

Objective B   **To define and describe geometric figures**

A **triangle** is a closed, three-sided plane figure. Figure *ABC* is a triangle. *AB* is called the **base.** The line *CD*, perpendicular to the base, is called the **height.**

The sum of the three angles in a triangle is 180°.

$$\angle A + \angle B + \angle C = 180°$$

➡ In triangle *DEF*, ∠D = 32° and ∠E = 88°. Find the measure of ∠F.

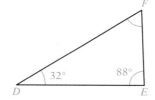

$$\angle D + \angle E + \angle F = 180°$$
$$32° + 88° + \angle F = 180°$$
$$120° + \angle F = 180°$$
$$120° - 120° + \angle F = 180° - 120°$$
$$\angle F = 60°$$

- The sum of the three angles in a triangle is 180°.
- ∠D = 32° and ∠E = 88°.
- Solve for ∠F.

A **right triangle** contains one right angle. The side opposite the right angle is called the **hypotenuse.** The other two sides are called **legs.** In a right triangle, the two acute angles are complementary.

$$\angle A + \angle B = 90°$$

➡ In the right triangle at the left, ∠A = 30°. Find the measure of ∠B.

$$\angle A + \angle B = 90°$$
$$30° + \angle B = 90°$$
$$30° - 30° + \angle B = 90° - 30°$$
$$\angle B = 60°$$

- The two acute angles are complementary.
- ∠A = 30°.
- Solve for ∠B.

A **quadrilateral** is a closed, four-sided plane figure. Three quadrilaterals with special characteristics are described here.

A **parallelogram** has opposite sides parallel and equal. The distance *AE* between the parallel sides is called the **height.**

Parallelogram

A **rectangle** is a parallelogram that has four right angles.

A **square** is a rectangle that has four equal sides.

Rectangle   Square

A **circle** is a plane figure in which all points are the same distance from point O, which is called the **center** of the circle.

The **diameter** (d) is a line segment through the center of the circle with endpoints on the circle. AB is a diameter of the circle.

The **radius** (r) is a line segment from the center to a point on the circle. OC is a radius of the circle.

$$d = 2r \quad \text{or} \quad r = \frac{1}{2}d$$

Circle

➡ The line segment AB is a diameter of the circle shown. Find the radius of the circle.

The radius is one-half the diameter. Therefore,

$$r = \frac{1}{2}d$$
$$= \frac{1}{2}(8 \text{ in.}) \quad \bullet \quad d = 8 \text{ in.}$$
$$= 4 \text{ in.}$$

Geometric solids are figures in space, or space figures. Four common space figures are the rectangular solid, cube, sphere, and cylinder.

A **rectangular solid** is a solid in which all six faces are rectangles.

Rectangular Solid

A **cube** is a rectangular solid in which all six faces are squares.

Cube

A **sphere** is a solid in which all points on the surface are the same distance from point O, which is called the **center** of the sphere.

The **diameter** of the sphere is a line segment going through the center with endpoints on the sphere. AB is a diameter of the sphere.

Sphere

The **radius** of the sphere is a line segment from the center to a point on the sphere. OC is a radius of the sphere.

$$r = \frac{1}{2}d \quad \text{or} \quad d = 2r$$

⇒ The radius of the sphere shown at the right is
5 cm. Find the diameter of the sphere.

$d = 2r$ • **The diameter equals twice the radius.**
$\quad = 2(5 \text{ cm})$ • $r = 5$ **cm.**
$\quad = 10 \text{ cm}$

The diameter is 10 cm.

The most common **cylinder** is one in which the
bases are circles and are perpendicular to the
height.

Cylinder

Example 4

One angle in a right triangle is equal to 50°.
Find the measure of the other angles.

Solution

In a right triangle, one angle measures 90° and
the two acute angles are complementary.

$$\angle A + \angle B = 90°$$
$$\angle A + 50° = 90°$$
$$\angle A + 50° - 50° = 90° - 50°$$
$$\angle A = 40°$$

The other angles measure 90° and 40°.

You Try It 4

A right triangle has one angle equal to 7°. Find
the measure of the other angles.

Your solution

Example 5

Two angles of a triangle measure 42° and 103°.
Find the measure of the third angle.

Solution

The sum of the three angles of a triangle is
180°.

$$\angle A + \angle B + \angle C = 180°$$
$$\angle A + 42° + 103° = 180°$$
$$\angle A + 145° = 180°$$
$$\angle A + 145° - 145° = 180° - 145°$$
$$\angle A = 35°$$

The measure of the third angle is 35°.

You Try It 5

Two angles of a triangle measure 62° and 45°.
Find the measure of the other angle.

Your solution

Example 6

A circle has a radius of 8 cm. Find the
diameter.

Solution

$d = 2r$
$\quad = 2 \cdot 8 \text{ cm} = 16 \text{ cm}$

The diameter is 16 cm.

You Try It 6

A circle has a diameter of 8 in. Find the radius.

Your solution

Solutions on pp. S30–S31

Objective C   **To solve problems involving angles formed by intersecting lines**

Four angles are formed by the intersection of two lines. If the two lines are perpendicular, then each of the four angles is a right angle. If the two lines are not perpendicular, then two of the angles formed are acute angles and two of the angles are obtuse angles. The two acute angles are always opposite each other, and the two obtuse angles are always opposite each other.

In the figure, $\angle w$ and $\angle y$ are acute angles. $\angle x$ and $\angle z$ are obtuse angles. Two angles that are on opposite sides of the intersection of two lines are called **vertical angles.** Vertical angles have the same measure. $\angle w$ and $\angle y$ are vertical angles. $\angle x$ and $\angle z$ are vertical angles.

$$\angle w = \angle y$$
$$\angle x = \angle z$$

Two angles that share a common side are called **adjacent angles.** In the previous figure, $\angle x$ and $\angle y$ are adjacent angles, as are $\angle y$ and $\angle z$, $\angle z$ and $\angle w$, and $\angle w$ and $\angle x$. Adjacent angles of intersecting lines are supplementary angles.

$$\angle x + \angle y = 180°$$
$$\angle y + \angle z = 180°$$
$$\angle z + \angle w = 180°$$
$$\angle w + \angle x = 180°$$

➡ In the figure at the left, $\angle c = 65°$. Find the measures of angles a, b, and d.

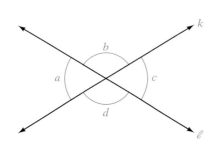

$\angle a = 65°$

- $\angle c = \angle a$ because $\angle c$ and $\angle a$ are vertical angles.

$$\angle b + \angle c = 180°$$

- $\angle c$ is supplementary to $\angle b$ because $\angle c$ and $\angle b$ are adjacent angles.

$$\angle b + 65° = 180°$$
$$\angle b + 65° - 65° = 180° - 65°$$
$$\angle b = 115°$$

- $\angle c = 65°$.

$\angle d = 115°$

- $\angle b = \angle d$ because $\angle b$ and $\angle d$ are vertical angles.

A line intersecting two other lines at two different points is called a **transversal.**

If the lines cut by a transversal are parallel lines and the transversal is perpendicular to the parallel lines, then all eight angles formed are right angles.

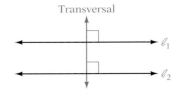

If the lines cut by a transversal are parallel lines and the transversal is not perpendicular to the parallel lines, then all four acute angles have the same measure and all four obtuse angles have the same measure. For the accompanying figure,

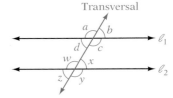

$$\angle a = \angle c = \angle w = \angle y \quad \text{and} \quad \angle b = \angle d = \angle x = \angle z$$

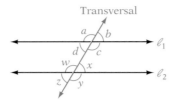

Alternate interior angles are two angles that are on opposite sides of the transversal and between the parallel lines. For the figure at the left, $\angle c$ and $\angle w$ are alternate interior angles. $\angle d$ and $\angle x$ are alternate interior angles. Alternate interior angles have the same measure.

Alternate exterior angles are two angles that are on opposite sides of the transversal and outside the parallel lines. For the figure at the left, $\angle a$ and $\angle y$ are alternate exterior angles. $\angle b$ and $\angle z$ are alternate exterior angles. Alternate exterior angles have the same measure.

Corresponding angles are two angles that are on the same side of the transversal and are both acute angles or are both obtuse angles. For the figure at the top left, the following pairs of angles are corresponding angles: $\angle a$ and $\angle w$, $\angle d$ and $\angle z$, $\angle b$ and $\angle x$, $\angle c$ and $\angle y$. Corresponding angles have the same measure.

➡ In the figure at the left, $\ell_1 \parallel \ell_2$ and $\angle c = 58°$. Find the measures of $\angle f$, $\angle h$, and $\angle g$.

$\angle f = 58°$ • $\angle c = \angle f$ because $\angle f$ and $\angle c$ are alternate interior angles.

$\angle h = 58°$ • $\angle c = \angle h$ because $\angle c$ and $\angle h$ are corresponding angles.

$\angle g + \angle h = 180°$ • $\angle g$ is supplementary to $\angle h$.
$\angle g + 58° = 180°$ • $\angle h = 58°$.
$\angle g = 122°$ • Subtract 58° from each side.

Example 7
In the figure,
$\angle a = 75°$.
Find $\angle b$.

You Try It 7
In the figure,
$\angle a = 125°$.
Find $\angle b$.

Solution
$\angle a + \angle b = 180°$ • $\angle a$ and $\angle b$ are supplementary.
$75° + \angle b = 180°$ • $\angle a = 75°$.
$\angle b = 105°$ • Subtract 75° from each side.

Your solution

Example 8
In the figure,
$\ell_1 \parallel \ell_2$ and
$\angle a = 70°$.
Find $\angle b$.

You Try It 8
In the figure,
$\ell_1 \parallel \ell_2$ and
$\angle a = 120°$.
Find $\angle b$.

Solution
$\angle c = \angle a = 70°$ • Corresponding angles are equal.
$\angle b + \angle c = 180°$ • $\angle b$ and $\angle c$ are supplementary.
$\angle b + 70° = 180°$ • $\angle c = 70°$.
$\angle b = 110°$ • Subtract 70° from each side.

Your solution

Solutions on p. S31

12.1 Exercises

Objective A

1. The measure of an acute angle is between ___ and ___.

2. The measure of an obtuse angle is between ___ and ___.

3. How many degrees are in a straight angle?

4. Two lines that intersect at right angles are ___ lines.

5. In the figure, *EF* = 20 and *FG* = 10. Find the length of *EG*.

6. In the figure, *EF* = 18 and *FG* = 6. Find the length of *EG*.

7. In the figure, it is given that *QR* = 7 and *QS* = 28. Find the length of *RS*.

8. In the figure, it is given that *QR* = 15 and *QS* = 45. Find the length of *RS*.

9. In the figure, it is given that *AB* = 12, *CD* = 9, and *AD* = 35. Find the length of *BC*.

10. In the figure, it is given that *AB* = 21, *BC* = 14, and *AD* = 54. Find the length of *CD*.

11. Find the complement of a 31° angle.

12. Find the complement of a 62° angle.

13. Find the supplement of a 72° angle.

14. Find the supplement of a 162° angle.

15. Find the complement of a 13° angle.

16. Find the complement of an 88° angle.

17. Find the supplement of a 127° angle.

18. Find the supplement of a 7° angle.

In Exercises 19 and 20, find the measure of angle *AOB*.

19.

20.

In Exercises 21 to 24, find the measure of angle *a*.

21.

22.

23.

24.

25. In the figure, it is given that ∠*LOM* = 53° and ∠*LON* = 139°. Find the measure of ∠*MON*.

26. In the figure, it is given that ∠*MON* = 38° and ∠*LON* = 85°. Find the measure of ∠*LOM*.

Objective B

27. What is the sum of the three angles of a triangle?

28. Name the side opposite the right angle in a right triangle.

29. Name a parallelogram with four right angles.

30. Name the rectangle with four equal sides.

31. Name a rectangular solid in which all six faces are squares.

32. Name the solid in which all points are the same distance from the center.

33. Name a quadrilateral in which opposite sides are parallel and equal.

34. Name the plane figure in which all points are the same distance from the center.

35. Name the solid in which the bases are circular and perpendicular to the height.

36. Name a solid in which all the faces are rectangles.

37. A triangle has a 13° angle and a 65° angle. Find the measure of the other angle.

38. A triangle has a 105° angle and a 32° angle. Find the measure of the other angle.

39. A right triangle has a 45° angle. Find the measure of the other two angles.

40. A right triangle has a 62° angle. Find the measure of the other two angles.

41. A triangle has a 62° angle and a 104° angle. Find the measure of the other angle.

42. A triangle has a 30° angle and a 45° angle. Find the measure of the other angle.

43. A right triangle has a 25° angle. Find the measure of the other two angles.

44. Two angles of a triangle are 42° and 105°. Find the measure of the other angle.

45. Find the radius of a circle with a diameter of 16 in.

46. Find the radius of a circle with a diameter of 9 ft.

47. Find the diameter of a circle with a radius of $2\frac{1}{3}$ ft.

48. Find the diameter of a circle with a radius of 24 cm.

49. The radius of a sphere is 3.5 cm. Find the diameter.

50. The radius of a sphere is $1\frac{1}{2}$ ft. Find the diameter.

51. The diameter of a sphere is 4 ft 8 in. Find the radius.

52. The diameter of a sphere is 1.2 m. Find the radius.

Objective C

In Exercises 53 to 56, find the measures of angles a and b.

53.

54.

55.

56.

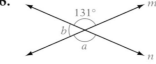

In Exercises 57 to 64, it is given that $\ell_1 \parallel \ell_2$. Find the measures of angles a and b.

57.

58.

59.

60.

61.

62.

63.

64.

APPLYING THE CONCEPTS

65. **a.** What is the smallest possible whole number of degrees in an angle of a triangle?
   b. What is the largest possible whole number of degrees in an angle of a right triangle?

66. Determine whether the statement is always true, sometimes true, or never true.
   a. Two lines that are both parallel to a third line are parallel to each other.
   b. A triangle contains at least two acute angles.
   c. Vertical angles are complementary angles.

67. If AB and CD intersect at point O, and $\angle AOC = \angle BOC$, explain why AB is perpendicular to CD.

12.2 Plane Geometric Figures

Objective A   **To find the perimeter of plane geometric figures**

A **polygon** is a closed figure determined by three or more line segments that lie in a plane. The line segments that form the polygon are called its **sides.** The figures below are examples of polygons.

    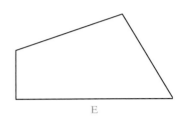

A   B   C   D   E

Point of Interest

Although a polygon is defined in terms of its *sides* (see the definition above), the word actually comes from the Latin word *polygonum*, which means having many *angles*. This is certainly the case for a polygon.

A **regular polygon** is one in which each side has the same length and each angle has the same measure. The polygons in Figures A, C, and D above are regular polygons.

The name of a polygon is based on the number of its sides. The table below lists the names of polygons that have from 3 to 10 sides.

Number of Sides	Name of the Polygon
3	Triangle
4	Quadrilateral
5	Pentagon
6	Hexagon
7	Heptagon
8	Octagon
9	Nonagon
10	Decagon

Triangles and quadrilaterals are two of the most common types of polygons. Triangles are distinguished by the number of equal sides and also by the measures of their angles.

An **isosceles triangle** has two sides of equal length. The angles opposite the equal sides are of equal measure.

$AC = BC$

$\angle A = \angle B$

The three sides of an **equilateral triangle** are of equal length. The three angles are of equal measure.

$AB = BC = AC$

$\angle A = \angle B = \angle C$

A **scalene triangle** has no two sides of equal length. No two angles are of equal measure.

An **acute triangle** has three acute angles.

An **obtuse triangle** has one obtuse angle.

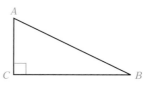

A **right triangle** has a right angle.

Quadrilaterals are also distinguished by their sides and angles, as shown below. Note that a rectangle, a square, and a rhombus are different forms of a parallelogram.

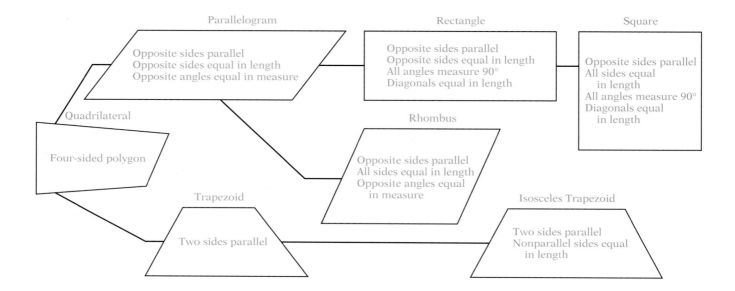

The **perimeter** of a plane geometric figure is a measure of the distance around the figure. Perimeter is used in buying fencing for a lawn or determining how much baseboard is needed for a room.

The perimeter of a triangle is the sum of the lengths of the three sides.

Perimeter of a Triangle

$P = a + b + c$

➡ Find the perimeter of the triangle shown at the right.

$P = a + b + c$
$ = 3 \text{ cm} + 5 \text{ cm} + 6 \text{ cm}$
$ = 14 \text{ cm}$

The perimeter of the triangle is 14 cm.

The perimeter of a square is the sum of the four equal sides.

Perimeter of a Square

$P = 4s$

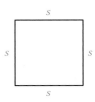

➡ Find the perimeter of the square shown at the right.

$P = 4s$
  $= 4(3 \text{ ft})$ • $s = 3$ **ft.**
  $= 12 \text{ ft}$

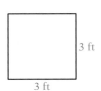

3 ft

3 ft

The perimeter of the square is 12 ft.

The perimeter of a quadrilateral is the sum of the lengths of the four sides.

A rectangle is a quadrilateral with opposite sides of equal length. The length of a rectangle refers to the longer side, and the width refers to the length of the shorter side.

Perimeter of a Rectangle

$P = 2L + 2W$

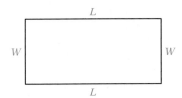

➡ Find the perimeter of the rectangle shown at the right.

$P = 2L + 2W$
  $= 2(6 \text{ m}) + 2(3 \text{ m})$ • $L = 6$ **m,** $W = 3$ **m.**
  $= 12 \text{ m} + 6 \text{ m}$
  $= 18 \text{ m}$

3 m

6 m

The perimeter of the rectangle is 18 m.

The distance around a circle is called the **circumference.** The circumference of a circle is equal to the product of π (pi) and the diameter.

Point of Interest

Archimedes (c. 287–212 B.C.) was the mathematician who gave us the approximate value of π as $\dfrac{22}{7} = 3\dfrac{1}{7}$. He actually showed that π was between $3\dfrac{10}{71}$ and $3\dfrac{1}{7}$. The approximation $3\dfrac{10}{71}$ is closer to the exact value of π, but it is more difficult to use.

Circumference of a Circle

$C = \pi d$

  or

$C = 2\pi r$ Because diameter = $2r$

The formula for circumference uses the number π (pi). The value of π can be approximated by a fraction or a decimal.

$$\pi \approx 3.14 \qquad \pi \approx \frac{22}{7}$$

The π key on a calculator gives a closer approximation of π than 3.14.

⇨ Find the circumference of the circle shown at the right.

$C = 2\pi r$
   $\approx 2 \cdot 3.14 \cdot 6$ in.   • $r = 6$ in.
   $= 37.68$ in.

The circumference of the circle is approximately 37.68 in.

Example 1

Find the perimeter of a rectangle with a width of $\frac{2}{3}$ ft and a length of 2 ft.

Solution

$P = 2L + 2W$

$= 2(2 \text{ ft}) + 2\left(\frac{2}{3} \text{ ft}\right)$ • $L = 2$ ft, $W = \frac{2}{3}$ ft

$= 4 \text{ ft} + \frac{4}{3} \text{ ft}$

$= 5\frac{1}{3} \text{ ft}$

The perimeter of the rectangle is $5\frac{1}{3}$ ft.

You Try It 1

Find the perimeter of a rectangle with a length of 2 m and a width of 0.85 m.

Your solution

Example 2

Find the perimeter of a triangle with sides 5 in., 7 in., and 8 in.

Solution

```
      5 in. /\  7 in.
           /  \
          /____\
          8 in.
```

$P = a + b + c$
   $= 5 \text{ in.} + 7 \text{ in.} + 8 \text{ in.}$
   $= 20 \text{ in.}$

The perimeter of the triangle is 20 in.

You Try It 2

Find the perimeter of a triangle with sides 12 cm, 15 cm, and 18 cm.

Your solution

Example 3

Find the circumference of a circle with a radius of 18 cm. Use 3.14 for π.

Solution

$C = 2\pi r$
   $\approx 2 \cdot 3.14 \cdot 18$ cm
   $= 113.04$ cm

The circumference is approximately 113.04 cm.

You Try It 3

Find the circumference of a circle with a diameter of 6 in. Use 3.14 for π.

Your solution

Solutions on p. S31

| *Objective B* | **To find the perimeter of composite geometric figures** | |

Composite geometric figures are figures made from two or more geometric figures. The following composite is made from part of a rectangle and part of a circle:

$\text{Perimeter of the composite figure} = 3$ sides of a rectangle $+ \frac{1}{2}$ the circumference of a circle

$\text{Perimeter of the composite figure} = 2L + W + \frac{1}{2}\pi d$

The perimeter of the composite figure below is found by adding the measures of twice the length plus the width plus one-half the circumference of the circle.

$$P = 2L + W + \frac{1}{2}\pi d$$

$$\approx 2(12 \text{ m}) + 4 \text{ m} + \frac{1}{2}(3.14)(4 \text{ m}) \qquad \bullet \ L = 12 \text{ m}, W = 4 \text{ m}, d = 4 \text{ m}.$$

$$= 34.28 \text{ m}$$

The perimeter is approximately 34.28 m.

- -

Example 4

Find the perimeter of the composite figure.

Use $\frac{22}{7}$ for π.

Solution

$$\begin{array}{c}\text{Perimeter}\\\text{of}\\\text{composite}\\\text{figure}\end{array} = \begin{array}{c}\text{sum of}\\\text{lengths}\\\text{of the 4}\\\text{sides}\end{array} + \begin{array}{c}\frac{1}{2}\text{ the cir-}\\\text{cumference}\\\text{of the}\\\text{circle}\end{array}$$

$$P = \quad 4s \quad + \quad \frac{1}{2}\pi d$$

$$\approx 4(5 \text{ cm}) + \frac{1}{2}\left(\frac{22}{7}\right)(7 \text{ cm})$$

$$= 20 \text{ cm} + 11 \text{ cm} = 31 \text{ cm}$$

The perimeter is approximately 31 cm.

You Try It 4

Find the perimeter of the composite figure. Use 3.14 for π.

Your solution

Solution on p. S31

Objective C   **To solve application problems**   ◄ 12 ►

Example 5

The dimensions of a triangular sail are 18 ft, 11 ft, and 15 ft. What is the perimeter of the sail?

Strategy

To find the perimeter, use the formula for the perimeter of a triangle.

Solution
$P = a + b + c$
$ = 18 \text{ ft} + 11 \text{ ft} + 15 \text{ ft}$
$ = 44 \text{ ft}$

The perimeter of the sail is 44 ft.

You Try It 5

What is the perimeter of a standard piece of computer paper that measures $8\frac{1}{2}$ in. by 11 in.?

Your strategy

Your solution

Example 6

If fencing costs $4.75 per foot, how much will it cost to fence a rectangular lot that is 108 ft wide and 240 ft long?

Strategy

To find the cost of the fence:
• Find the perimeter of the lot.
• Multiply the perimeter by the per-foot cost of fencing.

Solution
$P = 2L + 2W$
$ = 2(240 \text{ ft}) + 2(108 \text{ ft})$
$ = 480 \text{ ft} + 216 \text{ ft}$
$ = 696 \text{ ft}$

$\text{Cost} = 696 \times 4.75 = 3306$

The cost is $3306.

You Try It 6

Metal stripping is being installed around a workbench that is 0.74 m wide and 3 m long. At $2.76 per meter, find the cost of the metal stripping. Round to the nearest cent.

Your strategy

Your solution

Solutions on p. S31

12.2 Exercises

Objective A

In Exercises 1 to 8, find the perimeter or circumference of the given figures. Use 3.14 for π.

1.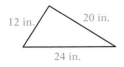
12 in.   20 in.
24 in.

2.
14 cm   13 cm
12 cm

3.
5 ft
5 ft

4.
2 m
2 m

5.
14 cm
32 cm

6.
5 ft
18 ft

7.
15 cm

8.
4 in.

9. Find the perimeter of a triangle with sides 2 ft 4 in., 3 ft, and 4 ft 6 in.

10. Find the perimeter of a rectangle with a length of 2 m and a width of 0.8 m.

11. Find the circumference of a circle with a radius of 8 cm. Use 3.14 for π.

12. Find the circumference of a circle with a diameter of 14 in. Use $\frac{22}{7}$ for π.

13. Find the perimeter of a square in which each side is equal to 60 m.

14. Find the perimeter of a triangle in which each side is $1\frac{2}{3}$ ft.

15. Find the perimeter of a five-sided figure with sides of 22 cm, 47 cm, 29 cm, 42 cm, and 17 cm.

16. Find the perimeter of a rectangular farm that is $\frac{1}{2}$ mi wide and $\frac{3}{4}$ mi long.

Objective B

In the figures accompanying Exercises 17 to 24, find the perimeter. Use 3.14 for π.

17.

18.

19.

20.

21.

22.

23.

24.

Objective C   *Application Problems*

25. How many feet of fencing should be purchased for a rectangular garden that is 18 ft long and 12 ft wide?

26. Wall-to-wall carpeting is installed in a room that is 12 ft long and 10 ft wide. The edges of the carpet are nailed to the floor. Along how many feet must the carpet be nailed down?

27. How many feet of binding are required to bind the edge of a rectangular quilt that measures 3.5 ft by 8.5 ft?

28. Find the length of molding needed to put around a circular table that is 3.8 ft in diameter. Use 3.14 for π.

29. The rectangular lot shown in the figure at the right is being fenced. The fencing along the road is to cost $6.20 per foot. The rest of the fencing costs $5.85 per foot. Find the total cost to fence the lot.

30. Bias binding is to be sewed around the edge of a rectangular tablecloth measuring 72 in. by 45 in. Each package of bias binding costs $3.50 and contains 15 ft of binding. How many packages of bias binding are needed for the tablecloth?

31. A bicycle tire has a diameter of 24 in. How many feet does the bicycle travel when the wheel makes 5 revolutions? Use 3.14 for π.

32. A tricycle tire has a diameter of 12 in. How many feet does the tricycle travel when the wheel makes 8 revolutions? Use 3.14 for π.

33. The floor plan of a roller rink is shown in the figure at the right.
 a. Use estimation to determine whether the perimeter of the rink is more than 70 m or less than 70 m.
 b. Calculate the perimeter of the roller rink. Use 3.14 for π.

34. A rain gutter is being installed on a home that has the dimensions shown in the figure at the right. At a cost of $11.30 per meter, how much will it cost to install the rain gutter?

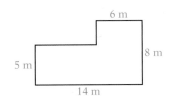

35. Find the length of weather stripping installed around the arched door shown in the figure at the right. Use 3.14 for π.

6 ft 6 in.

←— 3 ft —→

36. The distance from Earth to the sun is 93,000,000 mi. Approximate the distance Earth travels in making 1 revolution about the sun. Use 3.14 for π.

37. The distance from the surface to the center of Earth is 6356 km. Approximate the circumference of Earth. Use 3.14 for π.

APPLYING THE CONCEPTS

38. a. If the diameter of a circle is doubled, how many times larger is the resulting circumference?
   b. If the radius of a circle is doubled, how many times larger is the resulting circumference?

39. In the following pattern, the length of one side of a square is 1 unit. Find the perimeter of the eighth figure in the pattern.

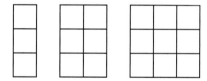

40. Remove six toothpicks from the figure at the right in such a way as to leave two squares.

41. An equilateral triangle is placed inside an equilateral triangle as shown at the right. Now three more equilateral triangles are placed inside the unshaded equilateral triangles. The process is repeated again. Determine the perimeter of all the shaded triangles in Figure C.

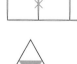

←2 cm→

Figure A   Figure B   Figure C

42. A wire whose length is given as x inches is bent into a square. Express the length of a side of the square in terms of x.

x

43. A forest ranger must determine the diameter of a redwood tree. Explain how the ranger could do this without cutting down the tree.

12.3 Area

Objective A   **To find the area of geometric figures**

Area is a measure of the amount of surface in a region. Area can be used to describe the size of a rug, a parking lot, a farm, or a national park. Area is measured in square units.

A square that measures 1 in. on each side has an area of 1 square inch, which is written 1 in^2.

A square that measures 1 cm on each side has an area of 1 square centimeter, which is written 1 cm^2.

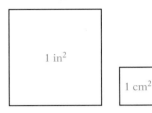

Larger areas can be measured in square feet (ft^2), square meters (m^2), square miles (mi^2), acres (43,560 ft^2), or any other square unit.

The area of a geometric figure is the number of squares that are necessary to cover the figure. In the figures below, two rectangles have been drawn and covered with squares. In the figure on the left, 12 squares, each of area 1 cm^2, were used to cover the rectangle. The area of the rectangle is 12 cm^2. In the figure on the right, 6 squares, each of area 1 in^2, were used to cover the rectangle. The area of the rectangle is 6 in^2.

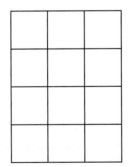

The area of the rectangle is 12 cm^2.

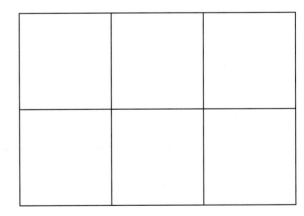

The area of the rectangle is 6 in^2.

Note from the above figures that the area of a rectangle can be found by multiplying the length of the rectangle by its width.

Area of a Rectangle

$A = LW$

➡ Find the area of the rectangle shown at the right.

$A = LW$
  $= (8 \text{ ft})(5 \text{ ft})$ • $L = 8 \text{ ft}, W = 5 \text{ ft}.$
  $= 40 \text{ ft}^2$

The area of the rectangle is 40 ft^2.

A square is a rectangle in which all sides are the same length. Therefore, both the length and width can be represented by a side.

Area of a Square

$A = s \cdot s = s^2$

➡ Find the area of the square shown at the right.

$A = s^2$
$\quad = (14 \text{ cm})^2 \qquad \bullet \ s = 14 \text{ cm.}$
$\quad = 196 \text{ cm}^2$

The area of the square is 196 cm².

The area of a circle is equal to the product of π and the square of the radius.

Area of a Circle

$A = \pi r^2$

Radius

➡ Find the area of the circle shown at the right.

$A = \pi r^2$
$\quad = \pi(8 \text{ in.})^2 = 64\pi \text{ in}^2$
$\quad \approx 64 \cdot 3.14 \text{ in}^2 = 200.96 \text{ in}^2$

The area is exactly 64π in².
The area is approximately 200.96 in².

8 in.

In the figure below, *AB* is the base of the triangle, and *CD*, which is perpendicular to the base, is the height. The area of a triangle is one-half the product of the base and the height.

Area of a Triangle

$A = \dfrac{1}{2}bh$

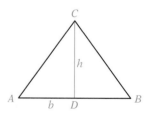

➡ Find the area of the triangle shown below.

$A = \dfrac{1}{2}bh$

$\quad = \dfrac{1}{2}(20 \text{ m})(5 \text{ m}) \qquad \bullet \ b = 20 \text{ m}, \ h = 5 \text{ m.}$

$\quad = 50 \text{ m}^2$

The area of the triangle is 50 m².

Example 1 Find the area of a circle with a diameter of 9 cm. Use 3.14 for π.

You Try It 1 Find the area of a triangle with a base of 24 in. and a height of 14 in.

Solution $r = \dfrac{1}{2}d = \dfrac{1}{2}(9 \text{ cm}) = 4.5 \text{ cm}$

$A = \pi r^2$
$\approx 3.14(4.5 \text{ cm})^2 = 63.585 \text{ cm}^2$

The area is approximately 63.585 cm².

Your solution

Solution on p. S31

Objective B   **To find the area of composite geometric figures**   ‹ 12 ›

The area of the composite figure shown below is found by calculating the area of the rectangle and then subtracting the area of the triangle.

$A = LW - \dfrac{1}{2}bh$

$= (20 \text{ in.})(8 \text{ in.}) - \dfrac{1}{2}(20 \text{ in.})(3 \text{ in.}) = 160 \text{ in}^2 - 30 \text{ in}^2 = 130 \text{ in}^2$

Example 2
Find the area of the shaded portion of the figure. Use 3.14 for π.

8 m

8 m

You Try It 2
Find the area of the composite figure.

6 in.

4 in.   10 in.

Solution

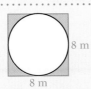

Your solution

Area of shaded portion = area of square − area of circle

$A = s^2 - \pi r^2$
$= (8 \text{ m})^2 - \pi(4 \text{ m})^2$
$\approx 64 \text{ m}^2 - 3.14(16 \text{ m}^2)$
$= 64 \text{ m}^2 - 50.24 \text{ m}^2 = 13.76 \text{ m}^2$

The area is approximately 13.76 m².

Solution on p. S31

Objective C   **To solve application problems**   ‹ 12 ›

....

Example 3

A walkway 2 m wide is built along the front and along both sides of a building, as shown in the figure. Find the area of the walkway.

```
|←— 50 m —→|
┌──────────────┐ ┬
│ ┌──────────┐ │ │
│ │ ○     °°° │ │ 35 m
│ │          │ │ │
│ │ ▯        │ │ │
│ └──────────┘ │ ┴
└──────────────┘
  ↕         →| |←—2 m
  2 m
```

Strategy

To find the area of the walkway, add the area of the front section (54 m · 2 m) and the area of the two side sections (each 35 m · 2 m).

Solution

$$\text{Area of walkway} = \underbrace{\text{area of front section}} + \underbrace{2(\text{area of one side section})}$$

$$A = (54\text{ m})(2\text{ m}) + 2(35\text{ m})(2\text{ m})$$
$$= 108\text{ m}^2 \quad + 140\text{ m}^2$$
$$= 248\text{ m}^2$$

The area of the walkway is 248 m².

You Try It 3

New carpet is installed in a room measuring 9 ft by 12 ft. Find the area of the room in square yards. ($9\text{ ft}^2 = 1\text{ yd}^2$)

Your strategy

Your solution

Solution on p. S31

12.3 Exercises

In Exercises 1 to 8, find the area of the given figures. Use 3.14 for π.

1.

6 ft
24 ft

2.

8 in.
18 in.

3.

9 in.
9 in.

4.

4 in.
4 in.

5.

4 ft

6.

3 cm

7.

4 in.
10 in.

8.

6 m
7 m

9. Find the area of a right triangle with a base of 3 cm and a height of 1.42 cm.

10. Find the area of a triangle with a base of 3 ft and a height of $\frac{2}{3}$ ft.

11. Find the area of a square with a side of 4 ft.

12. Find the area of a square with a side of 10 cm.

13. Find the area of a rectangle with a length of 43 in. and a width of 19 in.

14. Find the area of a rectangle with a length of 82 cm and a width of 20 cm.

15. Find the area of a circle with a radius of 7 in. Use $\frac{22}{7}$ for π.

16. Find the area of a circle with a diameter of 40 cm. Use 3.14 for π.

Objective B

Find the area. Use 3.14 for π.

17.

18.

19.

20.

21.

22.

23.

24.

Objective C   *Application Problems*

25. Artificial turf is being used to cover a playing field. If the field is rectangular with a length of 100 yd and a width of 75 yd, how much artificial turf must be purchased to cover the field?

26. The telescope lens of the Hale telescope at Mount Palomar, California, has a diameter of 200 in. Find the area of the lens. Leave the answer in terms of π.

27. An irrigation system waters a circular field that has a 50-foot radius. Find the area watered by the irrigation system. Use 3.14 for π.

28. A fabric wall hanging is to fill a space that measures 5 m by 3.5 m. Allowing for 0.1 m of the fabric to be folded back along each edge, how much fabric must be purchased for the wall hanging?

29. You plan to stain the wooden deck attached to your house. The deck measures 10 ft by 8 ft. A quart of stain will cost $9.95 and will cover 50 ft². How many quarts of stain should you buy?

30. A carpet is to be installed in one room and a hallway, as shown in the diagram at the right. At a cost of $18.50 per square meter, how much will it cost to carpet the area?

31. Find the area of a concrete driveway with the measurements shown in the figure.

32. You want to tile your kitchen floor. The floor measures 12 ft by 9 ft. How many tiles, each a square with side $1\frac{1}{2}$ ft, should you purchase for the job?

33. You are wallpapering two walls of a child's room. One wall measures 9 ft by 8 ft, and the other measures 11 ft by 8 ft. The wallpaper costs $18.50 per roll, and each roll of the wallpaper will cover 40 ft². What is the cost to wallpaper the two walls?

34. Find the area of the 2-meter boundary around the swimming pool shown in the figure.

35. An urban renewal project involves reseeding a park that is in the shape of a square, 60 ft on each side. Each bag of grass seed costs $5.75 and will seed 1200 ft². How much money should be budgeted for buying grass seed for the park?

36. The roller rink shown in the figure at the right is to be covered with hardwood floor.
  a. Without doing the calculations, indicate whether the area of the rink is more than 8000 ft² or less than 8000 ft².
  b. Calculate how much hardwood floor is needed to cover the roller rink. Use 3.14 for π.

37. Find the total area of the national park with the dimensions shown in the figure. Use 3.14 for π.

38. Find the cost of plastering the walls of a room 22 ft wide, 25 ft 6 in. long, and 8 ft high. Subtract 120 ft² for windows and doors. The cost is $2.50 per square foot.

39. **a.** A circle has a radius of 8 in. Find the increase in area when the radius is increased by 2 in. Use 3.14 for π.
  b. A circle has a radius of 5 cm. Find the increase in area when the radius is doubled. Use 3.14 for π.

APPLYING THE CONCEPTS

40. What fractional part of the area of the larger of the two squares is the shaded area? Write your answer as a fraction in simplest form. This problem appeared in *Math Teacher*, vol. 86, No. 3 (September 1993).

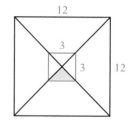

41. **a.** If both the length and width of a rectangle are doubled, how many times larger is the area of the resulting rectangle?
  b. If the radius of a circle is doubled, what happens to the area?
  c. If the diameter of a circle is doubled, what happens to the area?

42. The circles at the right are identical. Is the area in the circles to the left of the line equal to, less than, or greater than the area in the circles to the right of the line? Explain your answer.

43. Determine whether the statement is always true, sometimes true, or never true.
  a. If two triangles have the same perimeter, then they have the same area.
  b. If two rectangles have the same area, then they have the same perimeter.
  c. If two squares have the same area, then the sides of the squares have the same length.

44. All of the dots at the right are equally spaced, horizontally and vertically, 1 inch apart. What is the area of the triangle?

12.4 Volume

Objective A   **To find the volume of geometric solids**

Volume is a measure of the amount of space inside a figure in space. Volume can be used to describe the amount of heating gas used for cooking, the amount of concrete delivered for the foundation of a house, or the amount of water in storage for a city's water supply.

A cube that is 1 ft on each side has a volume of 1 cubic foot, which is written 1 ft³.

A cube that measures 1 cm on each side has a volume of 1 cubic centimeter, which is written 1 cm³.

The volume of a solid is the number of cubes that are necessary to fill the solid exactly. The volume of the rectangular solid at the right is 24 cm³ because it will hold exactly 24 cubes, each 1 cm on a side. Note that the volume can be found by multiplying the length times the width times the height.

Volume of a Rectangular Solid

$V = LWH$

⇒ Find the volume of a rectangular solid with a length of 9 in., a width of 3 in., and a height of 4 in.

$V = LWH$
$\quad = (9 \text{ in.})(3 \text{ in.})(4 \text{ in.})$ • $L = 9 \text{ in.}, W = 3 \text{ in.},$
$\quad = 108 \text{ in}^3$ $H = 4 \text{ in.}$

The volume of the rectangular solid is 108 in³.

The length, width, and height of a cube have the same measure. The volume of a cube is found by multiplying the side of the cube times itself three times (side cubed).

Volume of a Cube

$V = s^3$

➡ Find the volume of the cube shown at the right.

$V = s^3$
$= (3 \text{ ft})^3$ • $s = 3$ ft.
$= 27 \text{ ft}^3$

The volume of the cube is 27 ft³.

The volume of a sphere is found by multiplying four-thirds times pi (π) times the radius cubed.

Volume of a Sphere

$V = \dfrac{4}{3}\pi r^3$

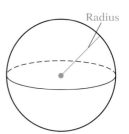

➡ Find the volume of the sphere shown below. Use 3.14 for π. Round to the nearest hundredth.

$V = \dfrac{4}{3}\pi r^3$

$\approx \dfrac{4}{3}(3.14)(2 \text{ in.})^3$ • $r = 2$ in.

$= \dfrac{4}{3}(3.14)(8 \text{ in}^3)$

$\approx 33.49 \text{ in}^3$

The volume is approximately 33.49 in³.

The volume of a cylinder is found by multiplying the area of the base of the cylinder (a circle) times the height.

Volume of a Cylinder

$V = \pi r^2 h$

⇒ Find the volume of the cylinder shown below. Use 3.14 for π.

$V = \pi r^2 h$
$\approx 3.14(3 \text{ cm})^2(8 \text{ cm})$ • $r = 3 \text{ cm}, h = 8 \text{ cm}.$
$= 3.14(9 \text{ cm}^2)(8 \text{ cm})$
$= 226.08 \text{ cm}^3$

The volume of the cylinder is approximately 226.08 cm³.

8 cm

Radius = 3 cm

Example 1

Find the volume of a rectangular solid with a length of 3 ft, a width of 1.5 ft, and a height of 2 ft.

Solution
$V = LWH$
$= (3 \text{ ft})(1.5 \text{ ft})(2 \text{ ft})$
$= 9 \text{ ft}^3$

The volume is 9 ft³.

You Try It 1

Find the volume of a rectangular solid with a length of 8 cm, a width of 3.5 cm, and a height of 4 cm.

Your solution

Example 2

Find the volume of a cube that has a side measuring 2.5 in.

Solution
$V = s^3$
$= (2.5 \text{ in.})^3$
$= 15.625 \text{ in}^3$

The volume is 15.625 in³.

You Try It 2

Find the volume of a cube with a side of 5 cm.

Your solution

Example 3

Find the volume of a cylinder with a radius of 12 cm and a height of 65 cm. Use 3.14 for π.

Solution
$V = \pi r^2 h$
$\approx 3.14(12 \text{ cm})^2(65 \text{ cm})$
$= 3.14(144 \text{ cm}^2)(65 \text{ cm})$
$= 29{,}390.4 \text{ cm}^3$

The volume is approximately 29,390.4 cm³.

You Try It 3

Find the volume of a cylinder with a diameter of 14 in. and a height of 15 in. Use $\frac{22}{7}$ for π.

Your solution

Solutions on p. S32

Example 4

Find the volume of a sphere with a diameter of 12 in. Use 3.14 for π.

Solution

$r = \dfrac{1}{2}d = \dfrac{1}{2}(12 \text{ in.}) = 6 \text{ in.}$

$V = \dfrac{4}{3}\pi r^3$

$\approx \dfrac{4}{3}(3.14)(6 \text{ in.})^3$

$= \dfrac{4}{3}(3.14)(216 \text{ in}^3)$

$= 904.32 \text{ in}^3$

The volume is approximately 904.32 in³.

You Try It 4

Find the volume of a sphere with a radius of 3 m. Use 3.14 for π.

Your solution

Solution on p. S32

Objective B   **To find the volume of composite geometric solids**

Composite geometric solids are solids made from two or more geometric solids. The solid shown is made from a cylinder and one-half of a sphere.

Volume of the composite solid = volume of the cylinder + $\dfrac{1}{2}$ the volume of the sphere

➡ Find the volume of the composite solid shown above if the radius of the base of the cylinder is 3 in. and the height of the cylinder is 10 in. Use 3.14 for π.

The volume equals the volume of a cylinder plus one-half the volume of a sphere. The radius of the sphere equals the radius of the base of the cylinder.

$V = \pi r^2 h + \dfrac{1}{2}\left(\dfrac{4}{3}\pi r^3\right)$

$\approx 3.14(3 \text{ in.})^2(10 \text{ in.}) + \dfrac{1}{2}\left(\dfrac{4}{3}\right)(3.14)(3 \text{ in.})^3$

$= 3.14(9 \text{ in}^2)(10 \text{ in.}) + \dfrac{1}{2}\left(\dfrac{4}{3}\right)(3.14)(27 \text{ in}^3)$

$= 282.6 \text{ in}^3 + 56.52 \text{ in}^3$

$= 339.12 \text{ in}^3$

The volume is approximately 339.12 in³.

Example 5

Find the volume of the solid in the figure. Use 3.14 for π.

Solution

| Volume of the solid | = | volume of rectangular solid | + | volume of cylinder |

$V = LWH + \pi r^2 h$
$\approx (8\ \text{cm})(8\ \text{cm})(2\ \text{cm}) + 3.14(1\ \text{cm})^2(2\ \text{cm})$
$= 128\ \text{cm}^3 + 6.28\ \text{cm}^3$
$= 134.28\ \text{cm}^3$

The volume is approximately 134.28 cm³.

You Try It 5

Find the volume of the solid in the figure. Use 3.14 for π.

Your solution

Example 6

Find the volume of the solid in the figure. Use 3.14 for π.

Solution

| Volume of the solid | = | volume of rectangular solid | − | volume of cylinder |

$V = LWH - \pi r^2 h$
$\approx (80\ \text{m})(40\ \text{m})(30\ \text{m}) - 3.14(14\ \text{m})^2(80\ \text{m})$
$= 96{,}000\ \text{m}^3 - 49{,}235.2\ \text{m}^3$
$= 46{,}764.8\ \text{m}^3$

The volume is approximately 46,764.8 m³.

You Try It 6

Find the volume of the solid in the figure. Use 3.14 for π.

Your solution

Solutions on p. S32

Objective C **To solve application problems** ‹12›

Example 7

An aquarium is 28 in. long, 14 in. wide, and 16 in. high. Find the volume of the aquarium.

Strategy

To find the volume of the aquarium, use the formula for the volume of a rectangular solid.

Solution

$V = LWH$
$\quad = (28 \text{ in.})(14 \text{ in.})(16 \text{ in.})$
$\quad = 6272 \text{ in}^3$

The volume of the aquarium is 6272 in³.

You Try It 7

Find the volume of a freezer that is 7 ft long, 3 ft high, and 2.5 ft wide.

Your strategy

Your solution

Example 8

Find the volume of the bushing shown in the figure below. Use 3.14 for π.

You Try It 8

Find the volume of the channel iron shown in the figure below.

Strategy

To find the volume of the bushing, subtract the volume of the half-cylinder from the volume of the rectangular solid.

Solution

Your strategy

Your solution

Volume of the bushing	=	volume of rectangular solid	−	½ the volume of cylinder

$V = LWH - \frac{1}{2}\pi r^2 h$

$\quad \approx (8 \text{ cm})(4 \text{ cm})(4 \text{ cm}) - \frac{1}{2}(3.14)(1 \text{ cm})^2(8 \text{ cm})$

$\quad = 128 \text{ cm}^3 - 12.56 \text{ cm}^3$

$\quad = 115.44 \text{ cm}^3$

The volume of the bushing is approximately 115.44 cm³.

Solutions on p. S32

12.4 Exercises

Objective A

In Exercises 1 to 8, find the volume. Round to the nearest hundredth. Use 3.14 for π.

1.

3 cm
12 cm
4 cm

2.

5 ft
6 ft
8 ft

3.

8 in.
8 in.
8 in.

4.

12 m
12 m
12 m

5.

8 in.

6.

7 in.

7.

12 cm
2 cm

8.

8 ft
5 ft

In Exercises 9 to 16, find the volume.

9. Find the volume of a rectangular solid with a length of 2 m, a width of 80 cm, and a height of 4 m.

10. Find the volume of a cylinder with a radius of 7 cm and a height of 14 cm. Use $\frac{22}{7}$ for π.

11. Find the volume of a sphere with an 11-millimeter radius. Use 3.14 for π. Round to the nearest hundredth.

12. Find the volume of a cube with a side of 2.14 m. Round to the nearest tenth.

13. Find the volume of a cylinder with a diameter of 12 ft and a height of 30 ft. Use 3.14 for π.

14. Find the volume of a sphere with a 6-foot diameter. Use 3.14 for π.

15. Find the volume of a cube with a side of $3\frac{1}{2}$ ft.

16. Find the volume of a rectangular solid with a length of 1.15 m, a width of 60 cm, and a height of 25 cm.

Objective B

In Exercises 17 to 22, find the volume. Use 3.14 for π.

17.

2 in.
1 in.
9 in.
6 in.

18.

6 ft
12 ft

19.

40 cm
80 cm
2 m
1.2 m

20.

1.5 m
1.5 m
50 cm
2 m
2 m

21.

22.

Objective C   *Application Problems*

Solve. Use 3.14 for π.

23. A rectangular tank at the fish hatchery is 9 m long, 3 m wide, and 1.5 m deep. Find the volume of the water in the tank when the tank is full.

24. A fuel tank in a booster rocket is a cylinder 10 ft in diameter and 52 ft high. Find the volume of the fuel tank.

25. A hot air balloon is in the shape of a sphere. Find the volume of a hot air balloon that is 32 ft in diameter. Round to the nearest hundredth.

26. A storage tank for propane is in the shape of a sphere that has a diameter of 9 m. Find the volume of the tank.

27. An oil tank, which is in the shape of a cylinder, is 4 m high and has a diameter of 6 m. The oil tank is two-thirds full. Find the number of cubic meters of oil in the tank. Round to the nearest hundredth.

28. A silo, which is in the shape of a cylinder, is 16 ft in diameter and has a height of 30 ft. The silo is three-fourths full. Find the volume of the portion of the silo that is not being used for storage. Round to the nearest hundredth.

29. An architect is designing the heating system for an auditorium and needs to know the volume of the structure. Find the volume of the auditorium with the measurements shown in the figure.

30. A swimming pool 50 ft long and 13 ft wide contains water to a depth of 10 ft. Find the total weight of the water in the swimming pool. (1 ft^3 weighs 62.4 lb.)

31. Find the volume of the bushing shown at the right.

32. How many gallons of water will fill an aquarium that is 12 in. wide, 18 in. long, and 16 in. high? Round to the nearest tenth. (1 gal = 231 in³)

33. How many gallons of water will fill a fish tank that is 12 in. long, 8 in. wide, and 9 in. high? Round to the nearest tenth. (1 gal = 231 in³)

34. A truck carrying an oil tank is shown in the figure at the right.
   a. Without doing the calculations, determine whether the volume of the oil tank is more than 240 ft³ or less than 240 ft³.
   b. If the tank is half full, how many cubic feet of oil is the truck carrying? Round to the nearest hundredth.

35. The concrete floor of a building is shown in the figure at the right. At a cost of $5.85 per cubic foot, find the cost of having the floor poured. Round to the nearest cent.

APPLYING THE CONCEPTS

36. Half a sphere is called a hemisphere. Derive a formula for the volume of a hemisphere.

37. **a.** If both the length and the width of a rectangular solid are doubled, how many times larger is the resulting rectangular solid?
   b. If the length, width, and height of a rectangular solid are all doubled, how many times larger is the resulting rectangular solid?
   c. If the side of a cube is doubled, how many times larger is the resulting cube?

38. Explain how you could cut through a cube so that the face of the resulting solid is **a.** a square, **b.** an equilateral triangle, **c.** a trapezoid, **d.** a hexagon.

39. Suppose a cylinder is cut into 16 equal pieces, which are then arranged as shown at the right. The figure resembles a rectangular solid. What variable expressions could be used to represent the length, width, and height of the rectangular solid? Explain how the formula for the volume of a cylinder is derived from this approach.

12.5 The Pythagorean Theorem

Objective A   **To find the square root of a number**

The area of a square is 36 in^2. What is the length of one side?

Area of the square = (side)^2
$$36 = \text{side} \cdot \text{side}$$

What number multiplied times itself equals 36?

$$36 = 6 \cdot 6$$

The side of the square is 6 in.

> Area =
> 36 in^2

The **square root** of a number is one of two identical factors of that number. The square root symbol is $\sqrt{}$.

The square root of 36 is 6.

$$\sqrt{36} = 6$$

A **perfect square** is the product of a whole number times itself.

1, 4, 9, 16, 25, and 36 are perfect squares.

$1 \cdot 1 = 1$	$\sqrt{1} = 1$
$2 \cdot 2 = 4$	$\sqrt{4} = 2$
$3 \cdot 3 = 9$	$\sqrt{9} = 3$
$4 \cdot 4 = 16$	$\sqrt{16} = 4$
$5 \cdot 5 = 25$	$\sqrt{25} = 5$
$6 \cdot 6 = 36$	$\sqrt{36} = 6$

The square root of a perfect square is a whole number.

Point of Interest

The square root of a number that is not a perfect square is an irrational number. There is evidence of irrational numbers as early as 500 B.C. These numbers were not very well understood, and they were given the name *numerus surdus*. This phrase comes from the Latin word *surdus*, which means deaf or mute. Thus irrational numbers were "inaudible numbers."

If a number is not a perfect square, its square root can only be approximated. The approximate square roots of numbers can be found using a calculator. For example:

Number	Square Root
33	$\sqrt{33} \approx 5.745$
34	$\sqrt{34} \approx 5.831$
35	$\sqrt{35} \approx 5.916$

Example 1   **a.** Find the square roots of the perfect squares 49 and 81.
b. Find the square roots of 27 and 108. Round to the nearest thousandth.

You Try It 1   **a.** Find the square roots of the perfect squares 16 and 169.
b. Find the square roots of 32 and 162. Round to the nearest thousandth.

Solution   **a.** $\sqrt{49} = 7$ $\sqrt{81} = 9$
b. $\sqrt{27} \approx 5.196$ $\sqrt{108} \approx 10.392$

Your solution

Solution on p. S32

Objective B   **To find the unknown side of a right triangle using the Pythagorean Theorem**

Point of Interest

The first known proof of the Pythagorean Theorem is in a Chinese textbook that dates from 150 B.C. The book is called *Nine Chapters on the Mathematical Art*. The diagram below is from that book and was used in the proof of the theorem.

The Greek mathematician Pythagoras is generally credited with the discovery that the square of the hypotenuse of a right triangle is equal to the sum of the squares of the two legs. This is called the **Pythagorean Theorem.** However, the Babylonians used this theorem more than 1000 years before Pythagoras lived.

Square of the hypotenuse equals sum of the squares of the two legs

$$5^2 = 3^2 + 4^2$$
$$25 = 9 + 16$$
$$25 = 25$$

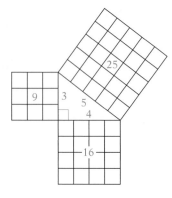

If the length of one side of a right triangle is unknown, one of the following formulas can be used to find it.

If the hypotenuse is unknown, use

$$\textbf{Hypotenuse} = \sqrt{\textbf{(leg)}^2 + \textbf{(leg)}^2}$$
$$= \sqrt{(3)^2 + (4)^2}$$
$$= \sqrt{9 + 16}$$
$$= \sqrt{25}$$
$$= 5$$

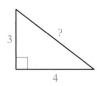

If the length of a leg is unknown, use

$$\textbf{Leg} = \sqrt{\textbf{(hypotenuse)}^2 - \textbf{(leg)}^2}$$
$$= \sqrt{(5)^2 - (4)^2}$$
$$= \sqrt{25 - 16}$$
$$= \sqrt{9}$$
$$= 3$$

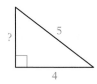

By using the Pythagorean Theorem and several facts from geometry, a relationship among the sides of two special right triangles can be found.

The first special right triangle has two 45° angles and is called a **45°–45°–90° triangle.** The sides opposite the 45° angles are equal.

In a 45°–45°–90° triangle:

$$\textbf{Hypotenuse} = \sqrt{\textbf{2}} \times \textbf{(length of a leg)}$$

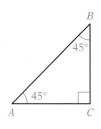

➡ Find the length of the hypotenuse for a 45°–45°–90° triangle in which the length of one leg is 26 m. Round to the nearest thousandth.

To calculate $\sqrt{2}\,(26\ m)$, enter

2 $\boxed{\sqrt{}}$ $\boxed{\times}$ 26 $\boxed{=}$

or

$\boxed{\sqrt{}}$ 2 $\boxed{\times}$ 26 $\boxed{=}$

Hypotenuse $= \sqrt{2} \times$ (length of a leg)

$= \sqrt{2}\,(26\ m)$

≈ 36.770

The hypotenuse is approximately 36.770 m.

The second special right triangle is the **30°–60°–90° triangle.**

For a 30°–60°–90° triangle, the length of the leg opposite the 30° angle is one-half the length of the hypotenuse.

$$BC = \frac{1}{2} \times AB$$

➡ Find the length of the two legs of a 30°–60°–90° triangle that has a hypotenuse 16 cm in length. Round to the nearest thousandth.

Leg $= \frac{1}{2} \times$ hypotenuse

$= \frac{1}{2}(16\ cm) = 8\ cm$

Leg $= \sqrt{(\text{hypotenuse})^2 - (\text{leg})^2}$

$= \sqrt{16^2 - 8^2}$

$= \sqrt{256 - 64}$

$= \sqrt{192} \approx 13.856$

- Because the right triangle is a 30°–60°–90° triangle, the leg opposite the 30° angle is one-half the hypotenuse.

- Use the Pythagorean Theorem to find the length of the other leg.

The lengths of the legs are 8 cm and 13.856 cm.

Example 2

Find the hypotenuse of the triangle in the figure. Round to the nearest thousandth.

Solution

Hypotenuse $= \sqrt{(\text{leg})^2 + (\text{leg})^2}$

$= \sqrt{8^2 + 4^2}$

$= \sqrt{64 + 16}$

$= \sqrt{80} \approx 8.944$

The hypotenuse is approximately 8.944 in.

You Try It 2

Find the hypotenuse of the triangle in the figure. Round to the nearest thousandth.

Your solution

Solution on p. S32

Example 3

Find the length of the leg of the triangle in the figure. Round to the nearest thousandth.

12 cm

9 cm

Solution

$$\text{Leg} = \sqrt{(\text{hypotenuse})^2 - (\text{leg})^2}$$
$$= \sqrt{12^2 - 9^2}$$
$$= \sqrt{144 - 81}$$
$$= \sqrt{63} \approx 7.937$$

The length of the leg is approximately 7.937 cm.

You Try It 3

Find the length of the leg of the triangle in the figure. Round to the nearest thousandth.

5 ft   12 ft

Your solution

Solution on p. S32

Objective C **To solve application problems**

Example 4

A 25-foot ladder is placed against a building at a point 21 ft from the ground, as shown in the figure. Find the distance from the base of the building to the base of the ladder. Round to the nearest thousandth.

25 ft

21 ft

Strategy

To find the distance from the base of the building to the base of the ladder, use the Pythagorean Theorem. The hypotenuse is the length of the ladder (25 ft). One leg is the distance along the building from the ground to the top of the ladder (21 ft). The distance from the base of the building to the base of the ladder is the unknown leg.

Solution

$$\text{Leg} = \sqrt{(\text{hypotenuse})^2 - (\text{leg})^2}$$
$$= \sqrt{25^2 - 21^2}$$
$$= \sqrt{625 - 441}$$
$$= \sqrt{184} \approx 13.565$$

The distance is approximately 13.565 ft.

You Try It 4

Find the distance between the centers of the holes in the metal plate in the figure. Round to the nearest thousandth.

3 cm

8 cm

Your strategy

Your solution

Solution on pp. S32–S33

12.5 Exercises

Objective A

Find the square root. Round to the nearest thousandth.

1. 7   **2.** 34   **3.** 42   **4.** 64

5. 165   **6.** 144   **7.** 189   **8.** 130

Objective B

In Exercises 9 to 26, round to the nearest thousandth. Find the unknown side of the triangle in the figures in Exercises 9 to 17.

9.

3 in.
4 in.

10.

5 in.
12 in.

11.

5 cm
7 cm

12.

7 cm
9 cm

13.

15 ft
10 ft

14.

20 ft
18 ft

15.

4 cm   6 cm

16.

9 m   12 m

17.

9 yd
9 yd

Find the lengths of the two legs in the figures in Exercises 18 to 20.

18.

20 cm
30°

19.

12 ft
30°

20.

16 cm
30°

Find the hypotenuse of the triangle in the figures in Exercises 21 to 26.

21.

15 cm
45°
15 cm

22.

6 in.   6 in.
45°   45°

23.

4 m
30°   60°

24.

4.3 cm
30°

25.

45°
8 yd

26.

8 ft
30°

27. Find the length of the ramp used to roll barrels up to the loading dock, which is 3.5 ft high. Round to the nearest hundredth.

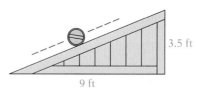

28. Find the distance between the holes in the metal plate in the figure at the right. Round to the nearest hundredth.

29. If you travel 18 mi east and then 12 mi north, how far are you from your starting point? Round to the nearest tenth.

30. A fence is built around the plot shown in the figure at the right. At $11.40 per meter, how much did it cost to fence the plot? (*Hint:* Use the Pythagorean Theorem to find the unknown length.)

31. Four holes are drilled in the circular plate shown at the right. The centers of the holes are 3 in. from the center. Find the distance between the centers of adjacent holes. Round to the nearest thousandth.

32. Find the offset distance, d, of the length of pipe shown in the diagram at the right. The total length of the pipe is 62 in.

APPLYING THE CONCEPTS

33. Determine whether the statement is always true, sometimes true, or never true.
   a. The sum of the lengths of two sides of a triangle is greater than the length of the third side of the triangle.
   b. The hypotenuse is the longest side of a right triangle.

34. What is a Pythagorean triple? Provide at least three examples of Pythagorean triples.

35. Buildings A and B are situated on opposite sides of a river. A construction company must lay a pipeline between the two buildings. The plan is to connect the buildings as shown. What is the total length of the pipe needed to connect the buildings?

To solve similar and congruent triangles

Similar objects have the same shape but not necessarily the same size. A baseball is similar to a basketball. A model airplane is similar to an actual airplane.

Similar objects have corresponding parts; for example, the propellers on the model airplane correspond to the propellers on the actual airplane. The relationship between the sizes of each of the corresponding parts can be written as a ratio, and all such ratios will be the same. If the propellers on the model plane are $\frac{1}{50}$ the size of the propellers on the actual plane, then the model wing is $\frac{1}{50}$ the size of the actual wing, the model fuselage is $\frac{1}{50}$ the size of the actual fuselage, and so on.

The two triangles *ABC* and *DEF* shown are similar. The ratios of corresponding sides are equal.

$$\frac{AB}{DE} = \frac{2}{6} = \frac{1}{3}, \frac{BC}{EF} = \frac{3}{9} = \frac{1}{3}, \text{ and } \frac{AC}{DF} = \frac{4}{12} = \frac{1}{3}$$

The ratio of corresponding sides $= \frac{1}{3}$.

Because the ratios of corresponding sides are equal, three proportions can be formed:

$$\frac{AB}{DE} = \frac{BC}{EF}, \frac{AB}{DE} = \frac{AC}{DF}, \text{ and } \frac{BC}{EF} = \frac{AC}{DF}$$

The ratio of corresponding heights equals the ratio of corresponding sides, as shown in the figure.

Ratio of corresponding sides $= \frac{1.5}{6} = \frac{1}{4}$

Ratio of heights $= \frac{2}{8} = \frac{1}{4}$

Congruent objects have the same shape *and* the same size.

The two triangles shown are congruent. They have exactly the same size.

For triangles, *congruent* means that the corresponding sides *and* angles of the triangle are equal (this contrasts with similar triangles, in which corresponding angles, but not necessarily corresponding sides, are equal).

Here are two major rules that can be used to determine whether two triangles are congruent.

> **Side-Side-Side Rule (SSS)**
>
> Two triangles are congruent if three sides of one triangle equal the corresponding sides of the second triangle.

In the two triangles at the right, $AB = DE$, $AC = DF$, and $BC = EF$. The corresponding sides of triangles ABC and DEF are equal. The triangles are congruent by the SSS rule.

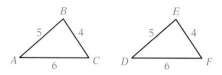

> **Side-Angle-Side Rule (SAS)**
>
> Two triangles are congruent if two sides and the included angle of one triangle equal the corresponding sides and included angle of the second triangle.

In the two triangles at the right, $AB = EF$, $AC = DE$, and $\angle CAB = \angle DEF$. The triangles are congruent by the SAS rule.

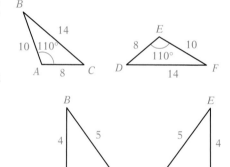

➡ Determine whether the two triangles in the figure at the right are congruent.

Because $AC = DF$, $AB = FE$, and $BC = DE$, all three sides of one triangle equal the corresponding sides of the second triangle. The triangles are congruent by the SSS rule.

Example 1

Find the ratio of corresponding sides for the similar triangles ABC and DEF in the figure.

Solution

$$\frac{7 \text{ m}}{12 \text{ m}} = \frac{7}{12}$$

You Try It 1

Find the ratio of corresponding sides for the similar triangles ABC and DEF in the figure.

D ⎯⎯ F
A ⎯ C
4 cm
B
7 cm
E

Your solution

Solution on p. S33

Example 2

Triangles *ABC* and *DEF* in the figure are similar. Find *x*, the length of side *EF*.

Solution

$$\frac{AB}{DE} = \frac{BC}{x}$$ • The ratios of corresponding sides of similar triangles are equal.

$$\frac{8 \text{ m}}{12 \text{ m}} = \frac{6 \text{ m}}{x}$$

$$8x = 12 \cdot 6 \text{ m}$$

$$8x = 72 \text{ m}$$

$$\frac{8x}{8} = \frac{72 \text{ m}}{8}$$

$$x = 9 \text{ m}$$

Side *EF* is 9 m.

You Try It 2

Triangles *ABC* and *DEF* in the figure are similar. Find *x*, the length of side *DF*.

Your solution

Example 3

Determine whether triangle *ABC* in the figure is congruent to triangle *FDE*.

Solution

Because *AB* = *DF*, *AC* = *EF*, and angle *BAC* = angle *DFE*, the triangles are congruent by the SAS rule.

You Try It 3

Determine whether triangle *ABC* in the figure is congruent to triangle *DEF*.

Your solution

Example 4

Triangles *ABC* and *DEF* in the figure are similar. Find *h*, the height of triangle *DEF*.

Solution

$$\frac{8 \text{ cm}}{12 \text{ cm}} = \frac{4 \text{ cm}}{h}$$

$$8h = 12 \cdot 4 \text{ cm}$$

$$8h = 48 \text{ cm}$$

$$\frac{8h}{8} = \frac{48 \text{ cm}}{8}$$

$$h = 6 \text{ cm}$$

• The ratios of corresponding sides of similar triangles equal the ratio of corresponding heights:

$$\frac{AB}{DE} = \frac{CH}{FG}.$$

The height of *DEF* is 6 cm.

You Try It 4

Triangles *ABC* and *DEF* in the figure are similar. Find *h*, the height of triangle *DEF*.

Your solution

Solutions on p. S33

To solve application problems

Example 5

Triangles *ABC* and *DEF* in the figure are similar. Find the area of triangle *DEF*.

You Try It 5

Triangles *ABC* and *DEF* in the figure are similar right triangles. Find the perimeter of triangle *ABC*.

Strategy

To find the area of triangle *DEF*:

• Solve a proportion to find the height of triangle *DEF*. Let h = the height.

• Use the formula $A = \frac{1}{2}bh$.

Your strategy

Solution

$$\frac{AB}{DE} = \frac{\text{height of triangle } ABC}{\text{height of triangle } DEF}$$

$$\frac{4 \text{ cm}}{12 \text{ cm}} = \frac{3 \text{ cm}}{h}$$

$$4h = 12 \cdot 3 \text{ cm}$$

$$4h = 36 \text{ cm}$$

$$\frac{4h}{4} = \frac{36 \text{ cm}}{4}$$

$$h = 9 \text{ cm}$$

$$A = \frac{1}{2}bh$$

$$= \frac{1}{2}(12 \text{ cm})(9 \text{ cm})$$

$$= 54 \text{ cm}^2$$

The area is 54 cm².

Your solution

Solution on p. S33

12.6 Exercises

Find the ratio of corresponding sides for the similar triangles in Exercises 1 to 4.

1.

2.

3.

4.

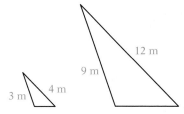

Determine whether the two triangles in Exercises 5 to 8 are congruent.

5.

6.

7.

8.

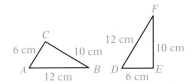

Triangles *ABC* and *DEF* in Exercises 9 to 12 are similar. Find the indicated distance. Round to the nearest tenth.

9. Find side *DE*.

10. Find side *DE*.

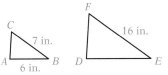

11. Find the height of triangle *DEF*.

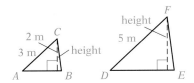

12. Find the height of triangle *ABC*.

Objective B *Application Problems*

The sun's rays, objects on Earth, and the shadows cast by those objects form similar triangles.

13. Find the height of the building shown.

8 m

8 m   4 m

14. Find the height of the building shown.

5.2 ft

1.3 ft   5.2 ft

In Exercises 15 to 18, triangles *ABC* and *DEF* are similar.

15. Find the perimeter of triangle *ABC*.

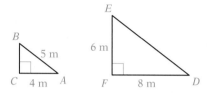

B
5 m
C   4 m   A

E
6 m
F   8 m   D

16. Find the perimeter of triangle *DEF*.

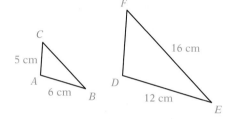

C
5 cm
A
6 cm   B

F
16 cm
D
12 cm   E

17. Find the area of triangle *ABC*.

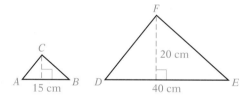

C
A   15 cm   B

F
20 cm
D   40 cm   E

18. Find the area of triangle *DEF*.

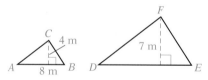

C   4 m
A   8 m   B

F
7 m
D   E

APPLYING THE CONCEPTS

19. Determine whether the statement is always true, sometimes true, or never true.
   a. If two angles of one triangle are equal to two angles of a second triangle, then the triangles are similar triangles.
   b. Two isosceles triangles are similar triangles.
   c. Two equilateral triangles are similar triangles.

20. Are all squares similar? Are all rectangles similar? Explain. Use a drawing in your explanation.

21. Figure *ABC* is a right triangle and *DE* is parallel to *AB*. What is the perimeter of the trapezoid *ABED*?

22. Explain how, by using only a yardstick, you could determine the approximate height of a tree without climbing it.

B
6
E   10
A   D   C
5
8

Focus on Problem Solving

Trial and Error   Some problems in mathematics are solved by using **trial and error.** The trial-and-error method of arriving at a solution to a problem involves repeated tests or experiments until a satisfactory conclusion is reached.

Many of the Applying the Concepts exercises in this text require a trial-and-error method of solution. For example, an exercise in Section 4 reads as follows:

Explain how you could cut through a cube so that the face of the resulting solid is **(a)** a square, **(b)** an equilateral triangle, **(c)** a trapezoid, **(d)** a hexagon.

There is no formula to apply to this problem; there is no computation to perform. This problem requires picturing a cube and the results after it is cut through at different places on its surface and at different angles. For part (a), cutting perpendicular to the top and bottom of the cube and parallel to two of its sides will result in a square. The other shapes may prove more difficult.

When solving problems of this type, keep an open mind. Sometimes when using the trial-and-error method, we are hampered by narrowness of vision; we cannot expand our thinking to include other possibilities. Then when we see someone else's solution, it appears so obvious to us! For example, for the Applying the Concepts question above, it is necessary to conceive of cutting through the cube at places other than the top surface; we need to be open to the idea of beginning the cut at one of the corner points of the cube.

A topic of the Projects and Group Activities in this chapter is symmetry. Here again, the trial-and-error method is used to determine the lines of symmetry inherent in an object. For example, in determining lines of symmetry for a square, begin by drawing a square. The horizontal line of symmetry and the vertical line of symmetry may be immediately obvious to you.

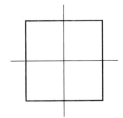

But there are two others. Do you see that a line drawn through opposite corners of the square is also a line of symmetry?

Many of the questions in this text that require an answer of "always true, sometimes true, or never true" are best solved by the trial-and-error method. For example, consider the statement presented in Section 3.

If two rectangles have the same area, then they have the same perimeter.

554 Chapter 12 / Geometry

Try some numbers. Each of two rectangles, one measuring 6 units by 2 units and another measuring 4 units by 3 units, has an area of 12 square units, but the perimeter of the first is 16 units and the perimeter of the second is 14 units, so the answer "always true" has been eliminated. We still need to determine whether there is a case when it is true. After experimenting with a lot of numbers, you may come to realize that we are trying to determine whether it is possible for two different pairs of factors of a number to have the same sum. Is it?

Don't be afraid to make many experiments, and remember that *errors,* or tests that "don't work," are a part of the trial-and-*error* process.

Projects and Group Activities

Investigating Perimeter

The perimeter of the square at the right is 4 units.

If two squares are joined along one of the sides, the perimeter is 6 units. Note that it does not matter which sides are joined; the perimeter is still 6 units.

If three squares are joined, the perimeter of the resulting figure is 8 units for each possible placement of the squares.

Four squares can be joined in five different ways as shown. There are two possible perimeters: 10 units for A, B, C, and D, and 8 for E.

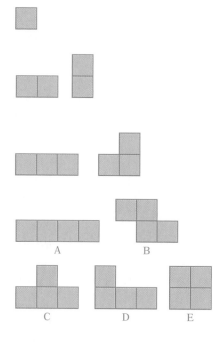

A B

C D E

1. If five squares are joined, what is the maximum perimeter possible?

2. If five squares are joined, what is the minimum perimeter possible?

3. If six squares are joined, what is the maximum perimeter possible?

4. If six squares are joined, what is the minimum perimeter possible?

Copyright © Houghton Mifflin Company. All rights reserved.

Symmetry

Look at the letter A printed at the left. If the letter were folded along line ℓ, the two sides of the letter would match exactly. This letter has **symmetry** with respect to line ℓ. Line ℓ is called the **axis of symmetry.**

Now consider the letter H printed below at the left. Both lines ℓ_1 and ℓ_2 are axes of symmetry for this letter; the letter could be folded along either line and the two sides would match exactly.

1. Does the letter A have more than one axis of symmetry?

2. Find axes of symmetry for other capital letters of the alphabet.

3. Which lowercase letters have one axis of symmetry?

4. Do any of the lowercase letters have more than one axis of symmetry?

5. Find the number of axes of symmetry for each of the plane geometric figures presented in this chapter.

6. There are other types of symmetry. Look up the meaning of *point symmetry* and *rotational symmetry*. Which plane geometric figures provide examples of these types of symmetry?

7. Find examples of symmetry in nature, art, and architecture.

Chapter Summary

Key Words

An *angle* is formed when two rays start from the same point. An angle is measured in *degrees*. [p. 502]

A 90° angle is called a *right angle*. [p. 502]

Perpendicular lines are intersecting lines that form right angles. [p. 502]

Complementary angles are two angles whose sum is 90°. [p. 503]

Supplementary angles are two angles whose sum is 180°. [p. 503]

A *right triangle* contains one right angle. The side opposite the right angle in a right triangle is called the *hypotenuse*. The other two sides are called *legs*. [p. 504]

A *45°–45°–90° triangle* is a special right triangle in which the sides opposite the 45° angles are equal and the hypotenuse is equal to $\sqrt{2} \times$ (length of a leg). [p. 542]

A *30°–60°–90° triangle* is a special right triangle in which the length of the leg opposite the 30° angle is one-half the length of the hypotenuse. [p. 543]

A *quadrilateral* is a four-sided plane figure. [p. 504]

A *rectangle* is a parallelogram that has four right angles. [p. 505]

A *square* is a rectangle that has four equal sides. [p. 505]

A *circle* is a plane figure in which all points are the same distance from the center of the circle. The *diameter* is a line segment across a circle going through the center. The *radius* is equal to one-half the diameter. [p. 505]

A *rectangular solid* is a solid in which all six faces are rectangles. [p. 505]

A *cube* is a rectangular solid in which all six faces are squares. [p. 505]

A *sphere* is a solid in which all points on the surface are the same distance from the center of the sphere. [p. 505]

Perimeter is the distance around a plane figure. [p. 514]

Area is a measure of the amount of surface in a region. [p. 523]

Volume is a measure of the amount of space inside a closed surface. [p. 531]

Composite geometric solids are solids made from two or more geometric solids. [p. 534]

The *square root* of a number is one of two identical factors of that number. [p. 541]

Similar objects have the same shape but not necessarily the same size. [p. 547]

Congruent objects have the same shape and the same size. [p. 547]

Essential Rules

Perimeter Equations

Triangle:	$P = a + b + c$ [p. 514]
Square:	$P = 4s$ [p. 515]
Rectangle:	$P = 2L + 2W$ [p. 515]
Circle:	$C = 2\pi r$ or $C = \pi d$ [p. 515]

Area Equations

Triangle:	$A = \frac{1}{2}bh$ [p. 524]
Square:	$A = s^2$ [p. 524]
Rectangle:	$A = LW$ [p. 523]
Circle:	$A = \pi r^2$ [p. 524]

Volume Equations

Rectangular solid:	$V = LWH$ [p. 531]
Cube:	$V = s^3$ [p. 532]
Sphere:	$V = \frac{4}{3}\pi r^3$ [p. 532]
Cylinder:	$V = \pi r^2 h$ [p. 532]

Pythagorean Theorem

Hypotenuse $= \sqrt{(\text{leg})^2 + (\text{leg})^2}$ Leg $= \sqrt{(\text{hypotenuse})^2 - (\text{leg})^2}$ [p. 542]

Side-Side-Side Rule (SSS)

Two triangles are congruent when three sides of one triangle equal the corresponding sides of the second triangle. [p. 548]

Side-Angle-Side Rule (SAS)

Two triangles are congruent when two sides and the included angle of one triangle equal the corresponding sides and included angle of the second triangle. [p. 548]

Chapter Review

1. The diameter of a sphere is 1.5 m. Find the radius of the sphere.

2. Find the circumference of a circle with a radius of 5 cm. Use 3.14 for π.

3. Find the perimeter of the rectangle in the figure below.

    5 ft

    8 ft

4. Given $AB = 15$, $CD = 6$, and $AD = 24$, find the length of BC.

5. Find the volume of the rectangular solid shown below.

    4 ft

    10 ft

    5 ft

6. Find the unknown side of the triangle in the figure below.

    10 cm

    24 cm

7. Find the supplement of a 105° angle.

8. Find the square root of 15. Round to the nearest thousandth.

9. Triangles ABC and DEF are similar. Find the height of triangle DEF.

    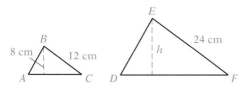

    E

    B

    8 cm   12 cm   24 cm

    h

    A C D F

10. Find the area of the circle shown below. Use 3.14 for π.

    9 cm

11. Given that $\ell_1 \parallel \ell_2$, find the measures of angles a and b.

    t

    b

    ℓ_1

    45° a

    ℓ_2

12. Find the area of the rectangle shown below.

    5 m

    11 m

13. Find the volume of the composite figure shown below.

    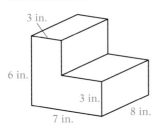

    3 in.

    6 in.

    3 in.

    8 in.

    7 in.

14. Find the area of the composite figure shown below. Use 3.14 for π.

    4 in.

    8 in.

15. Find the volume of a sphere with a diameter of 8 ft. Use 3.14 for π. Round to the nearest tenth.

16. Triangles *ABC* and *DEF* are similar. Find the area of triangle *DEF*.

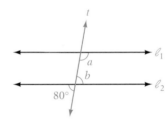

17. Find the perimeter of the composite figure shown below. Use 3.14 for π.

18. Given that $\ell_1 \parallel \ell_2$, find the measures of angles *a* and *b*.

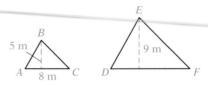

19. How high on a building will a 17-foot ladder reach when the bottom of the ladder is 8 ft from the building?

20. A right triangle has a 32° angle. Find the measures of the other two angles.

21. A bicycle tire has a diameter of 28 in. How many feet does the bicycle travel when the wheel makes 10 revolutions? Use 3.14 for π. Round to the nearest tenth of a foot.

22. New carpet is installed in a room measuring 18 ft by 14 ft. Find the area of the room in square yards. (9 ft² = 1 yd²)

23. A silo, which is in the shape of a cylinder, is 9 ft in diameter and has a height of 18 ft. Find the volume of the silo. Use 3.14 for π.

24. Find the area of a right triangle with a base of 8 m and a height of 2.75 m.

25. If you travel 20 mi west and then 21 mi south, how far are you from your starting point?

Chapter Test

1. Find the volume of a cylinder with a height of 6 m and a radius of 3 m. Use 3.14 for π.

2. Find the perimeter of a rectangle that has a length of 2 m and a width of 1.4 m.

3. Find the volume of the composite figure. Use 3.14 for π.

4. Triangles *ABC* and *FED* are congruent right triangles. Find the length of *FE*.

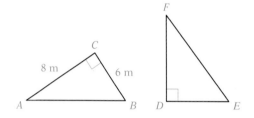

5. Find the complement of a 32° angle.

6. Find the area of a circle that has a diameter of 2 m. Use $\frac{22}{7}$ for π.

7. In the figure below, lines ℓ_1 and ℓ_2 are parallel. Angle x measures 30°. Find the measure of angle y.

8. Find the perimeter of the composite figure. Use 3.14 for π.

9. Find the square root of 189. Round to the nearest thousandth.

10. Find the unknown side of the triangle shown below. Round to the nearest thousandth.

11. Find the area of the composite figure.

12. In the figure below, lines ℓ_1 and ℓ_2 are parallel. Angle x measures 45°. Find the measures of angles a and b.

13. Triangles ABC and DEF are similar. Find side BC.

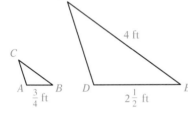

14. A right triangle has a 40° angle. Find the measures of the other two angles.

15. Find the width of the canal shown in the figure at the right.

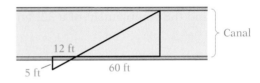

16. How much more pizza is contained in a pizza with radius 10 in. than in one with radius 8 in.? Use 3.14 for π.

17. A carpet is to be placed as shown in the diagram at the right. At $26.80 per square yard, how much will it cost to carpet the area? Round to the nearest cent. ($9\ \text{ft}^2 = 1\ \text{yd}^2$)

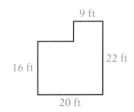

18. Find the cross-sectional area of a redwood tree that is 11 ft 6 in. in diameter. Use 3.14 for π. Round to the nearest hundredth.

19. Find the length of the rafter needed for the roof shown in the figure.

20. A toolbox is 1 ft 2 in. long, 9 in. wide, and 8 in. high. The sides and bottom of the toolbox are $\frac{1}{2}$ in. thick. Find the volume of the interior of the toolbox in cubic inches.

Cumulative Review

1. Find the GCF of 96 and 144.

2. Add: $3\frac{5}{12} + 2\frac{9}{16} + 1\frac{7}{8}$

3. Find the quotient of $4\frac{1}{3}$ and $6\frac{2}{9}$.

4. Simplify: $\left(\frac{2}{3}\right)^2 \div \left(\frac{1}{3} + \frac{1}{2}\right) - \frac{2}{5}$

5. Simplify: $-\frac{2}{3} - \left(-\frac{5}{8}\right)$

6. Write "$348.80 earned in 40 hours" as a unit rate.

7. Solve the proportion $\frac{3}{8} = \frac{n}{100}$.

8. Write $37\frac{1}{2}\%$ as a fraction.

9. Evaluate $a^2 - (b^2 - c)$ when $a = 2$, $b = -2$, and $c = -4$.

10. 30.94 is 36.4% of what number?

11. Solve: $\frac{x}{3} + 3 = 1$

12. Solve: $2(x - 3) + 2 = 5x - 8$

13. Convert 32.5 km to meters.

14. Subtract: 32 m − 42 cm

15. Solve: $\frac{2}{3}x = -10$

16. Solve: $2x - 4(x - 3) = 8$

17. You bought a car for $26,488 and made a down payment of $1000. You paid the balance in 36 equal monthly installments. Find the monthly payment.

18. The sales tax on a jacket costing $175 is $6.75. At the same rate, find the sales tax on a stereo system costing $1220.

19. A heavy-equipment operator receives an hourly wage of $16.06 an hour after receiving a 10% wage increase. Find the operator's hourly wage before the increase.

20. An after-Christmas sale has a markdown rate of 55%. Find the sale price of a dress that had a regular price of $120.

21. An IRA pays 10% annual interest compounded daily. What would be the value of an investment of $25,000 after 20 years? Use the table in the Appendix.

22. A square tile measuring 4 in. by 4 in. weighs 6 oz. Find the weight, in pounds, of a package of 144 such tiles.

23. Twenty-five rivets are used to fasten two steel plates together. The plates are 5.4 m long, and the rivets are equally spaced with a rivet at each end. Find the distance, in centimeters, between the rivets.

24. The total of four times a number and two is negative six. Find the number.

25. The lines ℓ_1 and ℓ_2 in the figure below are parallel. Find angles a and b.

26. Find the perimeter of the composite figure. Use 3.14 for π.

27. Find the area of the composite figure.

28. Find the volume of the composite figure. Use 3.14 for π.

29. Find the unknown side of the triangle shown in the figure below. Round to the nearest hundredth.

30. Triangles *ABC* and *FED* below are similar. Find the perimeter of *FED*.

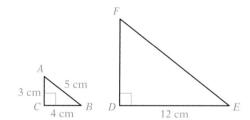

Final Examination

1. Subtract: $100,914 - 97,655$

2. Find 34,821 divided by 657.

3. Simplify: $3^2 \cdot (5 - 3)^2 \div 3 + 4$

4. Find the LCM of 9, 12, and 16.

5. Add: $\dfrac{3}{8} + \dfrac{5}{6} + \dfrac{1}{5}$

6. Subtract: $7\dfrac{5}{12} - 3\dfrac{13}{16}$

7. Find the product of $3\dfrac{5}{8}$ and $1\dfrac{5}{7}$.

8. Divide: $1\dfrac{2}{3} \div 3\dfrac{3}{4}$

9. Simplify: $\left(\dfrac{2}{3}\right)^3 \cdot \left(\dfrac{3}{4}\right)^2$

10. Simplify: $\left(\dfrac{2}{3}\right)^2 \div \left(\dfrac{3}{4} + \dfrac{1}{3}\right) - \dfrac{1}{3}$

11. Add:
$$\begin{array}{r} 4.972 \\ 28.6 \\ 1.88 \\ + 128.725 \\ \hline \end{array}$$

12. Find 90,001 decreased by 29,796.

13. Multiply:
$$\begin{array}{r} 2.97 \\ \times\ 0.0094 \\ \hline \end{array}$$

14. Divide: $0.062\overline{)0.0426}$
Round to the nearest hundredth.

15. Convert 0.45 to a fraction in simplest form.

16. Write "323.4 miles on 13.2 gallons of gas" as a unit rate.

17. Solve the proportion $\dfrac{12}{35} = \dfrac{n}{160}$.
Round to the nearest tenth.

18. Write $22\dfrac{1}{2}\%$ as a fraction.

19. Write 1.35 as a percent.

20. Write $\frac{5}{4}$ as a percent.

21. Find 120% of 30.

22. 12 is what percent of 9?

23. 42 is 60% of what number?

24. Convert $1\frac{2}{3}$ ft to inches.

25. Subtract: 3 ft 2 in. − 1 ft 10 in.

26. Convert 40 oz to pounds.

27. Find the sum of 3 lb 12 oz and 2 lb 10 oz.

28. Convert 18 pt to gallons.

29. Divide: $3\overline{)5\text{ gal 1 qt}}$

30. Convert 2.48 m to centimeters.

31. Convert 4 m 62 cm to meters.

32. Convert 1 kg 614 g to kilograms.

33. Convert 2 L 67 ml to milliliters.

34. Convert 55 mi to kilometers. Round to the nearest hundredth. (1.61 km ≈ 1 mi)

35. How much does it cost to run a 2400-watt air conditioner for 6 h at 8¢ per kilowatt-hour? Round to the nearest cent.

36. Write 0.0000000679 in scientific notation.

37. Find the perimeter of a rectangle with a length of 1.2 m and a width of 0.75 m.

38. Find the area of a rectangle with a length of 9 in. and a width of 5 in.

39. Find the volume of a box with a length of 20 cm, a width of 12 cm, and a height of 5 cm.

40. Add: $-2 + 8 + (-10)$

41. Subtract: $-30 - (-15)$

42. Multiply: $2\frac{1}{2} \times \left(-\frac{1}{5}\right)$

43. Find the quotient of $-1\frac{3}{8}$ and $5\frac{1}{2}$.

44. Simplify: $(-4)^2 \div (1 - 3)^2 - (-2)$

45. Simplify: $2x - 3(x - 4) + 5$

46. Solve: $\frac{2}{3}x = -12$

47. Solve: $3x - 5 = 10$

48. Solve: $8 - 3x = x + 4$

49. You have $872.48 in your checking account. You write checks of $321.88 and $34.23 and then make a deposit of $443.56. Find your new checking account balance.

50. On the basis of a pre-election survey, it is estimated that 5 out of 8 eligible voters will vote in an election. How many people will vote in an election with 102,000 eligible voters?

51. This month a company is paying its stockholders a dividend of $1.60 per share. This is 80% of what the dividend per share was one year ago. What was the dividend per share one year ago?

52. A sales executive received commissions of $4320, $3572, $2864, and $4420 during a 4-month period. Find the average income from commissions for the 4 months.

53. A contractor borrows $120,000 for 9 months at an annual interest rate of 10%. What is the simple interest due on the loan?

54. If two dice are tossed, what is the probability that the sum of the dots on the upward faces is divisible by 3?

55. The top four highest death counts, by country, in World War II are shown in the circle graph. Find what percent the death count of Chinese is of the total death count in all four countries. Round to the nearest tenth of a percent.

Japan 1100
China 1300
Germany 3300
USSR 13,600

Top Four Highest Death Counts in World War II (in thousands)

Source: U.S. Department of Defense

56. A compact disc player that regularly sells for $314.00 is on sale for $226.08. What is the discount rate?

57. A square tile measuring 8 in. by 8 in. weighs 9 oz. Find the weight, in pounds, of a box containing 144 tiles.

58. Find the perimeter of the composite figure. Use 3.14 for π.

8 in.   8 in.
8 in.

59. Find the area of the composite figure. Use 3.14 for π.

10 cm
2 cm

60. Five less than the quotient of a number and two is equal to three. Find the number.

Appendix

Compound Interest Table

Compounded Annually

	4%	5%	6%	7%	8%	9%	10%
1 year	1.04000	1.05000	1.06000	1.07000	1.08000	1.09000	1.10000
5 years	1.21665	1.27628	1.33823	1.40255	1.46933	1.53862	1.61051
10 years	1.48024	1.62890	1.79085	1.96715	2.15893	2.36736	2.59374
15 years	1.80094	2.07893	2.39656	2.75903	3.17217	3.64248	4.17725
20 years	2.19112	2.65330	3.20714	3.86968	4.66095	5.60441	6.72750

Compounded Semiannually

	4%	5%	6%	7%	8%	9%	10%
1 year	1.04040	1.05062	1.06090	1.07123	1.08160	1.09203	1.10250
5 years	1.21899	1.28008	1.34392	1.41060	1.48024	1.55297	1.62890
10 years	1.48595	1.63862	1.80611	1.98979	2.19112	2.41171	2.65330
15 years	1.81136	2.09757	2.42726	2.80679	3.24340	3.74531	4.32194
20 years	2.20804	2.68506	3.26204	3.95926	4.80102	5.81634	7.03999

Compounded Quarterly

	4%	5%	6%	7%	8%	9%	10%
1 year	1.04060	1.05094	1.06136	1.07186	1.08243	1.09308	1.10381
5 years	1.22019	1.28204	1.34686	1.41478	1.48595	1.56051	1.63862
10 years	1.48886	1.64362	1.81402	2.00160	2.20804	2.43519	2.68506
15 years	1.81670	2.10718	2.44322	2.83182	3.28103	3.80013	4.39979
20 years	2.21672	2.70148	3.29066	4.00639	4.87544	5.93015	7.20957

Compounded Monthly

	4%	5%	6%	7%	8%	9%	10%
1 year	1.04074	1.051162	1.061678	1.072290	1.083000	1.093807	1.104713
5 years	1.220997	1.283359	1.348850	1.417625	1.489846	1.565681	1.645309
10 years	1.490833	1.647009	1.819397	2.009661	2.219640	2.451357	2.707041
15 years	1.820302	2.113704	2.454094	2.848947	3.306921	3.838043	4.453920
20 years	2.222582	2.712640	3.310204	4.038739	4.926803	6.009152	7.328074

Compounded Daily

	4%	5%	6%	7%	8%	9%	10%
1 year	1.04080	1.05127	1.06183	1.07250	1.08328	1.09416	1.10516
5 years	1.22139	1.28400	1.34983	1.41902	1.49176	1.56823	1.64861
10 years	1.49179	1.64866	1.82203	2.01362	2.22535	2.45933	2.71791
15 years	1.82206	2.11689	2.45942	2.85736	3.31968	3.85678	4.48077
20 years	2.22544	2.71810	3.31979	4.05466	4.95217	6.04830	7.38703

To use this table:
1. Locate the section which gives the desired compounding period.
2. Locate the interest rate in the top row of that section.
3. Locate the number of years in the left-hand column of that section.
4. Locate the number where the interest-rate column and the number-of-years row meet. This is the compound interest factor.

Example   An investment yields an annual interest rate of 10% compounded quarterly for 5 years.
The compounding period is "compounded quarterly."
The interest rate is 10%.
The number of years is 5.
The number where the row and column meet is 1.63862. This is the compound interest factor.

Compound Interest Table

Compounded Annually

	11%	12%	13%	14%	15%	16%	17%
1 year	1.11000	1.12000	1.13000	1.14000	1.15000	1.16000	1.17000
5 years	1.68506	1.76234	1.84244	1.92542	2.01136	2.10034	2.19245
10 years	2.83942	3.10585	3.39457	3.70722	4.04556	4.41144	4.80683
15 years	4.78459	5.47357	6.25427	7.13794	8.13706	9.26552	10.53872
20 years	8.06239	9.64629	11.52309	13.74349	16.36654	19.46076	23.10560

Compounded Semiannually

	11%	12%	13%	14%	15%	16%	17%
1 year	1.11303	1.12360	1.13423	1.14490	1.15563	1.16640	1.17723
5 years	1.70814	1.79085	1.87714	1.96715	2.06103	2.15893	2.26098
10 years	2.91776	3.20714	3.52365	3.86968	4.24785	4.66096	5.11205
15 years	4.98395	5.74349	6.61437	7.61226	8.75496	10.06266	11.55825
20 years	8.51331	10.28572	12.41607	14.97446	18.04424	21.72452	26.13302

Compounded Quarterly

	11%	12%	13%	14%	15%	16%	17%
1 year	1.11462	1.12551	1.13648	1.14752	1.15865	1.16986	1.18115
5 years	1.72043	1.80611	1.89584	1.98979	2.08815	2.19112	2.29891
10 years	2.95987	3.26204	3.59420	3.95926	4.36038	4.80102	5.28497
15 years	5.09225	5.89160	6.81402	7.87809	9.10513	10.51963	12.14965
20 years	8.76085	10.64089	12.91828	15.67574	19.01290	23.04980	27.93091

Compounded Monthly

	11%	12%	13%	14%	15%	16%	17%
1 year	1.115719	1.126825	1.138032	1.149342	1.160755	1.172271	1.183892
5 years	1.728916	1.816697	1.908857	2.005610	2.107181	2.213807	2.325733
10 years	2.989150	3.300387	3.643733	4.022471	4.440213	4.900941	5.409036
15 years	5.167988	5.995802	6.955364	8.067507	9.356334	10.849737	12.579975
20 years	8.935015	10.892554	13.276792	16.180270	19.715494	24.019222	29.257669

Compounded Daily

	11%	12%	13%	14%	15%	16%	17%
1 year	1.11626	1.12747	1.13880	1.15024	1.16180	1.17347	1.18526
5 years	1.73311	1.82194	1.91532	2.01348	2.11667	2.22515	2.33918
10 years	3.00367	3.31946	3.66845	4.05411	4.48031	4.95130	5.47178
15 years	5.20569	6.04786	7.02625	8.16288	9.48335	11.01738	12.79950
20 years	9.02203	11.01883	13.45751	16.43582	20.07316	24.51534	29.94039

Monthly Payment Table

	4%	5%	6%	7%	8%	9%
1 year	0.0851499	0.0856075	0.0860664	0.0865267	0.0869884	0.0874515
2 years	0.0434249	0.0438714	0.0443206	0.0447726	0.0452273	0.0456847
3 years	0.0295240	0.0299709	0.0304219	0.0308771	0.0313364	0.0317997
4 years	0.0225791	0.0230293	0.0234850	0.0239462	0.0244129	0.0248850
5 years	0.0184165	0.0188712	0.0193328	0.0198012	0.0202764	0.0207584
15 years	0.0073969	0.0079079	0.0084386	0.0089883	0.0095565	0.0101427
20 years	0.0060598	0.0065996	0.0071643	0.0077530	0.0083644	0.0089973
25 years	0.0052784	0.0058459	0.0064430	0.0070678	0.0077182	0.0083920
30 years	0.0047742	0.0053682	0.0059955	0.0066530	0.0073376	0.0080462

	10%	11%	12%	13%
1 year	0.0879159	0.0883817	0.0888488	0.0893173
2 years	0.0461449	0.0466078	0.0470735	0.0475418
3 years	0.0322672	0.0327387	0.0332143	0.0336940
4 years	0.0253626	0.0258455	0.0263338	0.0268275
5 years	0.0212470	0.0217424	0.0222445	0.0227531
15 years	0.0107461	0.0113660	0.0120017	0.0126524
20 years	0.0096502	0.0103219	0.0110109	0.0117158
25 years	0.0090870	0.0098011	0.0105322	0.0112784
30 years	0.0087757	0.0095232	0.0102861	0.0110620

To use this table:
1. Locate the desired interest rate in the top row.
2. Locate the number of years in the left-hand column.
3. Locate the number where the interest-rate column and the number-of-years row meet. This is the monthly payment factor.

Example   A home has a 30-year mortgage at an annual interest rate of 12%.
The interest rate is 12%.
The number of years is 30.
The number where the row and column meet is 0.0102861. This is the monthly payment factor.

Solutions to Chapter 1 "You Try It"

SECTION 1.1

You Try It 1

$$\text{0 1 2 3 4 5 6 7 8 9 10 11 12 13 14}$$

You Try It 2   **a.** $45 > 29$ **b.** $27 > 0$

You Try It 3   Thirty-six million four hundred sixty-two thousand seventy-five

You Try It 4   452,007

You Try It 5 $60,000 + 8000 + 200 + 80 + 1$

You Try It 6 $100,000 + 9000 + 200 + 7$

You Try It 7

┌──── Given place value

368,492

└──── $8 > 5$

368,492 rounded to the nearest ten-thousand is 370,000.

You Try It 8

┌──── Given place value

3962

└──── $6 > 5$

3962 rounded to the nearest hundred is 4000.

SECTION 1.2

You Try It 1
$$
\begin{array}{r}
\overset{1\;1}{347} \\
+\,12,453 \\
\hline
12,800
\end{array}
$$

347 increased by 12,453 is 12,800.

You Try It 2
$$
\begin{array}{r}
\overset{2}{95} \\
88 \\
+\,67 \\
\hline
250
\end{array}
$$

You Try It 3
$$
\begin{array}{r}
\overset{1\;1\;\;2\;1}{392} \\
4,079 \\
89,035 \\
+\;\;4,992 \\
\hline
98,498
\end{array}
$$

You Try It 4

Strategy   To find the total amount spent on these items, add the four amounts (30, 22, 16, and 11).

Solution
$$
\begin{array}{r}
30 \\
22 \\
16 \\
+\,11 \\
\hline
79
\end{array}
$$

The total amount spent on these items is $79.

SECTION 1.3

You Try It 1
$$
\begin{array}{r}
8925 \\
-\,6413 \\
\hline
2512
\end{array}
\qquad
\begin{array}{r}
\textit{Check:}\;\;\; 6413 \\
+\,2512 \\
\hline
8925
\end{array}
$$

You Try It 2
$$
\begin{array}{r}
17,504 \\
-\;\;9,302 \\
\hline
8,202
\end{array}
\qquad
\begin{array}{r}
\textit{Check:}\;\;\;9,302 \\
+\;\;8,202 \\
\hline
17,504
\end{array}
$$

You Try It 3
$$
\begin{array}{r}
\overset{2\;\;14\;\;7\;\;11}{3\,4\,8\,1} \\
-\;\;\;\;8\,6\,5 \\
\hline
2\,6\,1\,6
\end{array}
\qquad
\begin{array}{r}
\textit{Check:}\;\;\;865 \\
+\,2616 \\
\hline
3481
\end{array}
$$

You Try It 4
$$
\begin{array}{r}
\overset{\;\;\;\;\;\;\;15}{\overset{4\;\;5\;\;12}{5\,4,3\,6\,2}} \\
-\,1\,4,4\,8\,5 \\
\hline
4\,0,0\,7\,7
\end{array}
\qquad
\begin{array}{r}
\textit{Check:}\;\;\;14,485 \\
+\,40,077 \\
\hline
54,562
\end{array}
$$

You Try It 5
$$
\begin{array}{r}
\overset{\;\;13\;\;9\;\;9}{\overset{5\;\;3\;\;10\;\;10\;\;13}{6\,4,0\,0\,3}} \\
-\,5\,4,9\,3\,6 \\
\hline
9,0\,6\,7
\end{array}
\qquad
\begin{array}{r}
\textit{Check:}\;\;\;54,936 \\
+\;\;9,067 \\
\hline
64,003
\end{array}
$$

You Try It 6

Strategy   To find the difference, subtract the number of personnel on active duty in the Air Force in 1945 (2,282,259) from the number of personnel on active duty in the Navy in 1945 (3,380,817).

Solution
$$
\begin{array}{r}
3,380,817 \\
-\,2,282,259 \\
\hline
1,098,558
\end{array}
$$

The difference between the number of personnel on active duty in the Navy and in the Air Force in 1945 was 1,098,558.

You Try It 7

Strategy   To find your take-home pay:
- Add to find the total of the deductions ($127 + 18 + 35$).
- Subtract the total of the deductions from your total salary (638).

Solution

$$\begin{array}{r} 127 \\ 18 \\ + \ 35 \\ \hline 180 \end{array} \text{ in deductions}$$

$$\begin{array}{r} 638 \\ - \ 180 \\ \hline 458 \end{array}$$

Your take-home pay is $458.

SECTION 1.4

You Try It 1

$$\begin{array}{r} {\scriptstyle 3\,5} \\ 648 \\ \times \quad 7 \\ \hline 4536 \end{array}$$

You Try It 2

$$\begin{array}{r} 756 \\ \times \ 305 \\ \hline 3780 \\ 22680 \\ \hline 230{,}580 \end{array}$$

You Try It 3

Strategy   To find the number of cars the dealer will receive in 12 months, multiply the number of months (12) by the number of cars received each month (37).

Solution

$$\begin{array}{r} 37 \\ \times \ \ 12 \\ \hline 74 \\ 37 \\ \hline 444 \end{array}$$

The dealer will receive 444 cars in 12 months.

You Try It 4

Strategy   To find the total cost of the order:
- Find the cost of the sports jackets by multiplying the number of jackets (25) by the cost for each jacket (23).
- Add the product to the cost for the suits (4800).

Solution

$$\begin{array}{r} 23 \\ \times \ 25 \\ \hline 115 \\ 46 \\ \hline 575 \end{array} \text{ cost for jackets}$$

$$\begin{array}{r} 4800 \\ + \ \ 575 \\ \hline 5375 \end{array}$$

The total cost of the order is $5375.

SECTION 1.5

You Try It 1

$$9\overline{)63} \quad \begin{array}{c} 7 \end{array}$$

Check: $7 \times 9 = 63$

You Try It 2

$$\begin{array}{r} 453 \\ 9{\overline{\smash{\big)}\,4077}} \\ \underline{-36} \\ 47 \\ \underline{-45} \\ 27 \\ \underline{-27} \\ 0 \end{array}$$

Check: $453 \times 9 = 4077$

You Try It 3

$$\begin{array}{r} 705 \\ 9{\overline{\smash{\big)}\,6345}} \\ \underline{-63} \\ 04 \\ \underline{- \ 0} \\ 45 \\ \underline{-45} \\ 0 \end{array}$$

Check: $705 \times 9 = 6345$

You Try It 4

$$\begin{array}{r} 870 \ \text{r5} \\ 6{\overline{\smash{\big)}\,5225}} \\ \underline{-48} \\ 42 \\ \underline{-42} \\ 05 \\ \underline{- \ 0} \\ 5 \end{array}$$

Check: $(870 \times 6) + 5 =$
$5220 + 5 = 5225$

You Try It 5

$$\begin{array}{r} 3{,}058 \ \text{r3} \\ 7{\overline{\smash{\big)}\,21{,}409}} \\ \underline{-21} \\ 0\,4 \\ \underline{- \ 0} \\ 40 \\ \underline{-35} \\ 59 \\ \underline{-56} \\ 3 \end{array}$$

Check: $(3058 \times 7) + 3 =$
$21{,}406 + 3 = 21{,}409$

You Try It 6

$$
\begin{array}{r}
109 \\
42)\overline{4578} \\
-42 \\
\hline
37 \\
-0 \\
\hline
378 \\
-378 \\
\hline
0
\end{array}
$$ *Check:* $109 \times 42 = 4578$

You Try It 7

$$
\begin{array}{r}
470 \text{ r}29 \\
39)\overline{18{,}359} \\
-15\ 6 \\
\hline
2\ 75 \\
-2\ 73 \\
\hline
29 \\
-0 \\
\hline
29
\end{array}
$$

Check: $(470 \times 39) + 29 =$
$18{,}330 + 29 = 18{,}359$

You Try It 8

$$
\begin{array}{r}
62 \text{ r}111 \\
534)\overline{33{,}219} \\
-32\ 04 \\
\hline
1\ 179 \\
-1\ 068 \\
\hline
111
\end{array}
$$

Check: $(62 \times 534) + 111 =$
$33{,}108 + 111 = 33{,}219$

You Try It 9

$$
\begin{array}{r}
421 \text{ r}33 \\
515)\overline{216{,}848} \\
-206\ 0 \\
\hline
10\ 84 \\
-10\ 30 \\
\hline
548 \\
-515 \\
\hline
33
\end{array}
$$

Check: $(421 \times 515) + 33 =$
$216{,}815 + 33 = 216{,}848$

You Try It 10

Strategy To find the number of tires that can
be stored on each shelf, divide the
number of tires (270) by the number
of shelves (15).

Solution

$$
\begin{array}{r}
18 \\
15)\overline{270} \\
-15 \\
\hline
120 \\
-120 \\
\hline
0
\end{array}
$$

Each shelf can store 18 tires.

You Try It 11

Strategy To find the number of cases
produced in 8 hours:
- Find the number of cases produced
  in 1 hour by dividing the number
  of cans produced (12,600) by the
  number of cans to a case (24).
- Multiply the number of cases
  produced in 1 hour by 8.

Solution

$$
\begin{array}{r}
525 \\
24)\overline{12{,}600} \\
-12\ 0 \\
\hline
60 \\
-48 \\
\hline
120 \\
-120 \\
\hline
0
\end{array}
$$ cases produced
in 1 hour

$$
\begin{array}{r}
525 \\
\times 8 \\
\hline
4200
\end{array}
$$

In 8 hours, 4200 cases are produced.

SECTION 1.6

You Try It 1 $2 \cdot 2 \cdot 2 \cdot 2 \cdot 3 \cdot 3 \cdot 3 = 2^4 \cdot 3^3$

You Try It 2 $10 \cdot 10 \cdot 10 \cdot 10 \cdot 10 \cdot 10 \cdot 10 = 10^7$

You Try It 3 $2^3 \cdot 5^2 = (2 \cdot 2 \cdot 2) \cdot (5 \cdot 5) = 8 \cdot 25$
$ = 200$

You Try It 4 $5 \cdot (8 - 4)^2 \div 4 - 2$
$ = 5 \cdot 4^2 \div 4 - 2$
$ = 5 \cdot 16 \div 4 - 2$
$ = 80 \div 4 - 2$
$ = 20 - 2$
$ = 18$

SECTION 1.7

You Try It 1 $40 \div 1 = 40$
$40 \div 2 = 20$
$40 \div 3 \qquad \text{Will not divide evenly}$
$40 \div 4 = 10$
$40 \div 5 = 8$
$40 \div 6 \qquad \text{Will not divide evenly}$
$40 \div 7 \qquad \text{Will not divide evenly}$
$40 \div 8 = 5$

1, 2, 4, 5, 8, 10, 20, and 40 are factors
of 40.

You Try It 2

$$\begin{array}{c} \underline{\quad 44 \quad} \\ 2 \mid 22 \\ 2 \mid 11 \\ 11 \mid 1 \\ 44 = 2 \cdot 2 \cdot 11 \end{array}$$

You Try It 3

$$\begin{array}{c} \underline{\quad 177 \quad} \\ 3 \mid 59 \\ 59 \mid 1 \\ 177 = 3 \cdot 59 \end{array}$$

Solutions to Chapter 2 "You Try It"

SECTION 2.1

You Try It 1

	2	3	5	7
50 =	2		5 · 5	
84 =	2 · 2	3		7
135 =		3 · 3 · 3	5	

The LCM $= 2 \cdot 2 \cdot 3 \cdot 3 \cdot 3 \cdot 5 \cdot 5 \cdot 7$
$= 18{,}900$

You Try It 2

	2	3	5
36 =	2 · 2	3 · 3	
60 =	2 · 2	3	5
72 =	2 · 2 · 2	3 · 3	

The GCF $= 2 \cdot 2 \cdot 3 = 12$.

You Try It 3

	2	3	5	11
11 =				11
24 =	2 · 2 · 2	3		
30 =	2	3	5	

Because no numbers are circled, the GCF $= 1$.

SECTION 2.2

You Try It 1 $\quad 4\dfrac{1}{4}$

You Try It 2 $\quad \dfrac{17}{4}$

You Try It 3

$$\begin{array}{r} 4 \\ 5\overline{)22} \\ -20 \\ \hline 2 \end{array} \qquad \frac{22}{5} = 4\frac{2}{5}$$

You Try It 4

$$\begin{array}{r} 4 \\ 7\overline{)28} \\ -28 \\ \hline 0 \end{array} \qquad \frac{28}{7} = 4$$

You Try It 5 $\quad 14\dfrac{5}{8} = \dfrac{112 + 5}{8} = \dfrac{117}{8}$

SECTION 2.3

You Try It 1 $\quad 45 \div 5 = 9 \qquad \dfrac{3}{5} = \dfrac{3 \cdot 9}{5 \cdot 9} = \dfrac{27}{45}$

$\dfrac{27}{45}$ is equivalent to $\dfrac{3}{5}$.

You Try It 2 $\quad$ Write 6 as $\dfrac{6}{1}$.

$18 \div 1 = 18 \qquad 6 = \dfrac{6 \cdot 18}{1 \cdot 18} = \dfrac{108}{18}$

$\dfrac{108}{18}$ is equivalent to 6.

You Try It 3 $\quad \dfrac{16}{24} = \dfrac{\overset{1}{\cancel{2}} \cdot \overset{1}{\cancel{2}} \cdot \overset{1}{\cancel{2}} \cdot 2}{\underset{1}{\cancel{2}} \cdot \underset{1}{\cancel{2}} \cdot \underset{1}{\cancel{2}} \cdot 3} = \dfrac{2}{3}$

You Try It 4 $\quad \dfrac{8}{56} = \dfrac{\overset{1}{\cancel{2}} \cdot \overset{1}{\cancel{2}} \cdot \overset{1}{\cancel{2}}}{\underset{1}{\cancel{2}} \cdot \underset{1}{\cancel{2}} \cdot \underset{1}{\cancel{2}} \cdot 7} = \dfrac{1}{7}$

You Try It 5 $\quad \dfrac{15}{32} = \dfrac{3 \cdot 5}{2 \cdot 2 \cdot 2 \cdot 2 \cdot 2} = \dfrac{15}{32}$

You Try It 6 $\quad \dfrac{48}{36} = \dfrac{\overset{1}{\cancel{2}} \cdot \overset{1}{\cancel{2}} \cdot 2 \cdot 2 \cdot \overset{1}{\cancel{3}}}{\underset{1}{\cancel{2}} \cdot \underset{1}{\cancel{2}} \cdot \underset{1}{\cancel{3}} \cdot 3} = \dfrac{4}{3} = 1\dfrac{1}{3}$

SECTION 2.4

You Try It 1

$$\begin{array}{r} \dfrac{3}{8} \\ +\dfrac{7}{8} \\ \hline \dfrac{10}{8} = \dfrac{5}{4} = 1\dfrac{1}{4} \end{array}$$

You Try It 2

$$\frac{5}{12} = \frac{20}{48}$$

$$+\ \frac{9}{16} = \frac{27}{48}$$

$$\frac{47}{48}$$

• The LCM of 12 and 16 is 48.

You Try It 3

$$\frac{7}{8} = \frac{105}{120}$$

$$+\ \frac{11}{15} = \frac{88}{120}$$

$$\frac{193}{120} = 1\frac{73}{120}$$

• The LCM of 8 and 15 is 120.

You Try It 4

$$\frac{3}{4} = \frac{30}{40}$$

$$\frac{4}{5} = \frac{32}{40}$$

$$+\ \frac{5}{8} = \frac{25}{40}$$

$$\frac{87}{40} = 2\frac{7}{40}$$

• The LCM of 4, 5, and 8 is 40.

You Try It 5

$$7 + \frac{6}{11} = 7\frac{6}{11}$$

You Try It 6

$$29$$

$$+\ 17\frac{5}{12}$$

$$46\frac{5}{12}$$

You Try It 7

$$7\frac{4}{5} = 7\frac{24}{30}$$

$$6\frac{7}{10} = 6\frac{21}{30}$$

$$+\ 13\frac{11}{15} = 13\frac{22}{30}$$

$$26\frac{67}{30} = 28\frac{7}{30}$$

• LCM = 30

You Try It 8

$$9\frac{3}{8} = 9\frac{45}{120}$$

$$17\frac{7}{12} = 17\frac{70}{120}$$

$$+\ 10\frac{14}{15} = 10\frac{112}{120}$$

$$36\frac{227}{120} = 37\frac{107}{120}$$

• LCM = 120

You Try It 9

Strategy To find the total time spent on the activities, add the three times $\left(4\frac{1}{2},\ 3\frac{3}{4},\ 1\frac{1}{3}\right)$.

Solution

$$4\frac{1}{2} = 4\frac{6}{12}$$

$$3\frac{3}{4} = 3\frac{9}{12}$$

$$+\ 1\frac{1}{3} = 1\frac{4}{12}$$

$$8\frac{19}{12} = 9\frac{7}{12}$$

The total time spent on the three activities was $9\frac{7}{12}$ hours.

You Try It 10

Strategy To find the overtime pay:
• Find the total number of overtime hours $\left(1\frac{2}{3} + 3\frac{1}{3} + 2\right)$.
• Multiply the total number of hours by the overtime hourly wage (36).

Solution

$$1\frac{2}{3}$$

$$3\frac{1}{3}$$

$$+\ 2$$

$$6\frac{3}{3} = 7 \text{ hours}$$

$$\begin{array}{r} 36 \\ \times\ \ 7 \\ \hline 252 \end{array}$$

Jeff earned $252 in overtime pay.

SECTION 2.5

You Try It 1

$$\frac{16}{27}$$

$$-\ \frac{7}{27}$$

$$\frac{9}{27} = \frac{1}{3}$$

You Try It 2

$$\frac{13}{18} = \frac{52}{72}$$

$$-\ \frac{7}{24} = \frac{21}{72}$$

$$\frac{31}{72}$$

• LCM = 72

You Try It 3

$$17\frac{5}{9} = 17\frac{20}{36} \quad \bullet \text{ LCM} = 36$$

$$-11\frac{5}{12} = 11\frac{15}{36}$$

$$6\frac{5}{36}$$

You Try It 4

$$8 = 7\frac{13}{13} \quad \bullet \text{ LCM} = 13$$

$$-2\frac{4}{13} = 2\frac{4}{13}$$

$$5\frac{9}{13}$$

You Try It 5

$$21\frac{7}{9} = 21\frac{28}{36} = 20\frac{64}{36} \quad \bullet \text{ LCM} = 36$$

$$-7\frac{11}{12} = 7\frac{33}{36} = 7\frac{33}{36}$$

$$13\frac{31}{36}$$

You Try It 6

Strategy To find the time remaining before the plane lands, subtract the number of hours already in the air $\left(2\frac{3}{4}\right)$ from the total time of the trip $\left(5\frac{1}{2}\right)$.

Solution

$$5\frac{1}{2} = 5\frac{2}{4} = 4\frac{6}{4}$$

$$-2\frac{3}{4} = 2\frac{3}{4} = 2\frac{3}{4}$$

$$2\frac{3}{4} \text{ hours}$$

The plane will land in $2\frac{3}{4}$ hours.

You Try It 7

Strategy To find the amount of weight to be lost during the third month:
- Find the total weight loss during the first two months $\left(7\frac{1}{2} + 5\frac{3}{4}\right)$.
- Subtract the total weight loss from the goal (24 pounds).

Solution

$$7\frac{1}{2} = 7\frac{2}{4}$$

$$+5\frac{3}{4} = 5\frac{3}{4}$$

$$12\frac{5}{4} = 13\frac{1}{4} \text{ pounds lost}$$

$$24 = 23\frac{4}{4}$$

$$-13\frac{1}{4} = 13\frac{1}{4}$$

$$10\frac{3}{4} \text{ pounds}$$

The patient must lose $10\frac{3}{4}$ pounds to achieve the goal.

SECTION 2.6

You Try It 1

$$\frac{4}{21} \times \frac{7}{44} = \frac{4 \cdot 7}{21 \cdot 44}$$

$$= \frac{\overset{1}{2} \cdot \overset{1}{2} \cdot \overset{1}{7}}{3 \cdot \underset{1}{7} \cdot \underset{1}{2} \cdot \underset{1}{2} \cdot 11} = \frac{1}{33}$$

You Try It 2

$$\frac{2}{21} \times \frac{10}{33} = \frac{2 \cdot 10}{21 \cdot 33}$$

$$= \frac{2 \cdot 2 \cdot 5}{3 \cdot 7 \cdot 3 \cdot 11} = \frac{20}{693}$$

You Try It 3

$$\frac{16}{5} \times \frac{15}{24} = \frac{16 \cdot 15}{5 \cdot 24}$$

$$= \frac{\overset{1}{2} \cdot \overset{1}{2} \cdot \overset{1}{2} \cdot 2 \cdot \overset{1}{3} \cdot \overset{1}{5}}{\underset{1}{5} \cdot \underset{1}{2} \cdot \underset{1}{2} \cdot \underset{1}{2} \cdot \underset{1}{3}} = 2$$

You Try It 4

$$5\frac{2}{5} \times \frac{5}{9} = \frac{27}{5} \times \frac{5}{9} = \frac{27 \cdot 5}{5 \cdot 9}$$

$$= \frac{\overset{1}{3} \cdot \overset{1}{3} \cdot 3 \cdot \overset{1}{5}}{\underset{1}{5} \cdot \underset{1}{3} \cdot \underset{1}{3}} = 3$$

You Try It 5

$$3\frac{2}{5} \times 6\frac{1}{4} = \frac{17}{5} \times \frac{25}{4} = \frac{17 \cdot 25}{5 \cdot 4}$$

$$= \frac{17 \cdot \overset{1}{5} \cdot 5}{\underset{1}{5} \cdot 2 \cdot 2} = \frac{85}{4} = 21\frac{1}{4}$$

You Try It 6

$$3\frac{2}{7} \times 6 = \frac{23}{7} \times \frac{6}{1} = \frac{23 \cdot 6}{7 \cdot 1}$$

$$= \frac{23 \cdot 2 \cdot 3}{7 \cdot 1} = \frac{138}{7} = 19\frac{5}{7}$$

You Try It 7

Strategy To find the value of the house today, multiply the old value of the house (70,000) by $2\frac{1}{2}$.

Solution
$$70,000 \times 2\frac{1}{2} = \frac{70,000}{1} \times \frac{5}{2}$$
$$= \frac{70,000 \cdot 5}{1 \cdot 2}$$
$$= 175,000$$

The value of the house today is $175,000.

You Try It 8

Strategy   To find the cost of the air compressor:
• Multiply to find the value of the drying chamber $\left(\frac{4}{5} \times 60,000\right)$.
• Subtract the value of the drying chamber from the total value of the two items (60,000).

Solution
$$\frac{4}{5} \times \frac{60,000}{1} = \frac{240,000}{5}$$
$$= 48,000 \quad \bullet \text{ Value of the drying chamber}$$

$$\begin{array}{r} 60,000 \\ -48,000 \\ \hline 12,000 \end{array}$$

The cost of the air compressor was $12,000.

SECTION 2.7

You Try It 1 $\dfrac{3}{7} \div \dfrac{2}{3} = \dfrac{3}{7} \times \dfrac{3}{2} = \dfrac{3 \cdot 3}{7 \cdot 2} = \dfrac{9}{14}$

You Try It 2 $\dfrac{3}{4} \div \dfrac{9}{10} = \dfrac{3}{4} \times \dfrac{10}{9}$

$$= \frac{3 \cdot 10}{4 \cdot 9} = \frac{\overset{1}{\cancel{3}} \cdot \overset{1}{\cancel{2}} \cdot 5}{2 \cdot 2 \cdot \underset{1}{\cancel{3}} \cdot 3} = \frac{5}{6}$$

You Try It 3 $\dfrac{5}{7} \div 6 = \dfrac{5}{7} \div \dfrac{6}{1}$

$$= \frac{5}{7} \times \frac{1}{6} = \frac{5 \cdot 1}{7 \cdot 6}$$
$$= \frac{5}{7 \cdot 2 \cdot 3} = \frac{5}{42}$$

You Try It 4 $12\dfrac{3}{5} \div 7 = \dfrac{63}{5} \div \dfrac{7}{1} = \dfrac{63}{5} \times \dfrac{1}{7}$

$$= \frac{63 \cdot 1}{5 \cdot 7} = \frac{3 \cdot 3 \cdot \overset{1}{\cancel{7}}}{5 \cdot \underset{1}{\cancel{7}}} = \frac{9}{5} = 1\frac{4}{5}$$

You Try It 5 $3\dfrac{2}{3} \div 2\dfrac{2}{5} = \dfrac{11}{3} \div \dfrac{12}{5}$

$$= \frac{11}{3} \times \frac{5}{12} = \frac{11 \cdot 5}{3 \cdot 12}$$
$$= \frac{11 \cdot 5}{3 \cdot 2 \cdot 2 \cdot 3} = \frac{55}{36} = 1\frac{19}{36}$$

You Try it 6 $2\dfrac{5}{6} \div 8\dfrac{1}{2} = \dfrac{17}{6} \div \dfrac{17}{2}$

$$= \frac{17}{6} \times \frac{2}{17} = \frac{17 \cdot 2}{6 \cdot 17}$$
$$= \frac{\overset{1}{\cancel{17}} \cdot 2}{2 \cdot 3 \cdot \underset{1}{\cancel{17}}} = \frac{1}{3}$$

You Try It 7 $6\dfrac{2}{5} \div 4 = \dfrac{32}{5} \div \dfrac{4}{1}$

$$= \frac{32}{5} \times \frac{1}{4} = \frac{32 \cdot 1}{5 \cdot 4}$$
$$= \frac{2 \cdot 2 \cdot 2 \cdot \overset{1}{\cancel{2}} \cdot \overset{1}{\cancel{2}}}{5 \cdot \underset{1}{\cancel{2}} \cdot \underset{1}{\cancel{2}}} = \frac{8}{5} = 1\frac{3}{5}$$

You Try It 8

Strategy   To find the cost of a similar coin weighing 1 ounce, divide the price of the coin (195) by the number of ounces $\left(\frac{1}{2}\right)$.

Solution
$$195 \div \frac{1}{2} = \frac{195}{1} \div \frac{1}{2}$$
$$= \frac{195}{1} \times \frac{2}{1} = \frac{195 \cdot 2}{1 \cdot 1} = 390$$

The cost of a similar coin weighing 1 ounce would be $390.

You Try It 9

Strategy   To find the length of the remaining piece:
• Divide the total length of the board (16) by the length of each shelf $\left(3\dfrac{1}{3}\right)$. This will give you the number of shelves cut, with a certain fraction of a shelf left over.
• Multiply the fractional part of the result in Step 1 by the length of one shelf to determine the length of the remaining piece.

Solution
$$16 \div 3\frac{1}{3} = 16 \div \frac{10}{3}$$
$$= \frac{16}{1} \times \frac{3}{10} = \frac{16 \cdot 3}{1 \cdot 10}$$
$$= \frac{\overset{1}{\cancel{2}} \cdot 2 \cdot 2 \cdot 2 \cdot 3}{\underset{1}{\cancel{2}} \cdot 5} = \frac{24}{5}$$
$$= 4\frac{4}{5}$$

There are 4 pieces $3\frac{1}{3}$ feet long. There is 1 piece that is $\frac{4}{5}$ of $3\frac{1}{3}$ feet long.

$$\frac{4}{5} \times 3\frac{1}{3} = \frac{4}{5} \times \frac{10}{3}$$
$$= \frac{4 \cdot 10}{5 \cdot 3} = \frac{2 \cdot 2 \cdot 2 \cdot \overset{1}{\cancel{5}}}{\underset{1}{\cancel{5}} \cdot 3}$$
$$= \frac{8}{3} = 2\frac{2}{3}$$

The length of the piece remaining is $2\frac{2}{3}$ feet.

SECTION 2.8

You Try It 1 $\quad \dfrac{9}{14} = \dfrac{27}{42} \qquad \dfrac{13}{21} = \dfrac{26}{42} \qquad \dfrac{9}{14} > \dfrac{13}{21}$

You Try It 2 $\quad \left(\dfrac{7}{11}\right)^2 \cdot \left(\dfrac{2}{7}\right) = \left(\dfrac{7}{11} \cdot \dfrac{7}{11}\right) \cdot \left(\dfrac{2}{7}\right)$
$$= \frac{\overset{1}{\cancel{7}} \cdot 7 \cdot 2}{11 \cdot 11 \cdot \underset{1}{\cancel{7}}} = \frac{14}{121}$$

You Try It 3 $\quad \left(\dfrac{1}{13}\right)^2 \cdot \left(\dfrac{1}{4} + \dfrac{1}{6}\right) \div \dfrac{5}{13} =$
$$\left(\frac{1}{13}\right)^2 \cdot \left(\frac{5}{12}\right) \div \frac{5}{13} =$$
$$\left(\frac{1}{169}\right) \cdot \left(\frac{5}{12}\right) \div \frac{5}{13} =$$
$$\left(\frac{1 \cdot 5}{13 \cdot 13 \cdot 12}\right) \div \frac{5}{13} =$$
$$\left(\frac{1 \cdot 5}{13 \cdot 13 \cdot 12}\right) \times \frac{13}{5} =$$
$$\frac{1 \cdot \overset{1}{\cancel{5}} \cdot \overset{1}{\cancel{13}}}{\underset{1}{\cancel{13}} \cdot 13 \cdot 12 \cdot \underset{1}{\cancel{5}}} = \frac{1}{156}$$

Solutions to Chapter 3 "You Try It"

SECTION 3.1

You Try It 1 Two hundred nine and five thousand eight hundred thirty-eight hundred-thousandths

You Try It 2 42,000.000207

You Try It 3
$$\text{4.349254} \quad \overset{\displaystyle \text{Given place value}}{\underset{\displaystyle 9 > 5}{}}$$

4.349254 rounded to the nearest hundredth is 4.35.

You Try It 4
$$\text{3.290532} \quad \overset{\displaystyle \text{Given place value}}{\underset{\displaystyle 2 < 5}{}}$$

3.290532 rounded to the nearest hundred-thousandth is 3.29053.

SECTION 3.2

You Try It 1

$$\begin{array}{r} {\scriptstyle 1\ 2} \\ 4.62 \\ 27.9 \\ +\ \ 0.62054 \\ \hline 33.14054 \end{array}$$

You Try It 2

$$1$$
$$6.05$$
$$12.$$
$$+ \ 0.374$$
$$\overline{18.424}$$

You Try It 3

Strategy   To find the amount he would be paid, add the amounts in the column headed "Boys."

Solution $2.61 + 1.64 + 2.86 + 1.29 + 9.02$
$+ \ 2.56 + 1.63 + 2.06 + 2.76 = 26.43$

For performing each one of the chores listed, a boy would be paid, on average, $26.43.

You Try It 4

Strategy   To find the total income, add the four commissions (985.80, 791.46, 829.75, and 635.42) to the salary (875).

Solution $875 + 985.80 + 791.46 + 829.75$
$+ \ 635.42 = 4117.43$

Anita's total income was $4117.43.

SECTION 3.3

You Try It 1

$$\begin{array}{r} {\scriptstyle 11\ \ 9} \\ {\scriptstyle 6\ \ X\ \ 10\ \ 13} \\ 7\,2.0\,3\,9 \\ -\ \ \ \ 8.4\,7 \\ \hline 6\,3.5\,6\,9 \end{array}$$

Check:
$$\begin{array}{r} {\scriptstyle 1\ 1\ \ 1} \\ 8.47 \\ +\ 63.569 \\ \hline 72.039 \end{array}$$

You Try It 2

$$\begin{array}{r} {\scriptstyle 14\ \ 9} \\ {\scriptstyle 2\ \ 4\ \ 10\ \ 10} \\ 3\,5.0\,0 \\ -\ \ \ 9.6\,7 \\ \hline 2\,5.3\,3 \end{array}$$

Check:
$$\begin{array}{r} {\scriptstyle 1\ 1\ \ 1} \\ 9.67 \\ +\ 25.33 \\ \hline 35.00 \end{array}$$

You Try It 3

$$\begin{array}{r} {\scriptstyle 16\ \ 9\ \ 9} \\ {\scriptstyle 2\ \ 6\ \ 10\ \ 10\ \ 10} \\ 3.7\,0\,0\,0 \\ -\ 1.9\,7\,1\,5 \\ \hline 1.7\,2\,8\,5 \end{array}$$

Check:
$$\begin{array}{r} {\scriptstyle 1\ 1\ 1\ 1} \\ 1.9715 \\ +\ 1.7285 \\ \hline 3.7000 \end{array}$$

You Try It 4

Strategy   To find the amount of change, subtract the amount paid (6.85) from 10.00.

Solution
$$\begin{array}{r} 10.00 \\ -\ \ 6.85 \\ \hline 3.15 \end{array}$$

Your change was $3.15.

You Try It 5

Strategy   To find the new balance:
 • Add to find the total of the three checks (1025.60 + 79.85 + 162.47).
 • Subtract the total from the previous balance (2472.69).

Solution
$$\begin{array}{r} 1025.60 \\ 79.85 \\ +\ 162.47 \\ \hline 1267.92 \end{array} \qquad \begin{array}{r} 2472.69 \\ -\ 1267.92 \\ \hline 1204.77 \end{array}$$

The new balance is $1204.77.

SECTION 3.4

You Try It 1

$$\begin{array}{r} 870 \\ \times\ \ \ 4.6 \\ \hline 522\ 0 \\ 3480 \\ \hline 4002.0 \end{array}$$

You Try It 2

$$\begin{array}{r} 0.000086 \\ \times\ \ \ \ 0.057 \\ \hline 602 \\ 430 \\ \hline 0.000004902 \end{array}$$

You Try It 3

$$\begin{array}{r} 4.68 \\ \times\ 6.03 \\ \hline 1404 \\ 28\ 080 \\ \hline 28.2204 \end{array}$$

You Try It 4 $6.9 \times 1000 = 6900$

You Try It 5 $4.0273 \times 10^2 = 402.73$

You Try It 6

Strategy   To find the total bill:
 • Find the number of gallons of water used by multiplying the number of gallons used per day (5000) by the number of days (62).
 • Find the cost of water by multiplying the cost per 1000 gallons (1.39) by the number of 1000-gallon units used.
 • Add the cost of the water to the meter fee (133.70).

Solution

Number of gallons = 5000(62) = 310,000

Cost of water = $\dfrac{310{,}000}{1000} \times 1.39 = 430.90$

Total cost = 430.90 + 133.70 = 564.60

The total bill is $564.60.

You Try It 7

Strategy   To find the cost of running the freezer for 210 hours, multiply the hourly cost (0.035) by the number of hours the freezer has run (210).

Solution
```
  0.035
×   210
  350
 7 0
 7.350
```

The cost of running the freezer for 210 hours is $7.35.

You Try It 8

Strategy   To find the total cost of the stereo:
- Multiply the monthly payment (37.18) by the number of months (18).
- Add that product to the down payment (175.00).

Solution
```
  37.18              175.00
×    18            + 669.24
 297 44              844.24
 371 8
 669.24
```

The total cost of the stereo is $844.24.

SECTION 3.5

You Try It 1
```
             2.7
0.052.)0.140.4
        −104
         36 4
        −36 4
            0
```

You Try It 2
```
       0.4873 ≈ 0.487
76)37.0420
  −30 4
    6 64
   −6 08
     562
    −532
     300
    −228
```

You Try It 3
```
            72.73 ≈ 72.7
5.09.)370.20.00
     −356 3
       13 90
       10 18
        3 720
       −3 563
        1570
       −1527
```

You Try It 4   309.21 ÷ 10,000 = 0.030921

You Try It 5 $42.93 \div 10^4 = 0.004293$

You Try It 6

Strategy   To find how many times greater the average hourly earnings were, divide the 1990 average hourly earnings (10.02) by the 1970 average hourly earnings (3.23).

Solution   10.02 ÷ 3.23 ≈ 3.1

The average hourly earnings in 1990 were about 3.1 times greater than in 1970.

You Try It 7

Strategy   To find the average number of people watching TV:
- Add the number of people watching each day of the week.
- Divide the total number of people watching by 7.

Solution   91.9 + 89.8 + 90.6 + 93.9 + 78.0 + 77.1 + 87.7 = 609

$\dfrac{609}{7} = 87$

An average of 87 million people watch television per day.

SECTION 3.6

You Try It 1
```
      0.56 ≈ 0.6
16)9.00
```

You Try It 2
$$4\dfrac{1}{6} = \dfrac{25}{6}$$
```
      4.166 ≈ 4.17
6)25.000
```

You Try It 3 $0.56 = \dfrac{56}{100} = \dfrac{14}{25}$

$5.35 = 5\dfrac{35}{100} = 5\dfrac{7}{20}$

You Try It 4

$$0.12\frac{7}{8} = \frac{12\frac{7}{8}}{100} = 12\frac{7}{8} \div 100$$

$$= \frac{103}{8} \times \frac{1}{100} = \frac{103}{800}$$

You Try It 5 $\dfrac{5}{8} = 0.625$

$0.63 > 0.625$

$0.63 > \dfrac{5}{8}$

Solutions to Chapter 4 "You Try It"

SECTION 4.1

You Try It 1 $\dfrac{20\ \text{pounds}}{24\ \text{pounds}} = \dfrac{20}{24} = \dfrac{5}{6}$

20 pounds : 24 pounds = 20 : 24 = 5 : 6

20 pounds to 24 pounds = 20 to 24
= 5 to 6

You Try It 2 $\dfrac{64\ \text{miles}}{8\ \text{miles}} = \dfrac{64}{8} = \dfrac{8}{1}$

64 miles : 8 miles = 64 : 8 = 8 : 1

64 miles to 8 miles = 64 to 8 = 8 to 1

You Try It 3

Strategy To find the ratio, write the ratio of board feet of cedar (12,000) to board feet of ash (18,000) in simplest form.

Solution $\dfrac{12,000}{18,000} = \dfrac{2}{3}$

The ratio is $\dfrac{2}{3}$.

You Try It 4

Strategy To find the ratio, write the ratio of the amount spent on radio advertising (15,000) to the amount spent on radio and television advertising (15,000 + 20,000) in simplest form.

Solution $\dfrac{\$15,000}{\$15,000 + \$20,000} = \dfrac{15,000}{35,000} = \dfrac{3}{7}$

The ratio is $\dfrac{3}{7}$.

SECTION 4.2

You Try It 1 $\dfrac{15\ \text{pounds}}{12\ \text{trees}} = \dfrac{5\ \text{pounds}}{4\ \text{trees}}$

You Try It 2 $\dfrac{260\ \text{miles}}{8\ \text{hours}}$

$$8)\overline{260.0}^{\,32.5}$$

32.5 miles/hour

You Try It 3

Strategy To find Erik's profit per ounce:
- Find the total profit by subtracting the cost (1625) from the selling price (1720).
- Divide the total profit by the number of ounces (5).

Solution $1720 - 1625 = 95$

$95 \div 5 = 19$

The profit was $19/ounce.

SECTION 4.3

You Try It 1 $\dfrac{6}{10} \bowtie \dfrac{9}{15}$ $\begin{array}{l} 10 \times 9 = 90 \\ 6 \times 15 = 90 \end{array}$

The cross products are equal. The proportion is true.

You Try It 2 $\dfrac{32}{6} \bowtie \dfrac{90}{8}$ $\begin{array}{l} 6 \times 90 = 540 \\ 32 \times 8 = 256 \end{array}$

The cross products are not equal. The proportion is not true.

You Try It 3 $\dfrac{n}{14} = \dfrac{3}{7}$

$n \times 7 = 14 \times 3$
$n \times 7 = 42$
$n = 42 \div 7$
$n = 6$

Check: $\dfrac{6}{14} \bowtie \dfrac{3}{7}$ $\begin{array}{l} 14 \times 3 = 42 \\ 6 \times 7 = 42 \end{array}$

You Try It 4

$$\frac{5}{7} = \frac{n}{20}$$

$$5 \times 20 = 7 \times n$$
$$100 = 7 \times n$$
$$100 \div 7 = n$$
$$14.3 \approx n$$

You Try It 5

$$\frac{15}{20} = \frac{12}{n}$$

$$15 \times n = 20 \times 12$$
$$15 \times n = 240$$
$$n = 240 \div 15$$
$$n = 16$$

Check: $\dfrac{15}{20} \asymp \dfrac{12}{16}$ ➝ $20 \times 12 = 240$ ➝ $15 \times 16 = 240$

You Try It 6

$$\frac{12}{n} = \frac{7}{4}$$

$$12 \times 4 = n \times 7$$
$$48 = n \times 7$$
$$48 \div 7 = n$$
$$6.86 \approx n$$

You Try It 7

$$\frac{n}{12} = \frac{4}{1}$$

$$n \times 1 = 12 \times 4$$
$$n \times 1 = 48$$
$$n = 48 \div 1$$
$$n = 48$$

Check: $\dfrac{48}{12} \asymp \dfrac{4}{1}$ ➝ $12 \times 4 = 48$ ➝ $48 \times 1 = 48$

You Try It 8

$$\frac{3}{8} = \frac{12}{n}$$

$$3 \times n = 8 \times 12$$
$$3 \times n = 96$$
$$n = 96 \div 3$$
$$n = 32$$

Check: $\dfrac{3}{8} \asymp \dfrac{12}{32}$ ➝ $8 \times 12 = 96$ ➝ $3 \times 32 = 96$

You Try It 9

Strategy   To find the number of jars that can be packed in 15 boxes, write and solve a proportion using n to represent the number of jars.

Solution $\dfrac{24 \text{ jars}}{6 \text{ boxes}} = \dfrac{n \text{ jars}}{15 \text{ boxes}}$

$$24 \times 15 = 6 \times n$$
$$360 = 6 \times n$$
$$360 \div 6 = n$$
$$60 = n$$

60 jars can be packed in 15 boxes.

You Try It 10

Strategy   To find the number of tablespoons of fertilizer needed, write and solve a proportion using n to represent the number of tablespoons of fertilizer.

Solution $\dfrac{3 \text{ tablespoons}}{4 \text{ gallons}} = \dfrac{n \text{ tablespoons}}{10 \text{ gallons}}$

$$3 \times 10 = 4 \times n$$
$$30 = 4 \times n$$
$$30 \div 4 = n$$
$$7.5 = n$$

For 10 gallons of water, 7.5 tablespoons of fertilizer are required.

Solutions to Chapter 5 "You Try It"

SECTION 5.1

You Try It 1 $125\% = 125 \times \dfrac{1}{100} = \dfrac{125}{100} = 1\dfrac{1}{4}$

$125\% = 125 \times 0.01 = 1.25$

You Try It 2 $33\dfrac{1}{3}\% = 33\dfrac{1}{3} \times \dfrac{1}{100}$

$$= \dfrac{100}{3} \times \dfrac{1}{100}$$

$$= \dfrac{100}{300} = \dfrac{1}{3}$$

You Try It 3 $0.25\% = 0.25 \times 0.01 = 0.0025$

You Try It 4 $0.048 = 0.048 \times 100\% = 4.8\%$

You Try It 5 $3.67 = 3.67 \times 100\% = 367\%$

You Try It 6 $0.62\frac{1}{2} = 0.62\frac{1}{2} \times 100\%$

$= 62\frac{1}{2}\%$

You Try It 7 $\frac{5}{6} = \frac{5}{6} \times 100\% = \frac{500}{6}\% = 83\frac{1}{3}\%$

You Try It 8 $1\frac{4}{9} = \frac{13}{9} = \frac{13}{9} \times 100\%$

$= \frac{1300}{9}\% \approx 144.4\%$

SECTION 5.2

You Try It 1 $n = 0.063 \times 150$
$n = 9.45$

You Try It 2 $n = \frac{1}{6} \times 66$

$n = 11$

You Try It 3

Strategy   To find the amount donated, write and solve the basic percent equation using n to represent the amount. The percent is 14%. The base is $190 billion.

Solution $14\% \times 190 = n$
$0.14 \times 190 = n$
$26.6 = n$

$26.6 billion was donated to educational organizations.

You Try It 4

Strategy   To find the new hourly wage:
- Find the amount of the raise. Write and solve the basic percent equation, using n to represent the amount of the raise (amount). The percent is 8%. The base is $23.50.
- Add the amount of the raise to the old wage (23.50).

Solution $8\% \times 23.50 = n$ 23.50
$0.08 \times 23.50 = n$ $\underline{+\ 1.88}$
$1.88 = n$ 25.38

The new hourly wage is $25.38.

SECTION 5.3

You Try It 1 $n \times 32 = 16$
$n = 16 \div 32$

$n = 0.50$
$n = 50\%$

You Try It 2 $n \times 15 = 48$
$n = 48 \div 15$
$n = 3.20$
$n = 320\%$

You Try It 3 $30 = n \times 45$
$30 \div 45 = n$
$\frac{2}{3} = n$
$66\frac{2}{3}\% = n$

You Try It 4

Strategy   To find what percent of the income the income tax is, write and solve the basic percent equation, using n to represent the percent. The base is $33,500 and the amount is $5025.

Solution $n \times 33,500 = 5025$
$n = 5025 \div 33,500$
$n = 0.15 = 15\%$

The income tax is 15% of the income.

You Try It 5

Strategy   To find the percent who were women:
- Subtract to find the number of enlisted personnel who were women (518,921 − 512,370).
- Write and solve the basic percent equation, using n to represent the percent. The base is 518,921, and the amount is the number of enlisted personnel who were women.

Solution $518,921 - 512,370 = 6551$

$n \times 518,921 = 6551$
$n = 6551 \div 518,921$
$n \approx 0.013$

In 1950, 1.3% of the enlisted personnel in the U.S. Army were women.

SECTION 5.4

You Try It 1 $0.86 \times n = 215$
$n = 215 \div 0.86$
$n = 250$

You Try It 2

$$15 = 0.025 \times n$$
$$15 \div 0.025 = n$$
$$600 = n$$

You Try It 3

$$\frac{1}{6} \times n = 5$$
$$n = 5 \div \frac{1}{6}$$
$$n = 30$$

You Try It 4

Strategy   To find the original value of the car, write and solve the basic percent equation, using n to represent the original value (base). The percent is 42%. The amount is $10,458.

Solution
$$42\% \times n = 10{,}458$$
$$0.42 \times n = 10{,}458$$
$$n = 10{,}458 \div 0.42$$
$$n = 24{,}900$$

The original value of the car was $24,900.

You Try It 5

Strategy   To find the difference between the original price and the sale price:
- Find the original price. Write and solve the basic percent equation, using n to represent the original price (base). The percent is 80%. The amount is $89.60.
- Subtract the sale price (89.60) from the original price.

Solution
$$80\% \times n = 89.60$$
$$0.80 \times n = 89.60$$
$$n = 89.60 \div 0.80$$
$$n = 112.00 \quad \text{original price}$$

$$112.00 - 89.60 = 22.40$$

The difference between the original price and the sale price is $22.40.

SECTION 5.5

You Try It 1

$$\frac{26}{100} = \frac{22}{n}$$
$$26 \times n = 100 \times 22$$
$$26 \times n = 2200$$
$$n = 2200 \div 26$$
$$n \approx 84.62$$

You Try It 2

$$\frac{16}{100} = \frac{n}{132}$$
$$16 \times 132 = 100 \times n$$
$$2112 = 100 \times n$$
$$2112 \div 100 = n$$
$$21.12 = n$$

You Try It 3

Strategy   To find the number of days it snowed, write and solve a proportion, using n to represent the number of days it snowed (amount). The percent is 64%. The base is 150.

Solution
$$\frac{64}{100} = \frac{n}{150}$$
$$64 \times 150 = 100 \times n$$
$$9600 = 100 \times n$$
$$9600 \div 100 = n$$
$$96 = n$$

It snowed 96 days.

You Try It 4

Strategy   To find the percent of pens that were not defective:
- Subtract to find the number of pens that were not defective $(200 - 5)$.
- Write and solve a proportion, using n to represent the percent of pens that were not defective. The base is 200, and the amount is the number of pens not defective.

Solution
$$200 - 5 = 195 \quad \begin{array}{l}\text{number of pens}\\ \text{not defective}\end{array}$$

$$\frac{n}{100} = \frac{195}{200}$$
$$n \times 200 = 100 \times 195$$
$$n \times 200 = 19{,}500$$
$$n = 19{,}500 \div 200$$
$$n = 97.5\%$$

97.5% of the pens were not defective.

Solutions to Chapter 6 "You Try It"

SECTION 6.1

You Try It 1

Strategy   To find the unit cost, divide the total cost by the number of units.

Solution   **a.** $7.67 \div 8 = 0.95875$
$.959$ for each battery
b. $2.29 \div 15 \approx 0.153$
$.153$ per ounce

You Try It 2

Strategy   To find the more economical purchase, compare the unit costs.

Solution $5.70 \div 6 = 0.95$
$3.96 \div 4 = 0.99$
$.95 < $.99$

The more economical purchase is 6 cans for $5.70.

You Try It 3

Strategy   To find the total cost, multiply the unit cost (9.96) by the number of units (7).

Solution $9.96 \times 7 = 69.72$

The total cost is $69.72.

SECTION 6.2

You Try It 1

Strategy   To find the percent increase:
• Find the amount of the increase.
• Solve the basic percent equation for *percent*.

Solution $1.83 - 1.46 = 0.37$

$n \times 1.46 = 0.37$
$n = 0.37 \div 1.46$
$n \approx 0.25 = 25\%$

The percent increase was 25%.

You Try It 2

Strategy   To find the new hourly wage:
• Solve the basic percent equation for *amount*.
• Add the amount of the increase to the original wage.

Solution $0.14 \times 12.50 = n$
$1.75 = n$

$12.50 + 1.75 = 14.25$

The new hourly wage is $14.25.

You Try It 3

Strategy   To find the markup, solve the basic percent equation for *amount*.

Solution $0.20 \times 8 = n$
$1.60 = n$

The markup is $1.60.

You Try It 4

Strategy   To find the selling price:
• Find the markup by solving the basic percent equation for *amount*.
• Add the markup to the cost.

Solution $0.55 \times 72 = n$
$39.60 = n$
$72 + 39.60 = 111.60$

The selling price is $111.60.

You Try It 5

Strategy   To find the percent decrease:
• Find the amount of the decrease.
• Solve the basic percent equation for *percent*.

Solution $261,000 - 215,000 = 46,000$

$n \times 261,000 = 46,000$
$n = 46,000 \div 261,000$
$n \approx 0.176$

The percent decrease is 17.6%.

You Try It 6

Strategy   To find the visibility:
• Find the amount of decrease by solving the basic percent equation for *amount*.
• Subtract the amount of decrease from the original visibility.

Solution $0.40 \times 5 = n$
$2 = n$

$5 - 2 = 3$

The visibility was 3 miles.

You Try It 7

Strategy To find the discount rate:
- Find the discount.
- Solve the basic percent equation for *percent*.

Solution $12.50 - 10.99 = 1.51$

$$n \times 12.50 = 1.51$$
$$n = 1.51 \div 12.50$$
$$n = 0.1208$$

The discount rate is 12.1%.

You Try It 8

Strategy To find the sale price:
- Find the discount by solving the basic percent equation for *amount*.
- Subtract to find the sale price.

Solution $0.15 \times 110 = n$
$$16.5 = n$$

$110 - 16.5 = 93.5$

The sale price is $93.50.

SECTION 6.3

You Try It 1

Strategy To find the simple interest due, multiply the principal (15,000) times the annual interest rate (8% = 0.08) times the time, in years
(18 months $= \dfrac{18}{12}$ years $= 1.5$ years).

Solution $15,000 \times 0.08 \times 1.5 = 1800$

The interest due is $1800.

You Try It 2

Strategy To find the maturity value:
- Use the simple interest formula to find the simple interest due.
- Find the maturity value by adding the principal and the interest.

Solution $3800 \times 0.06 \times \dfrac{90}{365} \approx 56.22$

$3800 + 56.22 = 3856.22$

The maturity value is $3856.22.

You Try It 3

Strategy To find the monthly payment:
- Find the maturity value by adding the principal and the interest.
- Divide the maturity value by the length of the loan in months (12).

Solution $1900 + 152 = 2052$

$2052 \div 12 = 171$

The monthly payment is $171.

You Try It 4

Strategy To find the finance charge, multiply the principal, or unpaid balance (1250), times the monthly interest rate (1.6%) times the number of months (1).

Solution $1250 \times 0.016 \times 1 = 20$

The simple interest due is $20.

You Try It 5

Strategy To find the interest earned:
- Find the new principal by multiplying the original principal (1000) by the factor found in the Compound Interest Table (3.29066).
- Subtract the original principal from the new principal.

Solution $1000 \times 3.29066 = 3290.66$

The new principal is $3290.66.

$3290.66 - 1000 = 2290.66$

The interest earned is $2290.66.

SECTION 6.4

You Try It 1

Strategy To find the mortgage:
- Find the down payment by solving the basic percent equation for *amount*.
- Subtract the down payment from the purchase price.

Solution $0.25 \times 216,000 = n$
$$54,000 = n$$

The down payment is $54,000.

$216,000 - 54,000 = 162,000$

The mortgage is $162,000.

You Try It 2

Strategy To find the loan origination fee, solve the basic percent equation for *amount*.

Solution $0.045 \times 80,000 = n$
$$3600 = n$$

The loan origination fee was $3600.

You Try It 3

Strategy   To find the monthly mortgage payment:
- Subtract the down payment from the purchase price to find the mortgage.
- Multiply the mortgage by the factor found in the Monthly Payment Table.

Solution   75,000 − 15,000 = 60,000

The mortgage is $60,000.

60,000 × 0.0089973 = 539.838

The monthly mortgage payment is $539.84.

You Try It 4

Strategy   To find the interest:
- Multiply the mortgage by the factor found in the Monthly Payment Table to find the monthly mortgage payment.
- Subtract the principal from the monthly mortgage payment.

Solution   125,000 × 0.0083920 = 1049

The monthly mortgage payment is $1049.

1049 − 492.65 = 556.35

The interest on the mortgage is $556.35.

You Try It 5

Strategy   To find the monthly payment:
- Divide the annual property tax by 12 to find the monthly property tax.
- Add the monthly property tax to the monthly mortgage payment.

Solution   744 ÷ 12 = 62

The monthly property tax is $62.

415.20 + 62 = 477.20

The total monthly payment is $477.20.

SECTION 6.5

You Try It 1

Strategy   To find the amount financed:
- Find the down payment by solving the basic percent equation for *amount*.

- Subtract the down payment from the purchase price.

Solution   0.20 × 19,200 = n
                      3840 = n

The down payment is $3840.

19,200 − 3840 = 15,360

The amount financed is $15,360.

You Try It 2

Strategy   To find the license fee, solve the basic percent equation for *amount*.

Solution   0.015 × 17,350 = n
                      260.25 = n

The license fee is $260.25.

You Try It 3

Strategy   To find the cost, multiply the cost per mile (0.31) by the number of miles driven (23,000).

Solution   23,000 × 0.31 = 7130

The cost is $7130.

You Try It 4

Strategy   To find the cost per mile for car insurance, divide the cost for insurance (360) by the number of miles driven (15,000).

Solution   360 ÷ 15,000 = 0.024

The cost per mile for insurance is $.024.

You Try It 5

Strategy   To find the monthly payment:
- Subtract the down payment from the purchase price to find the amount financed.
- Multiply the amount financed by the factor found in the Monthly Payment Table.

Solution   25,900 − 6475 = 19,425

19,425 × 0.0244129 ≈ 474.22

The monthly payment is $474.22.

SECTION 6.6

You Try It 1

Strategy   To find the worker's earnings:
- Find the worker's overtime wage by multiplying the hourly wage by 2.

• Multiply the number of overtime hours worked by the overtime wage.

Solution $18.50 \times 2 = 37$

The hourly wage for overtime is $37.

$37 \times 8 = 296$

The construction worker earns $296.

You Try It 2

Strategy   To find the salary per month, divide the annual salary by the number of months in a year (12).

Solution $48,228 \div 12 = 4019$

The contractor's monthly salary is $4019.

You Try It 3

Strategy   To find the total earnings:
• Find the sales over $50,000.
• Multiply the commission rate by sales over $50,000.
• Add the commission to the annual salary.

Solution $175,000 - 50,000 = 125,000$

Sales over $50,000 totaled $125,000.

$125,000 \times 0.095 = 11,875$

Earnings from commissions totaled $11,875.

$27,000 + 11,875 = 38,875$

The insurance agent earned $38,875.

SECTION 6.7

You Try It 1

Strategy   To find the current balance:
• Subtract the amount of the check from the old balance.
• Add the amount of each deposit.

Solution
```
   302.46
 - 20.59   check
  281.87
  176.86   first deposit
 + 94.73   second deposit
  553.46
```

The current checking account balance is $553.46.

You Try It 2   Current checkbook

```
balance:                  623.41
Check: 237              +  78.73
                          702.14
Interest:               +   2.11
                          704.25
Deposit:                - 523.84
                          180.41
```

Closing bank balance from bank statement: $180.41.

Checkbook balance: $180.41.

The bank statement and the checkbook balance.

Solutions to Chapter 7 "You Try It"

SECTION 7.1

You Try It 1

Strategy   To find what percent of the total number of cellular phones purchased the number purchased in March represents:
• Read the pictograph to determine the number of cellular phones purchased for each month.

• Find the total cellular phone purchases for the 4-month period.
• Solve the basic percent equation for percent (n). Amount = 3000; the base is the total sales for the 4-month period.

Solution
```
   4,500
   3,500
   3,000
   1,500
  12,500
```

$$n \times 12{,}500 = 3000$$
$$n = 3000 \div 12{,}500$$
$$n = 0.24$$

The number of cellular phones purchased in March represents 24% of the total number of cellular phones purchased.

You Try It 2

Strategy To find the ratio of the annual cost of fuel to the annual cost of maintenance:
- Locate the annual fuel cost and the annual maintenance cost in the circle graph.
- Write the ratio of the annual fuel cost to the annual maintenance cost in simplest form.

Solution Annual fuel cost: $700
Annual maintenance cost: $500

$$\frac{700}{500} = \frac{7}{5}$$

The ratio is $\frac{7}{5}$.

You Try It 3

Strategy To find the amount paid for medical/dental insurance:
- Locate the percent of the distribution that is medical/dental insurance.
- Solve the basic percent equation for amount.

Solution The percent of the distribution that is medical/dental insurance: 6%

Percent $\times$ base = amount
$$0.06 \times 2900 = n$$
$$174 = n$$

The amount paid for medical/dental insurance was $174.

SECTION 7.2

You Try It 1

Strategy To write the ratio:
- Read the graph to find the lung capacity of an inactive female and of an athletic female.
- Write the ratio in simplest form.

Solution Lung capacity of an inactive female: 25

Lung capacity of an athletic female: 55

$$\frac{25}{55} = \frac{5}{11}$$

The ratio is $\frac{5}{11}$.

You Try It 2

Strategy To determine between which two years the net income of Math Associates increased the most:
- Read the line graph to determine the net income of Math Associates for each of the years shown.
- Subtract to find the difference between each two years.
- Find the greatest difference.

Solution 2000: $1 million
2001: $3 million
2002: $5 million
2003: $8 million
2004: $12 million

Between 2000 and 2001: $3 - 1 = 2$
Between 2001 and 2002: $5 - 3 = 2$
Between 2002 and 2003: $8 - 5 = 3$
Between 2003 and 2004: $12 - 8 = 4$

$4 > 3 > 2$

The net income of Math Associates increased the most between the years 2003 and 2004.

SECTION 7.3

You Try It 1

Strategy To find the number of employees:
- Read the histogram to find the number of employees whose hourly wage is between $10 and $12 and the number whose hourly wage is between $12 and $14.
- Add the two numbers.

Solution Number whose wage is between $10 and $12: 7; between $12 and $14: 15.

$7 + 15 = 22$

22 employees earn between $10 and $14.

You Try It 2

Strategy   To write the ratio:
- Read the graph to find the number of states with a per capita income between $18,000 and $22,000 and the number with a per capita income between $26,000 and $30,000.
- Write the ratio in simplest form.

Solution   Number of states with a per capita income between $18,000 and $22,000: 13
Number of states with a per capita income between $26,000 and $30,000: 13

$$\frac{13}{13} = \frac{1}{1}$$

The ratio is $\frac{1}{1}$.

SECTION 7.4

You Try It 1

Strategy   To find the mean amount spent by the 12 customers:
- Find the sum of the numbers.
- Divide the sum by the number of customers (12).

Solution   6.26 + 8.23 + 5.09 + 8.11 + 7.50 + 6.69 + 5.66 + 4.89 + 5.25 + 9.36 + 6.75 + 7.05 = 80.84

$$\bar{x} = \frac{80.84}{12} \approx 6.74$$

The mean amount spent by the 12 customers was $6.74.

You Try It 2

Strategy   To find the median weight loss:
- Arrange the weight losses from smallest to largest.
- Because there is an even number of values, the median is the mean of the middle two numbers.

Solution   10, 14, 16, 16, 22, 27, 29, 31, 31, 40

$$\text{Median} = \frac{22 + 27}{2} = \frac{49}{2} = 24.5$$

The median weight loss was 24.5 pounds.

You Try It 3

Strategy   To draw the box-and-whiskers plot:
- Find the median, Q_1, and Q_3.
- Use the smallest value, Q_1, the median, Q_3, and the largest value to draw the box-and-whiskers plot.

Solution
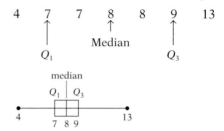

Answers about the spread of the data will vary. For example, in You Try It 3, the values in the interquartile range are all very close to the median. They are not so close to the median in Example 3. The whiskers are long with respect to the box in You Try It 3, whereas they are short with respect to the box in Example 3, showing that the data values outside the interquartile range are closer together in Example 3 than in You Try It 3.

SECTION 7.5

You Try It 1

Strategy   To find the probability:
- List the outcomes of the experiment in a systematic way. We will use a table.
- Use the table to count the number of possible outcomes of the experiment.
- Count the number of outcomes of the experiment that are favorable to the event of 2 true questions and 1 false question.
- Use the probability formula.

Solution

Question 1	Question 2	Question 3
T	T	T
T	T	F
T	F	T
T	F	F
F	T	T
F	T	F
F	F	T
F	F	F

There are 8 possible outcomes:

$S = \{TTT, TTF, TFT, TFF, FTT, FTF, FFT, FFF\}$

There are 3 outcomes favorable to the event:

$\{TTF, TFT, FTT\}$

Probability of an event =

$$\dfrac{\text{number of favorable outcomes}}{\text{number of possible outcomes}} = \dfrac{3}{8}$$

The probability of 2 true questions and 1 false question is $\dfrac{3}{8}$.

Solutions to Chapter 8 "You Try It"

SECTION 8.1

You Try It 1 $14 \text{ ft} = 14 \text{ ft} \times \dfrac{1 \text{ yd}}{3 \text{ ft}} = \dfrac{14 \text{ yd}}{3} = 4\dfrac{2}{3} \text{ yd}$

You Try It 2 $9240 \text{ ft} = 9240 \text{ ft} \times \dfrac{1 \text{ mi}}{5280 \text{ ft}}$

$= \dfrac{9240 \text{ mi}}{5280} = 1\dfrac{3}{4} \text{ mi}$

You Try It 3

$$\begin{array}{r} 3 \text{ ft } 6 \text{ in.} \\ 12\overline{)\ 42} \\ -36 \\ \hline 6 \end{array}$$ • 12 in. = 1 ft

$42 \text{ in.} = 3 \text{ ft } 6 \text{ in.}$

You Try It 4

$$\begin{array}{r} 4 \text{ yd } 2 \text{ ft} \\ 3\overline{)\ 14} \\ -12 \\ \hline 2 \end{array}$$ • 3 ft = 1 yd

$14 \text{ ft} = 4 \text{ yd } 2 \text{ ft}$

You Try It 5

$$\begin{array}{r} 3 \text{ ft } \ \ 5 \text{ in.} \\ + \ 4 \text{ ft } \ \ 9 \text{ in.} \\ \hline 7 \text{ ft } 14 \text{ in.} \end{array}$$ • 14 in. = 1 ft 2 in.

$7 \text{ ft } 14 \text{ in.} = 8 \text{ ft } 2 \text{ in.}$

You Try It 6

$$\begin{array}{r} {\scriptstyle 3 \text{ ft} \quad 14 \text{ in.}} \\ \cancel{4 \text{ ft } 2 \text{ in.}} \\ - \ 1 \text{ ft } 8 \text{ in.} \\ \hline 2 \text{ ft } 6 \text{ in.} \end{array}$$ • Borrow 1 ft (12 in.) from 4 ft and add it to 2 in.

You Try It 7

$$\begin{array}{r} 4 \text{ yd } 1 \text{ ft} \\ \times \qquad 8 \\ \hline 32 \text{ yd } 8 \text{ ft} \end{array}$$ • 8 ft = 2 yd 2 ft

$32 \text{ yd } 8 \text{ ft} = 34 \text{ yd } 2 \text{ ft}$

You Try It 8

$$\begin{array}{r} 3 \text{ yd } \quad 2 \text{ ft} \\ 2\overline{)\ 7 \text{ yd } \quad 1 \text{ ft}} \\ -6 \text{ yd} \\ \hline 1 \text{ yd } = 3 \text{ ft} \\ 4 \text{ ft} \\ -4 \text{ ft} \\ \hline 0 \end{array}$$

You Try It 9 $6\dfrac{1}{4} \text{ ft} = 6\dfrac{3}{12} \text{ ft} = 5\dfrac{15}{12} \text{ ft}$

$$\begin{array}{r} -3\dfrac{2}{3} \text{ ft} = 3\dfrac{8}{12} \text{ ft} = 3\dfrac{8}{12} \text{ ft} \\ \hline 2\dfrac{7}{12} \text{ ft} \end{array}$$

You Try It 10

Strategy   To find the width of the storage room:
- Multiply the number of tiles (8) by the width of each tile (9 in.).
- Divide the result by the number of inches in 1 foot (12) to find the width in feet.

Solution $9 \text{ in.} \times 8 = 72 \text{ in.}$

$72 \div 12 = 6$

The width is 6 ft.

You Try It 11

Strategy   To find the length of each piece, divide the total length (9 ft 8 in.) by the number of pieces (4).

Solution

$$\begin{array}{r} 2 \text{ ft } \quad 5 \text{ in.} \\ 4\overline{)\ 9 \text{ ft } \quad 8 \text{ in.}} \\ -8 \text{ ft} \\ \hline 1 \text{ ft } = 12 \text{ in.} \\ 20 \text{ in.} \\ -20 \text{ in.} \\ \hline 0 \text{ in.} \end{array}$$

Each piece is 2 ft 5 in. long.

SECTION 8.2

You Try It 1 $3 \text{ lb} = 3 \cancel{\text{ lb}} \times \dfrac{16 \text{ oz}}{1 \cancel{\text{ lb}}} = 48 \text{ oz}$

You Try It 2 $4200 \text{ lb} = 4200 \cancel{\text{ lb}} \times \dfrac{1 \text{ ton}}{2000 \cancel{\text{ lb}}}$

$\qquad\qquad = \dfrac{4200 \text{ tons}}{2000} = 2\dfrac{1}{10} \text{ tons}$

You Try It 3

$$\begin{array}{l} \overset{6\text{ lb}}{\cancel{7 \text{ lb}}} \ \overset{17\text{ oz}}{\cancel{1 \text{ oz}}} \\ -\ 3 \text{ lb} \ \ 4 \text{ oz} \\ \hline \ \ 3 \text{ lb} \ 13 \text{ oz} \end{array}$$

• Borrow 1 lb (16 oz) from 7 lb and add it to 1 oz.

You Try It 4

$$\begin{array}{r} 3 \text{ lb } 6 \text{ oz} \\ \times \qquad 4 \\ \hline 12 \text{ lb } 24 \text{ oz} = 13 \text{ lb } 8 \text{ oz} \end{array}$$

You Try It 5

Strategy   To find the weight of 12 bars of soap:
• Multiply the number of bars (12) by the weight of each bar (7 oz).
• Convert the number of ounces to pounds.

Solution

$$\begin{array}{r} 12 \\ \times \ \ 7 \text{ oz} \\ \hline 84 \text{ oz} \end{array} \qquad 84 \cancel{\text{ oz}} \times \dfrac{1 \text{ lb}}{16 \cancel{\text{ oz}}} = 5\dfrac{1}{4} \text{ lb}$$

The 12 bars of soap weigh $5\dfrac{1}{4}$ lb.

SECTION 8.3

You Try It 1 $18 \text{ pt} = 18 \cancel{\text{ pt}} \times \dfrac{1 \cancel{\text{ qt}}}{2 \cancel{\text{ pt}}} \times \dfrac{1 \text{ gal}}{4 \cancel{\text{ qt}}}$

$\qquad\qquad = \dfrac{18 \text{ gal}}{8} = 2\dfrac{1}{4} \text{ gal}$

You Try It 2

$$\begin{array}{r} 1 \text{ gal} \quad 2 \text{ qt} \\ 3\overline{)\ 4 \text{ gal} \quad 2 \text{ qt}} \\ -3 \text{ gal} \\ \hline 1 \text{ gal} = 4 \text{ qt} \\ \hline 6 \text{ qt} \\ -6 \text{ qt} \\ \hline 0 \end{array}$$

You Try It 3

Strategy   To find the number of gallons of water needed:
• Find the number of quarts required by multiplying the number of quarts one student needs per day (3) by the number of students (5) by the number of days (3).
• Convert the number of quarts to gallons.

Solution $3 \times 5 \times 3 = 45 \text{ qt}$

$$45 \cancel{\text{ qt}} \cdot \dfrac{1 \text{ gal}}{4 \cancel{\text{ qt}}} = \dfrac{45 \text{ gal}}{4} = 11\dfrac{1}{4} \text{ gal}$$

The students should take $11\dfrac{1}{4}$ gal of water.

SECTION 8.4

You Try It 1 $4.5 \text{ Btu} = 4.5 \cancel{\text{ Btu}} \times \dfrac{778 \text{ ft} \cdot \text{lb}}{1 \cancel{\text{ Btu}}}$

$\qquad\qquad = 3501 \text{ ft} \cdot \text{lb}$

You Try It 2 $800 \text{ lb} \times 16 \text{ ft} = 12{,}800 \text{ ft} \cdot \text{lb}$

You Try It 3 $56{,}000 \text{ Btu} =$

$56{,}000 \cancel{\text{ Btu}} \times \dfrac{778 \text{ ft} \cdot \text{lb}}{1 \cancel{\text{ Btu}}} =$

$43{,}568{,}000 \text{ ft} \cdot \text{lb}$

You Try It 4 $\text{Power} = \dfrac{90 \text{ ft} \times 1200 \text{ lb}}{24 \text{ s}}$

$\qquad\qquad = 4500 \dfrac{\text{ft} \cdot \text{lb}}{\text{s}}$

You Try It 5 $\dfrac{3300}{550} = 6 \text{ hp}$

Solutions to Chapter 9 "You Try It"

SECTION 9.1

You Try It 1   3.07 m = 307 cm

You Try It 2   750 m = 0.750 km

3 km 750 m = 3 km + 0.750 km
                        = 3.750 km

You Try It 3

Strategy   To find the cost of the shelves:
  • Multiply the length of the bookcase (175 cm) by the number of shelves (4).
  • Convert centimeters to meters.
  • Multiply the number of meters by the cost per meter ($15.75).

Solution
$$\begin{array}{r} 175 \text{ cm} \\ \times \quad 4 \\ \hline 700 \text{ cm} \end{array}$$

700 cm = 7 m

$$\begin{array}{r} \$15.75 \\ \times \quad 7 \\ \hline \$110.25 \end{array}$$

The cost of the shelves is $110.25.

SECTION 9.2

You Try It 1   42.3 mg = 0.0423 g

You Try It 2   54 mg = 0.054 g

3 g 54 mg = 3 g + 0.054 g
                      = 3.054 g

You Try It 3

Strategy   To find how much fertilizer is required:
  • Convert 300 g to kilograms.
  • Multiply the number of kilograms by the number of trees (400).

Solution   300 g = 0.3 kg

$$\begin{array}{r} 400 \\ \times \quad 0.3 \text{ kg} \\ \hline 120.0 \text{ kg} \end{array}$$

To fertilize the trees, 120 kg of fertilizer are required.

SECTION 9.3

You Try It 1   2 kl = 2000 L

2 kl 167 L = 2000 L + 167 L
                    = 2167 L

You Try It 2   325 cm³ = 325 ml = 0.325 L

You Try It 3

Strategy   To find the profit:
  • Convert 5 L to milliliters.
  • Find the number of jars by dividing the number of milliliters by 125 (the number of milliliters in each jar).
  • Multiply the number of jars by the cost per jar ($.85).
  • Find the total cost by adding the cost of the jars to the cost for the moisturizer ($299.50).
  • Find the income by multiplying the number of jars by the selling price per jar ($19.95).
  • Subtract the total cost from the income.

Solution   5 L = 5000 ml

5000 ÷ 125 = 40   This is the number of jars.

$$\begin{array}{r} \$.85 \\ \times \quad 40 \\ \hline \$34.00 \end{array}$$ This is the cost of the jars.

$$\begin{array}{r} \$299.50 \\ + \quad 34.00 \\ \hline \$333.50 \end{array}$$ This is the total cost.

$$\begin{array}{r} \$19.95 \\ \times \quad 40 \\ \hline \$798.00 \end{array}$$ This is the income from sales.

$$\begin{array}{r} \$798.00 \\ - \quad 333.50 \\ \hline \$464.50 \end{array}$$

The profit on the 5 L of moisturizer is $464.50.

SECTION 9.4

You Try It 1

Strategy To find the number of Calories burned off, multiply the number of hours spent doing housework $\left(4\frac{1}{2}\right)$ by the Calories used per hour (240).

Solution $4\frac{1}{2} \times 240 = \frac{9}{2} \times 240 = 1080$

Doing $4\frac{1}{2}$ h of housework burns off 1080 Calories.

You Try It 2

Strategy To find the number of kilowatt-hours used:
- Find the number of watt-hours used.
- Convert watt-hours to kilowatt-hours.

Solution $150 \times 200 = 30{,}000$
$30{,}000 \text{ Wh} = 30 \text{ kWh}$

30 kWh of energy are used.

You Try It 3

Strategy To find the cost:
- Convert 20 min to hours.
- Multiply to find the total number of hours the oven is used.
- Multiply the number of hours used by the number of watts to find the watt-hours.
- Convert watt-hours to kilowatt-hours.
- Multiply the number of kilowatt-hours by the cost per kilowatt-hour.

Solution $20 \text{ min} = 20 \text{ min} \times \dfrac{1 \text{ h}}{60 \text{ min}}$

$= \dfrac{20}{60} \text{ h} = \dfrac{1}{3} \text{ h}$

$\dfrac{1}{3} \text{ h} \times 30 = 10 \text{ h}$

$10 \text{ h} \times 500 \text{ W} = 5000 \text{ Wh}$

$5000 \text{ Wh} = 5 \text{ kWh}$

$5 \times 8.7¢ = 43.5¢$

The cost is 43.5¢.

SECTION 9.5

You Try It 1 $\dfrac{60 \text{ ft}}{\text{s}} \approx \dfrac{60 \text{ ft}}{\text{s}} \times \dfrac{1 \text{ m}}{3.28 \text{ ft}}$

$\approx \dfrac{60 \text{ m}}{3.28 \text{ s}} \approx 18.29 \text{ m/s}$

60 ft/s ≈ 18.29 m/s

You Try It 2 $\dfrac{\$2.69}{\text{gal}} \approx \dfrac{\$2.69}{\text{gal}} \times \dfrac{1 \text{ gal}}{3.79 \text{ L}}$

$\approx \dfrac{\$2.69}{3.79 \text{ L}} \approx \dfrac{\$.71}{\text{L}}$

\$2.69/gal ≈ \$.71/L

You Try It 3 $45 \text{ cm} = \dfrac{45 \text{ cm}}{1} \times \dfrac{1 \text{ in.}}{2.54 \text{ cm}}$

$= \dfrac{45 \text{ in.}}{2.54} \approx 17.72 \text{ in.}$

45 cm ≈ 17.72 in.

You Try It 4 $\dfrac{75 \text{ km}}{\text{h}} \approx \dfrac{75 \text{ km}}{\text{h}} \times \dfrac{1 \text{ mi}}{1.61 \text{ km}}$

$\approx \dfrac{75 \text{ mi}}{1.61 \text{ h}} \approx 46.58 \text{ mi/h}$

75 km/h ≈ 46.58 mi/h

You Try It 5 $\dfrac{\$1.75}{\text{L}} \approx \dfrac{\$1.75}{\text{L}} \times \dfrac{3.79 \text{ L}}{1 \text{ gal}}$

$\approx \dfrac{\$6.6325}{1 \text{ gal}} \approx \$6.63/\text{gal}$

\$1.75/L ≈ \$6.63/gal

Solutions to Chapter 10 "You Try It"

SECTION 10.1

You Try It 1 −232 ft

You Try It 2

You Try It 3 **a.** −12 < −8
b. −5 < 0

You Try It 4 $|-7| = 7$
$|21| = 21$

You Try It 5 $|2| = 2$
$|-9| = 9$

You Try It 6 $-|-12| = -12$

SECTION 10.2

You Try It 1 $-154 + (-37) = -191$

You Try It 2 $-5 + (-2) + 9 + (-3)$
$= -7 + 9 + (-3)$
$= 2 + (-3)$
$= -1$

You Try It 3 $-8 - 14$
$= -8 + (-14)$
$= -22$

You Try It 4 $3 - (-15)$
$= 3 + 15$
$= 18$

You Try It 5 $4 - (-3) - 12 - (-7) - 20$
$= 4 + 3 + (-12) + 7 + (-20)$
$= 7 + (-12) + 7 + (-20)$
$= -5 + 7 + (-20)$
$= 2 + (-20)$
$= -18$

You Try It 6

Strategy   To find the temperature, add the increase (12) to the previous temperature (-10).

Solution $-10 + 12 = 2$

After an increase of 12°C, the temperature is 2°C.

SECTION 10.3

You Try It 1 $(-3)(-5) = 15$

You Try It 2 $-38 \cdot 51 = -1938$

You Try It 3 $-7(-8)(9)(-2) = 56(9)(-2)$
$= 504(-2)$
$= -1008$

You Try It 4 $(-135) \div (-9) = 15$

You Try It 5 $84 \div (-6) = -14$

You Try It 6 $-72 \div 4 = -18$

You Try It 7   Division by zero is undefined.
$-39 \div 0$ is undefined.

You Try It 8

Strategy   To find the melting point of argon, multiply the melting point of mercury $(-38°)$ by 5.

Solution $5(-38) = -190$

The melting point of argon is $-190°$C.

You Try It 9

Strategy   To find the average daily low temperature:
- Add the seven temperature readings.
- Divide by 7.

Solution
$-6 + (-7) + 1 + 0 + (-5) + (-10) + (-1)$
$= -13 + 1 + 0 + (-5) + (-10) + (-1)$
$= -12 + 0 + (-5) + (-10) + (-1)$
$= -12 + (-5) + (-10) + (-1)$
$= -17 + (-10) + (-1)$
$= -27 + (-1)$
$= -28$

$-28 \div 7 = -4$

The average daily low temperature was $-4°$.

SECTION 10.4

You Try It 1   The LCM of 9 and 12 is 36.

$$\frac{5}{9} - \frac{11}{12} = \frac{20}{36} - \frac{33}{36} = \frac{20}{36} + \frac{-33}{36}$$

$$= \frac{20 + (-33)}{36} = \frac{-13}{36}$$

$$= -\frac{13}{36}$$

You Try It 2   The LCM of 8, 6, and 3 is 24.

$$-\frac{7}{8} - \frac{5}{6} + \frac{2}{3}$$

$$= -\frac{21}{24} - \frac{20}{24} + \frac{16}{24}$$

$$= \frac{-21}{24} + \frac{-20}{24} + \frac{16}{24}$$

$$= \frac{-21 + (-20) + 16}{24} = \frac{-25}{24}$$

$$= -\frac{25}{24} = -1\frac{1}{24}$$

You Try It 3 $16.127 - 67.91 = 16.127 + (-67.91)$

$$\begin{array}{r} 67.910 \\ -16.127 \\ \hline 51.783 \end{array}$$

$16.127 - 67.91 = -51.783$

You Try It 4 $2.7 + (-9.44) + 6.2$
$= -6.74 + 6.2$
$= -0.54$

You Try It 5 The product is positive.

$$\left(-\frac{2}{3}\right)\left(-\frac{9}{10}\right) = \frac{2 \cdot 9}{3 \cdot 10}$$

$$= \frac{3}{5}$$

You Try It 6 The quotient is negative.

$$-\frac{5}{8} \div \frac{5}{40} = -\left(\frac{5}{8} \div \frac{5}{40}\right)$$

$$= -\left(\frac{5 \cdot 40}{8 \cdot 5}\right)$$

$$= -5$$

You Try It 7 The product is negative.

$$\begin{array}{r} 5.44 \\ \times\ 3.8 \\ \hline 4352 \\ 1632\ \ \\ \hline 20.672 \end{array}$$

$-5.44 \times 3.8 = -20.672$

You Try It 8 $3.44 \times (-1.7) \times 0.6$
$= (-5.848) \times 0.6$
$= -3.5088$

You Try It 9

$$\begin{array}{r} 0.231 \\ 1.7{.}\overline{)0.3{.}940} \\ -3\ 4\ \ \ \\ \hline 54\ \ \\ -51\ \ \\ \hline 30 \\ -17 \\ \hline 13 \end{array}$$

$-0.394 \div 1.7 \approx -0.23$

You Try It 10

Strategy To find how many degrees the temperature fell, subtract the lower temperature (-13.33) from the higher temperature (12.78).

Solution $12.78 - (-13.33) = 12.78 + 13.33$
$= 26.11$

The temperature fell 26.11°C in the 15-minute period.

SECTION 10.5

You Try It 1 The number is less than 1. Move the decimal point 7 places to the right. The exponent on 10 is -7.

$0.000000961 = 9.61 \times 10^{-7}$

You Try It 2 The exponent on 10 is positive. Move the decimal point 6 places to the right.

$7.329 \times 10^6 = 7,329,000$

You Try It 3 $9 - 9 \div (-3)$
$= 9 - (-3)$
$= 9 + 3$
$= 12$

You Try It 4 $8 \div 4 \cdot 4 - (-2)^2$
$= 8 \div 4 \cdot 4 - 4$
$= 2 \cdot 4 - 4$
$= 8 - 4$
$= 8 + (-4)$
$= 4$

You Try It 5 $8 - (-15) \div (2 - 7)$
$= 8 - (-15) \div (-5)$
$= 8 - 3$
$= 8 + (-3)$
$= 5$

You Try It 6 $(-2)^2 \times (3 - 7)^2 - (-16) \div (-4)$
$= (-2)^2 \times (-4)^2 - (-16) \div (-4)$
$= 4 \times 16 - (-16) \div (-4)$
$= 64 - (-16) \div (-4)$
$= 64 - 4$
$= 64 + (-4)$
$= 60$

You Try It 7 $7 \div \left(\frac{1}{7} - \frac{3}{14}\right) - 9$

$= 7 \div \left(-\frac{1}{14}\right) - 9$

$= 7(-14) - 9$
$= -98 - 9$
$= -98 + (-9)$
$= -107$

Solutions to Chapter 11 "You Try It"

SECTION 11.1

You Try It 1 $6a - 5b$

$$6(-3) - 5(4) = -18 - 20$$
$$= -18 + (-20)$$
$$= -38$$

You Try It 2 $-3s^2 - 12 \div t$

$$-3(-2)^2 - 12 \div 4 = -3(4) - 12 \div 4$$
$$= -12 - 12 \div 4$$
$$= -12 - 3$$
$$= -12 + (-3)$$
$$= -15$$

You Try It 3 $-\dfrac{2}{3}m + \dfrac{3}{4}n^3$

$$-\dfrac{2}{3}(6) + \dfrac{3}{4}(2)^3 = -\dfrac{2}{3}(6) + \dfrac{3}{4}(8)$$
$$= -4 + 6$$
$$= 2$$

You Try It 4 $-3yz - z^2 + y^2$

$$-3\left(-\dfrac{2}{3}\right)\left(\dfrac{1}{3}\right) - \left(\dfrac{1}{3}\right)^2 + \left(-\dfrac{2}{3}\right)^2$$
$$= -3\left(-\dfrac{2}{3}\right)\left(\dfrac{1}{3}\right) - \dfrac{1}{9} + \dfrac{4}{9}$$
$$= \dfrac{2}{3} - \dfrac{1}{9} + \dfrac{4}{9} = \dfrac{6}{9} - \dfrac{1}{9} + \dfrac{4}{9} = \dfrac{9}{9}$$
$$= 1$$

You Try It 5 $5a^2 - 6b^2 + 7a^2 - 9b^2$

$$= 5a^2 + (-6)b^2 + 7a^2 + (-9)b^2$$
$$= 5a^2 + 7a^2 + (-6)b^2 + (-9)b^2$$
$$= 12a^2 + (-15)b^2$$
$$= 12a^2 - 15b^2$$

You Try It 6 $-6x + 7 + 9x - 10$

$$= (-6)x + 7 + 9x + (-10)$$
$$= (-6)x + 9x + 7 + (-10)$$
$$= 3x + (-3)$$
$$= 3x - 3$$

You Try It 7 $\dfrac{3}{8}w + \dfrac{1}{2} - \dfrac{1}{4}w - \dfrac{2}{3}$

$$= \dfrac{3}{8}w + \dfrac{1}{2} + \left(-\dfrac{1}{4}\right)w + \left(-\dfrac{2}{3}\right)$$
$$= \dfrac{3}{8}w + \left(-\dfrac{1}{4}\right)w + \dfrac{1}{2} + \left(-\dfrac{2}{3}\right)$$
$$= \dfrac{3}{8}w + \left(-\dfrac{2}{8}\right)w + \dfrac{3}{6} + \left(-\dfrac{4}{6}\right)$$
$$= \dfrac{1}{8}w + \left(-\dfrac{1}{6}\right)$$
$$= \dfrac{1}{8}w - \dfrac{1}{6}$$

You Try It 8 $5(a - 2) = 5a - 5(2)$
$$= 5a - 10$$

You Try It 9 $8s - 2(3s - 5)$
$$= 8s - 2(3s) - (-2)(5)$$
$$= 8s - 6s + 10$$
$$= 2s + 10$$

You Try It 10 $4(x - 3) - 2(x + 1)$
$$= 4x - 4(3) - 2x - 2(1)$$
$$= 4x - 12 - 2x - 2$$
$$= 4x - 2x - 12 - 2$$
$$= 2x - 14$$

SECTION 11.2

You Try It 1

$$\begin{array}{c|c} x(x + 3) = 4x + 6 \\ \hline (-2)(-2 + 3) & 4(-2) + 6 \\ (-2)(1) & (-8) + 6 \\ -2 = -2 \end{array}$$

Yes, -2 is a solution.

You Try It 2

$$\begin{array}{c|c} x^2 - x = 3x + 7 \\ \hline (-3)^2 - (-3) & 3(-3) + 7 \\ 9 + 3 & -9 + 7 \\ 12 \neq -2 \end{array}$$

No, -3 is not a solution.

You Try It 3
$$-2 + y = -5$$
$$-2 + 2 + y = -5 + 2$$
$$0 + y = -3$$
$$y = -3$$

The solution is -3.

You Try It 4
$$7 = y + 8$$
$$7 - 8 = y + 8 - 8$$
$$-1 = y + 0$$
$$-1 = y$$

The solution is -1.

You Try It 5
$$\dfrac{1}{5} = z + \dfrac{4}{5}$$
$$\dfrac{1}{5} - \dfrac{4}{5} = z + \dfrac{4}{5} - \dfrac{4}{5}$$
$$-\dfrac{3}{5} = z + 0$$
$$-\dfrac{3}{5} = z$$

The solution is $-\dfrac{3}{5}$.

You Try It 6

$$4z = -20$$
$$\frac{4z}{4} = \frac{-20}{4}$$
$$1z = -5$$
$$z = -5$$

The solution is -5.

You Try It 7

$$8 = \frac{2}{5}n$$
$$\left(\frac{5}{2}\right)(8) = \left(\frac{5}{2}\right)\frac{2}{5}n$$
$$20 = 1n$$
$$20 = n$$

The solution is 20.

You Try It 8

$$\frac{2}{3}t - \frac{1}{3}t = -2$$
$$\frac{1}{3}t = -2$$
$$\left(\frac{3}{1}\right)\frac{1}{3}t = \left(\frac{3}{1}\right)(-2)$$
$$1t = -6$$
$$t = -6$$

The solution is -6.

You Try It 9

Strategy To find the regular price, replace the variables S and D in the formula by the given values and solve for R.

Solution
$$S = R - D$$
$$44 = R - 16$$
$$44 + 16 = R - 16 + 16$$
$$60 = R$$

The regular price is $60.

You Try It 10

Strategy To find the monthly payment, replace the variables A and N in the formula by the given values and solve for M.

Solution
$$A = MN$$
$$6840 = M \cdot 24$$
$$6840 = 24M$$
$$\frac{6840}{24} = \frac{24M}{24}$$
$$285 = M$$

The monthly payment is $285.

SECTION 11.3

You Try It 1

$$5x + 8 = 6$$
$$5x + 8 - 8 = 6 - 8$$
$$5x = -2$$

$$\frac{5x}{5} = \frac{-2}{5}$$
$$x = -\frac{2}{5}$$

The solution is $-\frac{2}{5}$.

You Try It 2

$$7 - x = 3$$
$$7 - 7 - x = 3 - 7$$
$$-x = -4$$
$$(-1)(-x) = (-1)(-4)$$
$$x = 4$$

The solution is 4.

You Try It 3

Strategy To find the Celsius temperature, replace the variable F in the formula by the given value and solve for C.

Solution
$$F = \frac{9}{5}C + 32$$
$$-22 = \frac{9}{5}C + 32$$
$$-22 - 32 = \frac{9}{5}C + 32 - 32$$
$$-54 = \frac{9}{5}C$$
$$\left(\frac{5}{9}\right)(-54) = \left(\frac{5}{9}\right)\frac{9}{5}C$$
$$-30 = C$$

The Celsius temperature is $-30°$.

You Try It 4

Strategy To find the cost per unit, replace the variables T, N, and F in the formula by the given values and solve for U.

Solution
$$T = U \cdot N + F$$
$$4500 = U \cdot 250 + 1500$$
$$4500 = 250U + 1500$$
$$4500 - 1500 = 250U + 1500 - 1500$$
$$3000 = 250U$$
$$\frac{3000}{250} = \frac{250U}{250}$$
$$12 = U$$

The cost per unit is $12.

SECTION 11.4

You Try It 1

$$\frac{1}{5}x - 2 = \frac{2}{5}x + 4$$
$$\frac{1}{5}x - \frac{2}{5}x - 2 = \frac{2}{5}x - \frac{2}{5}x + 4$$
$$-\frac{1}{5}x - 2 = 4$$

$$-\frac{1}{5}x - 2 + 2 = 4 + 2$$
$$-\frac{1}{5}x = 6$$
$$(-5)\left(-\frac{1}{5}x\right) = (-5)6$$
$$x = -30$$

The solution is -30.

You Try It 2

$$4(x - 1) - x = 5$$
$$4x - 4 - x = 5$$
$$3x - 4 = 5$$
$$3x - 4 + 4 = 5 + 4$$
$$3x = 9$$
$$\frac{3x}{3} = \frac{9}{3}$$
$$x = 3$$

The solution is 3.

SECTION 11.5

You Try It 1 $8 - 2t$

You Try It 2 $\dfrac{5}{7x}$

You Try It 3   The product of a number and one-half of the number
The unknown number: x

One-half the number: $\dfrac{x}{2}$

$$(x)\left(\frac{x}{2}\right)$$

SECTION 11.6

You Try It 1   The unknown number: x

A number increased by four	equals	twelve

$$x + 4 = 12$$
$$x + 4 - 4 = 12 - 4$$
$$x = 8$$

The number is 8.

You Try It 2   The unknown number: x

The product of two and a number	is	ten

$$2x = 10$$
$$\frac{2x}{2} = \frac{10}{2}$$
$$x = 5$$

The number is 5.

You Try It 3   The unknown number: x

The sum of three times a number and six	equals	four

$$3x + 6 = 4$$
$$3x + 6 - 6 = 4 - 6$$
$$3x = -2$$
$$\frac{3x}{3} = \frac{-2}{3}$$
$$x = -\frac{2}{3}$$

The number is $-\dfrac{2}{3}$.

You Try It 4   The unknown number: x

Three more than one-half of a number	is	nine

$$\frac{1}{2}x + 3 = 9$$
$$\frac{1}{2}x + 3 - 3 = 9 - 3$$
$$\frac{1}{2}x = 6$$
$$2 \cdot \frac{1}{2}x = 2 \cdot 6$$
$$x = 12$$

The number is 12.

You Try It 5

Strategy   To find the regular price, write and solve an equation using R to represent the regular price of the slacks.

Solution

$38.95	is	$11 less than the regular price

$$38.95 = R - 11$$
$$38.95 + 11 = R - 11 + 11$$
$$49.95 = R$$

The regular price of the slacks is $49.95.

You Try It 6

Strategy To find the rpm of the engine when it is in third gear, write and solve an equation using R to represent the rpm of the engine in third gear.

Solution

| 2500 | is | two-thirds of the rpm of the engine in third gear |

$$2500 = \frac{2}{3}R$$

$$\frac{3}{2}(2500) = \left(\frac{3}{2}\right)\frac{2}{3}R$$

$$3750 = R$$

The rpm of the engine when in third gear is 3750.

You Try It 7

Strategy To find the total sales, write and solve an equation using S to represent the total sales.

Solution

| $2500 | is | the sum of $800 and an 8% commission on total sales |

$$2500 = 800 + 0.08S$$
$$2500 - 800 = 800 - 800 + 0.08S$$
$$1700 = 0.08S$$
$$\frac{1700}{0.08} = \frac{0.08S}{0.08}$$
$$21{,}250 = S$$

Natalie's total sales for the month were $21,250.

You Try It 8

Strategy To find the number of hours of labor, write and solve an equation using H to represent the number of hours of labor required.

Solution

| $300 | includes | $100 for materials plus $12.50 per hour of labor |

$$300 = 100 + 12.50H$$
$$300 - 100 = 100 - 100 + 12.50H$$
$$200 = 12.50H$$
$$\frac{200}{12.50} = \frac{12.50H}{12.50}$$
$$16 = H$$

The number of hours of labor required is 16.

Solutions to Chapter 12 "You Try It"

SECTION 12.1

You Try It 1
$$QT = QR + RS + ST$$
$$62 = 24 + RS + 17$$
$$62 = 41 + RS$$
$$62 - 41 = 41 - 41 + RS$$
$$21 = RS$$

You Try It 2 Let x represent the supplement of a 32° angle. The sum of supplementary angles is 180°.

$$x + 32° = 180$$
$$x + 32° - 32° = 180° - 32°$$
$$x = 148°$$

148° is the supplement of 32°.

You Try It 3
$$\angle a + 68° = 118°$$
$$\angle a + 68° - 68° = 118° - 68°$$
$$\angle a = 50°$$

You Try It 4 In a right triangle, one angle measures 90° and the two acute angles are complementary.

$$\angle A + \angle B = 90°$$
$$\angle A + 7° = 90°$$
$$\angle A + 7° - 7° = 90° - 7°$$
$$\angle A = 83°$$

The other angles measure 90° and 83°.

You Try It 5 The sum of the three angles of a triangle is 180°.

$$\angle A + \angle B + \angle C = 180°$$
$$\angle A + 62° + 45° = 180°$$
$$\angle A + 107° = 180°$$
$$\angle A = 73°$$

The measure of the other angle is 73°.

You Try It 6 $r = \frac{1}{2}d$

$r = \frac{1}{2}(8 \text{ in.}) = 4 \text{ in.}$

The radius is 4 in.

You Try It 7   Angles a and b are supplementary angles.

$$\angle a + \angle b = 180°$$
$$125° + \angle b = 180°$$
$$125° - 125° + \angle b = 180° - 125°$$
$$\angle b = 55°$$

You Try It 8 $\angle c$ and $\angle a$ are corresponding angles. Corresponding angles are equal.

$$\angle c = \angle a = 120°$$

$\angle b$ and $\angle c$ are supplementary angles.

$$\angle b + \angle c = 180°$$
$$\angle b + 120° = 180°$$
$$\angle b + 120° - 120° = 180° - 120°$$
$$\angle b = 60°$$

SECTION 12.2

You Try It 1 $P = 2L + 2W$
$= 2(2 \text{ m}) + 2(0.85 \text{ m})$
$= 4 \text{ m} + 1.7 \text{ m}$
$= 5.7 \text{ m}$

The perimeter of the rectangle is 5.7 m.

You Try It 2 $P = a + b + c$
$= 12 \text{ cm} + 15 \text{ cm} + 18 \text{ cm}$
$= 45 \text{ cm}$

The perimeter of the triangle is 45 cm.

You Try It 3 $C = \pi d$
$\approx 3.14 \cdot 6 \text{ in.}$
$= 18.84 \text{ in.}$

The circumference is approximately 18.84 in.

You Try It 4 $\begin{array}{ccc} \text{Perimeter} & & \text{the} \\ \text{of} & \text{two} & \text{circumference} \\ \text{composite} = \text{lengths} + \text{of a} \\ \text{figure} & \text{of a} & \text{circle} \\ & \text{rectangle} & \end{array}$

$P = 2L + \pi d$
$\approx 2(8 \text{ in.}) + 3.14(3 \text{ in.})$
$= 16 \text{ in.} + 9.42 \text{ in.}$
$= 25.42 \text{ in.}$

The perimeter is approximately 25.42 in.

You Try It 5

Strategy   To find the perimeter, use the formula for the perimeter of a rectangle.

Solution $P = 2L + 2W$
$= 2(11 \text{ in.}) + 2\left(8\frac{1}{2} \text{ in.}\right)$
$= 22 \text{ in.} + 17 \text{ in.}$
$= 39 \text{ in.}$

The perimeter of the computer paper is 39 in.

You Try It 6

Strategy   To find the cost:
• Find the perimeter of the workbench.
• Multiply the perimeter by the perimeter cost of the stripping.

Solution $P = 2L + 2W$
$= 2(3 \text{ m}) + 2(0.74 \text{ m})$
$= 6 \text{ m} + 1.48 \text{ m}$
$= 7.48 \text{ m}$

$\$2.76 \times 7.48 = \$20.6448$

The cost is $20.64.

SECTION 12.3

You Try It 1 $A = \frac{1}{2}bh = \frac{1}{2}(24 \text{ in.})(14 \text{ in.}) = 168 \text{ in}^2$

The area is 168 in².

You Try It 2

$A = \text{area of rectangle} - \text{area of triangle}$
$A = LW - \frac{1}{2}bh$

$= (10 \text{ in.} \times 6 \text{ in.}) - \left(\frac{1}{2} \times 6 \text{ in.} \times 4 \text{ in.}\right)$
$= 60 \text{ in}^2 - 12 \text{ in}^2$
$= 48 \text{ in}^2$

The area is 48 in².

You Try It 3

Strategy   To find the area of the room:
• Find the area in square feet.
• Convert to square yards.

Solution $A = LW$
$= 12 \text{ ft} \cdot 9 \text{ ft}$
$= 108 \text{ ft}^2$

$108 \text{ ft}^2 \times \frac{1 \text{ yd}^2}{9 \text{ ft}^2} = \frac{108}{9} \text{ yd}^2$
$= 12 \text{ yd}^2$

The area of the room is 12 yd².

SECTION 12.4

You Try It 1 $V = LWH$
$\quad = (8 \text{ cm})(3.5 \text{ cm})(4 \text{ cm})$
$\quad = 112 \text{ cm}^3$

The volume is 112 cm³.

You Try It 2 $V = s^3$
$\quad = (5 \text{ cm})^3$
$\quad = 125 \text{ cm}^3$

The volume is 125 cm³.

You Try It 3 $r = \frac{1}{2}d = \frac{1}{2}(14 \text{ in.}) = 7 \text{ in.}$

$V = \pi r^2 h$
$\quad = \frac{22}{7}(7 \text{ in.})^2(15 \text{ in.})$
$\quad = 2310 \text{ in}^3$

The volume is approximately 2310 in³.

You Try It 4 $V = \frac{4}{3}\pi r^3$

$\quad \approx \frac{4}{3}(3.14)(3 \text{ m})^3$
$\quad = 113.04 \text{ m}^3$

The volume is approximately 113.04 m³.

You Try It 5
V = volume of rectangular solid
  + volume of cylinder

$V = LWH + \pi r^2 h$
$\quad \approx (1.5 \text{ m})(0.4 \text{ m})(0.4 \text{ m}) + 3.14(0.8 \text{ m})^2(0.2 \text{ m})$
$\quad = 0.24 \text{ m}^3 + 0.40192 \text{ m}^3$
$\quad = 0.64192 \text{ m}^3$

The volume is approximately 0.64192 m³.

You Try It 6

V = volume of rectangular solid
  + $\frac{1}{2}$ the volume of cylinder

$V = LWH + \frac{1}{2}\pi r^2 h$

$\quad \approx (24 \text{ in.})(6 \text{ in.})(4 \text{ in.}) + \frac{1}{2}(3.14)(3 \text{ in.})^2(24 \text{ in.})$

$\quad = 576 \text{ in}^3 + 339.12 \text{ in}^3$
$\quad = 915.12 \text{ in}^3$

The volume is approximately 915.12 in³.

You Try It 7

Strategy   To find the volume of the freezer, use the formula for the volume of a rectangular solid.

Solution $V = LWH$
$\quad = (7 \text{ ft})(2.5 \text{ ft})(3 \text{ ft})$
$\quad = 52.5 \text{ ft}^3$

The volume of the freezer is 52.5 ft³.

You Try It 8

Strategy
To find the volume of the channel iron, add the volumes of the three rectangular solids.

Solution

$V = 2LWH + LWH$
$\quad = 2(10 \text{ ft})(0.5 \text{ ft})(0.3 \text{ ft}) + (10 \text{ ft})(0.2 \text{ ft})(0.2 \text{ ft})$
$\quad = 3 \text{ ft}^3 + 0.4 \text{ ft}^3$
$\quad = 3.4 \text{ ft}^3$

The volume of the channel iron is 3.4 ft³.

SECTION 12.5

You Try It 1   **a.** $\sqrt{16} = 4$
$\quad \sqrt{169} = 13$

b. $\sqrt{32} \approx 5.657$
$\quad \sqrt{162} \approx 12.728$

You Try It 2   Hypotenuse $= \sqrt{(\text{leg})^2 + (\text{leg})^2}$
$\quad = \sqrt{8^2 + 11^2}$
$\quad = \sqrt{64 + 121}$
$\quad = \sqrt{185}$
$\quad \approx 13.601$

The hypotenuse is approximately 13.601 in.

You Try It 3   Leg $= \sqrt{(\text{hypotenuse})^2 - (\text{leg})^2}$
$\quad = \sqrt{12^2 - 5^2}$
$\quad = \sqrt{144 - 25}$
$\quad = \sqrt{119}$
$\quad \approx 10.909$

The length of the leg is approximately 10.909 ft.

You Try It 4

Strategy   To find the distance between the holes, use the Pythagorean Theorem. The hypotenuse is the distance between the holes. The length of each leg is given (3 cm and 8 cm).

Solution

$$\text{Hypotenuse} = \sqrt{(\text{leg})^2 + (\text{leg})^2}$$
$$= \sqrt{3^2 + 8^2}$$
$$= \sqrt{9 + 64}$$
$$= \sqrt{73}$$
$$\approx 8.544$$

The distance is approximately 8.544 cm.

SECTION 12.6

You Try It 1 $\dfrac{4 \text{ cm}}{7 \text{ cm}} = \dfrac{4}{7}$

You Try It 2

$$\dfrac{AB}{DE} = \dfrac{AC}{x}$$
$$\dfrac{7 \text{ cm}}{14 \text{ cm}} = \dfrac{3 \text{ cm}}{x}$$
$$7x = 14 \cdot 3 \text{ cm}$$
$$7x = 42 \text{ cm}$$
$$\dfrac{7x}{7} = \dfrac{42 \text{ cm}}{7}$$
$$x = 6 \text{ cm}$$

Side *DF* is 6 cm.

You Try It 3 $AC = DF$, $\angle ACB = \angle DFE$, but $BC \neq EF$ because $\angle CAB \neq \angle FDE$. Therefore, the triangles are not congruent.

You Try It 4

$$\dfrac{AC}{DF} = \dfrac{\text{height } CH}{\text{height } FG}$$
$$\dfrac{10 \text{ m}}{15 \text{ m}} = \dfrac{7 \text{ cm}}{h}$$
$$10h = 15 \cdot 7 \text{ m}$$
$$10h = 105 \text{ m}$$
$$\dfrac{10h}{10} = \dfrac{105 \text{ m}}{10}$$
$$h = 10.5 \text{ m}$$

The height *FG* is 10.5 m.

You Try It 5

Strategy   To find the perimeter of triangle *ABC*:
- Solve a proportion to find the lengths of sides *BC* and *AC*.
- Use the formula Perimeter = side *AB* + side *BC* + side *AC*.

Solution

$$\dfrac{BC}{EF} = \dfrac{AB}{DE}$$
$$\dfrac{BC}{10 \text{ in.}} = \dfrac{4 \text{ in.}}{8 \text{ in.}}$$
$$8BC = 10 \text{ in.} \cdot 4$$
$$8BC = 40 \text{ in.}$$
$$\dfrac{8BC}{8} = \dfrac{40 \text{ in.}}{8}$$
$$BC = 5 \text{ in.}$$

$$\dfrac{AC}{DF} = \dfrac{AB}{DE}$$
$$\dfrac{AC}{6 \text{ in.}} = \dfrac{4 \text{ in.}}{8 \text{ in.}}$$
$$8AC = 6 \text{ in.} \cdot 4$$
$$8AC = 24 \text{ in.}$$
$$\dfrac{8AC}{8} = \dfrac{24 \text{ in.}}{8}$$
$$AC = 3 \text{ in.}$$

$$\text{Perimeter} = 4 \text{ in.} + 5 \text{ in.} + 3 \text{ in.}$$
$$= 12 \text{ in.}$$

The perimeter of triangle *ABC* is 12 in.

Answers to Chapter 1 Odd-Numbered Exercises

PREP TEST

1. 8   **2.** 1 2 3 4 5 6 7 8 9 10   **3.** a and D; b and E; c and A; d and B; e and F; f and C

SECTION 1.1

1.   **3.**   **5.** 37 < 49   **7.** 101 > 87

9. 245 > 158   **11.** 0 < 45   **13.** 815 < 928   **15.** millions   **17.** hundred-thousands   **19.** Three thousand seven hundred ninety   **21.** Fifty-eight thousand four hundred seventy-three   **23.** Four hundred ninety-eight thousand five hundred twelve   **25.** Six million eight hundred forty-two thousand seven hundred fifteen   **27.** 357   **29.** 63,780

31. 7,024,709   **33.** 6000 + 200 + 90 + 5   **35.** 400,000 + 50,000 + 3000 + 900 + 20 + 1

37. 300,000 + 1000 + 800 + 9   **39.** 3,000,000 + 600 + 40 + 2   **41.** 850   **43.** 4000   **45.** 53,000   **47.** 250,000

49. 999; 10,000

SECTION 1.2

1. 28   **3.** 125   **5.** 102   **7.** 154   **9.** 1489   **11.** 828   **13.** 1584   **15.** 1219   **17.** 102,317   **19.** 79,326

21. 1804   **23.** 1579   **25.** 19,740   **27.** 7420   **29.** 120,570   **31.** 207,453   **33.** 24,218   **35.** 11,974

37. 9323   **39.** 77,139   **41.** 14,383   **43.** 9473   **45.** 33,247   **47.** 5058   **49.** 1992   **51.** 68,263

53. Est.: 17,700   **55.** Est.: 2900   **57.** Est.: 101,000   **59.** Est.: 158,000   **61.** Est.: 260,000   **63.** Est.: 940,000
Cal.: 17,754   Cal.: 2872   Cal.: 101,712   Cal.: 158,763   Cal.: 261,595   Cal.: 946,718

65. Est.: 33,000,000   **67.** Est.: 34,000,000   **69.** There were 118,295 multiple births during the year.
Cal.: 32,691,621   Cal.: 34,420,922

71. The total income from the four movies was approximately $1,500,000,000.   **73. a.** The total income from the two movies with the lowest box-office incomes was $599,300,000.   **b.** Yes, the total income from the two movies with the lowest box-office incomes exceeded the income from the 1977 *Star Wars* production.   **75. a.** During the three days, 1285 miles will be driven.   **b.** The odometer will read 69,977 miles.   **77.** The total average amount for Americans aged 16 to 34 in checking accounts, savings accounts, and U.S. Savings Bonds is $1796.   **79.** The sum of the average amounts invested in home equity and retirement is greater for all Americans than that same sum for Americans between the ages of 16 and 34.   **81.** 11 different sums   **83.** No. 0 + 0 = 0   **85.** 10 numbers

SECTION 1.3

1. 4   **3.** 4   **5.** 10   **7.** 4   **9.** 9   **11.** 22   **13.** 60   **15.** 66   **17.** 31   **19.** 901   **21.** 791   **23.** 1125

25. 3131   **27.** 47   **29.** 925   **31.** 4561   **33.** 3205   **35.** 1222   **37.** 3021   **39.** 3022   **41.** 3040

43. 212   **45.** 60,245   **47.** 65   **49.** 17   **51.** 23   **53.** 456   **55.** 57   **57.** 375   **59.** 3139   **61.** 3621

63. 738   **65.** 3545   **67.** 749   **69.** 5343   **71.** 66,463   **73.** 16,590   **75.** 52,404   **77.** 38,777   **79.** 4638

81. 3612   **83.** 2913   **85.** 2583   **87.** 5268   **89.** 71,767   **91.** 11,239   **93.** 8482   **95.** 625   **97.** 76,725

99. 23   **101.** 4648   **103.** Est.: 30,000   **105.** Est.: 40,000   **107.** Est.: 100,000
Cal.: 29,837   Cal.: 36,668   Cal.: 101,998

109. IBM was granted 748 more patents than Canon.   **111.** The amount that remains to be paid is $10,275.

113. The difference is 15 feet.   **115. a.** The Rolling Stones grossed $466,000,000 more from their concerts.   **b.** Neil Diamond grossed $16,000,000 more from his concerts.   **117. a.** The 2-year period with the smallest increase is 2010 to 2012.   **b.** The 2-year period with the greatest increase is 2018 to 2020.

SECTION 1.4

1. 6 × 2 or 6 · 2   **3.** 4 × 7 or 4 · 7   **5.** 12   **7.** 35   **9.** 25   **11.** 0   **13.** 72   **15.** 198   **17.** 335

19. 2492   **21.** 5463   **23.** 4200   **25.** 6327   **27.** 1896   **29.** 5056   **31.** 1685   **33.** 46,963   **35.** 59,976

37. 19,120   **39.** 19,790   **41.** 108   **43.** 3664   **45.** 20,036   **47.** 71,656   **49.** 432   **51.** 1944   **53.** 41,832

55. 43,620   **57.** 335,195   **59.** 594,625   **61.** 321,696   **63.** 342,171   **65.** 279,220   **67.** 191,800

69. 463,712   **71.** 180,621   **73.** 478,800   **75.** 158,422   **77.** 4,696,714   **79.** 5,542,452   **81.** 198,423

83. 18,834   **85.** 260,178   **87.** 315,109,895   **89.** Est.: 450,000   **91.** Est.: 6,300,000   **93.** Est.: 18,000,000
Cal.: 440,076   Cal.: 6,491,166   Cal.: 18,728,744

95. Est.: 54,000,000   **97.** The plane used 5190 gallons of fuel on a 6-hour flight.   **99.** The area is 256 square miles.
Cal.: 57,691,192

101. Company A offers the lights for the lower total price.   **103.** The estimated cost for the electricians' labor is $5100.

105. The total cost is $2138.   **107.** Each hour 12 accidental deaths occurred, 288 occurred each day, and 105,120
throughout the year. Explanations will vary.

SECTION 1.5

1. 2   **3.** 6   **5.** 7   **7.** 16   **9.** 210   **11.** 44   **13.** 703   **15.** 910   **17.** 21,560   **19.** 3580   **21.** 482

23. 1075   **25.** 52   **27.** 5   **29.** 2 r1   **31.** 5 r2   **33.** 13 r1   **35.** 10 r3   **37.** 90 r2   **39.** 230 r1

41. 204 r3   **43.** 1347 r3   **45.** 1720 r2   **47.** 409 r2   **49.** 6214 r2   **51.** 8708 r2   **53.** 1080 r2   **55.** 4200

57. 19,600   **59.** 1 r38   **61.** 1 r26   **63.** 21 r21   **65.** 30 r22   **67.** 5 r40   **69.** 9 r17   **71.** 200 r21

73. 303 r1   **75.** 67 r13   **77.** 176 r13   **79.** 1086 r7   **81.** 403   **83.** 12 r456   **85.** 4 r160   **87.** 160 r27

89. 1669 r14   **91.** 7950   **93.** Est.: 2000   **95.** Est.: 10,000   **97.** Est.: 30,000   **99.** Est.: 500   **101.** Est.: 500
Cal.: 2225   Cal.: 11,016   Cal.: 26,656   Cal.: 504   Cal.: 541

103. Est.: 20,000   **105.** The average monthly expense for housing is $976.
Cal.: 20,621

107. The average monthly claim for theft was $25,000.   **109.** The average number of hours worked per week in Britain
is 35 hours.   **111.** It works out to 407 pennies for each person living in the United States.   **113.** 2112

SECTION 1.6

1. 2^3 **3.** $6^3 \cdot 7^4$ **5.** $2^3 \cdot 3^3$ **7.** $5 \cdot 7^5$ **9.** $3^3 \cdot 6^4$ **11.** $3^3 \cdot 5 \cdot 9^3$ **13.** 8   **15.** 400   **17.** 900   **19.** 972

21. 120   **23.** 360   **25.** 0   **27.** 90,000   **29.** 540   **31.** 4050   **33.** 11,025   **35.** 25,920   **37.** 4,320,000

39. 5   **41.** 10   **43.** 47   **45.** 8   **47.** 5   **49.** 8   **51.** 6   **53.** 53   **55.** 44   **57.** 19   **59.** 67   **61.** 168

63. 27   **65.** 14   **67.** 10   **69.** 9   **71.** 12   **73.** 32   **75.** 39   **77.** 1024

SECTION 1.7

1. 1, 2, 4   **3.** 1, 2, 5, 10   **5.** 1, 7   **7.** 1, 3, 9   **9.** 1, 13   **11.** 1, 2, 3, 6, 9, 18   **13.** 1, 2, 4, 7, 8, 14, 28, 56

15. 1, 3, 5, 9, 15, 45   **17.** 1, 29   **19.** 1, 2, 11, 22   **21.** 1, 2, 4, 13, 26, 52   **23.** 1, 2, 41, 82   **25.** 1, 3, 19, 57

27. 1, 2, 3, 4, 6, 8, 12, 16, 24, 48   **29.** 1, 5, 19, 95   **31.** 1, 2, 3, 6, 9, 18, 27, 54   **33.** 1, 2, 3, 6, 11, 22, 33, 66

35. 1, 2, 4, 5, 8, 10, 16, 20, 40, 80   **37.** 1, 2, 3, 4, 6, 8, 12, 16, 24, 32, 48, 96   **39.** 1, 2, 3, 5, 6, 9, 10, 15, 18, 30, 45, 90

41. $2 \cdot 3$ **43.** Prime   **45.** $2 \cdot 2 \cdot 2 \cdot 3$ **47.** $3 \cdot 3 \cdot 3$ **49.** $2 \cdot 2 \cdot 3 \cdot 3$ **51.** Prime   **53.** $2 \cdot 3 \cdot 3 \cdot 5$

55. $5 \cdot 23$ **57.** $2 \cdot 3 \cdot 3$ **59.** $2 \cdot 2 \cdot 7$ **61.** Prime   **63.** $2 \cdot 31$ **65.** $2 \cdot 11$ **67.** Prime   **69.** $2 \cdot 3 \cdot 11$

71. $2 \cdot 37$ **73.** Prime   **75.** $5 \cdot 11$ **77.** $2 \cdot 2 \cdot 2 \cdot 3 \cdot 5$ **79.** $2 \cdot 2 \cdot 2 \cdot 2 \cdot 2 \cdot 5$ **81.** $2 \cdot 2 \cdot 2 \cdot 3 \cdot 3 \cdot 3$

83. $5 \cdot 5 \cdot 5 \cdot 5$ **85.** 3 and 5, 5 and 7, 11 and 13. Other answers are possible.

CHAPTER REVIEW

1. 600 [1.6A]   **2.** 10,000 + 300 + 20 + 7 [1.1C]   **3.** 1, 2, 3, 6, 9, 18 [1.7A]   **4.** 12,493 [1.2A]

5. 1749 [1.3B]   **6.** 2135 [1.5A]   **7.** 101 > 87 [1.1A]   **8.** $5^2 \cdot 7^5$ [1.6A]   **9.** 619,833 [1.4B]

10. 5409 [1.3B]   **11.** 1081 [1.2A]   **12.** 2 [1.6B]   **13.** 45,700 [1.1D]   **14.** Two hundred seventy-six thousand
fifty-seven [1.1B]   **15.** 1306 r59 [1.5C]   **16.** 2,011,044 [1.1B]   **17.** 488 r2 [1.5B]   **18.** 17 [1.6B]

19. 32 [1.6B]   **20.** $2 \cdot 2 \cdot 2 \cdot 3 \cdot 3$ [1.7B]   **21.** 2133 [1.3A]   **22.** 22,761 [1.4B]   **23.** Vincent's total pay for the
week was $768. [1.4C]   **24.** He drove 27 miles per gallon of gasoline. [1.5D]   **25.** The monthly car payment is

Note: The numbers in brackets following the answers in the Chapter Review are a reference to the objective that
corresponds to that problem. For example, the reference [1.2A] stands for Section 1.2, Objective A. This notation will be
used for all Prep Tests, Chapter Reviews, Chapter Tests, and Cumulative Reviews throughout the text.

$310. [1.5D] **26.** The total income from commissions was $2567. [1.2B] **27.** The total amount deposited was $301. The new checking account balance is $817. [1.2B] **28.** The total of the car payments is $2952. [1.4C] **29.** The difference is $29,489. [1.3C] **30.** The difference is $19,937. [1.3C] **31.** A man earns $63,865 more in 5 years. [1.4C]

CHAPTER TEST

1. 432 [1.6A] **2.** Two hundred seven thousand sixty-eight [1.1B] **3.** 9333 [1.3A] **4.** 1, 2, 4, 5, 10, 20 [1.7A]
5. 6,854,144 [1.4B] **6.** 9 [1.6B] **7.** $900,000 + 6000 + 300 + 70 + 8$ [1.1C] **8.** 75,000 [1.1D]
9. 1121 r27 [1.5C] **10.** $3^3 \cdot 7^2$ [1.6A] **11.** 54,915 [1.2A] **12.** $2 \cdot 2 \cdot 3 \cdot 7$ [1.7B] **13.** 4 [1.6B]
14. 726,104 [1.4A] **15.** 1,204,006 [1.1B] **16.** 8710 r2 [1.5B] **17.** $21 > 19$ [1.1A] **18.** 703 [1.5A]
19. 96,798 [1.2A] **20.** 19,922 [1.3B] **21.** The difference is 1,469,000 students. [1.3C] **22.** The average projected enrollment is 4,108,500 students. [1.5D] **23.** The farmer needed 3000 boxes. [1.5D] **24.** The investor will receive $2844 over 12 months. [1.4C] **25. a.** They drove 855 miles. [1.2B] **b.** The odometer reading was 48,481 miles. [1.2B]

Answers to Chapter 2 Odd-Numbered Exercises

PREP TEST

1. 20 [1.4A] **2.** 120 [1.4A] **3.** 9 [1.4A] **4.** 10 [1.2A] **5.** 7 [1.3A] **6.** 2 r3 [1.5C] **7.** 1, 2, 3, 4, 6, 12 [1.5B] **8.** 59 [1.6B] **9.** 7 [1.3A] **10.** $44 < 48$ [1.1A]

SECTION 2.1

1. 40 **3.** 24 **5.** 30 **7.** 12 **9.** 24 **11.** 60 **13.** 56 **15.** 9 **17.** 32 **19.** 36 **21.** 660
23. 9384 **25.** 24 **27.** 30 **29.** 24 **31.** 576 **33.** 1680 **35.** 1 **37.** 3 **39.** 5 **41.** 25 **43.** 1
45. 4 **47.** 4 **49.** 6 **51.** 4 **53.** 1 **55.** 7 **57.** 5 **59.** 8 **61.** 1 **63.** 25 **65.** 7 **67.** 8
69. Relatively prime numbers are numbers with no common factors except the factor 1. Examples will vary.

SECTION 2.2

1. Improper fraction **3.** Proper fraction **5.** $\frac{3}{4}$ **7.** $\frac{7}{8}$ **9.** $1\frac{1}{2}$ **11.** $2\frac{5}{8}$ **13.** $3\frac{3}{5}$ **15.** $\frac{5}{4}$ **17.** $\frac{8}{3}$

19. $\frac{28}{8}$ **21.** **23.** **25.** **27.** $2\frac{3}{4}$ **29.** 5 **31.** $1\frac{1}{8}$ **33.** $2\frac{3}{10}$ **35.** 3

37. $1\frac{1}{7}$ **39.** $2\frac{1}{3}$ **41.** 16 **43.** $2\frac{1}{8}$ **45.** $2\frac{2}{5}$ **47.** 1 **49.** 9 **51.** $\frac{7}{3}$ **53.** $\frac{13}{2}$ **55.** $\frac{41}{6}$ **57.** $\frac{37}{4}$

59. $\frac{21}{2}$ **61.** $\frac{73}{9}$ **63.** $\frac{58}{11}$ **65.** $\frac{21}{8}$ **67.** $\frac{13}{8}$ **69.** $\frac{100}{9}$ **71.** $\frac{27}{8}$ **73.** $\frac{85}{13}$

SECTION 2.3

1. $\frac{5}{10}$ **3.** $\frac{9}{48}$ **5.** $\frac{12}{32}$ **7.** $\frac{9}{51}$ **9.** $\frac{12}{16}$ **11.** $\frac{27}{9}$ **13.** $\frac{20}{60}$ **15.** $\frac{44}{60}$ **17.** $\frac{12}{18}$ **19.** $\frac{35}{49}$ **21.** $\frac{10}{18}$

23. $\frac{21}{3}$ **25.** $\frac{35}{45}$ **27.** $\frac{60}{64}$ **29.** $\frac{21}{98}$ **31.** $\frac{30}{48}$ **33.** $\frac{15}{42}$ **35.** $\frac{102}{144}$ **37.** $\frac{153}{408}$ **39.** $\frac{340}{800}$ **41.** $\frac{1}{3}$ **43.** $\frac{1}{2}$

45. $\frac{1}{6}$ **47.** $1\frac{1}{9}$ **49.** 0 **51.** $\frac{9}{22}$ **53.** 3 **55.** $\frac{4}{21}$ **57.** $\frac{12}{35}$ **59.** $\frac{7}{11}$ **61.** $1\frac{1}{3}$ **63.** $\frac{3}{5}$ **65.** $\frac{1}{11}$

67. 4 **69.** $\frac{1}{3}$ **71.** $\frac{3}{5}$ **73.** $2\frac{1}{4}$ **75.** $\frac{1}{5}$ **77.** Answers will vary. For example, $\frac{3}{1}, \frac{6}{2}, \frac{9}{3}, \frac{12}{4}$, and $\frac{15}{5}$.

79. a. $\frac{4}{25}$ **b.** $\frac{4}{25}$

SECTION 2.4

1. $\dfrac{3}{7}$ **3.** 1 **5.** $1\dfrac{4}{11}$ **7.** $3\dfrac{2}{5}$ **9.** $2\dfrac{4}{5}$ **11.** $2\dfrac{1}{4}$ **13.** $1\dfrac{3}{8}$ **15.** $1\dfrac{7}{15}$ **17.** $\dfrac{15}{16}$ **19.** $1\dfrac{4}{11}$ **21.** 1

23. $1\dfrac{7}{8}$ **25.** $1\dfrac{1}{6}$ **27.** $\dfrac{13}{14}$ **29.** $\dfrac{53}{60}$ **31.** $1\dfrac{1}{56}$ **33.** $\dfrac{23}{60}$ **35.** $\dfrac{56}{57}$ **37.** $1\dfrac{17}{18}$ **39.** $1\dfrac{11}{48}$ **41.** $1\dfrac{9}{20}$

43. $1\dfrac{109}{180}$ **45.** $1\dfrac{91}{144}$ **47.** $2\dfrac{5}{72}$ **49.** $\dfrac{39}{40}$ **51.** $1\dfrac{19}{24}$ **53.** $1\dfrac{65}{72}$ **55.** $3\dfrac{2}{3}$ **57.** $10\dfrac{1}{12}$ **59.** $9\dfrac{2}{7}$ **61.** $6\dfrac{7}{40}$

63. $9\dfrac{47}{48}$ **65.** $8\dfrac{22}{39}$ **67.** $16\dfrac{29}{120}$ **69.** $24\dfrac{29}{40}$ **71.** $33\dfrac{7}{24}$ **73.** $10\dfrac{5}{36}$ **75.** $10\dfrac{5}{12}$ **77.** $14\dfrac{73}{90}$ **79.** $10\dfrac{13}{48}$

81. $42\dfrac{13}{54}$ **83.** $8\dfrac{1}{36}$ **85.** $14\dfrac{1}{12}$ **87.** $9\dfrac{37}{72}$ **89.** The length of the shaft is $1\dfrac{5}{16}$ inches. **91.** The total thickness

of the table top is $1\dfrac{5}{16}$ inches. **93. a.** You worked a total of 20 hours during the last five days. **b.** Your total wages

are $180. **95.** The total length is $55\dfrac{3}{4}$ feet. **97.** $\dfrac{1}{3}$ of those who changed homes moved outside the county they had

been living in.

SECTION 2.5

1. $\dfrac{2}{17}$ **3.** $\dfrac{1}{3}$ **5.** $\dfrac{1}{10}$ **7.** $\dfrac{5}{13}$ **9.** $\dfrac{1}{3}$ **11.** $\dfrac{4}{7}$ **13.** $\dfrac{1}{4}$ **15.** $\dfrac{9}{23}$ **17.** $\dfrac{1}{4}$ **19.** $\dfrac{1}{2}$ **21.** $\dfrac{19}{56}$ **23.** $\dfrac{1}{2}$

25. $\dfrac{11}{60}$ **27.** $\dfrac{15}{56}$ **29.** $\dfrac{37}{51}$ **31.** $\dfrac{17}{70}$ **33.** $\dfrac{49}{120}$ **35.** $\dfrac{8}{45}$ **37.** $\dfrac{11}{21}$ **39.** $\dfrac{23}{60}$ **41.** $\dfrac{1}{18}$ **43.** $5\dfrac{1}{5}$ **45.** $10\dfrac{9}{17}$

47. $4\dfrac{7}{8}$ **49.** $\dfrac{16}{21}$ **51.** $5\dfrac{1}{2}$ **53.** $5\dfrac{4}{7}$ **55.** $7\dfrac{5}{24}$ **57.** $48\dfrac{31}{70}$ **59.** $85\dfrac{2}{9}$ **61.** $9\dfrac{5}{13}$ **63.** $15\dfrac{11}{20}$ **65.** $4\dfrac{23}{24}$

67. $2\dfrac{11}{15}$ **69.** The missing dimension is $9\dfrac{1}{2}$ inches. **71.** Meyfarth jumped $9\dfrac{3}{8}$ inches farther than Coachman.

Kostadinova jumped $5\dfrac{1}{4}$ inches farther than Meyfarth. **73. a.** It is $7\dfrac{17}{24}$ miles from the starting point to the second

checkpoint. **b.** It is $4\dfrac{7}{24}$ miles from the second checkpoint to the finish line. **75. a.** Yes. **b.** The wrestler must

lose $3\dfrac{1}{4}$ pounds in the third week. **77.** $2\dfrac{5}{6}$ **79.**

$\frac{3}{8}$	$\frac{3}{4}$	$\frac{3}{4}$
1	$\frac{5}{8}$	$\frac{1}{4}$
$\frac{1}{2}$	$\frac{1}{2}$	$\frac{7}{8}$

SECTION 2.6

1. $\dfrac{7}{12}$ **3.** $\dfrac{7}{48}$ **5.** $\dfrac{1}{48}$ **7.** $\dfrac{11}{14}$ **9.** $\dfrac{1}{7}$ **11.** $\dfrac{1}{8}$ **13.** 6 **15.** $\dfrac{5}{12}$ **17.** 6 **19.** $\dfrac{2}{3}$ **21.** $\dfrac{3}{16}$ **23.** $\dfrac{3}{80}$

25. 10 **27.** $\dfrac{1}{15}$ **29.** $\dfrac{2}{3}$ **31.** $\dfrac{7}{26}$ **33.** 4 **35.** $\dfrac{100}{357}$ **37.** $\dfrac{5}{24}$ **39.** $\dfrac{1}{12}$ **41.** $\dfrac{4}{15}$ **43.** $1\dfrac{1}{2}$ **45.** 4

47. $\dfrac{4}{9}$ **49.** $\dfrac{1}{2}$ **51.** $16\dfrac{1}{2}$ **53.** 10 **55.** $6\dfrac{3}{7}$ **57.** $18\dfrac{1}{3}$ **59.** $1\dfrac{5}{7}$ **61.** $3\dfrac{1}{2}$ **63.** $25\dfrac{5}{8}$ **65.** $1\dfrac{11}{16}$ **67.** 16

69. 3 **71.** $8\dfrac{37}{40}$ **73.** $6\dfrac{19}{28}$ **75.** $7\dfrac{2}{5}$ **77.** 16 **79.** $32\dfrac{4}{5}$ **81.** $28\dfrac{1}{2}$ **83.** 9 **85.** $\dfrac{5}{8}$ **87.** $3\dfrac{1}{40}$

89. The cost of $2\dfrac{3}{4}$ pounds of salmon is $22. **91. a.** No **b.** The piece cut off is $3\dfrac{1}{12}$ feet long. **93.** The area is

$27\dfrac{9}{16}$ square feet. **95. a.** $1280 is budgeted for housing and utilities. **b.** $1920 remains for purposes other than

housing and utilities. **97.** The total cost is $504. **99.** The weight is $54\dfrac{19}{36}$ pounds. **101.** $\dfrac{3}{16}$ of the total portfolio is

invested in corporate bonds. **105.** *A*

SECTION 2.7

1. $\dfrac{5}{6}$ **3.** 1 **5.** 0 **7.** $\dfrac{1}{2}$ **9.** $2\dfrac{73}{256}$ **11.** $1\dfrac{1}{9}$ **13.** $\dfrac{1}{6}$ **15.** $\dfrac{7}{10}$ **17.** 2 **19.** 2 **21.** $\dfrac{1}{6}$ **23.** 6

25. $\dfrac{1}{15}$ **27.** 2 **29.** $2\dfrac{1}{2}$ **31.** 3 **33.** 6 **35.** $\dfrac{1}{2}$ **37.** $1\dfrac{1}{6}$ **39.** $3\dfrac{1}{3}$ **41.** $\dfrac{1}{30}$ **43.** $1\dfrac{4}{5}$ **45.** 13

47. $\dfrac{25}{288}$ **49.** 3 **51.** $\dfrac{1}{5}$ **53.** $\dfrac{11}{28}$ **55.** $\dfrac{5}{86}$ **57.** 120 **59.** $\dfrac{11}{40}$ **61.** $\dfrac{33}{40}$ **63.** $4\dfrac{4}{9}$ **65.** $\dfrac{13}{32}$ **67.** $10\dfrac{2}{3}$

69. $9\dfrac{39}{40}$ **71.** $\dfrac{12}{53}$ **73.** $4\dfrac{62}{191}$ **75.** 68 **77.** $8\dfrac{2}{7}$ **79.** $3\dfrac{13}{49}$ **81.** 4 **83.** $1\dfrac{3}{5}$ **85.** $\dfrac{9}{34}$ **87.** The box

contains twelve $1\dfrac{1}{3}$-ounce servings. **89.** Each acre cost \$24,000. **91.** The nut makes 12 turns to move $1\dfrac{7}{8}$ inches.

93. a. The total weight of the fat and bone is $1\dfrac{5}{12}$ pounds. **b.** There will be 28 servings. **95. a.** The actual length of

wall a is $12\dfrac{1}{2}$ feet. **b.** The actual length of wall b is 18 feet. **c.** The actual length of wall c is $15\dfrac{3}{4}$ feet. **97.** No. For

example, $4 \div \dfrac{1}{2} = 8$. Answers will vary. **99.** Each column is $2\dfrac{1}{4}$ inches wide. **101. a.** $2\dfrac{1}{2}$ **b.** $\dfrac{2}{3}$ **c.** $\dfrac{1}{3}$ **d.** $2\dfrac{1}{2}$

SECTION 2.8

1. $\dfrac{11}{40} < \dfrac{19}{40}$ **3.** $\dfrac{2}{3} < \dfrac{5}{7}$ **5.** $\dfrac{5}{8} > \dfrac{7}{12}$ **7.** $\dfrac{7}{9} < \dfrac{11}{12}$ **9.** $\dfrac{13}{14} > \dfrac{19}{21}$ **11.** $\dfrac{7}{24} < \dfrac{11}{30}$ **13.** $\dfrac{9}{64}$ **15.** $\dfrac{8}{729}$ **17.** $\dfrac{1}{24}$

19. $\dfrac{8}{245}$ **21.** $\dfrac{1}{121}$ **23.** $\dfrac{81}{625}$ **25.** $\dfrac{7}{36}$ **27.** $\dfrac{27}{49}$ **29.** $\dfrac{5}{6}$ **31.** $1\dfrac{5}{12}$ **33.** $\dfrac{7}{48}$ **35.** $\dfrac{29}{36}$ **37.** $\dfrac{55}{72}$ **39.** $\dfrac{35}{54}$

41. 2 **43.** $\dfrac{9}{19}$ **45.** $\dfrac{7}{32}$ **47.** $\dfrac{64}{75}$

CHAPTER REVIEW

1. $\dfrac{2}{3}$ [2.3B] **2.** $\dfrac{5}{16}$ [2.8B] **3.** $\dfrac{13}{4}$ [2.2A] **4.** $1\dfrac{13}{18}$ [2.4B] **5.** $\dfrac{11}{18} < \dfrac{17}{24}$ [2.8A] **6.** $14\dfrac{19}{42}$ [2.5C]

7. $\dfrac{5}{36}$ [2.8C] **8.** $9\dfrac{1}{24}$ [2.6B] **9.** 2 [2.7B] **10.** $\dfrac{25}{48}$ [2.5B] **11.** $3\dfrac{1}{3}$ [2.7B] **12.** 4 [2.1B]

13. $\dfrac{24}{36}$ [2.3A] **14.** $\dfrac{3}{4}$ [2.7A] **15.** $\dfrac{32}{44}$ [2.3A] **16.** $16\dfrac{1}{2}$ [2.6B] **17.** 36 [2.1A] **18.** $\dfrac{4}{11}$ [2.3B]

19. $1\dfrac{1}{8}$ [2.4A] **20.** $10\dfrac{1}{8}$ [2.5C] **21.** $18\dfrac{13}{54}$ [2.4C] **22.** 5 [2.1B] **23.** $3\dfrac{2}{5}$ [2.2B] **24.** $\dfrac{1}{15}$ [2.8C]

25. $5\dfrac{7}{8}$ [2.4C] **26.** 54 [2.1A] **27.** $\dfrac{1}{3}$ [2.5A] **28.** $\dfrac{19}{7}$ [2.2B] **29.** 2 [2.7A] **30.** $\dfrac{1}{15}$ [2.6A]

31. $\dfrac{1}{8}$ [2.6A] **32.** $1\dfrac{7}{8}$ [2.2A] **33.** The total rainfall was $21\dfrac{7}{24}$ inches. [2.4D] **34.** Each acre cost \$36,000. [2.7C]

35. The second checkpoint is $4\dfrac{3}{4}$ miles from the finish line. [2.5D] **36.** The car can travel 243 miles. [2.6C]

CHAPTER TEST

1. $\dfrac{4}{9}$ [2.6A] **2.** 8 [2.1B] **3.** $1\dfrac{3}{7}$ [2.7A] **4.** $\dfrac{7}{24}$ [2.8C] **5.** $\dfrac{49}{5}$ [2.2B] **6.** 8 [2.6B] **7.** $\dfrac{5}{8}$ [2.3B]

8. $\dfrac{3}{8} < \dfrac{5}{12}$ [2.8A] **9.** $\dfrac{5}{6}$ [2.8C] **10.** 120 [2.1A] **11.** $\dfrac{1}{4}$ [2.5A] **12.** $3\dfrac{3}{5}$ [2.2B] **13.** $2\dfrac{2}{19}$ [2.7B]

14. $\dfrac{45}{72}$ [2.3A] **15.** $1\dfrac{61}{90}$ [2.4B] **16.** $13\dfrac{81}{88}$ [2.5C] **17.** $\dfrac{7}{48}$ [2.5B] **18.** $\dfrac{1}{6}$ [2.8B] **19.** $1\dfrac{11}{12}$ [2.4A]

20. $22\dfrac{4}{15}$ [2.4C] **21.** $\dfrac{11}{4}$ [2.2A] **22.** The electrician's total earnings are \$840. [2.6C] **23.** There were 11 lots

available for sale. [2.7C] **24.** The value of one share is $27\dfrac{7}{8}$. [2.5D] **25.** The total rainfall was $21\dfrac{11}{24}$ inches. [2.4D]

CUMULATIVE REVIEW

1. 290,000 [1.1D]   **2.** 291,278 [1.3B]   **3.** 73,154 [1.4B]   **4.** 540 r12 [1.5C]   **5.** 1 [1.6B]

6. $2 \cdot 2 \cdot 11$ [1.7B]   **7.** 210 [2.1A]   **8.** 20 [2.1B]   **9.** $\frac{23}{3}$ [2.2B]   **10.** $6\frac{1}{4}$ [2.2B]   **11.** $\frac{15}{48}$ [2.3A]

12. $\frac{2}{5}$ [2.3B]   **13.** $1\frac{7}{48}$ [2.4B]   **14.** $14\frac{11}{48}$ [2.4C]   **15.** $\frac{13}{24}$ [2.5B]   **16.** $1\frac{7}{9}$ [2.5C]   **17.** $\frac{7}{20}$ [2.6A]

18. $7\frac{1}{2}$ [2.6B]   **19.** $1\frac{1}{20}$ [2.7A]   **20.** $2\frac{5}{8}$ [2.7B]   **21.** $\frac{1}{9}$ [2.8B]   **22.** $5\frac{5}{24}$ [2.8C]

23. There is $862 in the checking account at the end of the week. [1.3C]   **24.** The total income is $705. [1.4C]

25. The total weight is $12\frac{1}{24}$ pounds. [2.4D]   **26.** The length of the remaining piece is $4\frac{17}{24}$ feet. [2.5D]

27. The car traveled 225 miles. [2.6C]   **28.** 25 parcels can be sold from the remaining land. [2.7C]

Answers to Chapter 3 Odd-Numbered Exercises

PREP TEST

1. $\frac{3}{10}$ [2.2A]   **2.** 36,900 [1.1D]   **3.** four thousand seven hundred ninety-one [1.1B]   **4.** 6842 [1.1B]

5. 9394 [1.2A]   **6.** 1638 [1.3B]   **7.** 76,804 [1.4B]   **8.** 278 r18 [1.5C]

SECTION 3.1

1. hundredths   **3.** ten-thousandths   **5.** Twenty-seven hundredths   **7.** One and five thousandths   **9.** Thirty-six and four tenths   **11.** Thirty-five hundred-thousandths   **13.** Ten and seven thousandths   **15.** Fifty-two and ninety-five hundred-thousandths   **17.** Two hundred ninety-three ten-thousandths   **19.** Six and three hundred twenty-four thousandths   **21.** Two hundred seventy-six and three thousand two hundred ninety-seven ten-thousandths   **23.** Two hundred sixteen and seven hundred twenty-nine ten-thousandths   **25.** Four thousand six hundred twenty-five and three hundred seventy-nine ten-thousandths   **27.** One and one hundred-thousandth   **29.** 0.762   **31.** 0.000062
33. 8.0304   **35.** 304.07   **37.** 362.048   **39.** 3048.2002   **41.** 7.4   **43.** 23.0   **45.** 22.68   **47.** 7.073
49. 62.009   **51.** 0.0123   **53.** 2.07924   **55.** 0.100975   **59. a.** 5.5 and 6.4   **b.** 10.15 and 10.24

SECTION 3.2

1. 150.1065   **3.** 95.8446   **5.** 69.644   **7.** 92.883   **9.** 113.205   **11.** 0.69   **13.** 16.305   **15.** 110.7666
17. 104.4959   **19.** Est.: 234   **21.** Est.: 782   **23.** The length of the shaft is 5.65 inches.
        Cal.: 234.192   Cal.: 781.943

25. The amount in your checking account is $3664.20.   **27.** The combined populations of Asia and Africa are expected to be 7.1 billion.   **29.** There are 6.5 million self-employed people who earn $5000 or more.   **31.** No, the customer does not have enough money.   **33.** No, the rope cannot be wrapped around the box.

SECTION 3.3

1. 5.627   **3.** 113.6427   **5.** 6.7098   **7.** 215.697   **9.** 53.8776   **11.** 72.7091   **13.** 0.3142   **15.** 1.023
17. 261.166   **19.** 655.32   **21.** 342.9268   **23.** 8.628   **25.** 4.685   **27.** 10.383
29. Est.: 3   **31.** Est.: 7   **33.** The missing dimension is 7.55 inches.
        Cal.: 2.74506   Cal.: 7.14925

35. a. The total amount of the checks written is $607.36.   **b.** The new balance is $422.38.   **37.** Oakland's rainfall was 16.77 inches above normal.   **39.** The growth in online shopping is 22.6 million households.   **41. a.** 0.1   **b.** 0.01
c. 0.001

SECTION 3.4

1. 0.36 **3.** 0.30 **5.** 0.25 **7.** 0.45 **9.** 6.93 **11.** 1.84 **13.** 4.32 **15.** 0.74 **17.** 39.5 **19.** 2.72
21. 0.603 **23.** 0.096 **25.** 13.50 **27.** 79.80 **29.** 4.316 **31.** 1.794 **33.** 0.06 **35.** 0.072 **37.** 0.1323
39. 0.03568 **41.** 0.0784 **43.** 0.076 **45.** 34.48 **47.** 580.5 **49.** 20.148 **51.** 0.04255 **53.** 0.17686
55. 0.19803 **57.** 14.8657 **59.** 0.0006608 **61.** 53.9961 **63.** 0.536335 **65.** 0.429 **67.** 2.116 **69.** 0.476
71. 1.022 **73.** 37.96 **75.** 2.318 **77.** 3.2 **79.** 6.5 **81.** 6285.6 **83.** 3200 **85.** 35,700 **87.** 6.3
89. 3.9 **91.** 49,000 **93.** 6.7 **95.** 0.012075 **97.** 0.0117796 **99.** 0.31004 **101.** 0.082845 **103.** 5.175
105. Est.: 90 **107.** Est.: 0.8 **109.** Est.: 4.5 **111.** Est.: 12 **113.** Est.: 0.32 **115.** Est.: 30
Cal.: 91.2   Cal.: 1.0472   Cal.: 3.897   Cal.: 11.2406   Cal.: 0.371096   Cal.: 31.8528
117. Est.: 2000 **119.** Est.: 0.00005 **121.** The cost is $1.51.
Cal.: 1941.069459   Cal.: 0.00004351152
123. The estimate is $25. The actual amount received is $23.40.   **125.** The area is 23.625 square feet.   **127. a.** The
nurse's overtime pay is $425.25.   **b.** The nurse's total income for the week is $1181.25.   **129.** It would cost $306.25.
131. a. The total costs are $21.12 for Grade 1, $29.84 for Grade 2, and $201.66 for Grade 3.   **b.** The total cost of the
three pieces of steel is $252.62.   **133.** None of the cars would fail the emissions test.
137. $1.3 \times 2.31 = \dfrac{13}{10} \times \dfrac{231}{100} = \dfrac{3003}{1000} = 3.003$

SECTION 3.5

1. 0.82 **3.** 4.8 **5.** 89 **7.** 60 **9.** 84.3 **11.** 32.3 **13.** 5.06 **15.** 1.3 **17.** 0.11 **19.** 3.8 **21.** 6.3
23. 0.6 **25.** 2.5 **27.** 1.1 **29.** 130.6 **31.** 0.81 **33.** 0.09 **35.** 40.70 **37.** 0.46 **39.** 0.019
41. 0.087 **43.** 0.360 **45.** 0.103 **47.** 0.009 **49.** 1 **51.** 3 **53.** 1 **55.** 57 **57.** 0.407 **59.** 4.267
61. 0.01037 **63.** 0.008295 **65.** 0.82537 **67.** 0.032 **69.** 0.23627 **71.** 0.000053 **73.** 0.0018932
75. 18.42 **77.** 16.07 **79.** 0.0135 **81.** 0.023678 **83.** 0.112 **85.** Est.: 10 **87.** Est.: 1000
Cal.: 11.1632   Cal.: 884.0909
89. Est.: 1.5 **91.** Est.: 50 **93.** Est.: 100 **95.** Est.: 0.0025 **97.** Ramon averaged 6.23 yards per carry.
Cal.: 1.8269   Cal.: 58.8095   Cal.: 72.3053   Cal.: 0.0023
99. The cost is $.28 per can.   **101.** The monthly payment is $58.65.   **103.** The cost was $.04 per mile.   **109.** ÷
111. −   **113.** +   **115.** 5.217   **117.** 0.025

SECTION 3.6

1. 0.625 **3.** 0.667 **5.** 0.167 **7.** 0.417 **9.** 1.750 **11.** 1.500 **13.** 4.000 **15.** 0.003 **17.** 7.080
19. 37.500 **21.** 0.160 **23.** 8.400 **25.** $\dfrac{4}{5}$ **27.** $\dfrac{8}{25}$ **29.** $\dfrac{1}{8}$ **31.** $1\dfrac{1}{4}$ **33.** $16\dfrac{9}{10}$ **35.** $8\dfrac{2}{5}$ **37.** $8\dfrac{437}{1000}$
39. $2\dfrac{1}{4}$ **41.** $\dfrac{23}{150}$ **43.** $\dfrac{703}{800}$ **45.** $7\dfrac{19}{50}$ **47.** $\dfrac{57}{100}$ **49.** $\dfrac{2}{3}$ **51.** 0.6 > 0.45 **53.** 3.89 < 3.98
55. 0.025 < 0.105 **57.** $\dfrac{4}{5} < 0.802$ **59.** $0.85 < \dfrac{7}{8}$ **61.** $\dfrac{7}{12} > 0.58$ **63.** $\dfrac{11}{12} < 0.92$ **65.** 0.623 > 0.6023
67. 0.87 > 0.087 **69.** 0.033 < 0.3 **71.** The number of hearing-impaired individuals who are aged 65–74 is less than
4 times the number of hearing-impaired individuals who are aged 0–17. **73.** No. For example, 0.0402 rounded to the
nearest hundredth is 0.04 and rounded to the nearest thousandth is 0.040.

CHAPTER REVIEW

1. 54.5 [3.5A] **2.** 833.958 [3.2A] **3.** 0.055 < 0.1 [3.6C] **4.** Twenty-two and ninety-two ten-thousandths [3.1A]
5. 0.05678 [3.1B] **6.** 2.33 [3.6A] **7.** $\dfrac{3}{8}$ [3.6B] **8.** 36.714 [3.2A] **9.** 34.025 [3.1A] **10.** $\dfrac{5}{8} > 0.62$ [3.6C]
11. 0.778 [3.6A] **12.** $\dfrac{2}{3}$ [3.6B] **13.** 22.8635 [3.3A] **14.** 7.94 [3.1B] **15.** 8.932 [3.4A]
16. Three hundred forty-two and thirty-seven hundredths [3.1A] **17.** 3.06753 [3.1A] **18.** 25.7446 [3.4A]

19. 6.594 [3.5A]   **20.** 4.8785 [3.3A]   **21.** The new balance is $661.51. [3.3B]   **22.** The total number is 53.446 million children. [3.2B]   **23.** There are 40.49 million more children being educated in public school than in private school. [3.3B]   **24.** There are 9.5 million gallons of milk served in school cafeterias during a 5-day school week. [3.4B] **25.** The number of acres planted in Texas is 5.2 times greater than the number of acres planted in Mississippi.   [3.5B]

CHAPTER TEST

1. $0.66 < 0.666$ [3.6C]   **2.** 4.087 [3.3A]   **3.** Forty-five and three hundred two ten-thousandths [3.1A]

4. 0.692 [3.6A]   **5.** $\frac{33}{40}$ [3.6B]   **6.** 0.0740 [3.1B]   **7.** 1.538 [3.5A]   **8.** 27.76626 [3.3A]   **9.** 7.095 [3.1B]

10. 232 [3.5A]   **11.** 458.581 [3.2A]   **12.** 1.37 inches [3.3B]   **13.** 0.00548 [3.4A]   **14.** 255.957 [3.2A]

15. 209.07086 [3.1A]   **16.** The monthly payment is $395.40. [3.5B]   **17.** Your total income is $3087.14. [3.2B]

18. The cost is $4.63. [3.4B]   **19.** On average, a 10th-grade student uses a computer 348.4 hours per year. [3.4B]

20. A 2nd-grade student uses a computer 36.4 hours more per year than does a 5th-grade student. [3.4B]

CUMULATIVE REVIEW

1. 235 r17 [1.5C]   **2.** 128 [1.6A]   **3.** 3 [1.6B]   **4.** 72 [2.1A]   **5.** $4\frac{2}{5}$ [2.2B]   **6.** $\frac{37}{8}$ [2.2B]

7. $\frac{25}{60}$ [2.3A]   **8.** $1\frac{17}{48}$ [2.4B]   **9.** $8\frac{35}{36}$ [2.4C]   **10.** $5\frac{23}{36}$ [2.5C]   **11.** $\frac{1}{12}$ [2.6A]   **12.** $9\frac{1}{8}$ [2.6B]

13. $1\frac{2}{9}$ [2.7A]   **14.** $\frac{19}{20}$ [2.7B]   **15.** $\frac{3}{16}$ [2.8B]   **16.** $2\frac{5}{18}$ [2.8C]   **17.** Sixty-five and three hundred nine ten-thousandths [3.1A]   **18.** 504.6991 [3.2A]   **19.** 21.0764 [3.3A]   **20.** 55.26066 [3.4A]   **21.** 2.154 [3.5A]

22. 0.733 [3.6A]   **23.** $\frac{1}{6}$ [3.6B]   **24.** $\frac{8}{9} < 0.98$ [3.6C]   **25.** Sweden mandates 14 more vacation days than Germany. [1.3C]   **26.** The value of each share is $32\frac{7}{8}$. [2.5D]   **27.** The checking account balance is $617.38. [3.3B]   **28.** The resulting thickness of the bushing is 1.395 inches. [3.3B]   **29.** The amount of income tax paid last year was $6008.80. [3.4B]   **30.** The monthly payment is $23.87. [3.5B]

Answers to Chapter 4 Odd-Numbered Exercises

PREP TEST

1. $\frac{4}{5}$ [2.3B]   **2.** $\frac{1}{2}$ [2.3B]   **3.** 24.8 [3.6A]   **4.** 4×33 [1.1A; 1.4A]   **5.** 4 [1.5A]

SECTION 4.1

1. $\frac{1}{5}$ 1:5   1 to 5   **3.** $\frac{2}{1}$ 2:1   2 to 1   **5.** $\frac{3}{8}$ 3:8   3 to 8   **7.** $\frac{37}{24}$ 37:24   37 to 24   **9.** $\frac{1}{1}$ 1:1   1 to 1

11. $\frac{7}{10}$ 7:10   7 to 10   **13.** $\frac{1}{2}$ 1:2   1 to 2   **15.** $\frac{2}{1}$ 2:1   2 to 1   **17.** $\frac{3}{4}$ 3:4   3 to 4   **19.** $\frac{5}{3}$ 5:3   5 to 3

21. $\frac{2}{3}$ 2:3   2 to 3   **23.** $\frac{2}{1}$ 2:1   2 to 1   **25.** The ratio of housing cost to total expenses is $\frac{1}{3}$.   **27.** The ratio of utilities cost to food cost is $\frac{3}{8}$.   **29.** The ratio of first-year college basketball players to high school senior basketball players is $\frac{2}{77}$.   **31.** The ratio of the turns in the primary coil to the turns in the secondary coil is $\frac{1}{12}$.   **33. a.** The amount of increase is $20,000.   **b.** The ratio of the increase to the original value is $\frac{2}{9}$.   **35.** The debt–income ratio is $\frac{265}{778}$.

37. No; $\frac{830}{1389}$ is greater than $\frac{2}{5}$.

SECTION 4.2

1. $\dfrac{3 \text{ pounds}}{4 \text{ people}}$ **3.** $\dfrac{\$20}{3 \text{ boards}}$ **5.** $\dfrac{20 \text{ miles}}{1 \text{ gallon}}$ **7.** $\dfrac{5 \text{ children}}{2 \text{ families}}$ **9.** $\dfrac{8 \text{ gallons}}{1 \text{ hour}}$ **11.** 2.5 feet/second **13.** \$325/week

15. 110 trees/acre **17.** \$4.71/hour **19.** 52.4 miles/hour **21.** 28 miles/gallon **23.** \$1.65/pound **25.** The car was driven 28.4 miles per gallon of gas. **27.** The rocket uses 213,600 gallons of fuel per minute. **29. a.** There were 175 pounds of beef packaged. **b.** The cost was \$2.09 per pound. **31. a.** The camera goes through film at a rate of 336 feet per minute. **b.** It uses a 500-foot roll of 65-mm film at a rate of 89 seconds per roll. **33. a.** 42,484 euros would be paid for the order. **b.** The cost is 2,556,000 yen.

SECTION 4.3

1. True **3.** Not true **5.** Not true **7.** True **9.** True **11.** Not true **13.** True **15.** True **17.** True **19.** Not true **21.** True **23.** Not true **25.** 3 **27.** 6 **29.** 9 **31.** 5.67 **33.** 4 **35.** 4.38 **37.** 88 **39.** 3.33 **41.** 26.25 **43.** 96 **45.** 9.78 **47.** 3.43 **49.** 1.34 **51.** 50.4 **53.** There are 50 calories in the serving of cereal. **55.** There were 70 pounds of fertilizer used. **57.** There are 375 wooden bats produced. **59.** The distance between the two cities is 16 miles. **61.** 1.25 ounces of medication are required. **63.** There would be 160,000 people voting in the election. **65.** The monthly cost of the life insurance is \$176.75. **67.** You will own 400 shares of stock. **69.** A bowling ball would weigh 2.67 pounds on the moon. **71.** Carlos would receive a dividend of \$1071.

CHAPTER REVIEW

1. True [4.3A] **2.** $\dfrac{2}{5}$ 2:5 2 to 5 [4.1A] **3.** 62.5 miles/hour [4.2B] **4.** True [4.3A] **5.** 68 [4.3B]

6. \$7.50/hour [4.2B] **7.** \$1.75/pound [4.2B] **8.** $\dfrac{2}{7}$ 2:7 2 to 7 [4.1A] **9.** 36 [4.3B] **10.** 19.44 [4.3B]

11. $\dfrac{2}{5}$ 2:5 2 to 5 [4.1A] **12.** Not true [4.3A] **13.** $\dfrac{\$15}{4 \text{ hours}}$ [4.2A] **14.** 27.2 miles/gallon [4.2B]

15. $\dfrac{1}{1}$ 1:1 1 to 1 [4.1A] **16.** True [4.3A] **17.** 65.45 [4.3B] **18.** $\dfrac{100 \text{ miles}}{3 \text{ hours}}$ [4.2A] **19.** The ratio of the

decrease in price to the original price is $\dfrac{2}{5}$. [4.1B] **20.** The property tax is \$2400. [4.3C] **21.** The ratio of the high

temperature to the low temperature is $\dfrac{2}{1}$. [4.1B] **22.** The cost per radio is \$37.50. [4.2C] **23.** It would take 1344

blocks to build the wall. [4.3C] **24.** The ratio of TV advertising to newspaper advertising is $\dfrac{5}{2}$. [4.1B] **25.** The cost

is \$.68 per pound. [4.2C] **26.** Mahesh drove an average of 56.8 miles per hour. [4.2C] **27.** The cost of the

insurance is \$193.50. [4.3C] **28.** The cost was \$44.75 per share. [4.2C] **29.** She will have to use 22.5 pounds of

fertilizer. [4.3C] **30.** The ratio of the increase to the original value is $\dfrac{1}{2}$. [4.1B]

CHAPTER TEST

1. \$1836.40/month [4.2B] **2.** $\dfrac{1}{6}$ 1:6 1 to 6 [4.1A] **3.** $\dfrac{9 \text{ supports}}{4 \text{ feet}}$ [4.2A] **4.** Not true [4.3A]

5. $\dfrac{3}{2}$ 3:2 3 to 2 [4.1A] **6.** 144 [4.3B] **7.** 30.5 miles/gallon [4.2B] **8.** $\dfrac{1}{3}$ 1:3 1 to 3 [4.1A]

9. True [4.3A] **10.** 40.5 [4.3B] **11.** $\dfrac{\$27}{4 \text{ boards}}$ [4.2A] **12.** $\dfrac{3}{5}$ 3:5 3 to 5 [4.1A] **13.** The dividend would

be \$625. [4.3C] **14.** The ratio of the average city temperature to the average desert temperature is $\dfrac{43}{56}$. [4.1B]

15. The plane's speed is 538 miles per hour. [4.2C] **16.** The number of pounds of water is 132. [4.3C]

17. The cost is \$1.73 per foot. [4.2C] **18.** The amount of medication needed is 0.875 ounce. [4.3C] **19.** The ratio

of the cost of radio advertising to the total cost of advertising is $\dfrac{8}{13}$. [4.1B] **20.** The property tax is \$1200. [4.3C]

CUMULATIVE REVIEW

1. 9158 [1.3B] **2.** $2^4 \cdot 3^3$ [1.6A] **3.** 3 [1.6B] **4.** $2 \cdot 2 \cdot 2 \cdot 2 \cdot 2 \cdot 5$ [1.7B] **5.** 36 [2.1A] **6.** 14 [2.1B]

7. $\frac{5}{8}$ [2.3B] **8.** $8\frac{3}{10}$ [2.4C] **9.** $5\frac{11}{18}$ [2.5C] **10.** $2\frac{5}{6}$ [2.6B] **11.** $4\frac{2}{3}$ [2.7B] **12.** $\frac{23}{30}$ [2.8C]

13. Four and seven hundred nine ten-thousandths [3.1A] **14.** 2.10 [3.1B] **15.** 1.990 [3.5A] **16.** $\frac{1}{15}$ [3.6B]

17. $\frac{1}{8}$ [4.1A] **18.** $\frac{29¢}{2 \text{ pencils}}$ [4.2A] **19.** 33.4 miles/gallon [4.2B] **20.** 4.25 [4.3B]

21. 57.2 miles/hour [4.2C] **22.** 36 [4.3B] **23.** The new balance is $744. [1.3C] **24.** The monthly payment is $370. [1.5D] **25.** There are 105 pages remaining to be read. [2.6C] **26.** The cost for each acre was $36,000. [2.7C] **27.** The amount of change is $17.62. [3.3B] **28.** Your monthly salary is $3468.25. [3.5B] **29.** There will be 25 inches eroded. [4.3C] **30.** The person needs 1.6 ounces of medication. [4.3C]

Answers to Chapter 5 Odd-Numbered Exercises

PREP TEST

1. $\frac{19}{100}$ [2.6B] **2.** 0.23 [3.4A] **3.** 47 [3.4A] **4.** 2850 [3.4A] **5.** 4000 [3.5A] **6.** 32 [2.7B]

7. 62.5 [2.6B] **8.** $66\frac{2}{3}$ [2.2B] **9.** 1.75 [3.5A]

SECTION 5.1

1. $\frac{1}{4}$, 0.25 **3.** $1\frac{3}{10}$, 1.30 **5.** 1, 1.00 **7.** $\frac{73}{100}$, 0.73 **9.** $3\frac{83}{100}$, 3.83 **11.** $\frac{7}{10}$, 0.70 **13.** $\frac{22}{25}$, 0.88

15. $\frac{8}{25}$, 0.32 **17.** $\frac{2}{3}$ **19.** $\frac{5}{6}$ **21.** $\frac{1}{9}$ **23.** $\frac{5}{11}$ **25.** $\frac{3}{70}$ **27.** $\frac{1}{15}$ **29.** 0.065 **31.** 0.123 **33.** 0.0055

35. 0.0825 **37.** 0.0505 **39.** 0.02 **41.** 0.804 **43.** 0.049 **45.** 73% **47.** 1% **49.** 294% **51.** 0.6%

53. 310.6% **55.** 70% **57.** 37% **59.** 40% **61.** 12.5% **63.** 150% **65.** 166.7% **67.** 87.5% **69.** 48%

71. $33\frac{1}{3}\%$ **73.** $166\frac{2}{3}\%$ **75.** $87\frac{1}{2}\%$ **77. a.** False. For example, 4(200%) = 4(2) = 8. **b.** False. For example,

$\frac{4}{200\%} = \frac{4}{2} = 2.$ **c.** True. **79.** This represents $33\frac{1}{3}\%$ of the regular price. **81.** 60% of the population did not vote.

SECTION 5.2

1. 8 **3.** 10.8 **5.** 0.075 **7.** 80 **9.** 51.895 **11.** 7.5 **13.** 13 **15.** 3.75 **17.** 20 **19.** 210

21. 5% of 95 **23.** 79% of 16 **25.** 72% of 40 **27.** 25,805.0324 **29.** In the United States, 13.2 million people aged 18 to 24 do not have health insurance. **31.** Dell sold 1.87 million of the computers. **33. a.** The sales tax was $1170. **b.** The total cost of the car was $20,670. **35. a.** There was an increase of 21 eggs eaten annually from 1995 to 2000. **b.** Americans ate an average of 257 eggs during 2000. **37.** Employees spend 48.2 hours with family and friends. **39.** The difference is 3.4 hours.

SECTION 5.3

1. 32% **3.** $16\frac{2}{3}\%$ **5.** 200% **7.** 37.5% **9.** 18% **11.** 0.25% **13.** 20% **15.** 400% **17.** 2.5%

19. 37.5% **21.** 0.25% **23.** 2.4% **25.** 9.6% **27.** 70% of couples disagree about financial matters. **29.** 25.4% of the vegetables produced were wasted. **31.** 6% of the total amount spent on energy utilities is spent on lighting. **33.** 61.25% of the workday is not spent at a computer. **35.** 27.4% of the total amount is spent on food. **37.** 56.7% of the total bill was for home heating.

SECTION 5.4

1. 75   **3.** 50   **5.** 100   **7.** 85   **9.** 1200   **11.** 19.2   **13.** 7.5   **15.** 32   **17.** 200   **19.** 80   **21.** 9
23. 504   **25.** 108   **27.** There were 15.8 million travelers who allowed their children to miss school to go along on a trip.   **29.** The average cost of attending a four-year public college in 1998–1999 was $3246.   **31. a.** The number of computer boards tested was 3000.   **b.** The number of computer boards tested that were not defective was 2976 boards.
33. 70,982 of the Democratic votes were not for Al Gore.   **35.** The recommended daily amount of copper for an adult is 2 milligrams.   **37.** The buyer is paying 25% of the value of the company.

SECTION 5.5

1. 65   **3.** 25%   **5.** 75   **7.** 12.5%   **9.** 400   **11.** 19.5   **13.** 14.8%   **15.** 62.62   **17.** 5   **19.** 45
21. 15   **23.** The club collected $24,500.   **25.** The approximate total land area of the world is 58,750,000 square miles.
27. Of that amount, $23.45 billion came from the federal government.   **29.** There were 14,000,000 ounces mined in the United States that year.   **31.** 46.8% of the deaths were due to traffic accidents.   **33.** The potency has decreased by 93.75% after 60 hours.

CHAPTER REVIEW

1. 60 [5.2A]   **2.** 20% [5.3A]   **3.** 175% [5.1B]   **4.** 75 [5.4A]   **5.** $\frac{3}{25}$ [5.1A]   **6.** 19.36 [5.2A]
7. 150% [5.3A]   **8.** 504 [5.4A]   **9.** 0.42 [5.1A]   **10.** 5.4 [5.2A]   **11.** 157.5 [5.4A]   **12.** 0.076 [5.1A]
13. 77.5 [5.2A]   **14.** $\frac{1}{6}$ [5.1A]   **15.** 160% [5.5A]   **16.** 75 [5.5A]   **17.** 38% [5.1B]   **18.** 10.9 [5.4A]
19. 7.3% [5.3A]   **20.** 613.3% [5.3A]   **21.** Trent answered 85% of the questions correctly. [5.5B]
22. The company spent $4500 for TV advertising. [5.2B]   **23.** The population increased by 6.25%. [5.3B]
24. The total cost of the camera was $1041.25. [5.2B]   **25.** 78.6% of the women wore sunscreen often. [5.3B]
26. The world's population in 2000 was 6,100,000,000 people. [5.4B]   **27.** The cost of the computer 4 years ago was $3000. [5.5B]   **28.** The team scored 110 points. [5.4B]

CHAPTER TEST

1. 0.973 [5.1A]   **2.** $\frac{5}{6}$ [5.1A]   **3.** 30% [5.1B]   **4.** 163% [5.1B]   **5.** 150% [5.1B]   **6.** $66\frac{2}{3}\%$ [5.1B]
7. 50.05 [5.2A]   **8.** 61.36 [5.2A]   **9.** 76% of 13 [5.2A]   **10.** 212% of 12 [5.2A]   **11.** The company spends $4500 for advertising. [5.2B]   **12.** There were 1170 pounds of vegetables unspoiled. [5.2B]   **13.** One serving provides 14.7% of the daily recommended amount of potassium. [5.3B]   **14.** One serving provides 9.1% of the daily recommended number of calories. [5.3B]   **15.** The store's temporary employees were 16% of the permanent employees. [5.3B]
16. Conchita answered 91.3% of the questions correctly. [5.3B]   **17.** 80 [5.4A]   **18.** 28.3 [5.4A]   **19.** There were 32,000 transistors tested. [5.4B]   **20.** The increase was 60% of the original price. [5.3B]   **21.** 143.0 [5.5A]
22. 1000% [5.5A]   **23.** The increase in the hourly wage is $1.38. [5.5B]   **24.** The population now is 220% of the population 10 years ago. [5.5B]   **25.** The value of the car is $6500. [5.5B]

CUMULATIVE REVIEW

1. 4 [1.6B]   **2.** 240 [2.1A]   **3.** $10\frac{11}{24}$ [2.4C]   **4.** $12\frac{41}{48}$ [2.5C]   **5.** $12\frac{4}{7}$ [2.6B]   **6.** $\frac{7}{24}$ [2.7B]
7. $\frac{1}{3}$ [2.8B]   **8.** $\frac{13}{36}$ [2.8C]   **9.** 3.08 [3.1B]   **10.** 1.1196 [3.3A]   **11.** 34.2813 [3.5A]   **12.** 3.625 [3.6A]
13. $1\frac{3}{4}$ [3.6B]   **14.** $\frac{3}{8} < 0.87$ [3.6C]   **15.** 53.3 [4.3B]   **16.** $9.60/hour [4.2B]   **17.** $\frac{11}{60}$ [5.1A]
18. $83\frac{1}{3}\%$ [5.1B]   **19.** 19.56 [5.2A/5.5A]   **20.** $133\frac{1}{3}\%$ [5.3A/5.5A]   **21.** 9.92 [5.4A/5.5A]
22. 342.9% [5.3A/5.5A]   **23.** His take-home pay is $592. [2.6C]   **24.** The monthly payment is $204.25. [3.5B]
25. There were 420 gallons of gasoline used during the month. [3.5B]   **26.** The real estate tax is $3000. [4.3C]

27. The sales tax was 6% of the purchase price. [5.3B/5.5B]   **28.** 45% of the people surveyed did not favor the candidate. [5.3B/5.5B]   **29.** Cable households spend 61.3 hours per week watching TV. [5.2B/5.5B]   **30.** 18% of the children tested had levels of lead in the blood that exceeded federal standards. [5.3B/5.5B]

Answers to Chapter 6 Odd-Numbered Exercises

PREP TEST

1. 0.75 [3.5A]   **2.** 52.05 [3.4A]   **3.** 504.51 [3.3A]   **4.** 9750 [3.4A]   **5.** 45 [3.4A]   **6.** 1417.24 [3.2A]
7. 3.33 [3.5A]   **8.** 0.605 [3.5A]   **9.** 0.379 < 0.397 [3.6C]

SECTION 6.1

1. The unit cost is $.055 per ounce.   **3.** The unit cost is $.374 per ounce.   **5.** The unit cost is $.080 per tablet.
7. The unit cost is $6.975 per clamp.   **9.** The unit cost is $.199 per ounce.   **11.** The unit cost is $.119 per screw.
13. The more economical purchase is Sutter Home.   **15.** The more economical purchase is 20 ounces for $3.29.
17. The more economical purchase is 400 tablets for $12.99.   **19.** The more economical purchase is Land to Lake.
21. The more economical purchase is Maxwell House.   **23.** The more economical purchase is Friskies Chef's Blend.
25. The total cost is $13.77.   **27.** The total cost is $1.84.   **29.** The total cost is $6.37.   **31.** The total cost is $2.71.
33. The total cost is $5.96.

SECTION 6.2

1. The expected percent increase is 19.6%.   **3.** The percent increase was 117.6%.   **5.** The percent increase is 457.1%.
7. The population eight years later was 157,861 people.   **9.** In 2000, 7.6 hours were spent online each week.
11. The markup is $35.70.   **13.** The markup rate is 30%.   **15. a.** The markup is $77.76.   **b.** The selling price
is $239.76.   **17. a.** The markup is $1.10.   **b.** The selling price is $3.10.   **19.** The selling price is $74.
21. The percent decrease is 40%.   **23.** The decrease is 540 employees.   **25. a.** The amount of the decrease is
13 minutes.   **b.** The percent decrease is 25%.   **27. a.** The amount of the decrease was $.60.   **b.** The new dividend
is $1.00.   **29.** The percent decrease is 6.6%.   **31.** The discount rate is $33\frac{1}{3}$%.   **33.** The discount is $68.
35. The discount rate is 30%.   **37. a.** The discount is $.17 per pound.   **b.** The sale price is $.68 per pound.
39. a. The discount is $4.   **b.** The discount rate is 20%.   **41.** Yes, the new wage can be found by multiplying $12
by 1.10.

SECTION 6.3

1. a. $10,000   **b.** $2050   **c.** 10.25%   **d.** 2 years   **3.** The simple interest owed is $1440.   **5.** The simple
interest due is $6750.   **7.** The simple interest due is $1320.   **9.** The simple interest due is $92.47.   **11.** The simple
interest due is $146.40.   **13.** The maturity value of the loan is $5120.   **15.** The maturity value of the loan is $164,250.
17. The total amount due on the loan is $13,291.67.   **19.** The maturity value of the loan is $15,061.51.
21. The monthly payment is $6187.50.   **23. a.** The interest charged is $1080.   **b.** The monthly payment is $545.
25. The monthly payment is $1332.50.   **27.** The finance charge is $1.48.   **29.** The finance charge is $185.53.
31. The difference is $10.07.   **33.** The value after 1 year is $780.60.   **35.** The value after 15 years is $7281.78.
37. a. The value will be $111,446.25 in 5 years.   **b.** In 5 years, $36,446.25 will be earned in interest.   **39.** The amount
of interest earned over a 20-year period is $5799.48.   **41.** The interest owed is $16.

SECTION 6.4

1. The mortgage is $82,450.   **3.** The down payment is $7500.   **5.** The down payment was $212,500.   **7.** The loan
origination fee is $3750.   **9. a.** The down payment is $7500.   **b.** The mortgage is $142,500.   **11.** The mortgage is
$189,000.   **13.** The monthly mortgage payment is $1157.73.   **15.** No, the couple cannot afford to buy the home.

17. The monthly property tax is $112.35. **19. a.** The monthly mortgage payment is $1678.40. **b.** The interest is $736.68 during that month. **21.** The total monthly payment (mortgage and property tax) is $982.70. **23.** The monthly mortgage payment is $1977.83. **25.** $63,408 of interest can be saved.

SECTION 6.5

1. No, Amanda does not have enough money for the down payment. **3.** The sales tax was $1192.50. **5.** The license fee is $450. **7. a.** The sales tax is $1120. **b.** The total cost of the sales tax and the license fee is $1395.
9. a. The down payment is $4050. **b.** The amount financed is $12,150. **11.** The amount financed is $28,000.
13. The monthly truck payment is $348.39. **15.** The cost to operate the car is $5120. **17.** The cost per mile is $.11.
19. The interest is $74.75. **21. a.** The amount financed is $76,600. **b.** The monthly truck payment is $1590.09.
23. The monthly car payment is $709.88. **25.** The 10% loan with the application fee has a lower loan cost.

SECTION 6.6

1. Lewis earns $380. **3.** The commission earned is $3930. **5.** The commission received was $84. **7.** Keisha receives $3244 per month. **9.** The electrician's overtime hourly wage is $51.60. **11.** The commission earned is $112.50.
13. The typist earns $618.75. **15.** The consultant's hourly wage is $85. **17. a.** Mark's hourly wage is $23.85.
b. Mark earns $190.80 on Saturday. **19. a.** The increase in pay is $1.65 per hour. **b.** The nurse's hourly pay is $18.15. **21.** Nicole earned $475. **23.** The starting salary was $36,281. **25.** Liberal arts majors received the smaller amount of increase in starting salary.

SECTION 6.7

1. The new balance is $486.32. **3.** The balance in the account is $1222.47. **5.** The current balance is $825.27.
7. The current balance is $3000.82. **9.** Yes, there is enough money to purchase a refrigerator. **11.** Yes, there is enough money to make both purchases. **13.** The bank statement and checkbook balance. **15.** The bank statement and checkbook balance. **17.** added **19.** subtract

CHAPTER REVIEW

1. The unit cost is 19.5¢ per ounce. [6.1A] **2.** The cost per mile was 16.4¢. [6.5B] **3.** The percent increase was 30.4%. [6.2A] **4.** The markup is $72. [6.2B] **5.** The simple interest due on the loan is $6750. [6.3A] **6.** The value of the investment will be $45,550.75. [6.3C] **7.** The percent increase was 15%. [6.2A] **8.** The monthly payment for mortgage and property tax is $578.53. [6.4B] **9.** The monthly payment is $330.82. [6.5B]
10. The value of the investment will be $53,593. [6.3C] **11.** The down payment is $18,750. [6.4A] **12.** The total of the sales tax and license fee is $1471.25. [6.5A] **13.** The selling price is $2079. [6.2B] **14.** The amount of interest is $97.33. [6.5B] **15.** The commission is $3240. [6.6A] **16.** The sale price was $141. [6.2D] **17.** The current balance is $943.68. [6.7A] **18.** The maturity value of the loan is $31,200. [6.3A] **19.** The loan origination fee is $1875. [6.4A] **20.** The more economical purchase is 33 ounces for $6.99. [6.1B] **21.** The monthly mortgage payment is $934.08. [6.4B] **22.** The total income was $655.20. [6.6A] **23.** The current checkbook balance is $8866.58. [6.7A] **24.** The monthly payment is $14,093.75. [6.3A] **25.** The finance charge is $7.20. [6.3B]

CHAPTER TEST

1. The cost per foot is $6.92. [6.1A] **2.** The more economical purchase is 3 pounds for $7.49. [6.1B]
3. The cost is $14.53. [6.1C] **4.** The percent increase is 20%. [6.2A] **5.** The selling price is $301. [6.2B]
6. The percent decrease was 7.7%. [6.2C] **7.** The percent decrease was 20%. [6.2C] **8.** The sale price is $209.30. [6.2D] **9.** The discount rate is 40%. [6.2D] **10.** The simple interest due is $2000. [6.3A]
11. The maturity value of the loan is $26,725. [6.3A] **12.** The finance charge is $4.50. [6.3B] **13.** The interest earned was $24,420.60. [6.3C] **14.** The loan origination fee is $3350. [6.4A] **15.** The monthly mortgage payment is $1713.44. [6.4B] **16.** The amount financed is $19,000. [6.5A] **17.** The monthly truck payment is $482.68. [6.5B] **18.** Shaney earns $966. [6.6A] **19.** The current checkbook balance is $6612.25. [6.7A]
20. The bank statement and the checkbook balance. [6.7B]

CUMULATIVE REVIEW

1. 13 [1.6B] **2.** $8\frac{13}{24}$ [2.4C] **3.** $2\frac{37}{48}$ [2.5C] **4.** 9 [2.6B] **5.** 2 [2.7B] **6.** 5 [2.8C] **7.** 52.2 [3.5A]

8. 1.417 [3.6A] **9.** $51.25/hour [4.2B] **10.** 10.94 [4.3B] **11.** 62.5% [5.1B] **12.** 27.3 [5.2A]

13. 0.182 [5.1A] **14.** 42% [5.3A] **15.** 250 [5.4A] **16.** 154.76 [5.4A/5.5A] **17.** The total rainfall was $13\frac{11}{12}$

inches. [2.4D] **18.** The amount the family pays in taxes is $970. [2.6C] **19.** The ratio of decrease in price to

original price is $\frac{3}{5}$. [4.1B] **20.** The car was driven 33.4 miles per gallon. [4.2C] **21.** The unit cost is $.93/lb. [4.2C]

22. The dividend is $280. [4.3C] **23.** The sale price is $720. [6.2D] **24.** The selling price is $119. [6.2B]

25. The percent increase is 8%. [6.2A] **26.** The simple interest due is $6000. [6.3A] **27.** The monthly payment is

$791.81. [6.5B] **28.** The new balance is $2243.77. [6.7A] **29.** The cost per mile is $.20. [6.5B]

30. The monthly mortgage payment is $743.18. [6.4B]

Answers to Chapter 7 Odd-Numbered Exercises

PREP TEST

1. Approximately 48.0% of the first-class mail handled by the U.S. Postal Service was bill-related mail. [5.3B]

2. The increase in cost is greatest between 2009 and 2010. The increase is $5318. [1.3C] **3. a.** The ratio is $\frac{5}{3}$. [4.1B]

b. The ratio is 1:1. [4.1B] **4. a.** 3.9, 3.9, 4.2, 4.5, 5.2, 5.5, 7.1 [3.6C] **b.** The average number of viewers per night

is 4.9 million. [3.5B] **5. a.** Approximately 4500 women are in the Marine Corps. [5.2B] **b.** $\frac{1}{20}$ of the women in the

military are in the Marine Corps. [5.1A]

SECTION 7.1

1. The gross revenues were $750 million. **3.** *The Lion King* earned 40% of the total gross revenue. **5.** There were
50 more people who agreed that humanity should explore planets than agreed that space exploration impacts daily life.
7. 150 children hid vegetables under a napkin. **11.** The ratio of the number of units in finance to the number of units

in accounting is $\frac{1}{3}$. **13.** 9.4% of the units required to graduate are taken in mathematics. **15.** 150 people were

surveyed. **17.** 28% of the respondents said that people talking bothered them most. **19.** The amount spent on
portable game machines was $279,000,000. **21.** Yes. **23.** The number of homeless who are aged 25 to 34 is more
than twice the number who are under the age of 25. **25.** For every 100,000 homeless people, there are 8000 homeless
people over the age of 54. **27.** North America is 2,550,000 square miles larger than South America. **29.** Australia is
5.2% of the total world land area. **31.** $1700 of the family's income was spent on entertainment. **33.** Yes.

SECTION 7.2

1. There were 39 million passenger cars produced worldwide. **3.** 28% of the passenger cars were produced in Asia.
5. The Honda Insight gets 10 more miles per gallon while on the highway. **7.** The difference is $16,000. **9.** The
difference between the maximum salary in the suburbs and that in the city is the greatest in Philadelphia. **11.** There are
20 inches of snow during January. **13.** The total average snowfall during March and April is 25 inches.

15. The difference is 800 calories.   **17.** The ratio is $\frac{7}{6}$.   **19.**

Year	Difference Between the Amount Spent for Foreign and for Domestic Aid (in billions of dollars)
1991	$1.3
1992	$1.1
1993	$1.0
1994	$0.9
1995	$0.8
1996	$0.8
1997	$1.1
1998	$0.9
1999	$1.0

SECTION 7.3

1. There are 44 students whose tuition was between $3000 and $6000.   **3.** There are 18 students who paid more than $12,000 annually for tuition.   **5.** There are 410 cars between 6 and 12 years old.   **7.** There are 230 cars more than 12 years old.   **9.** There are 54 adults who spent between 1 and 2 hours at the mall.   **11.** 22% of the adults spend less than 1 hour at the mall.   **13.** 22 teams competed in the relay.   **15.** 72.7% of the teams completed the relay in less than 12 minutes.   **17.** 10.8% of the people purchased between 20 and 30 tickets each month.   **21.** 32.4% of the students scored between 800 and 1000.   **23.** There were 900,000 students who scored above 800.

SECTION 7.4

1. a. Median   **b.** Mean   **c.** Mode   **d.** Median   **e.** Mode   **f.** Mean   **3.** The mean is 381.5625 seats, and the median is 394.5 seats. There is no mode.   **5.** The mean is $45.615, and the median is $45.855.   **7.** The mean is $403.625, and the median is $404.50.   **9.** The mean is 70.8 years, and the median is 72 years.   **13. a.** 25% of the data lie above Q_3.   **b.** 75% of the data lie above Q_1.   **c.** 75% of the data lie below Q_3.   **d.** 25% of the data lie below Q_1.
15. The lowest value is $24,880. The highest value is $47,954. $Q_1 = $29,778, $Q_3 = $36,453, and the median is $33,858. The range is $23,074, and the interquartile range is $6675.   **17. a.** 40 adults had a cholesterol level above 217.
b. 60 adults had a cholesterol level below 254.   **c.** 20 cholesterol levels are represented in each quartile.   **d.** 25% of the adults had a cholesterol level of not more than 198.   **19. a.** In millions of metric tons per year, the range is 4.39 emissions, $Q_1 = 0.56$ emissions, $Q_3 = 2.10$ emissions, and the interquartile range is 1.54 emissions.
b.   **c.** The data value 4.80 is responsible for the long "whisker" at the right.

21. a. No   **b.** The difference is 0.3 inch.   **c.**

23. Answers will vary. For example, 55, 55, 55, 55, 55, or 50, 55, 55, 55, 60.

SECTION 7.5

1. {(HHHH), (HHHT), (HHTT), (HHTH), (HTTT), (HTHH), (HTTH), (HTHT), (TTTT), (TTTH), (TTHH), (THHH), (TTHT), (THHT), (THTT), (THTH)}   **3.** {(1, 1), (1, 2), (1, 3), (1, 4), (2, 1), (2, 2), (2, 3), (2, 4), (3, 1), (3, 2), (3, 3), (3, 4), (4, 1), (4, 2), (4, 3), (4, 4)}   **5. a.** {1, 2, 3, 4, 5, 6, 7, 8}   **b.** {1, 2, 3}   **7. a.** The probability is $\frac{1}{9}$.   **b.** The probability is 0.

c. The probablity is 1. **d.** The probability is $\frac{1}{36}$. **9. a.** The probability is $\frac{3}{16}$. **b.** The probability is $\frac{3}{16}$.

11. The probability of throwing a sum of 5 is greater. **13. a.** The probability is $\frac{4}{11}$. **b.** The probability of choosing an S is greater. **15. a.** The probability is $\frac{1}{3}$. **b.** The probability of choosing a red marble is greater. **17.** The probability is $\frac{8}{47}$. **19.** The probability is 0.81.

CHAPTER REVIEW

1. $349 million was spent on maintaining web sites. [7.1B] **2.** The ratio of the amount spent by the Department of Commerce to the amount spent by the EPA was $\frac{9}{8}$. [7.1B] **3.** 8.9% of the total was spent by NASA. [7.1B] **4.** In 1900, Texas had the larger population. [7.2B] **5.** In 2000, California's population was approximately 12.5 million people greater than the population of Texas. [7.2B] **6.** The population of Texas increased the least during the period from 1925 to 1950. [7.2B] **7.** There were 54 games in which fewer than 100 points were scored. [7.3B] **8.** The ratio of the number of games in which 90 to 100 points were scored to the number of games in which 110 to 120 points were scored was $\frac{31}{8}$. [7.3B] **9.** There were 110 or more points scored in 11.3% of the games. [7.3B]

10. 10 million more passengers pass through O'Hare each year. [7.1A] **11.** The ratio is 2 : 3. [7.1A] **12.** The difference between the numbers of days was 50. [7.2A] **13.** The Rocky Mountain ski areas were in full operation for 50% of the days. [7.2A] **14.** The Southeast had 30 days of full operation. [7.2A] **15.** The probability is $\frac{1}{4}$. [7.5A]

16. There were 15 people who slept 8 hours or more. [7.3A] **17.** 28.3% of the people surveyed slept 7 hours. [7.3A] **18.** The mean is 91.6 heartbeats per minute, the median is 93.5 heartbeats per minute, and the mode is 96 heartbeats per minute. [7.4A] **19.** The range is 36 heartbeats per minute, and the interquartile range is 15 heartbeats per minute. [7.4B]

CHAPTER TEST

1. There were 19 students who spent between $15 and $25 per week. [7.3B] **2.** The ratio of the number of students who spent between $10 and $15 to the number who spent between $15 and $20 is $\frac{2}{3}$. [7.3B] **3.** 45% of the students spent less than $15 per week. [7.3B] **4.** There were 36 people surveyed. [7.1A] **5.** The ratio of the number of people who gave their marriage a B to the number who gave it a C was $\frac{5}{2}$. [7.1A] **6.** 58.3% of the people surveyed gave their marriage an A. [7.1A] **7.** The numbers of fatalities were the same during 1995 and 1996. [7.2A] **8.** The total number of fatalities during 1991 through 1999 was 32. [7.2A] **9.** There were 4 more fatalities during 1995 through 1998. [7.2A] **10.** There were 355 more R-rated films released than PG films. [7.1B] **11.** There were 16 times more PG-13 films released than NC-17 films. [7.1B] **12.** 5.6% of the films released were rated G. [7.1B] **13.** There are 34 states with per capita incomes between $16,000 and $22,000. [7.3A] **14.** 38% of the states have a per capita income between $19,000 and $22,000. [7.3A] **15.** 14% of the states have a per capita income that is $25,000 or more. [7.3A] **16.** The probability is $\frac{3}{10}$. [7.5A] **17.** The student population increased the least during the 1990s. [7.2B] **18.** The increase in the college enrollment from 1960 to 2000 was approximately 11 million students. [7.2B] **19.** The mean is 2.53 days, and the median is 2.55 days. [7.4A] **20.**

2.0   2.35   2.55   2.8   3.1   [7.4B]

CUMULATIVE REVIEW

1. 540 [1.6A] **2.** 14 [1.6B] **3.** 120 [2.1A] **4.** $\frac{5}{12}$ [2.3B] **5.** $12\frac{3}{40}$ [2.4C] **6.** $4\frac{17}{24}$ [2.5C]

7. 2 [2.6B]   **8.** $\frac{64}{85}$ [2.7B]   **9.** $8\frac{1}{4}$ [2.8C]   **10.** 209.305 [3.1A]   **11.** 2.82348 [3.4A]   **12.** 16.67 [3.6A]

13. 26.4 miles/gallon [4.2B]   **14.** 3.2 [4.3B]   **15.** 80% [5.1B]   **16.** 80 [5.4A]   **17.** 16.34 [5.2A]

18. 40% [5.3A]   **19.** Tanim had $650 in income. [6.6A]   **20.** The life insurance costs $207.50. [4.3C]

21. The interest due is $6875. [6.3A]   **22.** The markup rate is 55%. [6.2B]   **23.** $570 is budgeted for food. [7.1B]

24. The difference between the numbers of problems answered correctly was 12 problems. [7.2B]   **25.** The mean high

temperature was 69.6°F. [7.4A]   **26.** The probability is $\frac{5}{36}$. [7.5A]

Answers to Chapter 8 Odd-Numbered Exercises

PREP TEST

1. 702 [1.2A]   **2.** 58 [1.3B]   **3.** 4 [2.6B]   **4.** 10 [2.6B]   **5.** 25 [2.6B]   **6.** 46 [2.6B]   **7.** 238 [1.5A]
8. 1.5 [3.5A]

SECTION 8.1

1. 72 in.   **3.** $2\frac{1}{2}$ ft   **5.** 39 ft   **7.** $5\frac{1}{3}$ yd   **9.** 84 in.   **11.** $3\frac{1}{3}$ yd   **13.** 10,560 ft   **15.** $\frac{5}{8}$ ft   **17.** 1 mi 1120 ft

19. 9 ft 11 in.   **21.** 2 ft 9 in.   **23.** 14 ft 6 in.   **25.** 2 ft 8 in.   **27.** $11\frac{1}{6}$ ft   **29.** 4 mi 2520 ft   **31.** 14 tiles can be

placed along one row.   **33.** The missing dimension is $1\frac{5}{6}$ ft.   **35.** The length of material needed is $7\frac{1}{2}$ in.   **37.** Each

piece is $1\frac{2}{3}$ ft long.   **39.** 6 ft 6 in. of framing is needed.   **41.** The wall is $33\frac{3}{4}$ ft in length.   **43.** Yes, the line would

reach around Earth at the equator.

SECTION 8.2

1. 4 lb   **3.** 128 oz   **5.** $1\frac{3}{5}$ tons   **7.** 12,000 lb   **9.** $4\frac{1}{8}$ lb   **11.** 24 oz   **13.** 2600 lb   **15.** $\frac{1}{4}$ ton

17. $11\frac{1}{4}$ lb   **19.** 4 tons 1000 lb   **21.** 2 lb 8 oz   **23.** 5 tons 400 lb   **25.** 1 ton 1700 lb   **27.** 33 lb

29. 14 lb   **31.** 9 oz   **33.** 1 lb 7 oz   **35.** The load will weigh 2000 lb.   **37.** The weight of 144 tiles is 63 lb.

39. The case of soft drink weighs 9 lb.   **41.** 1 lb 5 oz of shampoo is in each container.   **43.** The ham costs $18.

45. The cost of mailing the manuscript is $8.75.   **47.** Answers will vary.

SECTION 8.3

1. $7\frac{1}{2}$ c   **3.** 24 fl oz   **5.** 4 pt   **7.** 7 c   **9.** $5\frac{1}{2}$ gal   **11.** 9 qt   **13.** $3\frac{3}{4}$ qt   **15.** $1\frac{1}{4}$ pt   **17.** $4\frac{1}{4}$ qt

19. 3 gal 2 qt   **21.** 2 qt 1 pt   **23.** 7 qt   **25.** 6 fl oz   **27.** $17\frac{1}{2}$ pt   **29.** $\frac{7}{8}$ gal   **31.** 5 gal 1 qt

33. 1 gal 2 qt   **35.** $2\frac{3}{4}$ gal   **37.** $7\frac{1}{2}$ gal of coffee should be prepared.   **39.** There are $4\frac{1}{4}$ qt of solution.

41. The farmer used $8\frac{3}{4}$ gal of oil.   **43.** $1.59 for 1 qt is the more economical purchase.   **45.** $43.50 of profit

was made on each 5-quart package.   **47.** 1 grain = 0.002286 oz; 1 dram = 0.0625 oz; 1 furlong = $\frac{1}{8}$ mi; 1 rod = 16.5 ft

SECTION 8.4

1. 19,450 ft · lb   **3.** 19,450,000 ft · lb   **5.** 1500 ft · lb   **7.** 29,700 ft · lb   **9.** 30,000 ft · lb   **11.** 25,500 ft · lb

13. 35,010,000 ft · lb **15.** 9,336,000 ft · lb **17.** 2 hp **19.** 8 hp **21.** $2750 \dfrac{\text{ft} \cdot \text{lb}}{\text{s}}$ **23.** $3850 \dfrac{\text{ft} \cdot \text{lb}}{\text{s}}$

25. $500 \dfrac{\text{ft} \cdot \text{lb}}{\text{s}}$ **27.** $4800 \dfrac{\text{ft} \cdot \text{lb}}{\text{s}}$ **29.** $1440 \dfrac{\text{ft} \cdot \text{lb}}{\text{s}}$ **31.** 9 hp **33.** 12 hp

CHAPTER REVIEW

1. 48 in. [8.1A] **2.** 2 ft 6 in. [8.1B] **3.** 1600 ft · lb [8.4A] **4.** 40 fl oz [8.3A] **5.** $4\dfrac{2}{3}$ yd [8.1A]

6. $1\dfrac{1}{5}$ tons [8.2A] **7.** 2 lb 7 oz [8.2B] **8.** 54 oz [8.2A] **9.** 9 ft 3 in. [8.1B] **10.** 1 ton 1000 lb [8.2B]

11. 9 lb 3 oz [8.2B] **12.** 1 yd 2 ft [8.1B] **13.** 3 qt [8.3A] **14.** 13 ft 4 in. [8.1B] **15.** $1375 \dfrac{\text{ft} \cdot \text{lb}}{\text{s}}$ [8.4B]

16. 44 lb [8.2B] **17.** 38,900 ft · lb [8.4A] **18.** 7 hp [8.4B] **19.** A board 3 ft 6 in. long remains. [8.1C]

20. The cost of mailing the book is $6.30. [8.2C] **21.** There are $13\dfrac{1}{2}$ qt in a case of 24 cans. [8.3C]

22. There were 16 gal of milk sold. [8.3C] **23.** The furnace releases 27,230,000 ft · lb in 1 h. [8.4A] **24.** The power

is $480 \dfrac{\text{ft} \cdot \text{lb}}{\text{s}}$. [8.4B]

CHAPTER TEST

1. 30 in. [8.1A] **2.** 2 ft 5 in. [8.1B] **3.** Each piece is $1\dfrac{1}{3}$ ft long. [8.1C] **4.** The wall is 48 ft in length. [8.1C]

5. 46 oz [8.2A] **6.** 2 lb 8 oz [8.2B] **7.** 17 lb 1 oz [8.2B] **8.** 1 lb 11oz [8.2B] **9.** The books weigh

750 lb. [8.2C] **10.** The class earned $28.13. [8.2C] **11.** $3\dfrac{1}{4}$ gal [8.3A] **12.** 28 pt [8.3A]

13. $12\dfrac{1}{4}$ gal [8.3B] **14.** 8 gal 1 qt [8.3B] **15.** There are 60 c of juice in a case of 24 cans. [8.3C] **16.** Nick

made a profit of $144. [8.3C] **17.** 3750 ft · lb [8.4A] **18.** 31,120,000 ft · lb [8.4A] **19.** $160 \dfrac{\text{ft} \cdot \text{lb}}{\text{s}}$ [8.4B]

20. 4 hp [8.4B]

CUMULATIVE REVIEW

1. 180 [2.1A] **2.** $5\dfrac{3}{8}$ [2.2B] **3.** $3\dfrac{7}{24}$ [2.5C] **4.** 2 [2.7B] **5.** $4\dfrac{3}{8}$ [2.8C] **6.** 2.10 [3.1B]

7. 0.038808 [3.4A] **8.** 8.8 [4.3B] **9.** 1.25 [5.2A] **10.** 42.86 [5.4A] **11.** The unit cost is $4.15/lb. [6.1A]

12. $8\dfrac{11}{15}$ in. [8.1B] **13.** 1 lb 8 oz [8.2B] **14.** 31 lb 8 oz [8.2B] **15.** $2\dfrac{1}{2}$ qt [8.3B] **16.** 1 lb 12 oz [8.2B]

17. The dividend would be $280. [4.3C] **18.** The new checkbook balance is $642.79. [6.7A] **19.** The total monthly

income is $3100. [6.6A] **20.** 2425 lb of the shipment could be sold. [5.2B] **21.** 18% of the class received scores

between 80% and 90%. [7.3A] **22.** The selling price is $308. [6.2B] **23.** The interest paid was $14,666.67. [6.3A]

24. Each student received $1267. [8.2C] **25.** The cost to mail the books was $20.16. [8.2C] **26.** $2.98 for 36 oz is

the better buy. [6.1B] **27.** The probability is $\dfrac{1}{9}$. [7.5A] **28.** 3200 ft · lb [8.4A] **29.** $400 \dfrac{\text{ft} \cdot \text{lb}}{\text{s}}$ [8.4B]

Answers to Chapter 9 Odd-Numbered Exercises

1. 37,320 [4.3A] **2.** 659,000 [3.4A] **3.** 0.04107 [3.5A] **4.** 28.496 [3.5A] **5.** 5.125 [3.3A]
6. 5.96 [3.2A] **7.** 0.13 [3.4A] **8.** 56.35 [2.6B, 3.4A] **9.** 0.5 [2.6B, 3.5A] **10.** 675 [2.6B]

SECTION 9.1

1. 420 mm   **3.** 8.1 cm   **5.** 6.804 km   **7.** 2109 m   **9.** 4.32 m   **11.** 88 cm   **13.** 7.038 km   **15.** 3500 m
17. 2.60 m   **19.** 168.5 cm   **21.** 148 mm   **23.** 62.07 m   **25.** 31.9 cm   **27.** 8.075 km   **29.** The missing
dimension is 7.8 cm.   **31.** The distance between rivets is 17.9 cm.   **33.** 15.6 m of fencing is left on the roll.
35. It takes 500 s for light from the sun to reach Earth.   **37.** Light travels 25,920,000,000 km in 1 day.

SECTION 9.2

1. 0.420 kg   **3.** 0.127 g   **5.** 4200 g   **7.** 450 mg   **9.** 1.856 kg   **11.** 4.057 g   **13.** 1370 g   **15.** 45.6 mg
17. 18.000 kg   **19.** 3.922 kg   **21.** 7.891 g   **23.** 4.063 kg   **25.** There are 40 g in one serving.   **27. a.** There are
3.288 g of cholesterol in one dozen eggs.   **b.** There is 0.132 g of cholesterol in four glasses of milk.   **29. a.** There is
0.186 kg of mix in the package.   **b.** There is 0.42 g of sodium in two servings of the corn bread.   **31.** 1.6 kg of grass
seed are needed to cover 2000 m².   **33.** The profit on the container of nuts is $67.50.   **35.** 50.3% of the grain exports
was corn.

SECTION 9.3

1. 4.2 L   **3.** 3420 ml   **5.** 423 cm³   **7.** 642 ml   **9.** 0.042 L   **11.** 435 cm³   **13.** 4620 L   **15.** 1.423 kl
17. 1267 cm³   **19.** 3.042 L   **21.** 3.004 kl   **23.** 8.200 L   **25. a.** The amount of oxygen in 50 L of air is less than
25 L.   **b.** There are 10.5 L of oxygen in 50 L of air.   **27.** 24 L of chlorine are used in a month.   **29.** 4000 people can
be immunized.   **31.** The better buy is the case of 12 one-liter bottles of apple juice.   **33.** The profit on the gasoline
is $8465.   **35.** 2720 ml; 2.72 L; 2L 720 ml

SECTION 9.4

1. 3300 Calories can be eliminated from your diet.   **3. a.** There are 90 Calories in $1\frac{1}{2}$ servings.   **b.** In 6 slices of
bread, 30 Calories are from fat.   **5.** A 135-pound lightly active person should consume 2025 Calories per day.   **7.** You
burn 10,125 Calories in 30 days.   **9.** You would have to hike for 2.6 h.   **11.** 1250 Wh of energy are used.
13. 0.567 kWh of energy is used.   **15.** It costs $1.58 to run the air conditioner.   **17. a.** The light output of the Energy
Saver Bulb is less than half that of the Long Life Soft White Bulb.   **b.** The difference in cost is $.42.   **19.** The cost of
using the welder is $109.98.   **21.** Answers will vary.

SECTION 9.5

1. 91.74 m   **3.** 1.73 m   **5.** 6.82 kg   **7.** 235.85 ml   **9.** 104.65 km/h   **11.** $7.68/kg   **13.** $.39/L
15. The distance is 40,068.07 km.   **17.** 328 ft   **19.** 1.58 gal   **21.** 4920 ft   **23.** 6.33 gal   **25.** 49.69 mi/h
27. $1.46/gal   **29.** 4.62 lb   **31.** He will lose 1 lb.   **33.** The speed is 186,335.40 mi/s.   **35. a.** False.   **b.** False.
c. True.   **d.** True.   **e.** False.

CHAPTER REVIEW

1. 1250 m [9.1A]   **2.** 450 mg [9.2A]   **3.** 5.6 ml [9.3A]   **4.** 1090 yd [9.5B]   **5.** 7.9 cm [9.1A]
6. 5.34 m [9.1A]   **7.** 0.990 kg [9.2A]   **8.** 2.550 L [9.3A]   **9.** 4.870 km [9.1A]   **10.** 3.7 mm [9.1A]
11. 6.829 g [9.2A]   **12.** 1200 cm³ [9.3A]   **13.** 4050 g [9.2A]   **14.** 870 cm [9.1A]   **15.** 192 cm³ [9.3A]
16. 0.356 g [9.2A]   **17.** 3.72 m [9.1A]   **18.** 8300 L [9.3A]   **19.** 2.089 L [9.3A]   **20.** 5.410 L [9.3A]
21. 3.792 kl [9.3A]   **22.** 468 ml [9.3A]   **23.** 37.2 m of wire are left on the roll. [9.1B]   **24.** The cost is
$12.72. [9.2B]   **25.** The cost is $7.48/kg. [9.5A]   **26.** 50 L of coffee should be prepared. [9.3B]
27. 2700 Calories can be eliminated from your diet. [9.4A]   **28.** It costs $3.42 to run the set. [9.4A]
29. The weight is 4.18 lb. [9.5B]   **30.** 8.75 h of cycling are necessary to lose 1 lb. [9.4A]   **31.** The profit on the
soap is $52.80. [9.3B]   **32.** 1.120 kWh of energy are used during the week. [9.4A]   **33.** 125 kg of fertilizer are
required. [9.2B]

CHAPTER TEST

1. 2960 m [9.1A]   **2.** 378 mg [9.2A]   **3.** 46 ml [9.3A]   **4.** 919 ml [9.3A]   **5.** 4.26 cm [9.1A]

6. 7.96 m [9.1A] **7.** 0.847 kg [9.2A] **8.** 3.920 L [9.3A] **9.** 5.885 km [9.1A] **10.** 15 mm [9.1A]
11. 3.089 g [9.2A] **12.** 1600 cm³ [9.3A] **13.** 3290 g [9.2A] **14.** 420 cm [9.1A] **15.** 96 cm³ [9.3A]
16. 1.375 g [9.2A] **17.** 4.02 m [9.1A] **18.** 8920 L [9.3A] **19.** 5038 m [9.1A] **20.** 6.020 kl [9.3A]
21. The total length of the rafters is 114 m. [9.1B] **22.** The weight is 36 kg. [9.2B] **23.** 5.2 L of vaccine are needed. [9.3B] **24.** 35 mi/h is equivalent to 56.4 km/h. [9.5A] **25.** The distance between the rivets is 17.5 cm. [9.1B] **26.** It costs $660 to fertilize the orchard. [9.2B] **27.** It costs $16.32 to operate the air conditioner. [9.4A] **28.** 11 L of acid should be ordered. [9.3B] **29.** The distance is 449.36 ft. [9.5B]
30. The maximum length of the skis is 279.4 cm. [9.5A]

CUMULATIVE REVIEW

1. 6 [1.6B] **2.** $12\frac{13}{36}$ [2.4C] **3.** $\frac{29}{36}$ [2.5C] **4.** $3\frac{1}{14}$ [2.7B] **5.** 1 [2.8B] **6.** 2.0702 [3.3A]
7. 31.3 [4.3B] **8.** 175% [5.1B] **9.** 145 [5.4A] **10.** 2.25 gal [8.3A] **11.** 8.75 m [9.1A]
12. 3.420 km [9.1A] **13.** 5050 g [9.2A] **14.** 3.672 g [9.2A] **15.** 6000 ml [9.3A] **16.** 2400 L [9.3A]
17. There is $2433 left after the rent is paid. [2.6C] **18.** The amount of state income tax paid is $7207.20. [3.4B]
19. The property tax is $1500. [4.3C] **20.** The new car buyer would receive a $2820 rebate. [5.2B] **21.** The dividend is 6.5% of the investment. [5.3B] **22.** Your mean grade is 76. [7.4A] **23.** The salary next year will be $25,200. [5.2B] **24.** The discount rate is 22%. [6.2D] **25.** The length of the wall is 36 ft. [8.1C]
26. The number of quarts of apple juice is 12. [8.3C] **27.** The profit is $123.20. [8.3C] **28.** 24 L of chlorine are used during the month. [9.3B] **29.** It costs $1.89 to operate the hair dryer. [9.4A] **30.** The conversion is 96.6 km/h. [9.5A]

Answers to Chapter 10 Odd-Numbered Exercises

PREP TEST

1. 54 > 45 [1.1A] **2.** 4 units [1.2A] **3.** 15,847 [1.2A] **4.** 3779 [1.3B] **5.** 26,432 [1.4B] **6.** 6 [1.5C]
7. $1\frac{4}{15}$ [2.4B] **8.** $\frac{7}{16}$ [2.5B] **9.** 11.058 [3.2A] **10.** 3.781 [3.3A] **11.** $\frac{2}{5}$ [2.6A] **12.** $\frac{5}{9}$ [2.7A]
13. 9.4 [3.4A] **14.** 0.4 [3.5A] **15.** 31 [1.6B]

SECTION 10.1

1. −120 ft **3.** +2 dollars **5.** **7.**
9. −2 > −5 **11.** −16 < 1 **13.** 3 > −7 **15.** −11 < −8 **17.** 35 > 28 **19.** −42 < 27 **21.** 21 > −34
23. −27 > −39 **25.** −87 < 63 **27.** 86 > −79 **29.** −62 > −84 **31.** −131 < 101 **33.** −4 **35.** 2
37. −22 **39.** 31 **41.** −70 **43.** 2 **45.** 6 **47.** 8 **49.** 9 **51.** −1 **53.** 0 **55.** 19 **57.** 22
59. −20 **61.** −18 **63.** 23 **65.** −27 **67.** 25 **69.** 74 **71.** −88 **73.** The state of New York has the lowest recorded temperature. **75. a.** −6 **b.** −8 **c.** −9 or −6

SECTION 10.2

1. −14, −364 **3.** −2 **5.** 20 **7.** −11 **9.** −9 **11.** −3 **13.** 1 **15.** −5 **17.** −30 **19.** 9 **21.** 1
23. −10 **25.** −28 **27.** −41 **29.** −392 **31.** −20 **33.** −23 **35.** −2 **37.** −6 **39.** −6
41. Negative six minus positive four **43.** Positive six minus negative four **45.** 9 + 5 **47.** 1 + (−8) **49.** 8
51. −7 **53.** −9 **55.** 36 **57.** −3 **59.** 18 **61.** −9 **63.** 11 **65.** 0 **67.** 11 **69.** 2 **71.** −138
73. 86 **75.** −337 **77.** 4 **79.** −12 **81.** −12 **83.** 3 **85.** The temperature is −1°C. **87.** Nick's score was −15 points. **89.** The change in the price of the stock was −9 dollars. **91.** The difference is 115°F. **93.** The

difference in elevation is 22,965 ft.   **95.** The difference is 82°C.   **97.** The difference is 13°C.   **99.**

−3	2	1
4	0	−4
−1	−2	3

SECTION 10.3

1. Subtraction   **3.** Multiplication   **5.** 42   **7.** −24   **9.** 6   **11.** 18   **13.** −20   **15.** −16   **17.** 25   **19.** 0
21. −72   **23.** −102   **25.** 140   **27.** −228   **29.** −320   **31.** −156   **33.** −70   **35.** 162   **37.** 120
39. 36   **41.** 192   **43.** −108   **45.** −2100   **47.** 20   **49.** −48   **51.** 140   **53.** $3(-12) = -36$
55. $-5(11) = -55$ **57.** −2   **59.** 8   **61.** 0   **63.** −9   **65.** −9   **67.** 9   **69.** −24   **71.** −12   **73.** 31
75. 17   **77.** 15   **79.** −13   **81.** −18   **83.** 19   **85.** 13   **87.** −19   **89.** 17   **91.** 26   **93.** 23   **95.** 25
97. −34   **99.** 11   **101.** −13   **103.** 13   **105.** 12   **107.** −14   **109.** −14   **111.** −6   **113.** −4
115. The average daily high temperature was −4°.   **117.** The average score of the ten golfers was −2.   **119.** The wind
chill factor at −15°F with a 20-mph wind is −60°F.   **121. a.** 25   **b.** −17   **123. a.** True   **b.** True

SECTION 10.4

1. $-\dfrac{5}{24}$ **3.** $-\dfrac{19}{24}$ **5.** $\dfrac{5}{26}$ **7.** $\dfrac{7}{24}$ **9.** $-\dfrac{19}{60}$ **11.** $-1\dfrac{3}{8}$ **13.** $\dfrac{3}{4}$ **15.** $-\dfrac{47}{48}$ **17.** $\dfrac{3}{8}$ **19.** $-\dfrac{7}{60}$ **21.** $\dfrac{13}{24}$
23. −3.4   **25.** −8.89   **27.** −8.0   **29.** −0.68   **31.** −181.51   **33.** 2.7   **35.** −20.7   **37.** −37.19
39. −34.99   **41.** 6.778   **43.** −8.4   **45.** $\dfrac{1}{21}$ **47.** $\dfrac{2}{9}$ **49.** $-\dfrac{2}{9}$ **51.** $-\dfrac{45}{328}$ **53.** $\dfrac{2}{3}$ **55.** $\dfrac{15}{64}$ **57.** $-\dfrac{3}{7}$
59. $-1\dfrac{1}{9}$ **61.** $\dfrac{5}{6}$ **63.** $-2\dfrac{2}{7}$ **65.** $-13\dfrac{1}{2}$ **67.** 31.15   **69.** −112.97   **71.** 0.0363   **73.** 97   **75.** 2.2
77. −4.14   **79.** −3.8   **81.** −22.70   **83.** 0.07   **85.** 0.55   **87.** The temperature fell 32.22°C.   **89.** The
difference is 35.438°C.   **91. a.** The closing price on the previous day for Kellogg Co. was $\$25\dfrac{5}{16}$.   **b.** The closing price
on the previous day for Unilever PLC was $\$26\dfrac{5}{8}$.   **93.** Answers will vary. For example, $-\dfrac{17}{24}$.

SECTION 10.5

1. 2.37×10^6 **3.** 4.5×10^{-4} **5.** 3.09×10^5 **7.** 6.01×10^{-7} **9.** 5.7×10^{10} **11.** 1.7×10^{-8} **13.** 710,000
15. 0.000043   **17.** 671,000,000   **19.** 0.00000713   **21.** 5,000,000,000,000   **23.** 0.00801   **25.** 1.6×10^{10} mi
27. $\$3.1 \times 10^{12}$ **29.** 3.7×10^{-6} m   **31.** 4   **33.** 0   **35.** 12   **37.** −6   **39.** −5   **41.** 2   **43.** −3   **45.** 1
47. 2   **49.** −3   **51.** 14   **53.** −4   **55.** 33   **57.** −13   **59.** −12   **61.** 17   **63.** 0   **65.** 30   **67.** 94
69. −8   **71.** 39   **73.** 22   **75.** 0.21   **77.** −0.96   **79.** −0.29   **81.** −1.76   **83.** 2.1   **85.** $\dfrac{3}{16}$ **87.** $\dfrac{7}{16}$
89. $-\dfrac{5}{16}$ **91.** $-\dfrac{5}{8}$ **93. a.** $3.45 \times 10^{-14} > 3.45 \times 10^{-15}$ **b.** $5.23 \times 10^{18} > 5.23 \times 10^{17}$
c. $3.12 \times 10^{12} > 3.12 \times 10^{11}$ **95. a.** 100   **b.** −100   **c.** 225   **d.** −225   **99.** Sun: 1.99×10^{30} kg;
neutron: 1.67×10^{-27} kg

CHAPTER REVIEW

1. −22 [10.1B]   **2.** 1 [10.2B]   **3.** $-\dfrac{5}{24}$ [10.4A]   **4.** 0.21 [10.4A]   **5.** $\dfrac{8}{25}$ [10.4B]   **6.** −1.28 [10.4B]

7. 10 [10.5B]   **8.** $-\dfrac{7}{18}$ [10.5B]   **9.** 4 [10.1B]   **10.** $0 > -3$ [10.1A]   **11.** −6 [10.1B]   **12.** 6 [10.3B]

13. $\dfrac{17}{24}$ [10.4A]   **14.** $-\dfrac{1}{4}$ [10.4B]   **15.** $1\dfrac{5}{8}$ [10.4B]   **16.** 24 [10.5B]   **17.** −26 [10.2A]   **18.** 2 [10.5B]

19. 3.97×10^{-5} [10.5A]   **20.** −0.08 [10.4B]   **21.** $\dfrac{1}{36}$ [10.4A]   **22.** $\dfrac{3}{40}$ [10.4B]   **23.** −0.042 [10.4B]

24. $\frac{1}{3}$ [10.5B]   **25.** 5 [10.1B]   **26.** $-2 > -40$ [10.1A]   **27.** -26 [10.3A]   **28.** 1.33 [10.5B]

29. $-\frac{1}{4}$ [10.4A]   **30.** -7.4 [10.4A]   **31.** $\frac{15}{32}$ [10.4B]   **32.** 240,000 [10.5A]   **33.** $-4°$ [10.2C]

34. The student scored 98 points. [10.3C]   **35.** The difference is 395.45°C. [10.4C]

CHAPTER TEST

1. 3 [10.2B]   **2.** -2 [10.1B]   **3.** $\frac{1}{15}$ [10.4A]   **4.** -0.0608 [10.4B]   **5.** $-8 > -10$ [10.1A]

6. -1.88 [10.4A]   **7.** -26 [10.5B]   **8.** 90 [10.3A]   **9.** -4.014 [10.4A]   **10.** -9 [10.3B]   **11.** -7 [10.2A]

12. $\frac{7}{24}$ [10.4A]   **13.** 8.76×10^{10} [10.5A]   **14.** -48 [10.3A]   **15.** 0 [10.3B]   **16.** 10 [10.2B]

17. $-\frac{4}{5}$ [10.4B]   **18.** $0 > -4$ [10.1A]   **19.** -14 [10.2A]   **20.** -4 [10.5B]   **21.** $\frac{3}{10}$ [10.4A]

22. 0.00000009601 [10.5A]   **23.** 3.4 [10.4B]   **24.** $\frac{1}{12}$ [10.4A]   **25.** $\frac{1}{12}$ [10.4B]   **26.** 3.213 [10.4A]

27. The temperature is 7°C. [10.2C]   **28.** The melting point of oxygen is -213°C. [10.3C]   **29.** The temperature fell 46.62°C. [10.4C]   **30.** The average low temperature was $-2°$. [10.3C]

CUMULATIVE REVIEW

1. 0 [1.6B]   **2.** $4\frac{13}{14}$ [2.5C]   **3.** $2\frac{7}{12}$ [2.7B]   **4.** $1\frac{2}{7}$ [2.8C]   **5.** 1.80939 [3.3A]   **6.** 18.67 [4.3B]

7. 13.75 [5.4A]   **8.** 1 gal 3 qt [8.3A]   **9.** 6.692 L [9.3A]   **10.** 1.28 m [9.5A]   **11.** 57.6 [5.2A]

12. 340% [5.1B]   **13.** -3 [10.2A]   **14.** $-3\frac{3}{8}$ [10.4A]   **15.** $-10\frac{13}{24}$ [10.4A]   **16.** 19 [10.5B]

17. 3.488 [10.4B]   **18.** $31\frac{1}{2}$ [10.4B]   **19.** -7 [10.3B]   **20.** $\frac{25}{42}$ [10.4B]   **21.** -4 [10.5B]   **22.** -4 [10.5B]

23. The length of board remaining is $2\frac{1}{3}$ ft. [2.5D]   **24.** Nimisha's new checkbook balance is $803.31. [6.7A]

25. The percent decrease in price is 27.3%. [6.2C]   **26.** There should be 10 gal of coffee prepared. [8.3C]

27. The dividend per share is $1.68. [6.2A]   **28.** The median hourly pay is $10.40. [7.4A]   **29.** There would be 600,000 people voting in the city election. [4.3C]   **30.** The average high temperature was $-4°$. [10.3C]

Answers to Chapter 11 Odd-Numbered Exercises

PREP TEST

1. -7 [10.2B]   **2.** -20 [10.3A]   **3.** 0 [10.2A]   **4.** 1 [10.3B]   **5.** 1 [10.4B]   **6.** $\frac{1}{15}$ [2.8B]

7. $\frac{19}{24}$ [2.8C]   **8.** 4 [10.5B]   **9.** -21 [10.5B]

SECTION 11.1

1. -33 **3.** -12 **5.** -4 **7.** -3 **9.** -22 **11.** -40 **13.** -1 **15.** 14   **17.** 32   **19.** 6   **21.** 7

23. 49   **25.** -24 **27.** 63   **29.** 9   **31.** 4   **33.** $-\frac{2}{3}$ **35.** 0   **37.** $-3\frac{1}{4}$ **39.** 8.5023

41. -18.950658 **43.** $2x^2, 3x, \underline{-4}$ **45.** $3a^2, -4a, \underline{8}$ **47.** $\underline{3x^2}, \underline{-4x}$ **49.** $\underline{1y^2}, 6a$ **51.** $16z$ **53.** $9m$ **55.** $12at$

57. $3yt$ **59.** Unlike terms   **61.** $-2t^2$ **63.** $13c - 5$ **65.** $-2t$ **67.** $3y^2 - 2$ **69.** $14w - 8u$

71. $-23xy + 10$ **73.** $-11v^2$ **75.** $-8ab - 3a$ **77.** $4y^2 - y$ **79.** $-3a - 2b^2$ **81.** $7x^2 - 8x$ **83.** $-3s + 6t$

85. $-3m + 10n$ **87.** $5ab + 4ac$ **89.** $\frac{2}{3}a^2 + \frac{3}{5}b^2$ **91.** $6.994x$ **93.** $1.56m - 3.77n$ **95.** $5x + 20$ **97.** $4y - 12$

99. $-2a - 8$ **101.** $15x + 30$ **103.** $15c - 25$ **105.** $-3y + 18$ **107.** $7x + 14$ **109.** $4y - 8$ **111.** $5x + 24$
113. $y - 6$ **115.** $6n - 3$ **117.** $6y + 20$ **119.** $10x + 4$ **121.** $-6t - 6$ **123.** $z - 2$ **125.** $y + 6$
127. $9t + 15$ **129.** $5t + 24$
131. a.

| 1 | 1 | 1 | x | x | | x | x | x | x | 1 | 1 | 1 | 1 | 1 | 1 | | x | x | x | 1 | 1 | | x | x | 1 | 1 | 1 | 1 |

b.

| x | x | x | 1 | 1 | 1 | $3x + 3$ **133.** Answers will vary.

SECTION 11.2

1. Yes **3.** No **5.** Yes **7.** Yes **9.** Yes **11.** Yes **13.** Yes **15.** Yes **17.** No **19.** No **21.** Yes
23. -2 **25.** 14 **27.** 2 **29.** -4 **31.** -5 **33.** 0 **35.** 2 **37.** -3 **39.** 10 **41.** -5 **43.** -12
45. -7 **47.** $\frac{2}{3}$ **49.** -1 **51.** $\frac{1}{3}$ **53.** $-\frac{1}{8}$ **55.** $-\frac{3}{4}$ **57.** $-1\frac{1}{12}$ **59.** 6 **61.** -9 **63.** -5 **65.** 14
67. 8 **69.** -3 **71.** 20 **73.** -21 **75.** -15 **77.** 16 **79.** -42 **81.** 15 **83.** $-38\frac{2}{5}$ **85.** $9\frac{3}{5}$
87. $-10\frac{4}{5}$ **89.** $1\frac{17}{18}$ **91.** $-2\frac{2}{3}$ **93.** 3 **95.** The amount of the original investment was \$23,610. **97.** The
increase in value of the investment was \$1190. **99.** You used 18.5 gal of gasoline. **101.** Your car averages
34.2 mi/gal. **103.** The markup on each shirt is \$16.30. **105.** The cost of a compact disc is \$14.50. **107.** Answers
will vary. **109.** Answers will vary. For example, $x + 5 = 1$.

SECTION 11.3

1. 3 **3.** 5 **5.** -1 **7.** -4 **9.** 1 **11.** 3 **13.** -2 **15.** -3 **17.** -2 **19.** -1 **21.** 3 **23.** -1
25. 7 **27.** 2 **29.** 2 **31.** 0 **33.** 2 **35.** 0 **37.** $\frac{2}{3}$ **39.** $2\frac{5}{6}$ **41.** $-2\frac{1}{2}$ **43.** $1\frac{1}{2}$ **45.** 2 **47.** -2
49. $2\frac{1}{3}$ **51.** $-1\frac{1}{2}$ **53.** 10 **55.** 5 **57.** 36 **59.** -6 **61.** -9 **63.** $-4\frac{4}{5}$ **65.** $-8\frac{3}{4}$ **67.** 36 **69.** $1\frac{1}{3}$
71. 2 **73.** 4 **75.** 9 **77.** 4 **79.** -3.125 **81.** 1 **83.** -3 **85.** -9 **87.** 4.95 **89.** -0.0862069
91. The Celsius temperature is $-40°C$. **93.** The time required is 14.5 s. **95.** There were 2500 units made.
97. The mechanic's monthly income is \$1800. **99.** The total sales were \$32,000. **101.** Miguel's commission rate was
4.5%. **103.** Answers will vary.

SECTION 11.4

1. $\frac{1}{2}$ **3.** -1 **5.** -4 **7.** -1 **9.** -4 **11.** 3 **13.** -1 **15.** 2 **17.** 3 **19.** 0 **21.** -2 **23.** -1
25. $-\frac{2}{3}$ **27.** -4 **29.** $\frac{1}{3}$ **31.** $-\frac{3}{4}$ **33.** $\frac{1}{4}$ **35.** 0 **37.** $-1\frac{3}{4}$ **39.** 1 **41.** -1 **43.** -5 **45.** $\frac{5}{12}$
47. $-\frac{4}{7}$ **49.** $2\frac{1}{3}$ **51.** 21 **53.** 21 **55.** 2 **57.** -1 **59.** -4 **61.** $-\frac{6}{7}$ **63.** 5 **65.** 3 **67.** 0
69. -1 **71.** 2 **73.** $1\frac{9}{10}$ **75.** 13 **77.** $2\frac{1}{2}$ **79.** 2 **81.** 0 **83.** $3\frac{1}{5}$ **85.** 4 **87.** $-6\frac{1}{2}$ **89.** $-\frac{3}{4}$
91. $-\frac{2}{3}$ **93.** -3 **95.** $-1\frac{1}{4}$ **97.** $2\frac{5}{8}$ **99.** 1.9936196 **101.** 13.383119 **103.** 48

SECTION 11.5

1. $y - 9$ **3.** $z + 3$ **5.** $\frac{2}{3}n + n$ **7.** $\frac{m}{m-3}$ **9.** $9(x + 4)$ **11.** $n - (-5)n$ **13.** $c\left(\frac{1}{4}c\right)$ **15.** $m^2 + 2m^2$
17. $2(t + 6)$ **19.** $\frac{x}{9+x}$ **21.** $3(b + 6)$ **23.** x^2 **25.** $\frac{x}{20}$ **27.** $4x$ **29.** $\frac{3}{4}x$ **31.** $4 + x$ **33.** $5x - x$

35. $x(x + 2)$ **37.** $7(x + 8)$ **39.** $x^2 + 3x$ **41.** $(x + 3) + \frac{1}{2}x$ **43.** $\frac{3x}{x}$ **45.** Answers will vary. For example:

3 more than twice x; twice the sum of x and 3   **49.** carbon: $\frac{1}{2}x$; oxygen: $\frac{1}{2}x$

SECTION 11.6

1. $x + 7 = 12$; 5   **3.** $3x = 18$; 6   **5.** $x + 5 = 3$; -2 **7.** $6x = 14$; $2\frac{1}{3}$ **9.** $\frac{5}{6}x = 15$; 18   **11.** $3x + 4 = 8$; $1\frac{1}{3}$

13. $\frac{1}{4}x - 7 = 9$; 64   **15.** $\frac{x}{9} = 14$; 126   **17.** $\frac{x}{4} - 6 = -2$; 16   **19.** $7 - 2x = 13$; -3 **21.** $9 - \frac{x}{2} = 5$; 8

23. $\frac{3}{5}x + 8 = 2$; -10 **25.** $\frac{x}{4.186} - 7.92 = 12.529$; 85.599514   **27.** Target's price is $76.75.   **29.** The value of the

sports utility vehicle last year was $20,000.   **31.** The total quick-service restaurant market was $260 billion.
33. Their monthly income is $3440.   **35.** The monthly output last year was 5000 computers.   **37.** The recommended
daily intake of sodium is 2.5 g.   **39.** Budget Plumbing charged for 3 h of labor.   **41.** There are 11,065 plants and
animals at risk of extinction in the world.   **43.** 3.3% of the total is spent in New York.   **45.** The original rate was
5 gal/min.   **47.** The total sales were $42,540.   **49.** The percent subsidized is 21.5%.

CHAPTER REVIEW

1. $-2a + 2b$ [11.1C]   **2.** Yes [11.2A]   **3.** -4 [11.2B]   **4.** 7 [11.3A]   **5.** 13 [11.1A]   **6.** -9 [11.2C]
7. -18 [11.3A]   **8.** $-3x + 4$ [11.1C]   **9.** 5 [11.4A]   **10.** -5 [11.2B]   **11.** No [11.2A]   **12.** 6 [11.1A]
13. 5 [11.4B]   **14.** 10 [11.3A]   **15.** $-4bc$ [11.1B]   **16.** $-2\frac{1}{2}$ [11.4A]   **17.** $1\frac{1}{4}$ [11.2C]   **18.** $\frac{71}{30}x^2$ [11.1B]
19. $-\frac{1}{3}$ [11.4B]   **20.** $10\frac{4}{5}$ [11.3A]   **21.** The car averaged 23 mi/gal. [11.2D]   **22.** The Celsius temperature

is 37.8°C. [11.3B]   **23.** $n + \frac{n}{5}$ [11.5A]   **24.** $(n + 5) + \frac{1}{3}n$ [11.5B]   **25.** The number is 2. [11.6A]   **26.** The

number is 10. [11.6A]   **27.** The regular price is $380. [11.6B]   **28.** The farmer harvested 25,300 bushels of corn last
year. [11.6B]

CHAPTER TEST

1. 95 [11.3A]   **2.** 26 [11.2B]   **3.** $-11x - 4y$ [11.1B]   **4.** $3\frac{1}{5}$ [11.4A]   **5.** -2 [11.3A]   **6.** -38 [11.1A]

7. No [11.2A]   **8.** $-14ab + 9$ [11.1B]   **9.** $-2\frac{4}{5}$ [11.2C]   **10.** $8y - 7$ [11.1C]   **11.** 0 [11.4B]

12. $-\frac{1}{2}$ [11.3A]   **13.** $-5\frac{5}{6}$ [11.1A]   **14.** -16 [11.2C]   **15.** -3 [11.3A]   **16.** 11 [11.4B]   **17.** The monthly

payment is $137.50. [11.2D]   **18.** There were 4000 clocks made during the month. [11.3B]   **19.** The time required

is 11.5 s. [11.3B]   **20.** $x + \frac{1}{3}x$ [11.5A]   **21.** $5(x + 3)$ [11.5B]   **22.** $2x - 3 = 7$; 5 [11.6A]   **23.** The number

is $-3\frac{1}{2}$. [11.6A]   **24.** His total sales for the month were $40,000. [11.6B]   **25.** The mechanic worked on your car for

3 h. [11.6B]

CUMULATIVE REVIEW

1. 41 [1.6B]   **2.** $1\frac{7}{10}$ [2.5C]   **3.** $\frac{11}{18}$ [2.8C]   **4.** 0.047383 [3.4A]   **5.** $4.20/h [4.2B]   **6.** 26.67 [4.3B]

7. $\frac{4}{75}$ [5.1A]   **8.** 140% [5.3A]   **9.** 6.4 [5.4A]   **10.** 18 ft 9 in. [8.1B]   **11.** 22 oz [8.2A]

12. 0.282 g [9.2A]   **13.** -1 [10.2A]   **14.** 19 [10.2B]   **15.** 6 [10.5B]   **16.** -48 [11.1A]

17. $-10x + 8z$ [11.1B]   **18.** $3y + 23$ [11.1C]   **19.** -1 [11.3A]   **20.** $-6\frac{1}{4}$ [11.4B]   **21.** $-7\frac{1}{2}$ [11.2C]

22. −21 [11.3A]   **23.** The percent was 17.6%. [5.3B]   **24.** The price of the pottery was $39.90. [6.2B]
25. a. The discount is $81.   **b.** The discount rate is 18%. [6.2D]   **26.** The simple interest due is $2933.33. [6.3A]
27. 3n + 4 [11.5B]   **28.** The probability is $\frac{1}{8}$. [7.5A]   **29.** The total monthly sales are $32,500. [11.3B]
30. The number is 2. [11.6A]

Answers to Chapter 12 Odd-Numbered Exercises

PREP TEST

1. 43 [11.2B]   **2.** 51 [11.2B]   **3.** 56 [1.6B]   **4.** 56.52 [11.1A]   **5.** 113.04 [11.1A]   **6.** 14.4 [4.3B]

SECTION 12.1

1. 0°, 90°   **3.** 180°   **5.** 30   **7.** 21   **9.** 14   **11.** 59°   **13.** 108°   **15.** 77°   **17.** 53°   **19.** 77°   **21.** 118°
23. 133°   **25.** 86°   **27.** 180°   **29.** Rectangle or square   **31.** Cube   **33.** Parallelogram, rectangle, or square
35. Cylinder   **37.** 102°   **39.** 90° and 45°   **41.** 14°   **43.** 90° and 65°   **45.** 8 in.   **47.** $4\frac{2}{3}$ ft   **49.** 7 cm
51. 2 ft 4 in.   **53.** a = 106°, b = 74°   **55.** a = 112°, b = 68°   **57.** a = 38°, b = 142°   **59.** a = 58°, b = 58°
61. a = 152°, b = 152°   **63.** a = 130°, b = 50°   **65. a.** 1°   **b.** 90°

SECTION 12.2

1. 56 in.   **3.** 20 ft   **5.** 92 cm   **7.** 47.1 cm   **9.** 9 ft 10 in.   **11.** 50.24 cm   **13.** 240 m   **15.** 157 cm
17. 121 cm   **19.** 50.56 m   **21.** 3.57 ft   **23.** 139.3 m   **25.** 60 ft of fencing should be purchased.   **27.** 24 ft of
binding are required.   **29.** The total cost is $24,422.50.   **31.** The bicycle travels 31.4 ft.   **33. a.** The perimeter is
more than 70 m.   **b.** The perimeter is 81.4 m.   **35.** The length of weather stripping is 20.71 ft.   **37.** The
circumference of Earth is approximately 39,915.68 km.   **39.** The perimeter of the eighth figure is 22 units.   **41.** The
perimeter of all the shaded triangles is $14\frac{1}{4}$ cm.

SECTION 12.3

1. 144 ft²   **3.** 81 in²   **5.** 50.24 ft²   **7.** 20 in²   **9.** 2.13 cm²   **11.** 16 ft²   **13.** 817 in²   **15.** 154 in²
17. 26 cm²   **19.** 2220 cm²   **21.** 150.72 in²   **23.** 8.851323 ft²   **25.** 7500 yd² of artificial turf must be purchased.
27. The area watered is 7850 ft².   **29.** You should buy 2 qt of stain.   **31.** The area is 1250 ft².   **33.** The cost is $74.
35. $17.25 should be budgeted.   **37.** The total area of the park is 125.1492 mi².   **39. a.** The increase in area is
113.04 in².   **b.** The increase in area is 235.5 cm².   **41. a.** 4 times   **b.** Quadrupled   **c.** Quadrupled
43. a. Sometimes true   **b.** Sometimes true   **c.** Always true

SECTION 12.4

1. 144 cm³   **3.** 512 in³   **5.** 2143.57 in³   **7.** 150.72 cm³   **9.** 6.4 m³   **11.** 5572.45 mm³   **13.** 3391.2 ft³
15. $42\frac{7}{8}$ ft³   **17.** 82.26 in³   **19.** 1.6688 m³   **21.** 69.08 in³   **23.** The volume is 40.5 m³.   **25.** The volume
is 17,148.59 ft³.   **27.** There are 75.36 m³ of oil in the tank.   **29.** The volume is 809,516.25 ft³.   **31.** The volume
is 212.64 in³.   **33.** The fish tank will be filled by 3.7 gal of water.   **35.** The cost is $6526.41.   **37. a.** 4 times larger
b. 8 times larger   **c.** 8 times larger

SECTION 12.5

1. 2.646   **3.** 6.481   **5.** 12.845   **7.** 13.748   **9.** 5 in.   **11.** 8.602 cm   **13.** 11.180 ft   **15.** 4.472 cm
17. 12.728 yd   **19.** 6 ft and 10.392 ft   **21.** 21.213 cm   **23.** 8 m   **25.** 11.314 yd   **27.** The length of the ramp
is 9.66 ft.   **29.** You are 21.6 mi from your starting point.   **31.** The distance is 4.243 in.   **33. a.** Always true
b. Always true   **35.** A total of 8 m of pipe is needed.

SECTION 12.6

1. $\frac{1}{2}$ **3.** $\frac{3}{4}$ **5.** Yes **7.** Yes **9.** 7.2 cm **11.** 3.3 m **13.** The height of the building is 16 m.
15. The perimeter is 12 m. **17.** The area is 56.25 cm². **19. a.** Always true **b.** Sometimes true **c.** Always true
21. 16.5 units

CHAPTER REVIEW

1. 0.75 m [12.1B] **2.** 31.4 cm [12.2A] **3.** 26 ft [12.2A] **4.** 3 [12.1A] **5.** 200 ft³ [12.4A]
6. 26 cm [12.5B] **7.** 75° [12.1A] **8.** 3.873 [12.5A] **9.** 16 cm [12.6A] **10.** 63.585 cm² [12.3A]
11. $a = 135°, b = 45°$ [12.1C] **12.** 55 m² [12.3A] **13.** 240 in³ [12.4B] **14.** 57.12 in² [12.3B]
15. 267.9 ft³ [12.4A] **16.** 64.8 m² [12.6B] **17.** 47.7 in. [12.2B] **18.** $a = 100°, b = 80°$ [12.1C]
19. The ladder reaches 15 ft up the building. [12.5C] **20.** The other two angles measure 90° and 58°. [12.1B]
21. The bicycle travels 73.3 ft. [12.2C] **22.** The area of the room is 28 yd². [12.3C] **23.** The volume is
1144.53 ft³. [12.4C] **24.** The area is 11 m². [12.3A] **25.** You are 29 mi from your starting point. [12.5C]

CHAPTER TEST

1. 169.56 m³ [12.4A] **2.** 6.8 m [12.2A] **3.** 1406.72 cm³ [12.4B] **4.** 10 m [12.6A] **5.** 58° [12.1A]
6. $3\frac{1}{7}$ m² [12.3A] **7.** 150° [12.1C] **8.** 15.85 ft [12.2B] **9.** 13.748 [12.5A] **10.** 9.747 ft [12.5B]
11. 10.125 ft² [12.3B] **12.** $\angle a = 45°, \angle b = 135°$ [12.1C] **13.** $1\frac{1}{5}$ ft [12.6A] **14.** The other two angles measure
90° and 50°. [12.1B] **15.** The width is 25 ft. [12.6B] **16.** The larger pizza contains 113.04 in² more. [12.3C]
17. The cost will be $1113.69. [12.3C] **18.** The cross-sectional area is 103.82 ft². [12.3C] **19.** The length of the
rafter is 15 ft. [12.5C] **20.** The volume is 780 in³. [12.4C]

CUMULATIVE REVIEW

1. 48 [2.1B] **2.** $7\frac{41}{48}$ [2.4C] **3.** $\frac{39}{56}$ [2.7B] **4.** $\frac{2}{15}$ [2.8C] **5.** $-\frac{1}{24}$ [10.4A] **6.** $8.72/h [4.2B]
7. 37.5 [4.3B] **8.** $\frac{3}{8}$ [5.1A] **9.** −4 [11.1A] **10.** 85 [5.4A] **11.** −6 [11.3A] **12.** $\frac{4}{3}$ [11.4B]
13. 32,500 m [9.1A] **14.** 31.58 m [9.1A] **15.** −15 [11.2C] **16.** 2 [11.4B] **17.** The monthly payment
is $708. [1.5D] **18.** The sales tax is $47.06. [4.3C] **19.** The operator's hourly wage was $14.60. [5.4B]
20. The sale price is $54. [6.2D] **21.** The value after 20 years would be $184,675.75. [6.3C] **22.** The weight of
the package is 54 lb. [8.2C] **23.** The distance between the rivets is 22.5 cm. [9.1B] **24.** The number is
−2. [11.6A] **25.** $a = 74°, b = 106°$ [12.1C] **26.** 29.42 cm [12.2B] **27.** 50 in² [12.3B]
28. 92.86 in³ [12.4B] **29.** 10.63 ft [12.5B] **30.** 36 cm [12.6B]

FINAL EXAMINATION

1. 3259 [1.3B] **2.** 53 [1.5C] **3.** 16 [1.6B] **4.** 144 [2.1A] **5.** $1\frac{49}{120}$ [2.4B] **6.** $3\frac{29}{48}$ [2.5C]
7. $6\frac{3}{14}$ [2.6B] **8.** $\frac{4}{9}$ [2.7B] **9.** $\frac{1}{6}$ [2.8B] **10.** $\frac{1}{13}$ [2.8C] **11.** 164.177 [3.2A] **12.** 60,205 [1.3B]
13. 0.027918 [3.4A] **14.** 0.69 [3.5A] **15.** $\frac{9}{20}$ [3.6B] **16.** 24.5 mi/gal [4.2B] **17.** 54.9 [4.3B]
18. $\frac{9}{40}$ [5.1A] **19.** 135% [5.1B] **20.** 125% [5.1B] **21.** 36 [5.2A] **22.** $133\frac{1}{3}$% [5.3A] **23.** 70 [5.4A]
24. 20 in. [8.1A] **25.** 1 ft 4 in. [8.1B] **26.** 2.5 lb [8.2A] **27.** 6 lb 6 oz [8.2B] **28.** 2.25 gal [8.3A]
29. 1 gal 3 qt [8.3B] **30.** 248 cm [9.1A] **31.** 4.62 m [9.1A] **32.** 1.614 kg [9.2A] **33.** 2067 ml [9.3A]
34. 88.55 km [9.5A] **35.** It costs $1.15 to run the air conditioner. [9.4A] **36.** 6.79×10^{-8} [10.5A]

37. The perimeter is 3.9 m. [12.2A]   **38.** The area is 45 in². [12.3A]   **39.** The volume is 1200 cm³. [12.4A]

40. −4 [10.2A]   **41.** −15 [10.2B]   **42.** $-\frac{1}{2}$ [10.4B]   **43.** $-\frac{1}{4}$ [10.4B]   **44.** 6 [10.5B]

45. $-x + 17$ [11.1C]   **46.** −18 [11.2C]   **47.** 5 [11.3A]   **48.** 1 [11.4A]   **49.** Your new balance is $959.93. [6.7A]   **50.** There will be 63,750 people voting in the election. [4.3C]   **51.** One year ago the dividend per share was $2.00. [5.4B]   **52.** The average income was $3794. [7.4A]   **53.** The simple interest due is $9000. [6.3A]

54. The probability is $\frac{1}{3}$. [7.5A]   **55.** The percent is 6.7%. [7.1B]   **56.** The discount rate is 28%. [6.2D]

57. The weight of the tiles is 81 lb. [8.2C]   **58.** The perimeter is 28.56 in. [12.2B]   **59.** The area is 16.86 cm². [12.3B]   **60.** The number is 16. [11.6A]

Glossary

absolute value of a number The distance between zero and the number on the number line. (Sec. 10.1)

acute angle An angle whose measure is between 0° and 90°. (Sec. 12.1)

acute triangle A triangle that has three acute angles. (Sec. 12.2)

addend In addition, one of the numbers added. (Sec. 1.2)

addition The process of finding the total of two numbers. (Sec. 1.2)

Addition Property of Zero Zero added to a number does not change the number. (Sec. 1.2)

adjacent angles Two angles that share a common side. (Sec. 12.1)

alternate exterior angles Two angles that are on opposite sides of the transversal and outside the parallel lines. (Sec. 12.1)

alternate interior angles Two angles that are on opposite sides of the transversal and between the parallel lines. (Sec. 12.1)

angle An angle is formed when two rays start at the same point; it is measured in degrees. (Sec. 12.1)

area A measure of the amount of surface in a region. (Sec. 12.3)

Associative Property of Addition Numbers to be added can be grouped (with parentheses, for example) in any order; the sum will be the same. (Sec. 1.2)

Associative Property of Multiplication Numbers to be multiplied can be grouped (with parentheses, for example) in any order; the product will be the same. (Sec. 1.4)

average value The sum of all values divided by the number of those values; also known as the mean value. (Sec. 7.4)

balancing a checkbook Determining if the checking account balance is accurate. (Sec. 6.7)

bar graph A graph that represents data by the height of the bars. (Sec. 7.2)

basic percent equation Percent times base equals amount. (Sec. 5.2)

box-and-whiskers plot A graph that shows the smallest value in a set of numbers, the first quartile, the median, the third quartile, and the greatest value. (Sec. 7.4)

British Thermal Unit A unit of energy. 1 British Thermal Unit = 778 foot-pounds. (Sec. 8.4)

broken-line graph A graph that represents data by the position of the lines and shows trends and comparisons. (Sec. 7.2)

Calorie A unit of energy in the metric system. (Sec. 9.4)

capacity A measure of liquid substances. (Sec. 8.3)

circle A plane figure in which all points are the same distance from point *O*, which is called the center of the circle. (Sec. 12.1)

circle graph A graph that represents data by the size of the sectors. (Sec. 7.1)

circumference The distance around a circle. (Sec. 12.2)

class frequency The number of occurrences of data in a class interval on a histogram; represented by the height of each bar. (Sec. 7.3)

class interval Range of numbers represented by the width of a bar on a histogram. (Sec. 7.3)

class midpoint The center of a class interval in a frequency polygon. (Sec. 7.3)

commission That part of the pay earned by a salesperson that is calculated as a percent of the salesperson's sales. (Sec. 6.6)

common factor A number that is a factor of two or more numbers is a common factor of those numbers. (Sec. 2.1)

common multiple A number that is a multiple of two or more numbers is a common multiple of those numbers. (Sec. 2.1)

Commutative Property of Addition Two numbers can be added in either order; the sum will be the same. (Sec. 1.2)

Commutative Property of Multiplication Two numbers can be multiplied in either order; the product will be the same. (Sec. 1.4)

complementary angles Two angles whose sum is 90°. (Sec. 12.1)

composite geometric figure A figure made from two or more geometric figures. (Sec. 12.2)

composite geometric solid A solid made from two or more geometric solids. (Sec. 12.4)

composite number A number that has whole-number factors besides 1 and itself. For instance, 18 is a composite number. (Sec. 1.7)

compound interest Interest computed not only on the original principal but also on interest already earned. (Sec. 6.3)

congruent objects Objects that have the same shape and the same size. (Sec. 12.6)

constant term A term that has no variables. (Sec. 11.1)

corresponding angles Two angles that are on the same side of the transversal and are both acute angles or are both obtuse angles. (Sec. 12.1)

cost The price that a business pays for a product. (Sec. 6.2)

cube A rectangular solid in which all six faces are squares. (Sec. 12.1)

cubic centimeter A unit of capacity equal to 1 milliliter. (Sec. 9.3)

cup A U.S. Customary measure of capacity. 2 cups = 1 pint. (Sec. 8.3)

data Numerical information. (Sec. 7.1)

decimal A number written in decimal notation. (Sec. 3.1)

decimal notation Notation in which a number consists of a whole-number part, a decimal point, and a decimal part. (Sec. 3.1)

decimal part In decimal notation, that part of the number that appears to the right of the decimal point. (Sec. 3.1)

decimal point In decimal notation, the point that separates the whole-number part from the decimal part. (Sec. 3.1)

degree Unit used to measure angles; one complete revolution is 360°. (Sec. 12.1)

denominator The part of a fraction that appears below the fraction bar. (Sec. 2.2)

diameter of a circle A line segment with endpoints on the circle and going through the center. (Sec. 12.1)

diameter of a sphere A line segment with endpoints on the sphere and going through the center. (Sec. 12.1)

difference In subtraction, the result of subtracting two numbers. (Sec. 1.3)

discount The difference between the regular price and the sale price. (Sec. 6.2)

discount rate The percent of a product's regular price that is represented by the discount. (Sec. 6.2)

dividend In division, the number into which the divisor is divided to yield the quotient. (Sec. 1.5)

division The process of finding the quotient of two numbers. (Sec. 1.5)

divisor In division, the number that is divided into the dividend to yield the quotient. (Sec. 1.5)

empirical probability The ratio of the number of observations of an event to the total number of observations. (Sec. 7.5)

energy The ability to do work. (Sec. 8.4)

equation A statement of the equality of two mathematical expressions. (Sec. 11.2)

equilateral triangle A triangle that has three sides of equal length; the three angles are also of equal measure. (Sec. 12.2)

equivalent fractions Equal fractions with different denominators. (Sec. 2.3)

evaluating a variable expression Replacing the variable or variables with numbers and then simplifying the resulting numerical expression. (Sec. 11.1)

event One or more outcomes of an experiment. (Sec. 7.5)

expanded form The number 46,208 can be written in expanded form as $40,000 + 6000 + 200 + 0 + 8$. (Sec. 1.1)

experiment Any activity that has an observable outcome. (Sec. 7.5)

exponent In exponential notation, the raised number that indicates how many times the number to which it is attached is taken as a factor. (Sec. 1.6)

exponential notation The expression of a number to some power, indicated by an exponent. (Sec. 1.6)

factors In multiplication, the numbers that are multiplied. (Sec. 1.4)

favorable outcomes The outcomes of an experiment that satisfy the requirements of a particular event. (Sec. 7.5)

finance charges Interest charges on purchases made with a credit card. (Sec. 6.3)

first quartile In a set of numbers, the number below which one-quarter of the data lie. (Sec. 7.4)

fluid ounce A U.S. Customary measure of capacity. 8 fluid ounces = 1 cup. (Sec. 8.3)

foot A U.S. Customary unit of length. 3 feet = 1 yard. (Sec. 8.1)

foot-pound A U.S. Customary unit of energy. One foot-pound of energy is required to lift 1 pound a distance of 1 foot. (Sec. 8.4)

foot-pounds per second A U.S. Customary unit of power. (Sec. 8.4)

fraction The notation used to represent the number of equal parts of a whole. (Sec. 2.2)

fraction bar The bar that separates the numerator of a fraction from the denominator. (Sec. 2.2)

frequency polygon A graph that displays information similarly to a histogram. A dot is placed above the center of each class interval at a height corresponding to that class's frequency. (Sec. 7.3)

gallon A U.S. Customary measure of capacity. 1 gallon = 4 quarts. (Sec. 8.3)

geometric solid A figure in space. (Sec. 12.1)

gram The basic unit of mass in the metric system. (Sec. 9.2)

graph A display that provides a pictorial representation of data. (Sec. 7.1)

graph of a whole number A heavy dot placed directly above that number on the number line. (Sec. 1.1)

greater than A number that appears to the right of a given number on the number line is greater than the given number. (Sec. 1.1)

greatest common factor The largest common factor of two or more numbers. (Sec. 2.1)

histogram A bar graph in which the width of each bar corresponds to a range of numbers called a class interval. (Sec. 7.3)

horsepower The U.S. Customary unit of power. 1 horsepower = 550 foot-pounds per second. (Sec. 8.4)

hourly wage Pay calculated on the basis of a certain amount for each hour worked. (Sec. 6.6)

hypotenuse The side opposite the right angle in a right triangle. (Sec. 12.1)

improper fraction A fraction greater than or equal to 1. (Sec. 2.2)

inch A U.S. Customary unit of length. 12 inches = 1 foot. (Sec. 8.1)

integers The integers are $\ldots, -3, -2, -1, 0, 1, 2, 3, \ldots$. (Sec 10.1)

interest Money paid for the privilege of using someone else's money. (Sec. 6.3)

interest rate The percent used to determine the amount of interest. (Sec. 6.3)

interquartile range The difference between the third quartile and the first quartile. (Sec. 7.4)

intersecting lines Lines that cross at a point in the plane. (Sec. 12.1)

inverting a fraction Interchanging the numerator and denominator. (Sec. 2.7)

isosceles triangle A triangle that has two sides of equal length; the angles opposite the equal sides are of equal measure. (Sec. 12.2)

least common denominator The least common multiple of denominators. (Sec. 2.4)

least common multiple The smallest common multiple of two or more numbers. (Sec. 2.1)

legs The two shortest sides of a right triangle. (Sec. 12.1)

less than A number that appears to the left of a given number on the number line is less than the given number. (Sec. 1.1)

like terms Terms of a variable expression that have the same variable part. (Sec. 11.1)

line A line extends indefinitely in two directions in a plane; it has no width. (Sec. 12.1)

line segment Part of a line; it has two endpoints. (Sec. 12.1)

liter The basic unit of capacity in the metric system. (Sec. 9.3)

markup The difference between selling price and cost. (Sec. 6.2)

markup rate The percent of a product's cost that is represented by the markup. (Sec. 6.2)

mass The amount of material in an object. On the surface of Earth, mass is the same as weight. (Sec. 9.2)

mean value The sum of all values divided by the number of those values; also known as the average value. (Sec. 7.4)

measurement A measurement has both a number and a unit. Examples include 7 feet, 4 ounces, and 0.5 gallon. (Sec. 8.1)

median value The value that separates a list of values in such a way that there is the same number of values below the median as above it. (Sec. 7.4)

meter The basic unit of length in the metric system. (Sec. 9.1)

metric system A system of measurement based on the decimal system. (Sec. 9.1)

mile A U.S. Customary unit of length. 5280 feet = 1 mile. (Sec. 8.1)

minuend In subtraction, the number from which another number (the subtrahend) is subtracted. (Sec. 1.3)

mixed number A number greater than 1 that has a whole-number part and a fractional part. (Sec. 2.2)

mode In a set of numbers, the value that occurs most frequently. (Sec. 7.4)

mortgage The amount borrowed to buy real estate. (Sec. 6.4)

multiples of a number The products of that number and the numbers 1, 2, 3, ... (Sec. 2.1)

multiplication The process of finding the product of two numbers. (Sec. 1.4)

Multiplication Property of One The product of a number and one is the number. (Sec. 1.4)

Multiplication Property of Zero The product of a number and zero is zero. (Sec. 1.4)

negative integers The numbers ..., -5, -4, -3, -2, -1. (Sec. 10.1)

negative numbers Numbers less than zero. (Sec. 10.1)

number line A line on which a number can be graphed. (Sec. 1.1)

numerator The part of a fraction that appears above the fraction bar. (Sec. 2.2)

numerical coefficient The number part of a variable term. When the numerical coefficient is 1 or -1, the 1 is usually not written. (Sec. 11.1)

obtuse angle An angle whose measure is between 90° and 180°. (Sec. 12.1)

obtuse triangle A triangle that has one obtuse angle. (Sec. 12.2)

opposites Two numbers that are the same distance from zero on the number line, but on opposite sides. (Sec. 10.1)

Order of Operations Agreement A set of rules that tell us in what order to perform the operations that occur in a numerical expression. (Sec. 1.6)

ounce A U.S. Customary unit of weight. 16 ounces = 1 pound. (Sec. 8.2)

parallel lines Lines that never meet; the distance between them is always the same. (Sec. 12.1)

parallelogram A quadrilateral that has opposite sides equal and parallel. (Sec. 12.1)

percent Parts per hundred. (Sec. 5.1)

percent decrease A decrease of a quantity expressed as a percent of its original value. (Sec. 6.2)

percent increase An increase of a quantity expressed as a percent of its original value. (Sec. 6.2)

perfect square The product of a whole number and itself. (Sec. 12.5)

perimeter The distance around a plane figure. (Sec. 12.2)

period In a number written in standard form, each group of digits separated by a comma. (Sec. 1.1)

perpendicular lines Intersecting lines that form right angles. (Sec. 12.1)

pictograph A graph that uses symbols to represent information. (Sec. 7.1)

pint A U.S. Customary measure of capacity. 2 pints = 1 quart. (Sec. 8.3)

place value The position of each digit in a number written in standard form determines that digit's place value. (Sec. 1.1)

plane A flat surface. (Sec. 12.1)

plane figure A figure that lies totally in a plane. (Sec. 12.1)

points A term banks use to mean percent of a mortgage; used to express the loan origination fee. (Sec. 6.4)

polygon A closed figure determined by three or more line segments that lie in a plane. (Sec. 12.2)

positive integers The numbers 1, 2, 3, 4, 5, ...; also called the natural numbers. (Sec. 10.1)

positive numbers Numbers greater than zero. (Sec. 10.1)

pound A U.S. Customary unit of weight. 1 pound = 16 ounces. (Sec. 8.2)

power The rate at which work is done or energy is released. (Sec. 8.4)

prime factorization The expression of a number as the product of its prime factors. (Sec. 1.7)

prime number A number whose only whole-number factors are 1 and itself. For instance, 13 is a prime number. (Sec. 1.7)

principal The amount of money originally deposited or borrowed. (Sec. 6.3)

probability A number from 0 to 1 that tells us how likely it is that a certain outcome of an experiment will happen. (Sec. 7.5)

product In multiplication, the result of multiplying two numbers. (Sec. 1.4)

proper fraction A fraction less than 1. (Sec. 2.2)

proportion An expression of the equality of two ratios or rates. (Sec. 4.3)

Pythagorean Theorem The square of the hypotenuse of a right triangle is equal to the sum of the squares of the two legs. (Sec. 12.5)

quadrilateral A four-sided closed figure. (Sec. 12.1)

quart A U.S. Customary measure of capacity. 4 quarts = 1 gallon. (Sec. 8.3)

quotient In division, the result of dividing the divisor into the dividend. (Sec. 1.5)

radius of a circle A line segment going from the center to a point on the circle. (Sec. 12.1)

radius of a sphere A line segment going from the center to a point on the sphere. (Sec. 12.1)

range In a set of numbers, the difference between the largest and smallest values. (Sec. 7.4)

rate A comparison of two quantities that have different units. (Sec. 4.2)

ratio A comparison of two quantities that have the same units. (Sec. 4.1)

rational number A number that can be written as the ratio of two integers, where the denominator is not zero. (Sec. 10.4)

ray A ray starts at a point and extends indefinitely in one direction. (Sec. 12.1)

reciprocal of a fraction The fraction with the numerator and denominator interchanged. (Sec. 2.7)

rectangle A parallelogram that has four right angles. (Sec. 12.1)

rectangular solid A solid in which all six faces are rectangles. (Sec. 12.1)

regular polygon A polygon in which each side has the same length and each angle has the same measure. (Sec. 12.2)

remainder In division, the quantity left over when it is not possible to separate objects or numbers into a whole number of equal groups. (Sec. 1.5)

right angle A 90° angle. (Sec. 12.1)

right triangle A triangle that contains one right angle. (Sec. 12.1)

rounding Giving an approximate value of an exact number. (Sec. 1.1)

salary Pay based on a weekly, biweekly, monthly, or annual time schedule. (Sec. 6.6)

sale price The reduced price. (Sec. 6.2)

sample space All the possible outcomes of an experiment. (Sec. 7.5)

scalene triangle A triangle that has no sides of equal length; no two of its angles are of equal measure. (Sec. 12.2)

scientific notation Notation in which a number is expressed as a product of two factors, one a number between 1 and 10 and the other a power of 10. (Sec. 10.5)

selling price The price for which a business sells a product to a customer. (Sec. 6.2)

similar objects Objects that have the same shape but not necessarily the same size. (Sec. 12.6)

simple interest Interest computed on the original principal. (Sec. 6.3)

simplest form of a fraction A fraction is in simplest form when there are no common factors in the numerator and the denominator. (Sec. 2.3)

solid An object in space. (Sec. 12.1)

solution of an equation A number that, when substituted for the variable, results in a true equation. (Sec. 11.2)

solving an equation Finding a solution of the equation. (Sec. 11.2)

sphere A solid in which all points are the same distance from point O, which is called the center of the sphere. (Sec. 12.1)

square A rectangle that has four equal sides. (Sec. 12.1)

square root A square root of a number is one of two identical factors of that number. (Sec. 12.5)

standard form A whole number is in standard form when it is written using the digits 0, 1, 2, ..., 9. An example is 46,208. (Sec. 1.1)

statistics The branch of mathematics concerned with data, or numerical information. (Sec. 7.1)

straight angle A 180° angle. (Sec. 12.1)

subtraction The process of finding the difference between two numbers. (Sec. 1.3)

subtrahend In subtraction, the number that is subtracted from another number (the minuend). (Sec. 1.3)

sum In addition, the total of the numbers added. (Sec. 1.2)

supplementary angles Two angles whose sum is 180°. (Sec. 12.1)

terms of a variable expression The addends of the expression. (Sec. 11.1)

theoretical probability A fraction with the number of favorable outcomes of an experiment in the numerator and the total number of possible outcomes of the experiment in the denominator. (Sec. 7.5)

third quartile In a set of numbers, the number above which one-quarter of the data lie. (Sec. 7.4)

ton A U.S. Customary unit of weight. 1 ton = 2000 pounds. (Sec. 8.2)

transversal A line intersecting two other lines at two different points. (Sec. 12.1)

triangle A three-sided closed figure. (Sec. 12.1)

true proportion A proportion in which the fractions are equal. (Sec. 4.3)

unit cost The cost of one item. (Sec. 6.1)

unit rate A rate in which the number in the denominator is 1. (Sec. 4.2)

units In the quantity 3 feet, feet are the units in which the measurement is made. (Sec. 4.1)

variable A letter used to stand for a quantity that is unknown or that can change. (Sec. 11.1)

variable expression An expression that contains one or more variables. (Sec. 11.1)

variable part In a variable term, the variable or variables and their exponents. (Sec. 11.1)

variable term A term composed of a numerical coefficient and a variable part. (Sec. 11.1)

vertex The common endpoint of two rays that form an angle. (Sec. 12.1)

vertical angles Two angles that are on opposite sides of the intersection of two lines. (Sec. 12.1)

volume A measure of the amount of space inside a closed surface. (Sec. 12.4)

watt-hour A unit of electrical energy in the metric system. (Sec. 9.4)

weight A measure of how strongly Earth is pulling on an object. (Sec. 8.2)

whole numbers The whole numbers are 0, 1, 2, 3, ... (Sec. 1.1)

whole-number part In decimal notation, that part of the number that appears to the left of the decimal point. (Sec. 3.1)

yard A U.S. Customary unit of length. 36 inches = 1 yard. (Sec. 8.1)

Index